# Smart Connection Systems

## Design and Seismic Analysis

*A man of genius cannot exceed a hard worker and a hard worker cannot exceed a person who enjoys his work*

# Smart Connection Systems

## Design and Seismic Analysis

Jong Wan Hu

*Department of Civil and Environmental Engineering,*
*Incheon National University, Incheon, Republic of Korea*

CRC Press
Taylor & Francis Group
Boca Raton London New York Leiden

CRC Press is an imprint of the
Taylor & Francis Group, an **informa** business
A BALKEMA BOOK

*CRC Press/Balkema is an imprint of the Taylor & Francis Group, an informa business*



Typeset by MPS Limited, Chennai, India
Printed and Bound by CPI Group (UK) Ltd, Croydon, CR0 4YY



*Library of Congress Cataloging-in-Publication Data*
Applied for

Published by: CRC Press/Balkema
P.O. Box 11320, 2301 EH Leiden, The Netherlands
e-mail: Pub.NL@taylorandfrancis.com
www.crcpress.com – www.taylorandfrancis.com

ISBN: 978-1-138-02778-7 (Hbk)
ISBN: 978-1-315-68705-6 (eBook PDF)

For my wife
Se Won Kim,
and
For my two daughters
Jeong Won Hu
Jeong A Hu

# Table of contents

# List of Figures

# List of Tables

# Preface

The major fields referred to in this book are architectural engineering, civil engineering, building construction, and structural engineering. The manuscript complies with the current design code. The author (J.W. Hu) intends to write this book that can be utilized as technical teaching aids for undergraduate and graduate lectures. The author has continuously developed finite element models to accurately predict the response of smart structures under seismic loading. The contents of this book were written based on the author's Ph.D. dissertation as well as journal papers in the references, and is made up of three parts with ten chapters.

Seismic design and analyses for smart PR-CFT (Partially restrained-concrete filled tube) connections are treated in Part 1. This part includes four chapters: (a) Introduction, (b) Design and modeling for smart PR-CFT connection system, (c) Analyses for smart PR-CFT connections – end plate connection type, and (d) Analyses for smart PR-CFT connections – T-stub connection type. Design and analyses for several bolted PR connections are investigated in Part 2. This part includes three chapters: (a) Refined FE (finite element) models for bolted connections, (b) Component spring models for bolted connections, and (c) Component models for recentering bolted connections. Finally, design and analyses for other smart connection systems are treated in Part 3. This part includes three chapters: (a) FE models for gusset plate connections, (b) recentering slit damper connections, and (c) smart damping connectors. All chapters are related to design and analyses for smart connection systems.

The author would like to express his appreciation to Dr. Roberto T. Leon and Dr. Won-sup Hwang who have helped the author to obtain his Ph.D. degree and to learn seismic design. The helpful comments of Dr. Dong-keon Kim, Dr. Taehyo Park, Dr. Eunsoo Choi, and Dr. Dongho Choi are also deeply acknowledged. They used to be the author's co-workers. The author would also thank Mr. Ji-woong Park for his assistance. Finally, many chapters of this book were researched and written over several years in the laboratory of Civil Engineering Department at Incheon National University. A special thanks to the author's colleagues and students.

Prof. Jong Wan Hu
Department of Civil and Environmental Engineering
Incheon National University
Republic of Korea

# About the Author

**Prof. Jong Wan Hu** received his BS degree from the Department of Civil and Environmental Engineering at Inha University in South Korea. After his BS degree, he received three MS degrees from (1) the Department of Civil and Environmental Engineering at Inha University (2) the G.W.W. School of Mechanical Engineering at the Georgia Institute of Technology, and (3) the School of Civil and Environmental Engineering at the Georgia Institute of Technology in the USA. He then received his Ph.D. degree from the School of Civil and Environmental Engineering at the Georgia Institute of Technology in the USA. Dr. Hu has been a Post-Doctorate Research Fellow at the Structural, Mechanics, and Material Research Group at the Georgia Institute of Technology. Dr. Hu also worked as an Associate Research Fellow at the Korea Institute of S&T Evaluation and Planning (KISTEP) and as Assistant Administrator at the National S&T Council (NSTC) for two years. He is currently an Assistant Professor at Incheon National University (INU) and head of the Incheon Disaster Prevention Research Center (IDPRC). He has been an active member of ASME and ASCE. He has two professional licenses such as Civil and Structural Professional Engineers (PE). His research interests are in the area of smart structures, seismic design, composite structures, and finite element modeling. He has published over 60 SCI-indexed and peer-reviewed journal papers.

Part 1

# Design and analyses for smart PR-CFT connections

Chapter 1

# Preliminary study

## 1.1 INTRODUCTION

In this study, three structural design concepts are integrated: the use of composite concrete-filled tube columns, the use of partially restrained connections, and the introduction of innovative materials (shape memory alloys) in the connection area. To understand the integration of these concepts and the scope of this dissertation, a brief description of each of these three topics will be given first.

In recent years, concrete filled steel tube (CFT) columns have become widely accepted and used in multistory buildings as well as bridges. These elements provide the synergetic advantages of ductility and toughness associated with steel structures and high compressive strength associated with confined concrete components. The advantages of CFT columns over other so-called mixed or hybrid systems (fully encased or partially encased systems) include the fact that the concrete prevents local buckling of the steel tube wall and the confinement action of the steel tube extends the usable strain of the concrete. In other words, the advantages of two materials (steel and concrete) can be utilized while their disadvantages can be compensated or offset. In addition, a CFT column has improved fire resistance (if properly reinforced) and significant cost reductions in comparison with traditional steel construction. Moreover, the steel tubes can be utilized as the formwork for casting concrete, giving CFT structures improved constructability over conventional reinforced concrete structures.

Composite CFT columns are especially efficient as the vertical elements in moment resisting frames in high seismic areas because they have a high strength to weight ratio, provide excellent monotonic and dynamic resistance under biaxial bending plus axial force, and improve damping behavior (Tsai *et al.*, 2004). A typical composite frame consisting of steel I shape girder and either circular or rectangular CFT (CCFT or RCFT) columns tested by Tsai *et al.*, (Tsai *et al.* 2004) is illustrated in Fig. 1.1.

Typical details of moment connections to RCFT or CCFT columns for this type of structure as constructed in the Far East and the USA are shown in Fig. 1.2. The external diaphragm plates are intended to alleviate the severe distortions of the steel tube skin during fabrication and provide a simple location for making a welded or bolted connection in the field.

To evaluate the performance of a moment frame subjected to lateral-loads, the flexural effects on the rotational deformation at the connections are the critical issue. Therefore, connection behavior can be generally represented by a moment-rotation curve as shown in Fig. 1.3. Connections are classified by three main parameters:

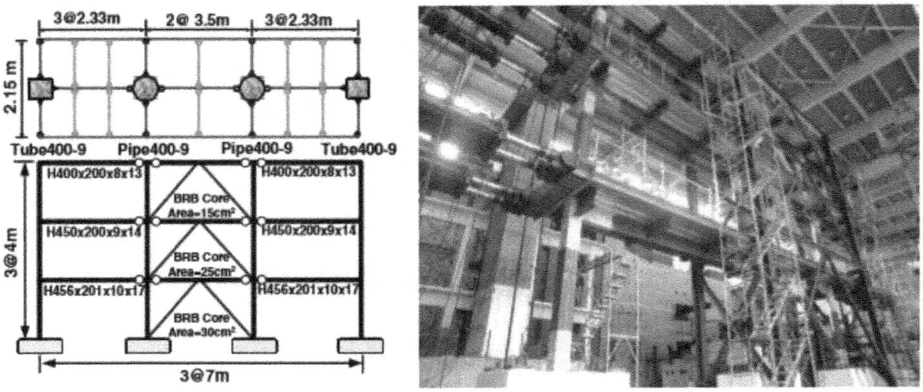


(a) Overview of the composite test frames

(b) Details of steel beam-to-CFT column connections

*Figure 1.1* 3 story by 3 bay CFT composite frame with buckling restrained braces.

stiffness, strength, and ductility (Leon, 1997). For stiffness, connections are classified as fully restrained (FR), partially restrained (PR) or simple pinned connections. An ideal pinned connection only transmits shear force from the beam to columns. For strength, connections are classified as either full strength (FS) or partial strength (PS) depending on whether they can transmit the full plastic moment ($M_p$) of the beam. Finally, connections are classified as brittle or ductile connections based on their ability to achieve a certain plastic rotational demand. The rotational demands at the connections vary according to whether they are in use in ordinary, intermediate, or special moment frames. For example, in the aftermath of Northridge earthquake, the capacity to undergo an elastic rotation of 0.01 radian and a plastic rotation of 0.03 radians under cyclic loading has been accepted as the rotational limit between ductile and brittle connections for special moment resisting frames. This limit accepts up to a 20% decrease from peak bending resistance at the rotational limit.

Major failures of fully welded moment connections during the 1994 Northridge and 1995 Kobe earthquakes have led to the conclusion that the traditional fully welded

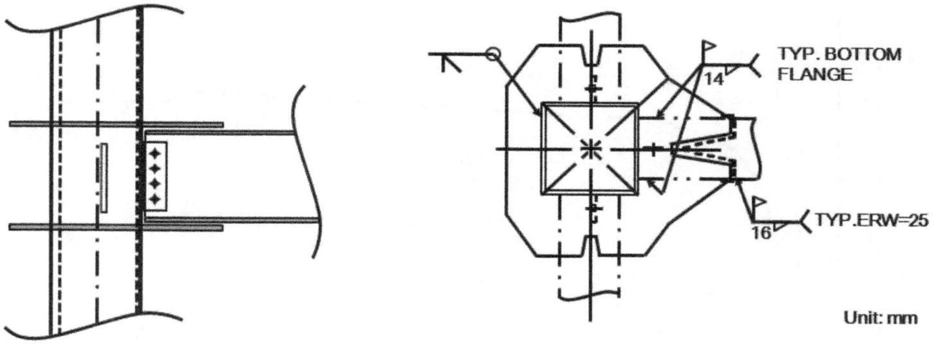


(a) Moment connection details with diaphragm plates (RCFT, Tasi, K.C. et al., 2003)

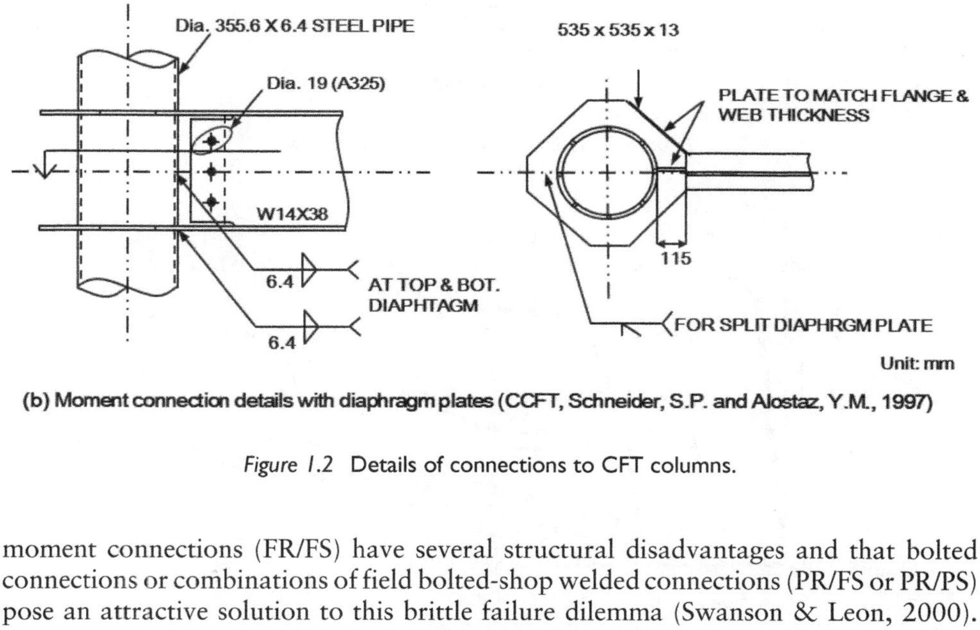


(b) Moment connection details with diaphragm plates (CCFT, Schneider, S.P. and Alostaz, Y.M., 1997)

*Figure 1.2* Details of connections to CFT columns.

moment connections (FR/FS) have several structural disadvantages and that bolted connections or combinations of field bolted-shop welded connections (PR/FS or PR/PS) pose an attractive solution to this brittle failure dilemma (Swanson & Leon, 2000). It also has been demonstrated that well-detailed PR structures can provide similar or superior seismic behavior to their FR counterparts (Rassati *et al.*, 2004). The improved performance is derived from the combination of both (a) the decrease in seismic forces stemming from the additional flexibility of the component members owing to the PR nature of the connections and (b) the increase in the structural strength reserve capacity owing to the lack of brittle connection failure modes.

More recently, work at GT on shape memory alloys (SMAs) has explored the applications of this material to the design of connections in steel structures subjected to large cyclic loads. SMA undergo large deformations with little permanent residual strain through either the shape memory effect or the super-elastic effect due to changes in either temperature or stress. Super-elastic Nitinol (NiTi) is a type of SMA with the

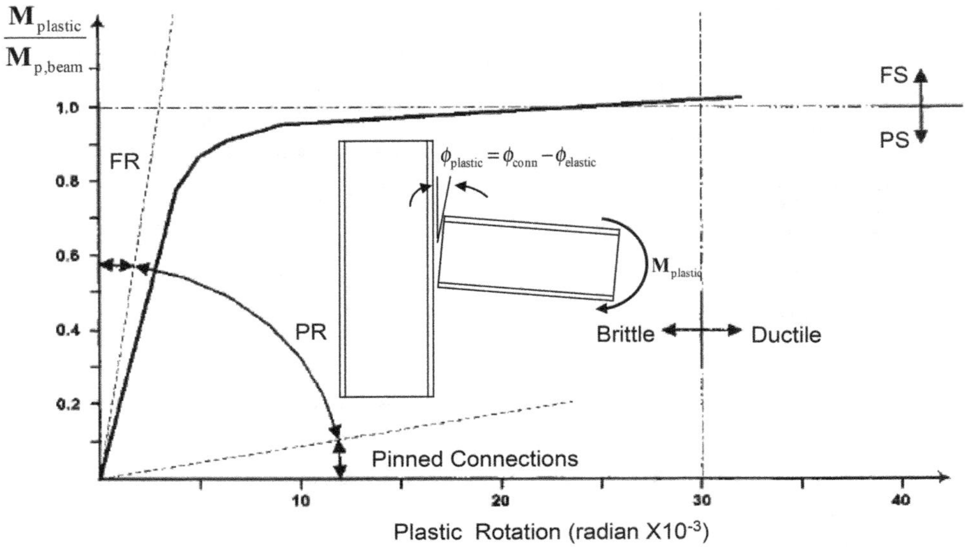


*Figure 1.3* Typical moment-rotation curve.

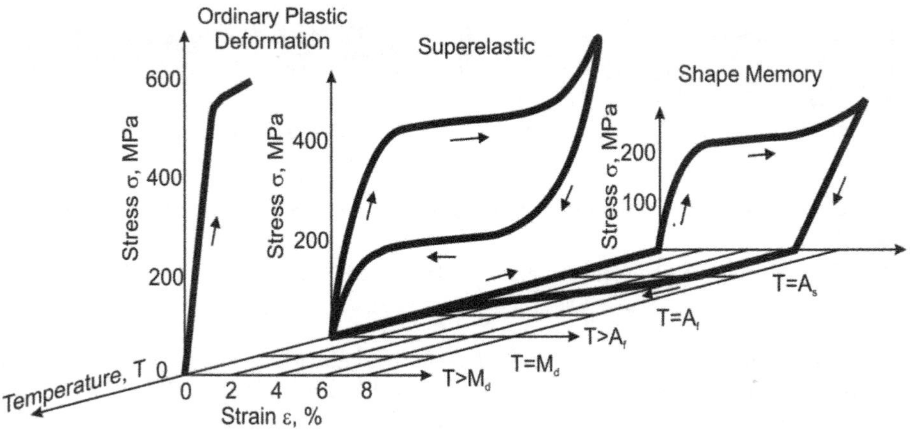


*Figure 1.4* Stress-strain-temperature relationships in SMA.

unique ability to sustain large strains (e.g. 6 to 8 percents) that are crystallographically reversible, thereby maintaining the material without residual deformation as illustrated in Fig. 1.4 (DesRoches *et al.*, 2004).

Utilizing super-elastic Nitinol tendons as the moment transfer elements in a steel beam-column connection will create smart structures that automatically adjust to seismic activity (Ocel *et al.*, 2004; Penar, 2005). This type of connection (See Fig. 1.5) not only contains all the advantages of bolted PR connections mentioned above, but also

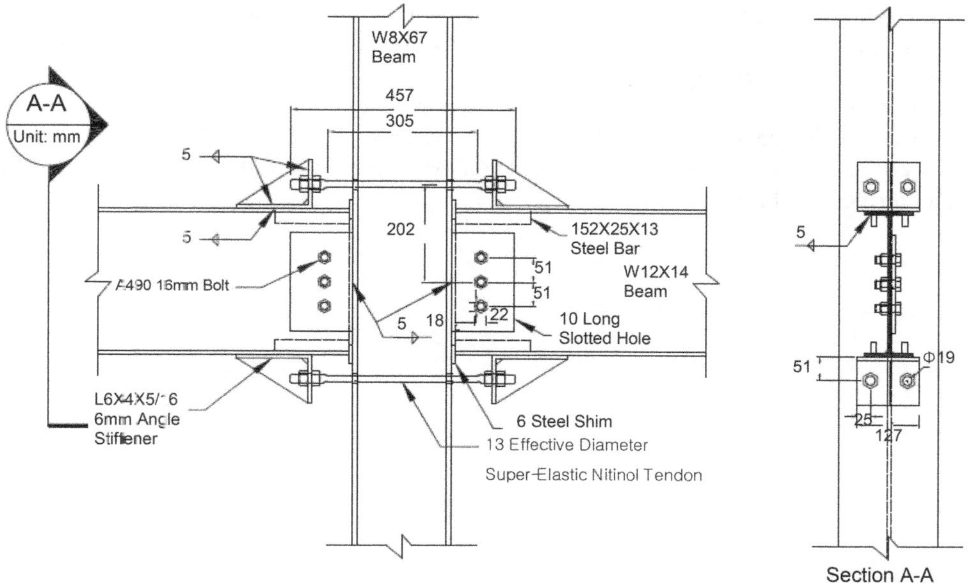


(a) Auto-CAD detail of the Super-elastic Nitionl tendon connection

(b) Picture of Nitinol tendon connection

*Figure 1.5* Super-elastic Nitinol tendon connection (Penar, 2005).

provides a recentering capacity because of the lack of permanent residual deformation in the tendons due to the SMA material characteristics. The ultimate purpose of this research is to develop suitable new design criteria for incorporating composite CFT structures into a partially restrained, partially strength connections. In addition, this research intends to explore a mixture of steel bars and super-elastic Nitinol bars as connecting elements to CFT columns. It is hypothesized that such combinations of CFT columns and SMA connections will achieve excellent ductility, high strength, and recentering capability.

Chapter 2

# Design and modeling for smart PR-CFT connection systems

## 2.1 HISTORY OF RELATED SYSTEMS

### 2.1.1 Related systems

This research is intended to take advantage of the synergetic characteristics of steel and SMA tendon bars to develop a flexible (PR) moment resisting connection with recentering capabilities. This chapter briefly describes some past experimental and analytical research on traditional PR/CFT connections relevant to the innovative types of connection models to be developed in this study. This chapter does not intend to provide a complete and systematic literature search on that topic but rather just present some examples of how connections have been investigated in the past. In addition, previous research on recentering connections as well as prior practical uses of shape memory alloys for seismic applications will be reviewed.

### 2.1.2 PR connection systems

Numerous investigations on a wide range of bolted connection types has been performed since the early 1900s both to understand behavior of various PR connection types and to model the connection behavior in the analysis of entire frames. Table 2.1 summarizes some of the data available for each type of PR connection. This review of the literature will describe the behavior and modeling of typical PR connections in steel moment frame construction (Fig. 2.1 and Fig. 2.2) as the goal of present work is to apply PR connections to composite structures.

Since the earliest tests aimed at determining the rotational stiffness of PR connections by Wilson and Moore (1917), hundreds of tests have been performed to establish the relationship between moments and relative rotations in beam-to-column connections. Before 1950s, tests of riveted connections were performed by Young and Jackson (1934) and Rathbun (1936). PR connections with high strength bolts as structural fasteners were first tested by Bell *et al.* (1958). Thereafter, behavior of header plate (or end plate) connections was investigated in twenty tests by Sommer (1969).

Extended end-plate and flush end-plate connections have been extensively accepted since the late 1960s. Flush end-plate and extended end-plate connections with performance close to that of rigid connections were tested by Ostrander (1970) and Johnstone and Walpole (1981), respectively. A series of tests on a variety of beam-to-column

*Table 2.1* Available experimental moment-rotational data for several connection types (Summarized by Chan and Chui, 2000).

| *Connection type* | *Reference* | *Fastener* | *Column axis restrained* | *Number of test* |
|---|---|---|---|---|
| Single Web Cleat | Lipson (1977) | Bolts | Major | 43 |
| | Lipson and Antonio (1980) | Bolts | Major | 33 |
| Double Web Cleat | Lathbun (1936) | Rivets | Minor | 7 |
| | Tompson *et al.* (1970) | Bolts | Major | 24 |
| | Bjorhovde (1984) | Bolts | Major | 10 |
| Header Plate | Sommer (1969) | Bolts | Major | 16 |
| | Kennedy and Hafez (1986) | Bolts | Major | 19 |
| Top and Seat Angle | Hechtman and Johnston (1947) | Rivets | 14 Major<br>5 Minor | 19 |
| | Azizinamini *et al.* (1985) | Rivets | Major | 20 |
| Flush End Plate | Zoetemeijer and Kolstein (1975) | Bolts | Major | 12 |
| | Ostrander (1970) | Bolts | Major | 24 |
| Extended End Plate | Sherbourne (1961) | Bolts | Major | 4 |
| | Bailey (1970) | Bolts | Major | 13 |

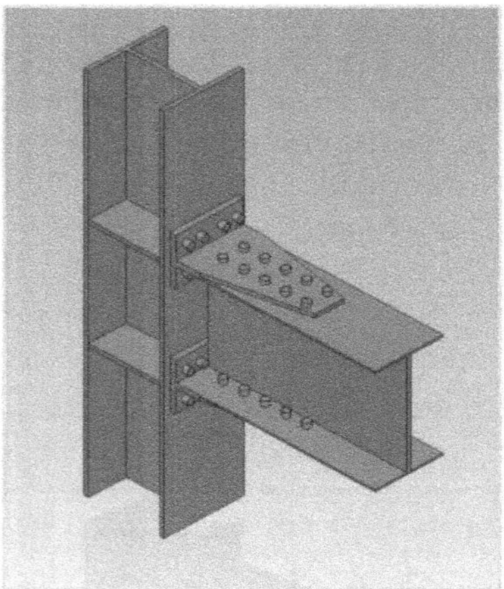

(a) T-Stub connection

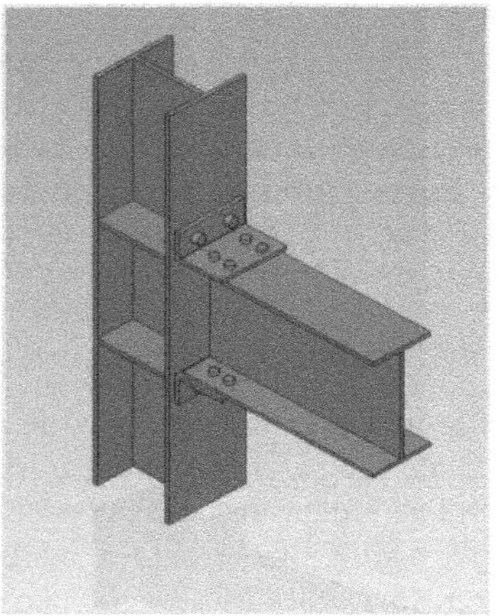

(b) Clip Angle connection

*Figure 2.1* T-Stub and Seat Angle connection configurations.

connections containing the web-cleat, flange cleat, seating cleat and web cleat, flush end-plate and extended end-plate connections were performed by Davison *et al.* (1987).

The earliest relevant T-stub connection research available was conducted by Batho and Rowan (1934). Eighteen beam-to-column tests were performed by Rathburn (1936). The work included the result of web angle, clip angle and T-stub connection tests. Following this work, forty seven nominally pinned connections were tested by Hechtman and Johnston (1947), who concluded that the connection slip contributes greatly to the overall rotation of a bolted or riveted connection. Dulty and McGuire

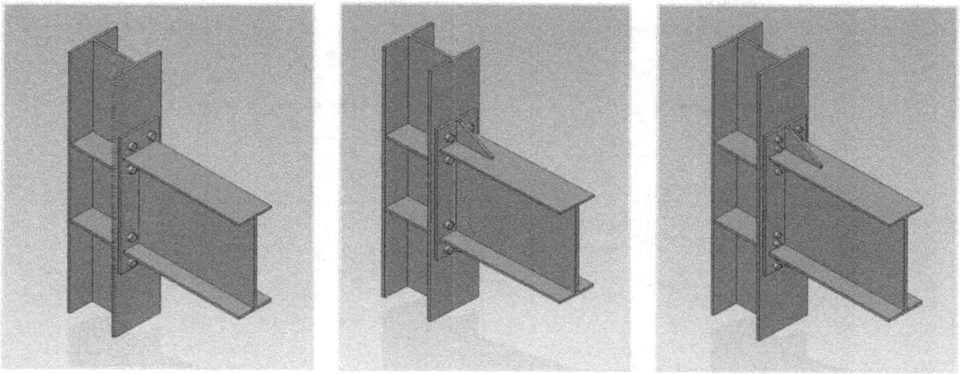

(a) Four bolt unstiffened, 4E (b) Four bolt stiffened, 4ES (c) Eight bolt stiffened, 8ES

*Figure 2.2* Extended End-Plate connection configurations.

(1964, 1965) carried out twenty seven component tests of wide flange and built up T-stubs in addition to fifteen splice plate tests intended to replicate the interaction between the T-stem and beam flange. Azizinamini (1982, 1985) performed an extensive and detailed experimental study for top and bottom seat angle connections with double web angles along with pull tests. Recently, Swanson and Leon (2000) performed tests on forty eight T-stub specimens in order to provide insight into failure modes, deformations, and ductility of these components. Smallidge (1999) and Schrauben (2000) also conducted tests on ten full scale T-stub and thick clip angle connection specimens and compared the results to the component tests performed by the SAC project (Swanson 1999).

The available data on cyclic behavior of PR connections was reviewed by Leon (1997) Similar surveys for monotonic load cases have been given by Bjorhovde (1984), Nethercot (1986), Chen and Lui (1991), and Chan and Chui (2000), among others. The reader is referred to those sources for more detailed descriptions.

### 2.1.3 Steel beam to CFT column connection

Experimental research on CFT connection details has been conducted on a wide variety of configurations depending on the tube shape and the desired connection performance. The beam-to-column connections used with CFT columns can be classified broadly into two categories. In the first connection category, the most convenient connection method is to weld the steel beam directly to the skin of the steel tube (Fig. 2.3(a)) or through the diaphragm (Fig. 2.3(b)). For this type of connection, the very large stresses and strains due to welding will lead to severe distortions of the tube wall. Shakir-Khalil (1992) tested structural steel girders connected to CFT columns using shear tabs which were fillet welded to the wall of circular steel tube columns. Many configurations for continuity diaphragms were tested by this research group in order to reduce severe distortions on the tube skin. Morino *et al.* (1992) used diaphragm plates continuous

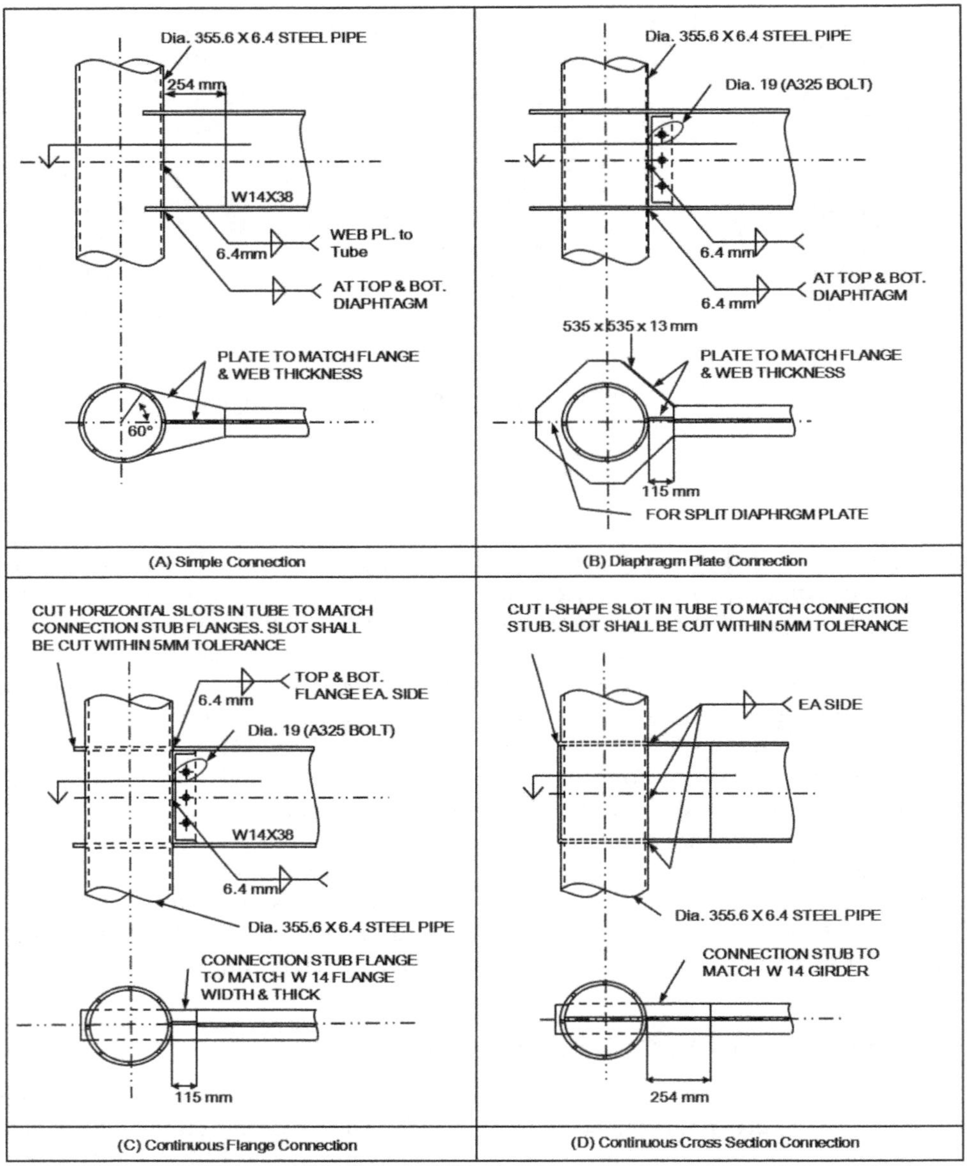


*Figure 2.3* Typical welded connections to composite columns (Alostaz and Schneider, 1997).

through square CFT columns at each girder flange location. The steel tube column was spliced and re-welded above and below each diaphragm.

In the second connection method, the beam flange, fastener, web or entire cross section goes though the steel tube (See Fig. 2.3(c)) or the girder end-plate is welded with embedded elements in the CFT column (See Fig. 2.3(d)). Embedding connection components into the concrete core reduces the high shear demand on the tube skin. Several tests were conducted by Kanatani *et al.* (1987) and Prion and McLellan (1992) on penetrated bolted connections to square tube columns. Kanatani used T-stub connection element by bolting the stem of the T-stub to the girder flanges and attaching the T-stub flanges to the column with bolts continuous through the CFT. Prion tested similar bolted connections but using end-plates fully welded to the girder. Azizinamini and Prakesh (1993) examined behavior of a beam-to-column connection in which the steel beam extended continuously through the CFT.

Alostaz and Scheider (1996, 1997) investigated six types of connection details with circular CFTs. These details were arranged in approximate order of increased fabrication difficulty. Alostaz and Scheider suggested four kinds of fabrication methods. The first one was embedding weldable deformed bars. It was shown through experimental and analytical results that the deformed bars could transfer the beam flange force to the concrete core. In the second method, headed studs were welded to the inside wall at the beam flange to alleviate severe distortion of the steel tube wall. In the third method, a configuration extending the web plate into the concrete core with attached headed studs was investigated. In the fourth method, continuing the beam through the depth of the CFT column was considered to be the most rigid connection type. The last connection type had the best seismic resistance behavior, but the fabrication difficulties are the main disadvantage of this connection type.

### 2.1.4 Application of SMA in structures

Smart structures for civil engineering are described as systems that can automatically adjust structural characteristics in response to external disturbances and unexpected severe loading. The idea is that the structure can be coaxed towards performance that results in improved structural safety, serviceability and extension of structural life (Otani, 2000). The focus in seismic design and retrofit has been towards performance-based design, often leading to structural solutions that make use of passive energy dissipation devices in order to mitigate inter-story drift and structural damage. One key avenue to achieve these goals is the development and implementation of smart materials. These materials exhibit synergetic functions such as sensing, actuating, self-recovery and healing. One example of smart material is a class of metals known as shape memory alloys (SMAs). When SMA are integrated within structures, SMA can act as passive, semi-active or active components to reduce damage under strong ground motions. SMA exhibit high power density, solid state actuation, high damping capacity, durability, and fatigue resistance.

The widest use of SMA for seismic applications is for passive structural control and self recentering applications in order to reduce the response to external disturbances and the resulting residual deformation. Saadat *et al.* (2002) suggested that SMA could be effectively used as the devices for passive structural control through two systems: a ground isolation system and an energy dissipation system.

With regard to a ground isolation system, SMA for isolators which connect a super-structure to the ground foundation can screen the seismic energy transferred from the ground acceleration to the superstructure. These systems have the ability to reduce the damage to the superstructure. Wilde *et al.* (2000) applied a base isolation

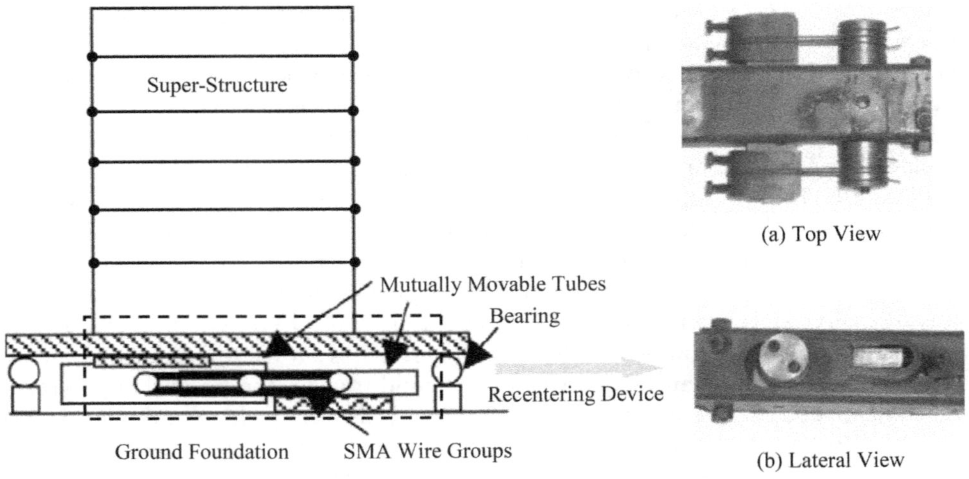


*Figure 2.4* Isolator systems for buildings.

system using super-elastic SMA bars to elevated highway bridges. Choi *et al.* (2005) proposed an isolation device in which SMA wires were incorporated in an elastomeric bearing to improve conventional lead-rubber bearings, which have problems related to instability and residual deformation under a strong ground motion. As the part of MANSIDE (Memory Alloys for New Seismic Isolation and Energy Dissipation Devices) project to study the feasible use of Nitinol wire for vibration isolation, Dolce *et al.* (2001) proposed and tested the three types of Nitinol wire based devices: supplemental recentering devices (SRCD), non recentering devices (NRD), and recentering devices (RCD). Uses of SRDC and SMA isolation systems in buildings are illustrated in Fig. 2.4. Khan and Lagoudas (2002) analytically studied SMA springs to isolate a single degree of freedom (SDOF) system from a ground excitation simulated by a shake table, while Corbi (2003) proposed SMA tendon to isolate a multi-story shear frame from the ground excitation.

With regard to an energy dissipation system, martensite or super-elastic SMA materials integrated into structures absorb vibration energy through hysteretic behavior. Some SMA energy dissipation devices that have been used include: braces for framed structures, dampers for simply supported bridges, and tendons for connection elements and retrofitting devices for historical building.

Casciati *et al.* (1998) studied the application of the large martensite Nitinol bar in seismic protection devices for bridges. They used a finite element model to analyze the device under both static and dynamic response. Dolce *et al.* (2000, 2001) proposed Nitinol wire recentering braces and tested several different scale prototypes of the devices. Tamai and Kitagawa (2002) proposed a combined steel-SMA type brace as their seismic resistance device. DesRoches and Delemont (2002) performed testing on a full scale super-elastic SMA bar and applied it to seismic retrofit of simply supported bridges. They also performed analytical simulations on a multi-span simply supported

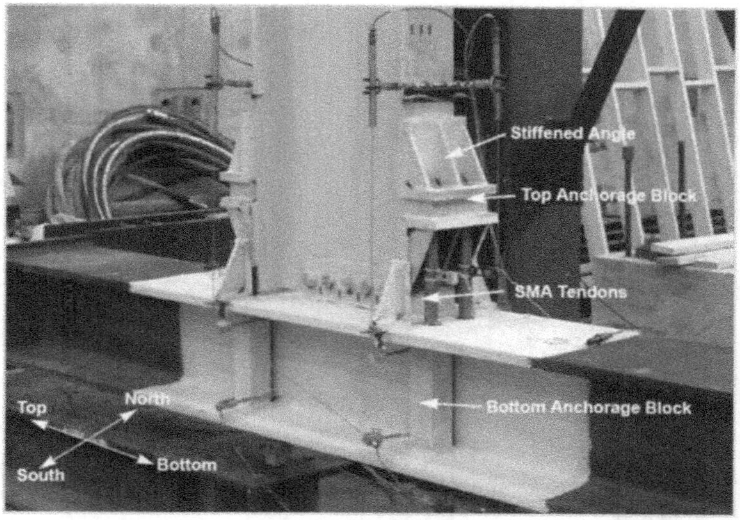


*Figure 2.5* Test specimen of beam to column connection using martensite Nitinol tendons (Ocel, 2002).

bridge with the SMA restrainer. Leon *et al.* (2002) applied martensite SMA tendons to the primary bending elements in a steel beam-column connections as shown in Fig. 2.5. Tami and Kitagawa (2002) suggested an exposed type column base with a SMA anchorage made of Nitinol SMA bars in order to resist seismic loading.

SMAs have been used to retrofit damaged structures. Super-elastic SMA bars were used by Indirli *et al.* (2001) so that they rehabilitated the S. Giorgio Chruch Bell Tower seriously damaged by the earthquake of Oct. 15th 1996 as shown in Fig. 2.6.

## 2.2 UNIQUE CHARACTERISTICS FOR PR-CFT CONNECTION STUDY

Several new types of connection details are proposed in this research. Contrasting these proposed configurations with those deecribed briefly in this chapter, it can be said that the characteristics of this research that set it apart from previous studies are:

- Complete Design Details for Several Innovative PR-CFT Connections: Several types of CFT connections with fasteners penetrating through the steel tube or the beam end are designed based on step by step procedures consistent with current code provisions. Strength models for connection components such as shear bolts, shear tab, web bolts, component angles, and plates are investigated in this research. The strength of each connection component obtained from the design strength model shold exceed the demand strength based on the full plastic capacity of the beam element in order to induce the ideal yielding/failure modes.
- Smart Structure Systems: The combined use of steel and SMA through bars in assemblies that connect beams to columns is examined. Super-elastic SMA bars

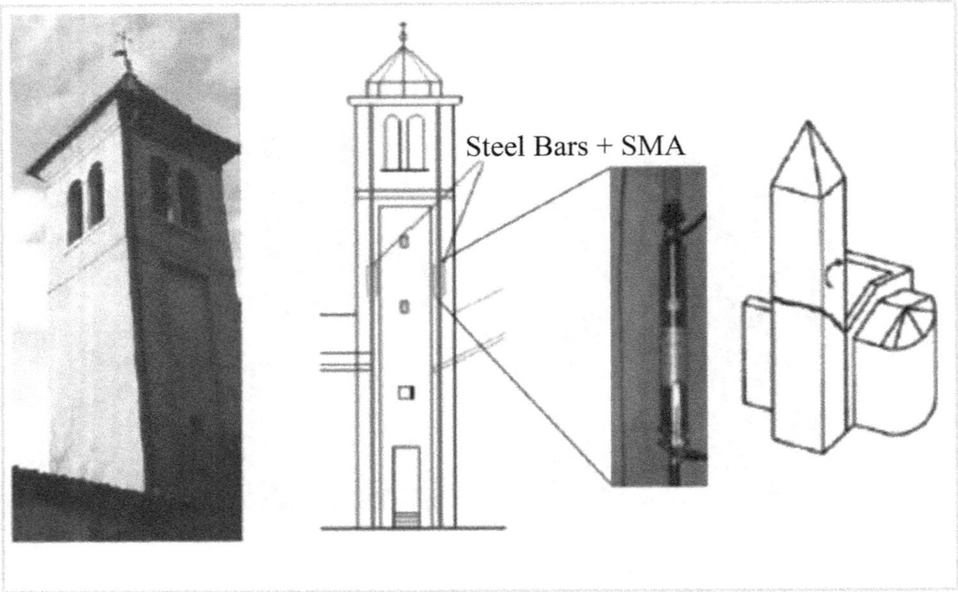


*Figure 2.6* Retrofit application using Nitinol devices – The bell tower of the S. Giorgio Church in Trignano (Indirli *et al.*, 2001).

have a strong recentering force to restitute the structure to its original position with little residual strain. Steel bars are used to help dissipate more energy and increase damping capacity.

- Refined FE Analysis Using Three-Dimensional (3-D) Elements: Nonlinear 3-D FE studies on a variety of details for smart SMA PR-CFT connections are presented in this thesis. Different types of 3-D elements include material inelastic and geometric nonlinear behavior. In order to formulate FE models very close to real connections in buildings, advanced modeling tools such as surface interactions, interface elements and initial pretensions are introduced into the FE models.
- Connection Behavior under Static and Cyclic Response: Behavior of connection components has a significant influence on that of full connections. First, behavior models for bolted components under either uniaxial static or uniaxial cyclic deformations are developed. These models intend to consider interactions between angles and bolts such as prying action, slip distance, bearing effects and various other possible failure modes. For the computational simulations, the behavior of the entire connections can be formulated based on superposition of the behavior models of individual components.
- Frame Analyses and Performance Evaluations: Two-dimensional (2-D) and 3-D composite frame structures will be designed and their performance assessed using typical PR moment-rotation behavior. Both building performance and new findings to estimate the building damage are developed from frame analyses consisting of both linear elastic analyses and second order inelastic analyses.

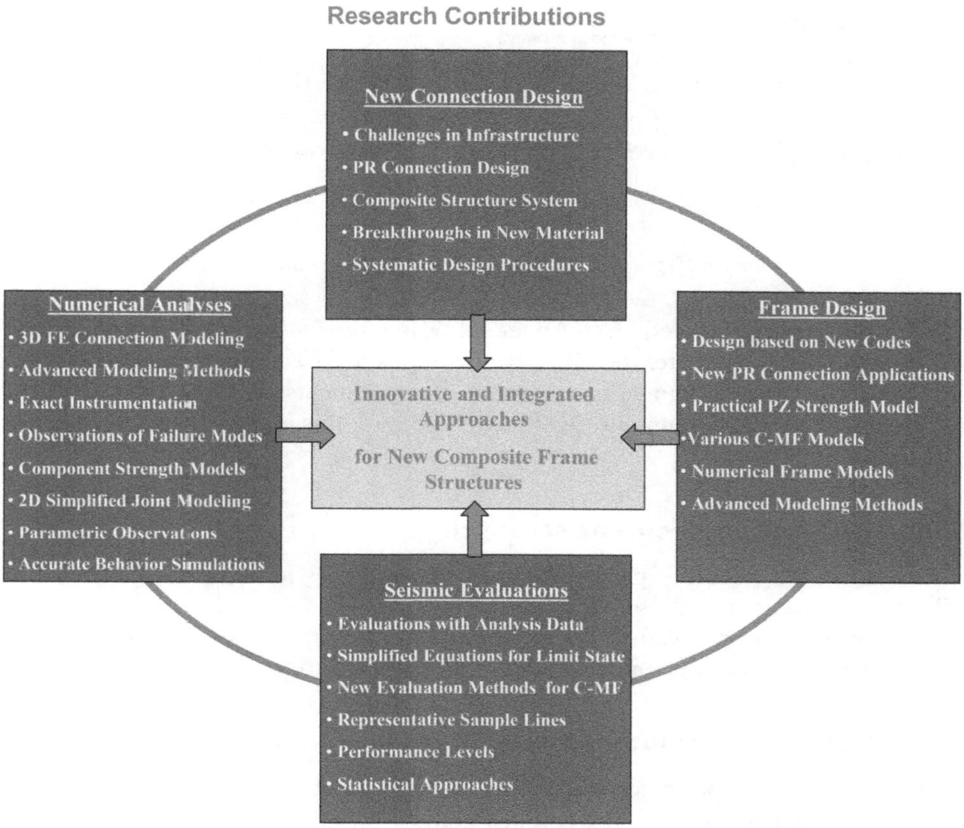


*Figure 2.7* Original contributions for this research.

## 2.3 DESIGN PROCEDURE FOR PROTOTYPE STRUCTURES

Several PR moment connections were designed including detailed designs of the panel zone, angle, web shear plate, stiffener, shear bolts, and tension fasteners. These connections are meant to be pre-qualified to meet the strength and ductility requirements in applicable design codes. The pre-qualification will be accomplished through the advanced analyses to be discussed in later chapters of this dissertation. The term "pre-qualified" implies that the connections will not be subject to the stringent physical testing requirements that connections used in steel and composite structures must currently undergo (2005 AISC Seismic Provisions).

The design of these connections is divided into two parts. In Section 2.3 of this chapter, the calculation of the resistance of the connection components and the composite column are described. In Section 2.4 a series of step-by-step seismic design procedures are proposed for PR-CFT connection systems Appendix A (Interaction Strength for CFT Columns) and Appendix B (Design Examples for PR Connections)

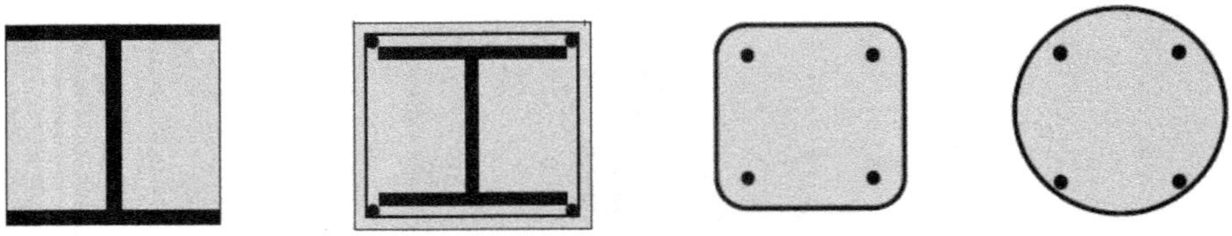


*Figure 2.8* Types of cross sections for composite column system.

are associated with this section. In Appendix A, it is shown that the *2005 AISC Specification* accurately evaluates the capacity of composite CFT columns when comparing its results with those obtained from numerical experiments. The SI unit systems (kN and m) will be used throughout this research because this is the current construction practice in the world.

### 2.3.1 Design requirement strength

This section describes the prequalification design limits and required strength for PR-CFT connections. The scope of this section includes overall models, geometric configurations, and rational formulas to estimate the capacity of these connections. These are shown, when possible, as step-by-step procedures.

### 2.3.2 Composite column strength

In steel-concrete composite structures, the advantages of two materials can be added while their disadvantages can are compensated by the composite effect. Two systems are widely utilized in the vertical members of composite construction: steel reinforced concrete (SRC), where a steel section is encased in concrete, and concrete filled tube (CFT) columns. Typical cross sections for the composite columns are illustrated in Fig. 2.8.

The connection developed in this research are intended to connect CFT columns to steel I-shape beams so only CFT column systems will be considered. The current applicable design code, the *2005 AISC Specification* (AISC 2005 Specification) includes design guidelines for composite columns consisting of rolled or built-up structural steel shapes, pipe or hollow steel section (HSS) and structural concrete acting together as a composite member. To qualify as a composite CFT column, the following limits listed in the *2005 AISC Specification* should be satisfied:

- The cross sectional area of HSS shall be at least 1 percent of the total composite cross section.
- The maximum width-thickness ratio for a rectangular HSS shall be equal or less than $2.26\sqrt{E/F_y}$.
- The maximum diameter-thickness ratio for a circular HSS filled with concrete shall be less than or equal to $0.15E/F_y$. Higher ratios are accepted when their use is verified by testing or analysis.

The design compressive strength, $\phi_c P_n$, for an axially loaded composite CFT columns should be determined for the limit state of flexural buckling based on the column slenderness ratio as shown below:

$\phi_c = 0.75$ (LRFD resistance factor for axially loaded columns)

(a) If $P_e \geq 0.44P_c$,

$$P_n = P_o[0.658]^{(P_o/P_e)} \tag{2.1}$$

otherwise,

$$P_n = 0.877P_\epsilon \tag{2.2}$$

where,

$$P_o = A_s F_y + A_{sr} F_{yr} + C_2 A_c f_c' \tag{2.3}$$

$C_2 = 0.85$ for rectangular CFT sections and $C_2 = 0.95$ for circular CFT sections

$$P_e = \pi^2 (EI_{eff})/(KL)^2 \tag{2.4}$$

where,

$$EI_{eff} = E_s I_s - 0.5 E_s I_{sr} + C_1 E_c I_c \tag{2.5}$$

$$C_1 = 0.6 + 2\left(\frac{A_s}{A_c + A_s}\right) \leq 0.9 \tag{2.6}$$

$EI_{eff}$ = effective stiffness of the composite section
$A_s$ = area of the steel section
$A_c$ = area of the concrete section
$A_{sr}$ = area of continuous reinforcing bars
$E_c$ = elastic modulus of concrete = $w_c^{1.5}\sqrt{f_c'}$ ksi
$E_s$ = elastic modulus of steel = 29000 ksi
$f_c'$ = specific compressive strength of concrete
$F_y$ = specific minimum yield stress of steel
$F_{yr}$ = specific minimum yield stress of reinforcing bars
$I_c$ = moment of the inertia of the concrete section
$I_s$ = moment of the inertia of the steel section
$I_{sr}$ = moment of the inertia of the reinforcing bars
$K$ = effective length factor determined in the boundary conditions
$L$ = laterally untraced length of the members
$w_c$ = weight of concrete per unit volume

The design tensile strength, $\phi_t P_n$, for filled composite columns should be determined for the limit state of yielding, neglecting the tensile strength of concrete, as shown below:

$\phi_t = 0.90$ (LRFD resistance factor for columns under tension)

$$P_o = A_s F_y + A_{sr} F_{yr} \quad (2.7)$$

In addition to the available axial strength, the flexural strength also needs to be calculated. The *2005 AISC Specification* adopts a full plastic stress distribution based on the assumption of linear strain across the section and perfect elasto-plastic material behavior. With these simple assumptions, the nominal strength can be estimated by assuming that the steel has reached its yield stress under either tension or compression and that the concrete has reached the crushing strength under compression (Fig. 2.9).

The P-M interaction diagram illustrated in Fig. 2.10 for a composite section is based on a full plastic stress distribution and can be approximated by a conservative linear interpolation between five points (Galambos, 1998). Points (A) and (B) correspond to the crushing axial strength and the flexural strength of the section, respectively. Point (C) is anchored to the same plastic neutral axis (PNA) position from corresponding to that of Point (B) but on the other side of the centerline, so it contains the same flexural capacity as Point (B) and the same magnitude of axial resistance from the concrete alone. For Point (D), the PNA is located at the centerline. As a result, this point corresponds to the maximum flexural strength and one half of axial strength of that determined for Point (C). Point (E) is an additionally arbitrary point to better describe the curvature of the interaction diagram at high axial loads. The five points can be easily calculated as shown in Table 2.2. For design, a simplified bilinear interpolation may be used between Point (A), (C), and (B) as shown in Fig. 2.10. The simplified interaction equations shown as Eq. 2.8 and Eq. 2.9 can be used to check the design of composite beam-columns. They are reasonably accurate and conservative. Exact expressions will be too cumbersome to use in everyday design practice.

If $P_r \leq P_D$

$$\frac{M_{rx}}{M_{Bx}} + \frac{M_{ry}}{M_{By}} \leq 1.0 \quad (2.8)$$

otherwise

$$\frac{P_r - P_D}{P_A - P_D} + \frac{M_{rx}}{M_{Bx}} + \frac{M_{ry}}{M_{By}} \leq 1.0 \quad (2.9)$$

where,
$M_A = \cdots = M_E$ = allowable flexural strength (Capital subscript indicates the observed point)
$M_r$ = required flexural strength
$P_A = \cdots = P_E$ = allowable axial force (Capital subscript indicates the observed point)
$P_r$ = required axial strength
$x$ = subscript referring to symbol related to strong axis bending
$y$ = subscript referring to symbol related to weak axis bending

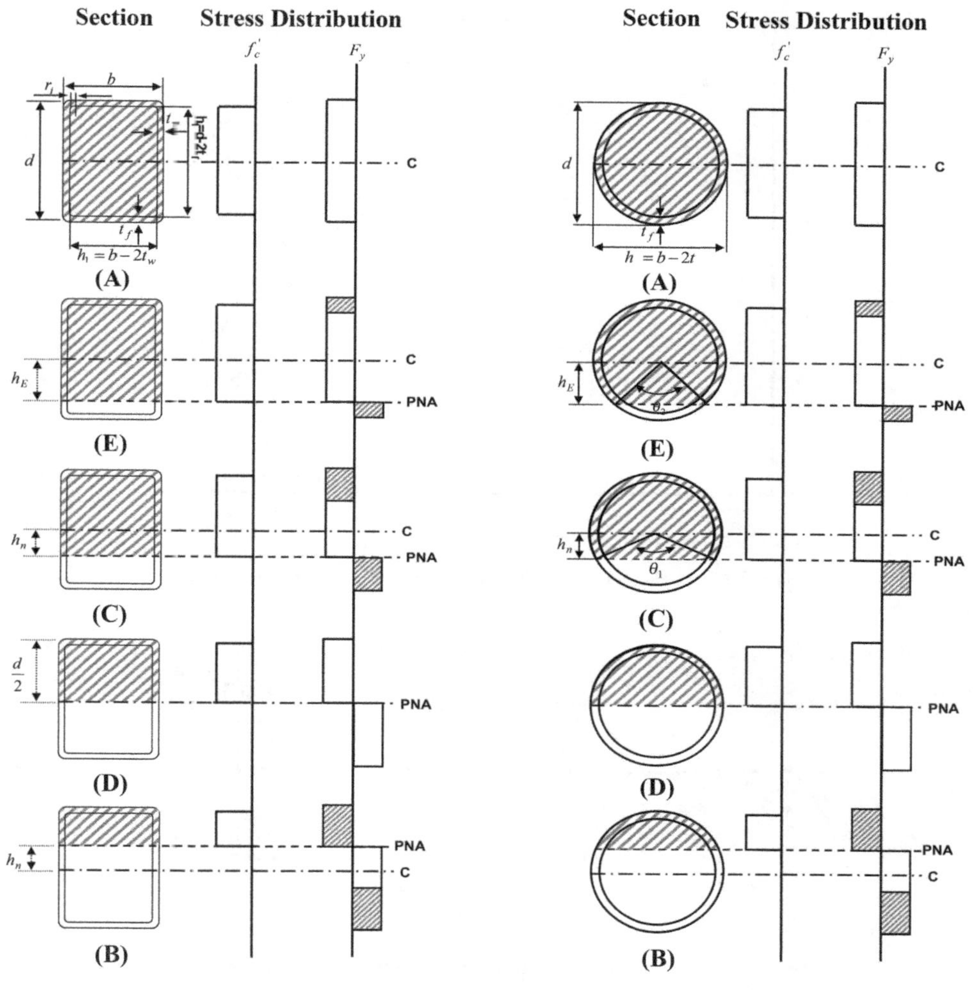


*Figure 2.9* Full plastic stress distributions for RCFT and CCFT at point A, E, C, D, and B.

### 2.3.3 Component member strength

This section presents the determination of the strength for the components which connect the I-shape or wide flange beam to the column. Force transfer components include tension bolts/bars, shear bolts, shear tabs, web bolts, plates, and angles. The prediction of the ultimate strength for connections is a quite complex process because the yielding and failure modes of the components interact with one another and are tied to uncertainties in material properties and fabrication/construction tolerances. Typical T-stub, clip angle and end-plate connections, respectively, were shown in Figs. 2.1 and 2.2.

A complicating factor for both T-stub connection and clip angle connections (See Fig. 2.1) is the need to include slip in the moment-rotational behavior due to the shear

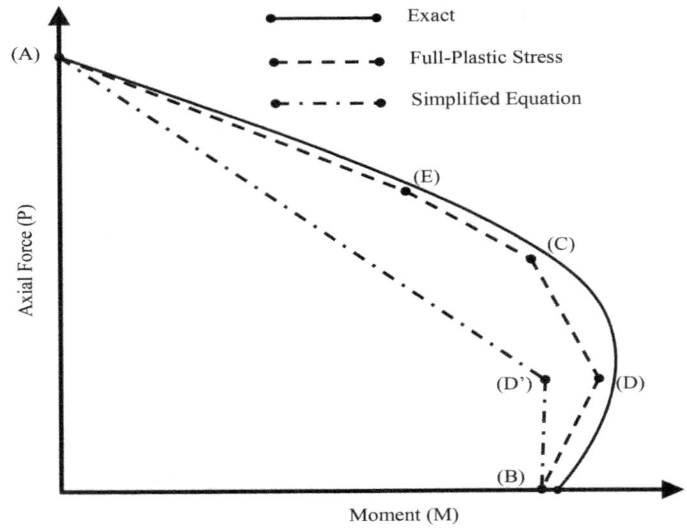


*Figure 2.10* P-M interaction diagrams for composite beam-columns.

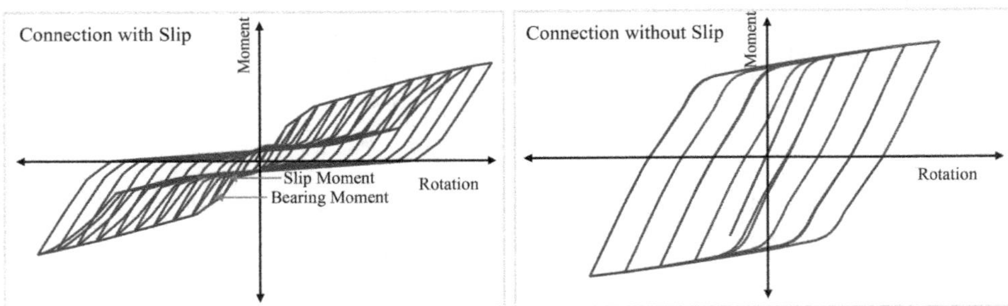


*Figure 2.11* Behavior of PR moment connections used in this study.

force acting on the faying surface between the beam flange and the corresponding T-stub stem/clip angle leg. On the other hand, end-plate connections (See Fig. 2.2) which are fabricated by directly welding the beam to the moment plate, do not produce slip for lack of the faying surface. Fig. 2.11 shows schematically the influence of slip on the monotonic and cyclic behavior of bolted connections. Component member strength will be classified according to the existence of slip.

### 2.3.3.1 *Components with slip*

Slip between surfaces of bolted components occurs as the shear force on the slip plane exceeds that provided by the clamping force from the pretensioned bolts. Slip occurs because construction tolerances require that bolt holes be at least 1.6 mm larger than

*Table 2.2* Equations for the specific 5 points in the P-M interaction diagram.

| *Point* | *Equation for RCFT* | *Equations for CCFT* |
|---|---|---|
| (A) | $P_A = A_sF_y + A_c(0.85f'_c)$<br>$M_A = 0$<br>$A_s$ = area of steel shape<br>$A_c = h_1h_2 - 0.858r_i^2$ | $P_A = A_sF_y + 0.95f'_cA_c^*$ $\quad r_m = \frac{d-t}{2}$<br>$M_A = 0 \quad A_c = \frac{\pi h^2}{4}$<br>$A_s = \pi r_m t$ |
| (E) | $P_E = \frac{1}{2}(0.85f'_c)A_c + 0.85f'_c h_1 h_E + 4F_yt_wh_E$<br>$M_E = M_{max} - \Delta M_E$<br>$Z_{sE} = bh_E^2 - Z_{cE} \quad z_{cE} = h_1h_E^2$<br>$\Delta M_E = Z_{sE}F_y + 1/2Z_{cE}(0.85f'_c)$<br>$h_E = \frac{h_n}{2} + \frac{d}{4}$ | $P_E = (0.95f'_cA_c + F_yA_s) - 1/2$ $[F_y(d^2 - h^2) + 1/2(0.95f'_c)h^2]\,[\theta_2/2 - \sin\theta_2/2]$<br>$M_E = Z_{sE}F_y + 1/2Z_{cE}(0.95f'_c) \quad h_E = \frac{h_n}{2} + \frac{d}{4}$<br>$Z_{sE} = \frac{d^3\sin^3(\theta^2/2)}{6} - z_{cE} \quad Z_{cE} = \frac{h^3\sin^3(\theta^2/2)}{6}$<br>$\theta_2 = \pi - 2\arcsin(2h_E/h)$ |
| (C) | $P_C = A_C(0.85f'_c)$<br>$M_C = M_B$ | $P_C = 0.95f'_cA_c$<br>$M_C = M_B$ |
| (D) | $P_D = \frac{0.85f'_cA_c}{2}$<br>$M_D = Z_sF_y + 1/2Z_c(0.85f'_c)$<br>$Z_s$ = full y-axis plastic section modulus of steel shape<br>$Z_c = \frac{h_1h_2^2}{4} - 0.192r_i^3$ | $P_D = \frac{0.95f'_cA_c}{2}$<br>$M_D = Z_sF_y + 1/2Z_c(0.95f'_c)$<br>$Z_s$ = plastic section modulus of steel shape $= \frac{d^3}{6} - Z_c$<br>$Z_c = \frac{h^3}{6}$ |
| (B) | $P_B = 0$<br>$M_B = M_D - Z_{sn}F_y - 1/2Z_{cn}(0.85f'_c)$<br>$Z_{sn} = 2t_wh_n^2$<br>$Z_{cn} = h_1h_n^2$<br>$h_n = \frac{0.85f'_cA_c}{2[0.85f'_ch_1 + 4t_wF_y]} \le \frac{h_2}{2}$ | $P_B = 0$<br>$M_B = Z_{sB}F_y + 1/2Z_{cB}(0.95f'_c)$<br>$Z_{sB} = \frac{d^3\sin^3(\theta/2)}{6\,\theta_1} - Z_{cB} \quad K_c = f'_ch^2$<br>$Z_{cB} = \frac{h^3\sin^3(\theta/2)}{6} \quad K_s = F_yr_mt$ ("thin" HSS)<br>$\theta_1 = \frac{0.0260K_c - 2K_s}{0.0848K_c} = \frac{\sqrt{(0.0260K_C + 2K_S)^2 + 0.857K_CK_S}}{0.0848K_C} \quad h_n = \frac{h}{2}\sin\left(\frac{\pi-\theta}{2}\right) \le \frac{h}{2}$ |

the nominal bolt diameter. Once this tolerance is exceeded, the bolt begins to bear on the plates and the stiffness and strength increase again. The amount of rotation that will result from a 1.6 mm slip on a 24 inch deep connection is about 0.005 radian, a non-trivial value if one assumes that typical connections are assumed to reach their yield strength at about 0.01 radian.

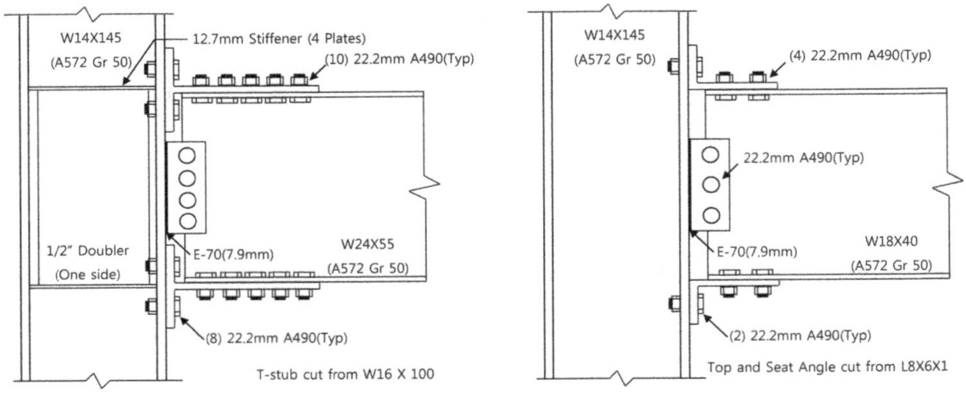


(a) Typical T-stub connection

(b) Typical thick clip angle connection

*Figure 2.12* Typical connection types with friction slippage (SAC Project).

*Table 2.3* Current LRFD design resistance factors.

| *Failure modes* | *The value of design resistance factor* |
|---|---|
| Member yielding | 0.9 |
| Bending and plastic moment | 0.9 |
| Compression buckling | 0.85 |
| Bolt fracture | 0.75 |
| Net section and block shear failure | 0.75 |
| Bolt bearing failure | 0.75 |
| Member rupture | 0.75 |

Slip gives rise to temporary loss of stiffness that acts as a fuse during reversed cyclic loading (Astaneh-Asl 1995). Slip limits the force which is transmitted though the bolts at a given deformation and produces significant energy dissipation and damping. Fig. 2.12 shows typical T-stub and clip angle connections that will exhibit slip. This section discusses the existing strength models for components which are cut from standard rolled shapes.

(A) Design Resistance Factor
The design strength must equal or exceed the required strength ($P_r$, $M_r$, and $V_r$). The design strength is computed as the product of the resistance factor ($\phi$) and the nominal strength ($P_n$, $M_n$, and $V_n$) per the current AISC LRFD provisions (AISC 2001). In general, the resistant factor is less than unity. The value usually depends on the accuracy of the models used to estimate nominal capacities, the desirability of specific failure modes and the scatter on material properties. Representative resistance factors for connection design are shown in Table 2.3.

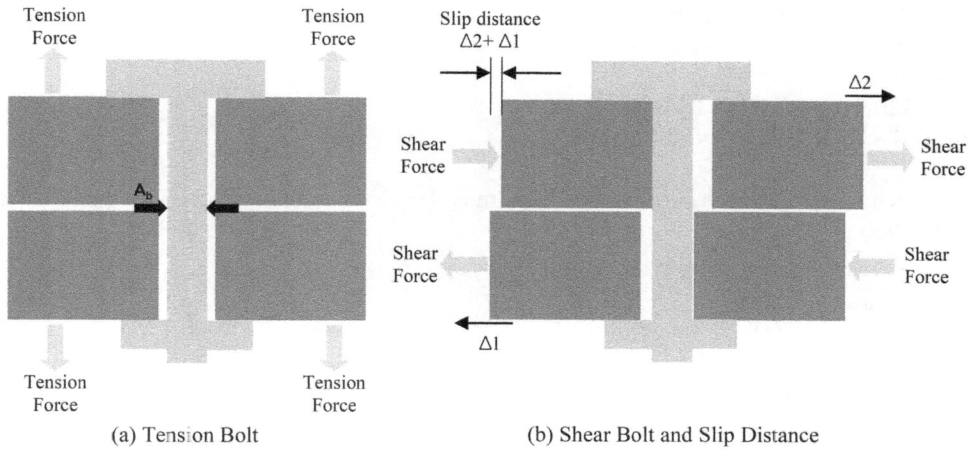


*Figure 2.13* View of a connection with a bolts in tension and shear.

(B) Slip Resistance
Slip is produced by the shear force parallel acting to the faying surface as shown in Fig. 2.13. Expressions for the nominal resistance to slip are based on simple friction models calibrated to numerous tests on bolted splices. The design slip resistance ($\phi R_{n,slip}$) is given in Section 16.4–34 of the LRFD (AISC 2005) as follows:

$\phi_f = 1.0$ for the slip

$$R_{n,slip} = 1.13uh_{sc}T_bN_s\left(1 - \frac{T_u}{1.13T_bN_b}\right) \tag{2.10}$$

where,
$h_{sc} = 1.00$ for standard holes
$h_{sc} = 0.85$ for oversized and short-slotted holes
$h_{sc} = 0.70$ for long-slotted holes transverse to the direction of the load
where,
$N_b$ = the number of bolts
$N_s$ = the number of shear plane
$T_b$ = specified minimum bolt pretension (i.e. A490 bolt with $d_b = 25.4$ mm has 284.4 kN)
$T_u$ = the required strength in tension
and where,
$u$ = the mean slip coefficient
$u = 0.33$ for Class A faying surfaces
$u = 0.50$ for Class B surfaces

(C) Tensile Strength of Bolts
The design tensile strength of a bolt ($\phi_f$ $B_{n,tension}$) can be taken as (AISC 2001):

$$B_{n,tension} = A_bF_t \tag{2.11}$$

where,
$\phi_f = 0.75$ for the bolt fracture
$A_b =$ nominal area of a bolt shank
and where,
$F_t = 620.1$ MPa for A325 bolts
$F_t = 778.57$ MPa for A490 bolts

Alternatively, the strength of a bolt may be determined by another procedure (Swanson 1999):

$$B_{n,tension} = A_{be}F_u \tag{2.12}$$

$$A_{be} = \frac{\pi}{4}\left(d_b - \frac{0.9743}{n_{th}}\right)^2 \tag{2.13}$$

where,
$A_{be} =$ effective tensile area of the bolt's threaded portion
$d_b =$ diameter of the bolt's shank
$n_{th} =$ number of treads per inch for the bolt (Table 8.7 in LRFD)
$F_u =$ ultimate strength of the bolt
$F_u = 723.45$ MPa for A325 bolts larger than 25.4 mm in diameter
$F_u = 826.8$ MPa for A325 bolts up to 25.4 mm in diameter
$F_u = 1033.5$ MPa for A490 bolts

(D) Shear Strength of Bolts
The design shear strength of a bolt ($\phi B_{v,shear}$) is given by (AISC 2001) as:

$$B_{v,shear} = A_bF_v \tag{2.14}$$

where,
$\phi_f = 0.75$ for the bolt fracture
$A_b =$ nominal area of a bolt shank
$F_v =$ ultimate shear strength
$F_v = 330.72$ MPa for A325 bolts with threads included form the shear plane
$F_v = 413.4$ MPa for A325 bolts with threads excluded from the shear plane
$F_v = 413.4$ MPa for A490 bolts with threads included form the shear plane
$F_v = 516.75$ MPa for A490 bolts with threads excluded form the shear plane

(E) Prying Action Mechanism
Fig. 2.14 shows the flange of components prior to the tension bolt failure. Prying action refers to the additional forces due to the reactions at the tip of the uplifting plate shown in these photographs; these additional forces increase the tension in the bolts and can lead to premature failure.

The basic prying mechanism and the fundamental equilibrium equation ($B = T + Q$) are shown in Fig. 2.15(a). In general, the prying force (Q) acting on the tip of the flange can be minimized by either increasing the member thickness or reducing the tension bolt gauge length.

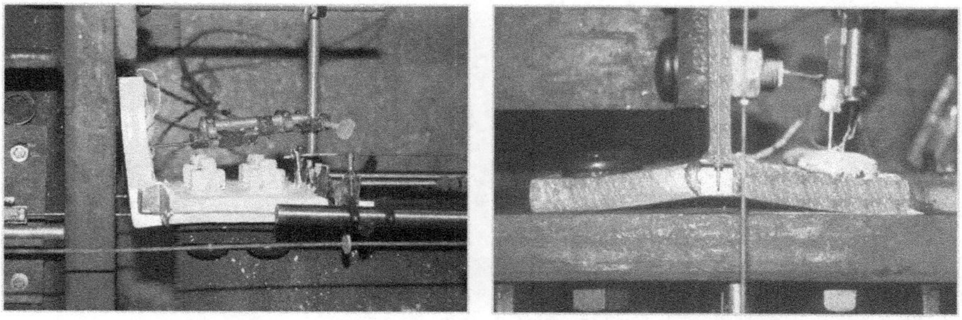

(a) Flange prying of clip angle component

(B) Flange prying of T-stub component

*Figure 2.14* Components before tension bolt fracture (SAC Project).

The prying model used in the LRFD (AISC 2001) is based upon one of the most widely used models developed by Kulak, Fisher and Struik (1987). In this model, the bolt force is assumed to act at the inside edge of the bolt shank instead of acting at the centerline of the bolt shank. For this model, the moment is shown in Fig. 2.15(b) and the geometry in Fig. 2.15(c).

The ultimate capacity of the component members is computed based upon considering three possible failure modes (Fig. 2.16). These three failure modes can be expressed by Eq. 2.15 to Eq. 2.17. They correspond to formation of a plastic mechanism on the flange, bolt prying mixed with flange yielding (Thornton 1985), and tension bolt fracture without the prying force, respectively.

Plastic mechanism formation:

$\phi_y = 0.9$ for the plastic flange

$$\mathrm{T} = \frac{(1+\delta)\mathrm{pF_y}t_f^2}{4b'} \tag{2.15}$$

Bolt prying mixed with the flange yielding:

$\phi_f = 0.75$ for the bolt fracture and $\phi_y = 0.9$ for the plastic flange

$$\mathrm{T} = \frac{\mathrm{B_{n,tension}a'}}{\mathrm{a'+b'}} + \frac{\mathrm{pF_y}t_f^2}{\mathrm{4b'}} \tag{2.16}$$

Tension bolt fracture without the prying force:

$\phi_f = 0.75$ for the bolt fracture

$$\mathrm{T} = \mathrm{B_{n,tension}} \tag{2.17}$$

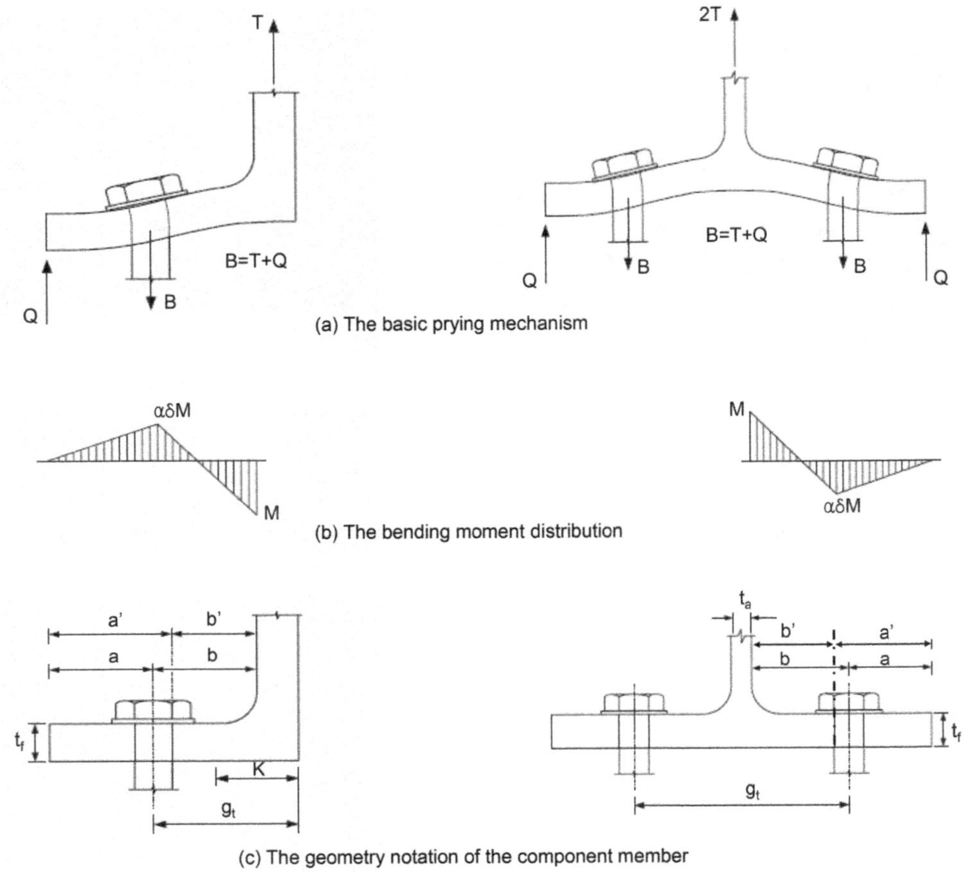


*Figure 2.15* Typical flange prying action.

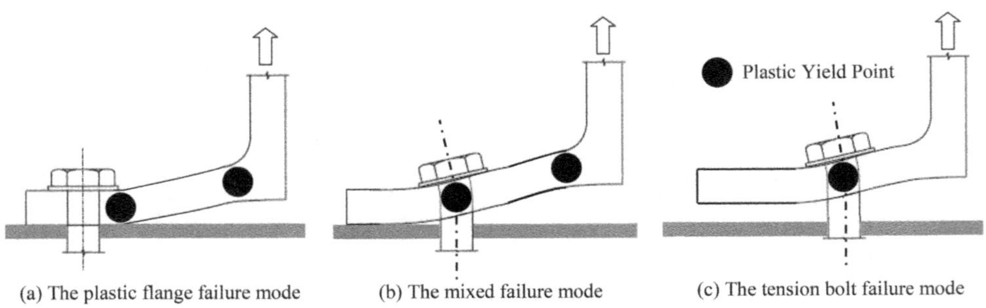


*Figure 2.16* Three possible failure modes.

where,

$$p = \frac{2W_{T\text{-stub}}}{n_{tb}} = \frac{W_{Clip}}{n_{tb}} \tag{2.18}$$

$n_{tb}$ = number of tension bolts connecting the component member
p = effective width per tension bolt
$W_{T\text{-stub}}$ = width of the T-stub at the flange
$W_{Clip}$ = width of the clip angle normal to the section area
a = distance from the centerline of the tension bolt to the edge of the flange
b = distance from the centerline of the tension bolt to the surface of the clip leg/T-stem
B = bolt reaction force
M = moment capacity of the flange
$t_a$ = thickness of T-stem
$t_f$ = thickness of the flange
$g_t$ = gauge length
T = applied tension force equivalent to one tension bolt
Q = prying force per bolt

and where,

$$\delta = 1 - \frac{d_h}{p} \tag{2.19}$$

$$a' = a + \frac{d_b}{2} \tag{2.20}$$

$$b' = b - \frac{d_b}{2} \tag{2.21}$$

$$\alpha = \frac{1}{\delta}\left(\frac{Tb'}{M} - 1\right) \tag{2.22}$$

$\delta$ = ratio of the net section area to the gross flange area
$d_h$ = diameter of the bolt hole
$\alpha$ = parameter for the level of prying present

The parameter ($\alpha$) is an indicator of the level of prying present (Kulak *et al.*, 1987). When $\alpha$ excess unity ($\alpha \geq 1.0$), the thickness of the flange is sufficient to cause the plastic flange mechanism to form as if the flange were a fixed-fixed beam (Fig. 2.16(a)) When $\alpha \leq 0$, the flange can separate from the contact surface and the combination of bolt fracture due to tension and flange yielding is dominant (Fig. 2.16(b)).

(F) Bearing Strength at Bolt Holes
The bearing strength can be determined by the sum of the strengths of the connected material at the individual bolt holes (AISC 2001, Section J3). The design resistance ($\phi R_{n,bearing}$) due to bearing in a standard bolt hole, oversized bolt hole, or short-slotted bolt hole is taken as either:

$$R_{n,bearing} = 1.2L_c tF_u \leq 2.4d_b tF_u \tag{2.23}$$

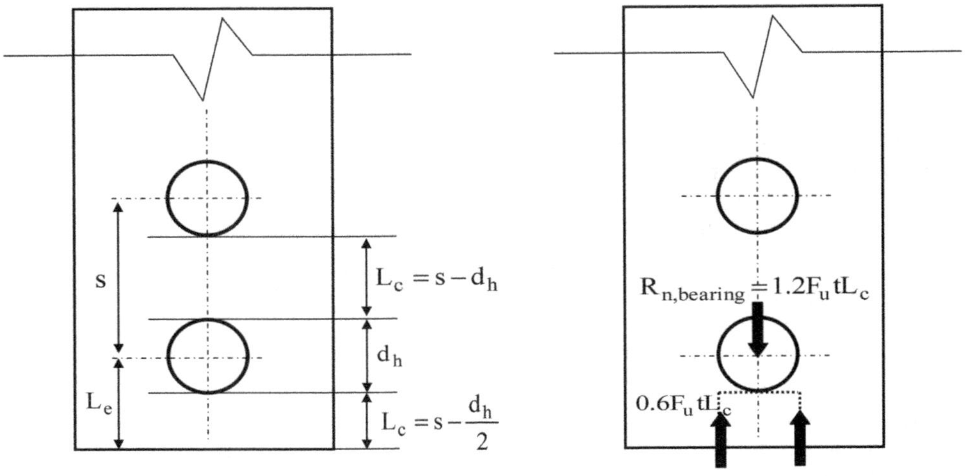


(a) The definition of dimension (b) The bearing strength

*Figure 2.17* Bolt bearing strength.

if deformation of the bolt hole under the service load is a design consideration, or

$$R_{n,bearing} = 1.5L_c tF_u \leq 3.0d_b tF_u \tag{2.24}$$

if deformation of the bolt hole under the service load is not a design consideration. For both cases, $\phi_f = 0.75$

The design resistance ($\phi_f\ R_{n,bearing}$) due to bearing in a long-slotted bolt hole is taken as:

$$R_{n,bearing} = L_c tF_u \leq 2.0d_b tF_u \tag{2.25}$$

with $\phi_f = 0.75$ for the resistant factor for the bearing failure.
In Eqs. (2.23) thru (2.25),

$F_u$ = specified minimum tensile strength of the connected material
$L_c$ = clear distance (Fig. 2.17)
s = bolt spacing
t = thickness of the connected material

(G) Net Section Strength
In general, the net section is defined as the cross sectional area excluding any area lost to drilling or punching of the bolt holes. A stress concentration occurs around the edges of bolt holes as shown in Fig. 2.18(a). Yielding is concentrated along the lines connected by the shear bolt holes and the tapered edge in the stem. The yielding will lead to a fracture of the T-stem along the lines shown in Fig. 2.18(b). The resistance

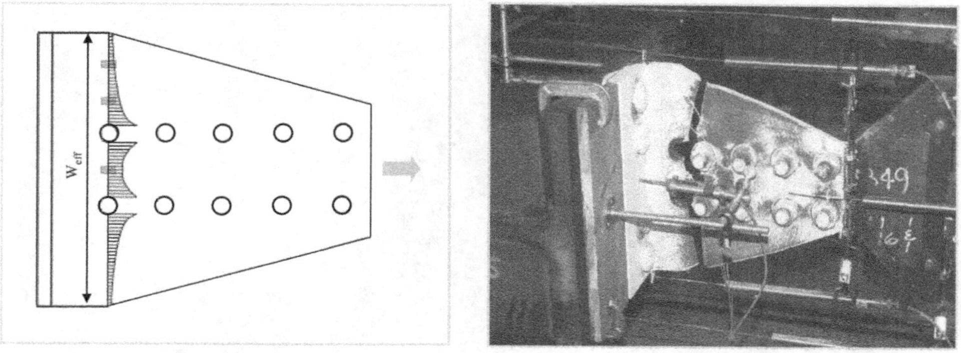


(a) The stress distribution in the T-stub member (b) The typical net section fracture (Swanson, 1999)

*Figure 2.18* **Stress distribution and net section fracture in T-stub members.**

of the stem against the net section fracture ($\phi_t R_{n,net}$) can be taken as follows (AISC 2001):

$$R_{n,net} = F_u A_{net,stem} \tag{2.26}$$

where,

$$A_{net,stem} = (W_{eff} - n_{sb} d_h)t \tag{2.27}$$

$A_{net,stem}$ = stem net section area
$n_{sb}$ = number of shear bolts along the effective width
$W_{eff}$ = effective width
$\phi_t$ = for the resistant factor for fracture on the net section.

(H) Block Shear Failure
Block shear failure is a combined failure, with one surface fracturing as the other one yields. It consists of either shear yielding plus tensile fracture as shown in Eq. 2.28 or shear fracture plus tensile yielding as shown in Eq. 2.29. Components that have failed by block shear are illustrated in Fig. 2.19. The design resistance against the block shear failure ($\phi_f R_{n,block}$) is determined in accordance with the model suggested by the LRFD (AISC 2001):

$$\phi_f = 0.75 \text{ for the resistant factor for the fracture}$$

If $F_u A_{nt} \geq 0.6 F_u A_{nv}$

$$R_{n,block} = 0.6 F_y A_{gv} + F_u A_{nt} \tag{2.28}$$

otherwise

$$R_{n,block} = 0.6 F_u A_{nv} + F_y A_{gt} \tag{2.29}$$

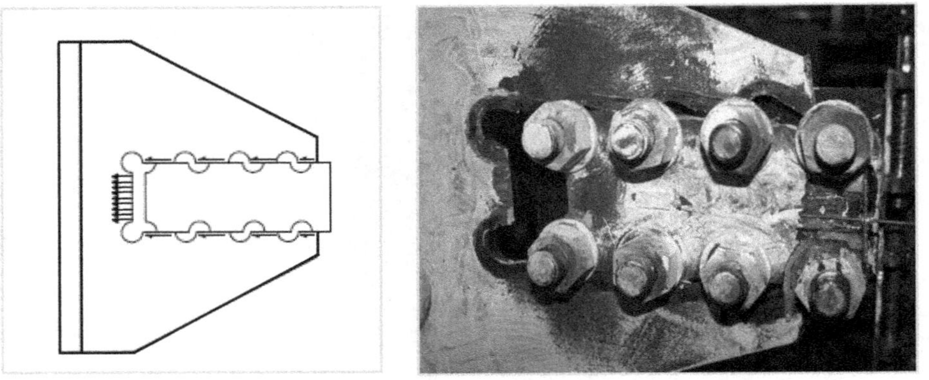

(a) The force distribution in T-stub member (b) The typical block shear failure of T-stub member

*Figure 2.19* Block shear failure mechanism (Swanson, 1999).

where,
$A_{gv}$ = gross section area under shear force
$A_{gt}$ = gross section area under tension force
$A_{nv}$ = net section area under shear force
$A_{nt}$ = net section area under tension force

#### 2.3.3.2 *Components without slip*

The end-plate connection is composed of a plate shop welded to the tip of the beam and tension bolted to the member. End-plate moment connections such as that shown in Fig. 2.20 are generally referred to as four tension bolt type without slip. Reliable welding can be achieved with end plate connections because they are generally fillet welds executed in the shop as compared to complete joint penetration welds executed in the field for traditional full moment welded ones (Adey *et al.*, 2000). End-plate connections also have advantages such as easy fabrication and fast erection when compared to welded connections. Moreover, end-plate connections provide enhanced ductility they share some of the deformation modes of typical semi-rigid connection.

As a result of the absence of faying surfaces, end-plate connections exhibit moment-rotational behavior without slippage and a different main failure mechanism (either a ductile yielding of the connected beam. Formation of a plate yield mechanism in the plate or an undesirable tension fracture of the connecting bolts). The design, detailing, fabrication and quality criteria for end-plate connections in this research shall conform to the requirement of the *2005 AISC Seismic Provisions* (2005 AISC Seismic Provisions). The design procedures for finding the adequate size of end-plate and tension fastener are as follows:

(A) Required bolt diameter
The *2005 AISC Seismic Provisions* require that the tension fasteners at the end-plate connection should be strong enough to resist the maximum probable moment ($M_{pr}$).

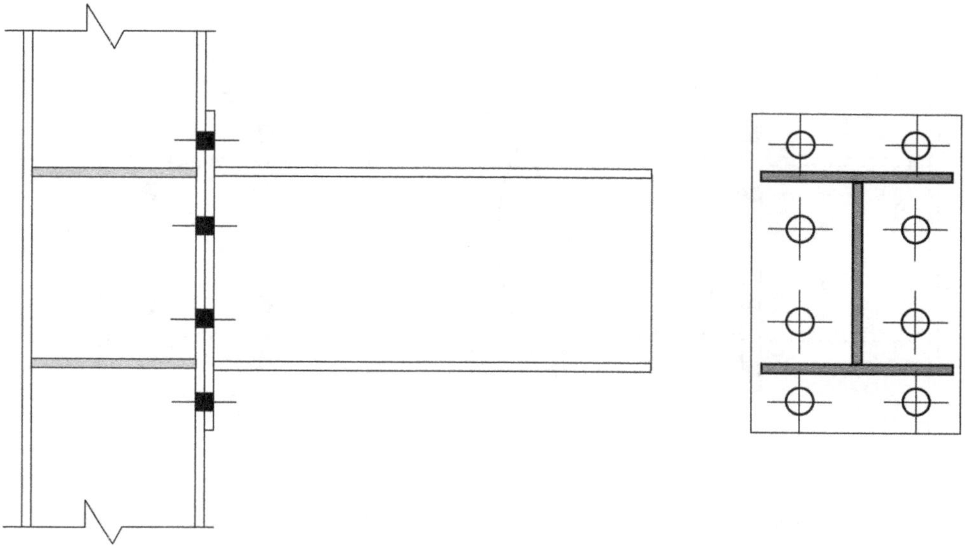

*Figure 2.20* The detail of end-plate connection (4E, Four-bolt unstiffened type).

The maximum probable moment intends to account for both the material over-strength and the fact that the material will likely begin to strain harden before failure occurs. The bolt force and bending moment mechanism according are shown in Fig. 2.21(a) to Fig. 2.23(a) for three types of end plate connections (4E, 4ES, and 8ES). To determine the required bolt diameter ($d_{b,req}$), the following equations are used:

$\phi = 0.90$ for the non-ductile limit state (2005 AISC Seismic Provisions)

$$\text{4E and 4ES Connection: } d_{b,req} = \sqrt{\frac{2M_{pr}}{\pi\phi F_{nt}(h_1 + h_2)}} \tag{2.30}$$

$$\text{8ES Connection: } d_{b,req} = \sqrt{\frac{2M_{pr}}{\pi\phi F_{nt}(h_1 + h_2 + h_3 + h_4)}} \tag{2.31}$$

where,

$$M_{pr} = C_{pr} R_y F_y Z_e \tag{2.32}$$

$$C_{pr} = \frac{F_y + F_u}{2F_y} \leq 1.20 \tag{2.33}$$

$C_{pr}$ = factor to account for the peak connection strength including strain hardening, local restraint, additional reinforcement, and other connection conditions.

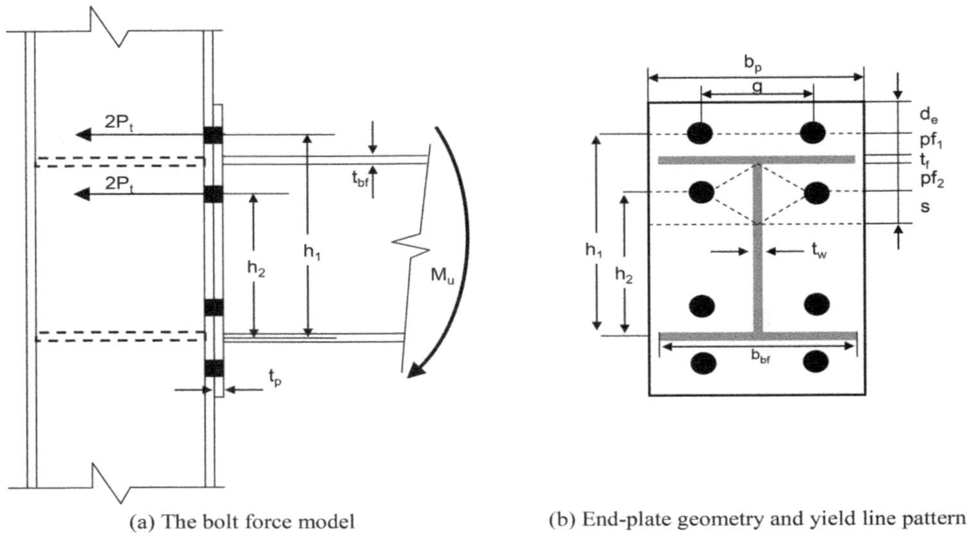


*Figure 2.21* Parameters for four bolt extended unstiffened end-plate (4E) yield line mechanism.

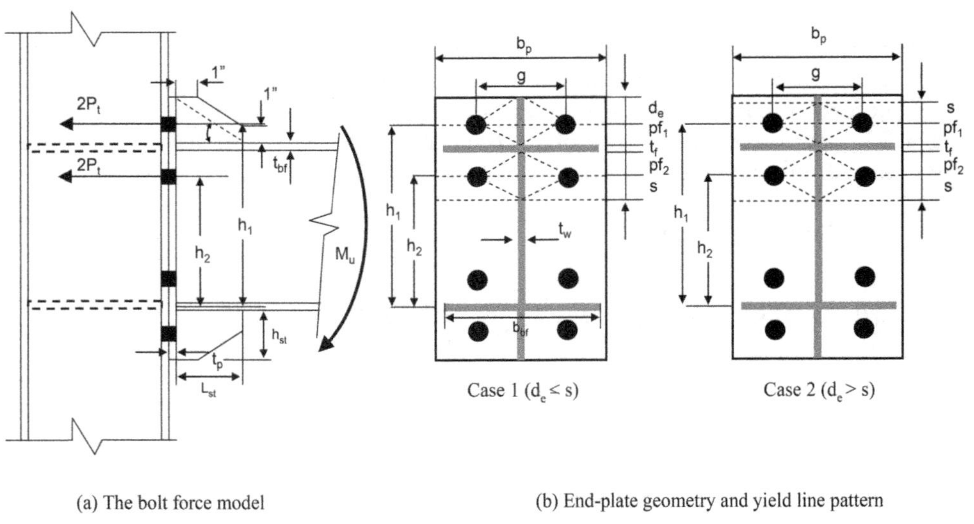


*Figure 2.22* Parameters for four bolt extended stiffened end-plate (4ES) yield line mechanism.

$F_{nt}$ = nominal tensile stress of bolt (620.1 MPa for A325 bolts and 826.8 MPa for A 490 bolts)
$F_u$ = specified minimum tensile stress of the type of steel, MPa
$F_y$ = specified minimum yield stress of the type of steel, MPa

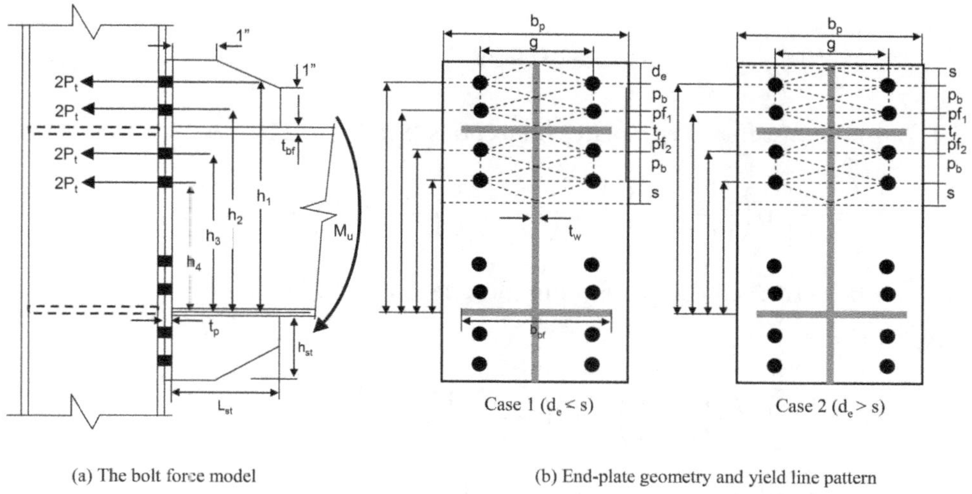


*Figure 2.23* Parameters for eight bolt extended stiffened end-plate (8ES) yield line mechanism.

$h_i$ = distance from the centerline of the beam compression flange to the centerline of the ith tension bolt holes.
$R_y$ = material over strength factor (2005 AISC Seismic Provisions)
$R_y = 1.5$ for A36 steel
$R_y = 1.3$ for A572-42 steel
$R_y = 1.1$ for other types of rolled shapes and bars
$Z_e$ = effective plastic section modulus

(B) Required end-plate thickness
The behavior of this type of connection is controlled by the thickness of the end plate. Typically end plate connections in seismic areas are designed such that beam hinging will occur before a plastic mechanism (yield lines) forms in the plate. The controlling yield line mechanisms for end-plates ($Y_p$) are illustrated in Figs. 2.21(b) to Fig. 2.23(b). The required end-plate thickness ($t_{p,req}$) is:

$\phi = 1.0$ for the ductile limit state (2005 AISC Seismic Provisions)

$$t_{p,req} = \sqrt{\frac{1.11M_{pr}}{\phi F_{yp} Y_p}} \tag{2.34}$$

where,
$F_{yp}$ = specified minimum yield stress of the end-plate material, MPa
$Y_p$ = end-plate yield line mechanism parameter from Eq. 2.36 to Eq. 2.40

The basic spacing parameter (s) is given by:

$$s = \frac{1}{2}\sqrt{b_p g} \quad (\text{If } pf_2 \geq s, \text{ use } pf_2 = s) \tag{2.35}$$

For four-bolt unstiffened end-plate connections (See Fig. 2.21):

$$Y_P = \frac{b_p}{2}\left[h_2\left(\frac{1}{pf_2} + \frac{1}{s}\right) + h_1\left(\frac{1}{pf_1} + \frac{1}{s}\right) - \frac{1}{2}\right] + \frac{2}{g}[h_2(pf_2 + s)] \tag{2.36}$$

For four-bolt stiffened end-plate connections (See Fig. 2.22):

Case 1 ($d_e \leq s$)

$$Y_P = \frac{b_p}{2}\left[h_2\left(\frac{1}{pf_2} + \frac{1}{s}\right) + h_1\left(\frac{1}{pf_1} + \frac{1}{s}\right) - \frac{1}{2}\right] + \frac{2}{g}[h_2(pf_2 + s) + h_1(pf_1 + d_e)] \tag{2.37}$$

Case 2 ($d_e > s$)

$$Y_P = \frac{b_p}{2}\left[h_2\left(\frac{1}{pf_2} + \frac{1}{s}\right) + h_1\left(\frac{1}{pf_1} + \frac{1}{s}\right) - \frac{1}{2}\right] + \frac{2}{g}[h_2(pf_2 + s) + h_1(pf_1 + s)] \tag{2.38}$$

For eight-bolt stiffened end-plate connection (See Fig. 2.21)

Case 1 ($d_e \leq s$)

$$Y_P = \frac{b_p}{2}\left[h_1\left(\frac{1}{2d_e}\right) + h_2\left(\frac{1}{pf_1}\right) + h_3\left(\frac{1}{pf_2}\right) + h_4\left(\frac{1}{s}\right)\right] + \frac{2}{g}\left[h_1\left(d_e + \frac{p_b}{4}\right) + h_2\left(pf_1 + \frac{3p_b}{4}\right) + h_3\left(pf_2 + \frac{p_b}{4}\right) + h_4\left(s + \frac{3p_b}{4}\right) + p_b^2\right] + g \tag{2.39}$$

Case 2 ($d_e > s$)

$$Y_P = \frac{b_p}{2}\left[h_1\left(\frac{1}{2d_e}\right) + h_2\left(\frac{1}{pf_1}\right) + h_3\left(\frac{1}{pf_2}\right) + h_4\left(\frac{1}{s}\right)\right] + \frac{2}{g}\left[h_1\left(s + \frac{p_b}{4}\right) + h_2\left(pf_1 + \frac{3p_b}{4}\right) + h_3\left(pf_2 + \frac{p_b}{4}\right) + h_4\left(s + \frac{3p_b}{4}\right) + p_b^2\right] + g \tag{2.40}$$

Yield line analysis has proven to be a useful instrument in calculating the ultimate capacity of end-plates. This analysis focus on the plastic deformation caused by the formation of the plastic hinges or yield lines and neglects the elastic deformation. The yield lines are selected from any valid pattern that results in a mechanism. The yield lines are assumed as straight and the moment along each line is constant and equal to the plastic moment capacity of the plate. The beam web and stiffeners are considered as rigid sections that provide boundaries to the yield lines along the end-plate width ($b_p$).

(C) Welds
Welding procedures in this research are assumed to satisfy the requirements of Section 7.3 and Appendix W of the *2005 AISC Seismic Provisions*. Welding of the beam to the end-plate shall be in accordance with the following limitations:

- Welding near to the holes shall not be used.
- The beam web to end-plate connection shall be fabricated using either fillet welds or complete joint penetration (CJP) groove welds.
- The beam flange to end-plate connection shall be fabricated using a CJP groove weld without backing. The CJP groove weld shall be fabricated in that the root of the weld lies on the beam web side of the flange.
- All end-plate stiffener connections shall be fabricated using CJP groove welds.

### 2.3.4 Composite panel zone strength

The composite panel zone treated in this research is composted of the steel tube and concrete core. In general, the shear strength of the composite panel zone can be calculated as the superposition of the shear strengths of the steel and concrete components. A mechanical model proposed by Wu *et al.* (2005 and 2007) is used in this research in order to compute the stiffness, the yielding shear strength, and the ultimate shear strength of the composite panel zone. The theoretical equations for this mechanical model are driven by using the shear stiffness contributions of both materials.

(A) Steel Tube
The contribution of the rectangular steel tube to the shear resistance is composed of two parts: (a) the column flanges deforming in a flexural mode and (b) the webs deforming in a shear mode (Fig. 2.24). Therefore, the shear strength and stiffness of the panel zone are affected by both deformation mechanisms.

The two column flanges subjected to shear force can be modeled as columns with fixed supports resisting flexural deformations. The shear stiffness of the two column flanges ($K_f$) is:

$$K_f = 2\frac{12E_sI_f}{(d_b - t_{bf})^2} \tag{2.41}$$

where,

$$I_f = \frac{b_c t_f^3}{12} \tag{2.42}$$

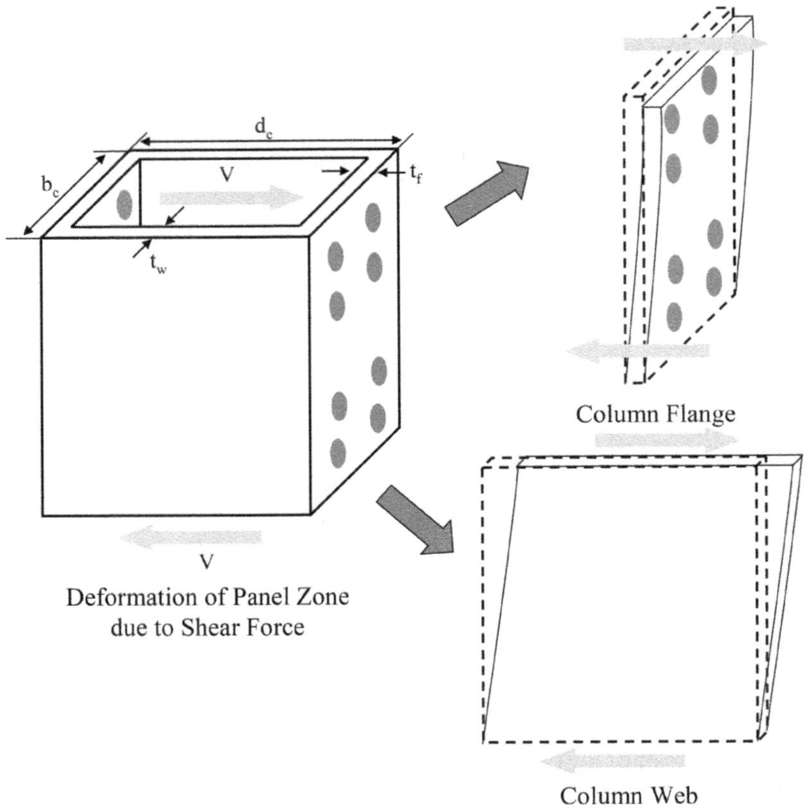


*Figure 2.24* Schematic figures for deformation of steel tube in the panel zone region.

and where,
$b_c$ = width of the column
$d_b$ = depth of the beam
$K_f$ = shear stiffness for the column flange at the panel zone
$I_f$ = moment of inertia of the column flange
$t_{bf}$ = thickness of the beam flange
$t_f$ = thickness of the column flange

The existence of bolt holes can reduce the shear strength of the tube. As a result, the steel tube column is divided into two regions which are a column web without bolt holes and a column flange without bolt holes. The original mechanical model by Wu *et al.* (2005 and 2007) considers this effect. However, only column webs without bolt holes will be treated in this study because the specimens shown in the next chapters do not contain any bolt holes in the column webs. The shear stiffness of the two column webs ($K_w$) is as follows:

$$K_w = 2(d_c - 2t_f)t_w G_s \tag{2.43}$$

where,
$d_c$ = depth of the column
$G_s$ = shear modulus of steel
$K_w$ = shear stiffness in the two column webs
$t_w$ = thickness of the column web

The shear stiffness ($K_{s1}$) of the rectangular steel tube at the panel zone is the superposition of the shear stiffness of the column webs ($K_w$) and the shear stiffness of the column flanges ($K_f$).

$$K_{s1} = K_f + K_w \tag{2.44}$$

The resulting shear yield strength ($V_{wy}$) and yield strain ($\gamma_{wy}$) at the column web are as follows:

$$V_{wy} = 2(d_c - 2t_f)t_w\tau_{sy} \tag{2.45}$$

$$\gamma_{wy} = \frac{V_{wy}}{K_w} \tag{2.46}$$

where,

$$\tau_{sy} = \frac{F_y}{\sqrt{3}} \tag{2.47}$$

$\tau_{sy}$ = yield shear stress of steel

When the shear strain in the panel zone reaches the yield shear strain ($\gamma_{wy}$) of the column web, the yield shear strength ($V_{sy}$) is the superposition of the shear strengths of the column webs and that of the column flanges.

$$V_{sy} = K_{s1}\gamma_{wy} \tag{2.48}$$

As the external loads increase, the column webs will yield and the stiffness of the column webs vanishes. After this yielding occurs, the shear stiffness ($K_{s2}$) is the shear stiffness of the column flange ($K_f$) alone:

$$K_{s2} = K_f \tag{2.49}$$

The stress in the column flanges subsequently arrives at the yielding point as the external load increases. The resulting shear yield strength ($V_{fy}$) and yield strain ($\gamma_{fy}$) at the column web are as follows:

$$V_{fy} = \frac{4M_{fy}}{(d_b - t_{bf})} \tag{2.50}$$

$$\gamma_{fy} = \frac{V_{fy}}{K_f} \tag{2.51}$$

where,

$$M_{fy} = \frac{b_c t_f^2 F_y}{6} \tag{2.52}$$

$M_{fy}$ = yielding flexural strength of the column flanges

The ultimate shear strength ($V_{su}$) of the panel zone is the summation of the shear strengths of the column webs and column flanges when the shear strain of the panel zone arrives at the yield strain ($\gamma_{fy}$) for the column flange.

$$V_{su} = V_{wy} + V_{fy} \tag{2.53}$$

(B) Confined Concrete Core

A Mohr-Coulomb failure criterion is adopted to estimate the ultimate shear strength of the concrete. The ultimate shear stress ($\tau_{cu}$) of the concrete in the panel zone can be obtained as:

$$\tau_{cu} = \sqrt{\left[\frac{f_c'}{11} + \frac{9(f_{cp} + f_{ct})}{22}\right]^2 - \left(\frac{f_{cp} - f_{ct}}{2}\right)^2} \tag{2.54}$$

where,

$$f_{cp} = \frac{PE_c}{E_s A_s + E_c A_c} \tag{2.55}$$

$$f_{ct} = \frac{-T}{b_c d_b} \tag{2.56}$$

$A_c$ = cross section area of concrete core
$A_s$ = cross section area of steel tube
P = axial compression loaded on CFT
T = sum of the forces of the pre-stressed bolts

The shear stiffness ($K_c$) and the ultimate shear strength ($V_{cu}$) of the concrete core in the panel zone are as follows:

$$K_c = G_c A_c \tag{2.57}$$

$$V_{cu} = \tau_{cu} A_c \tag{2.58}$$

The corresponding ultimate shear strain of the concrete in the panel zone is:

$$\gamma_{cu} = \frac{\tau_{cu}}{G_c} \tag{2.59}$$

(C) Combined Steel Tube and Concrete Core

The steel tube at the panel zone is divided structurally into webs and flanges. The strain and strength at which the steel webs yield is defined as the yielding shear strain and strength of the panel zone, while the strain and strength at which the steel flanges yield is defined as the ultimate shear strain and strength of the panel zone. For the composite panel zone, the shear stiffness (K), the yielding shear strength ($V_y$), and the ultimate shear strength ($V_u$) is the summation of those of the steel tube and the concrete core respectively, as follows:

$$K = K_{s1} + K_c \tag{2.60}$$

$$V_y = V_{sy} + V_{cu} \tag{2.61}$$

$$V_u = V_{wy} + V_{fy} + V_{cu} \tag{2.62}$$

## 2.4 PRELIMINARY DESIGN PROCEDURE FOR CONNECTION COMPONENTS

Practical methods for ductile design of components will be presented in this section. The ductility and strength of the moment connections should be greater than the demand due to the external response in order to satisfy the general requirements of seismic design. Strength requirements can be met by satisfying the strength equations discussed in the previous sections and this is a relatively straight forward process based on well-established limit theories. On the other hand, satisfying ductility (or drift) requirements is a difficult task and one not amenable to simplification. As shown by the calculations for the shear capacity of the panel zones, strength and deformation capacity need to be carefully considered for each component. When several components are merged to make a connection, these relationships become complex and simple equations can generally not be used to insure that available ductility exceeds demand. In this dissertation an approach that emphasizes strength design will be used for the preliminary design, with the deformation checks carried out with the aid of advanced analyses.

In the seismic design of steel components, limit states can be classified as either ductile (yielding) or brittle (fractures) modes. The dominant failure modes for steel or composite components should be ductile failure modes such as slip, yielding of steel and minor local buckling in order to avoid the entire collapse and ensure the survival of the structure (Astaneh-Asl 1995). An approach to fulfill this requirement is to require that the minimum capacity of all brittle failure modes be greater than that of the strongest yielding modes. In this way, several ductile mechanisms will be activated before any brittle one can occur. Suitable use of capacity reduction factors and reliability approaches will lead to connection designs with a suitable low probability of not achieving the desired ductility. Seismic design procedures considering this approach for the design of composite PR-CFT connection frames are discussed in the next sections.

### 2.4.1 Composite column design

The *2005 AISC Specification* contains revised rules for the design of composite columns. As the design of columns is governed by stability effects, the first step is to determine the factored design demand including the dead and the live load on the floor systems. In the real composite frame structures, the composite column systems behave as members subjected simultaneously to axial load and bending moment (So called beam-columns). Beam-column behavior as related to the composite frame performance will be treated in the following chapters. The second step is to determine the slenderness effects. Finally, the design capacity can be determined by applying the design resistance factor to the nominal strength. A detail step-by-step design procedure for the composite columns under the axially compressive load is as follows:

**Step 1) Compute the design demand**
The first step for the design is to determine the required design strength for the composite columns. The initial design will be based on the highest compressive axial load, which is given by the load combination:

$$P_u = P_r = 1.2P_D + 1.6P_L \tag{2.63}$$

**Step 2) Select the cross section area**
To qualify as a composite filled column, all column elements shall be within the limitations as mentioned Section 2.3.2. Note that this step includes a check for local buckling that ensures that the plastic capacity of the section can be achieved.

**Step 3) Compute effective moment of inertia for the composite section**
Once the cross section selected has been shown to exceed the required design strength ($P_r$), the modified stiffness ($EI_{eff}$) is calculated for the CFT:

$$EI_{eff} = E_s I_s + 0.5 E_s I_{sr} + C_1 E_c I_c \tag{2.64}$$

$$C_1 = 0.6 + 2\left(\frac{A_s}{A_c + A_s}\right) \leq 0.9 \tag{2.65}$$

**Step 4) Compute the slenderness ratio**
The limit state global buckling ($P_o$) is based upon the slenderness ratio ($\alpha$) as shown below:

$$\alpha = \sqrt{\frac{P_o}{P_e}} \tag{2.66}$$

$$P_o = A_s F_y + A_{sr} F_{yr} + C_2 A_c f'_c \tag{2.67}$$

$$P_e = \pi^2 (EI_{eff})/(KL)^2 \tag{2.68}$$

**Step 5) Compute the factored design capacity**
To fulfill the general axiom of design, the factored design strength should be equal or greater than the design demand as shown below.

$$\phi_c P_n \geq P_u \tag{2.69}$$

The nominal compressive strength should be determined based on the slenderness ratio.

When $\alpha \leq 1.5$,

$$\phi_c P_n = P_o(0.658^{\alpha^2}) \tag{2.70}$$

When $\alpha > 1.5$,

$$\phi_c P_n = P_o \frac{0.877}{\alpha^2} \tag{2.71}$$

Extensive numerical examples of composite column design are given in Appendix A. In that Appendix, comparisons between the simplified AISC procedure used here and more "exact" fiber models are also given.

### 2.4.2 End plate connection

The *ANSI standard* (ANSI 358-05) specifies design, detailing, fabrication, limitation, and quality criteria for end-plate connections that satisfy the *2005 AISC Seismic Provisions* (2005 AISC Seismic Provisions) for use with composite moment frame structures. The behavior of this type of connections is generally controlled by a number of different limit states closely related to the ductile modes such as local buckling of the beam flanges, flexural yielding of the beam section, flexural yielding of the end-plate, and yielding of the column panel zone. The resistance to brittle failure modes such as tension failure of the end-plate bolts, shear failure of the end-plate bolts or tearing of various welded connections should be greater than those computed for the ductile modes. Example 1 in Appendix B is a complete example of the application of the procedure described in the next subsections.

**Step 1) Compute the design strength**

The first step is to determine the design strength for the end-plate connection. In case of a full strength connection, i.e., one where all the plastic deformations will be confined to the framing beam, the required design strength shall be determined by the plastic moment specified in the *2005 AISC Seismic Provisions*.

$$M_{design} = C_{pr} F_{ye} Z_x \tag{2.72}$$

where,

$$C_{pr} = \frac{F_y + F_u}{2F_y} \leq 1.2 \tag{2.73}$$

$$F_{ye} = R_y F_y \tag{2.74}$$

$C_{pr}$ = factor to consider the peak connection strength, typically taken as 1.1
$F_{ye}$ = factored yield stress at the beam
$R_y$ = material over strength factor (See Eq. 2.32)
$Z_x$ = plastic section modulus of the beam

*Table 2.4* Prequalification limitations for geometric parameters (ANSI 385-05).

| | 4E | | 4ES | | 8ES | |
|---|---|---|---|---|---|---|
| *Parameter* | *Max.* | *Min.* | *Max.* | *Min.* | *Max.* | *Min.* |
| $t_p$ | 57.2 | 12.7 | 38.1 | 12.7 | 63.5 | 19.1 |
| $b_p$ | 273.1 | 177.8 | 273.1 | 273.1 | 381 | 228.6 |
| g | 152.4 | 101.6 | 152.4 | 82.55 | 152.4 | 127 |
| $P_{f1}, P_{f2}$ | 114.3 | 38.1 | 139.7 | 44.45 | 50.8 | 44.45 |
| $P_b$ | – | – | – | – | 95.25 | 88.9 |
| d | 1397 | 635 | 609.6 | 349.25 | 914.4 | 469.9 |
| $T_{bf}$ | 19.1 | 9.5 | 19.1 | 9.5 | 25.4 | 15.1 |
| $b_{bf}$ | 234.9 | 152.4 | 228.6 | 152.4 | 311.2 | 196.9 |

Unit: mm

On the other hand, for a partial strength connection requires the advanced analysis including second order effect and overall frame stability in order to find the design strength.

**Step 2) Select the geometric parameters for connection components**
One of the three end-plate moment connection configurations, preliminary value for the connection geometric parameters and bolt grades should be selected in this step. All connection components shall satisfy the following limitations as shown in Table 2.4 associated with Fig. 2.21 to 2.23 in order to avoid the stress concentrations and possible brittle fractures.

**Step 3) Determine the tension bar/bolt diameter**
For an end plate connection, the dominant brittle failure modes that needed to be avoided are the tension/shear fracture of the welds or tension fracture of the bolts. For any kind of full strength bolted connections, the connection must insure that the ductile limit state given by beam yielding occurs well before tension bolts or welds fail. The required bolt diameter ($d_{b,req}$) can be determined by using Eq. 2.30 or 2.31. The tension bars designed in this structure consist of the super-elastic SMA tension bars and the A490 type tension bars. The nominal strength of the individual tension bar should reflect the difference in material properties between steel and SMA as well as the different strain demands based on the placement of the bars relative to the centerline of the bottom beam flange. The procedure about the averaged strength according to the different connection types is illustrated in Fig. 2.25.

**Step 4) Determine the required end-plate thickness**
After the size of the bolt diameter has been determined, the geometric parameters for the end-plate should be established. The most significant parameter is the end-plate thickness ($t_p$). It is necessary to understand the effect of the end-plate geometry on the response of the connection as the behavior can be highly nonlinear and counterintuitive. Based on these considerations, the end plate thickness can be computed from Eq. 2.34 to 2.40. The plate deformations should be consistent with the selected yield line mechanism as shown in Fig. 2.23 to Fig. 2.25.

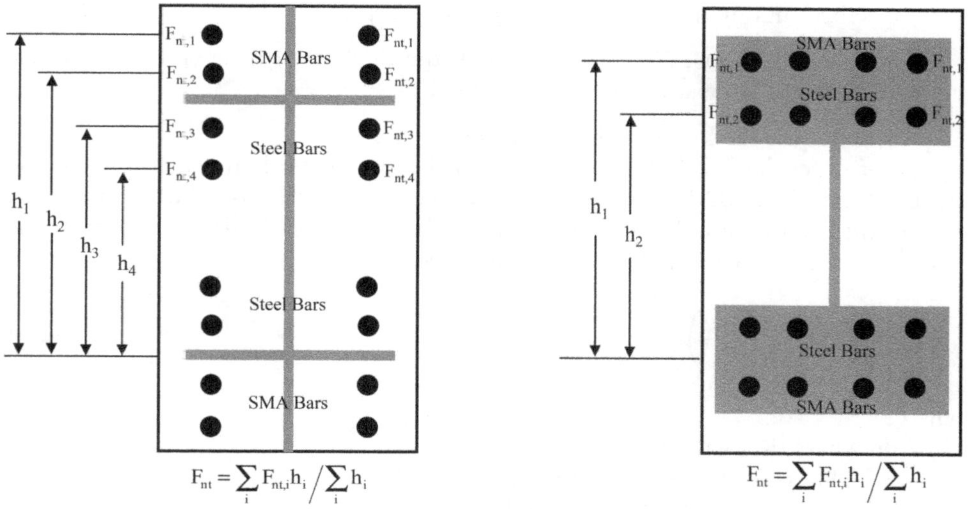


*Figure 2.25* Average tensile strength for a bar.

**Step 5) Check the shear resistance of the end-plate**

For a full strength connection, the design axial force ($F_{fu}$) at the beam flange can be backcalculated by a simple model from the factored plastic moment of the beam as mentioned in the *2005 AISC Seismic Provisions*. After finding of the factored design axial force for the beam flange, that the designer must insure that the shear yielding resistance and shear rupture resistance at the end-plate should be equal or larger than this design force (See Eqs. 2.75 to 2.76).

$\phi = 0.90$ for the non-ductile limit state (2005 AISC Seismic Provisions)

$$F_{fu} = \frac{M_{design}}{d - t_{bf}} \tag{2.75}$$

$$\frac{F_{fu}}{2} \leq \phi R_{n,shear} = 0.6\phi F_{yp} b_p t_p \tag{2.76}$$

$$\frac{F_{fu}}{2} \leq \phi R_{n,shear} = 0.6\phi F_{up} A_n \tag{2.77}$$

where,

$$A_n = \left[ b_p - 2\left( d_b + \frac{1}{16} \right) \right] t_p \tag{2.78}$$

$A_n$ = net area of the end-plate when standard holes are used
$b_p$ = width of the end-plate
$d$ = depth of the beam

$d_b$ = diameter of bolts
$F_{fu}$ = specified minimum tensile strength of the end-plate

If either Eq. 2.75 or 2.76 is not satisfied, the thickness of the end-plate must be increased until these relationships are satisfied.

**Step 6) Select the thickness of the end-plate stiffener**
In case of either the four bolt extended stiffened end-plate (4ES) or eight bolt extended stiffened end-plate (8ES) connection, the end-plate stiffener needs to be designed. After the thickness of the end-plate stiffener ($t_{s,min}$) is determined, the stiffener to beam-flange and the stiffener to end-plate welds are modeled.

$$t_{s,min} = t_{bw}\left(\frac{F_{yb}}{F_{ys}}\right) \tag{2.79}$$

where,
$t_{bw}$ = thickness the beam web
$t_{s,min}$ = thickness of the end-plate stiffener
$F_{yb}$ = specified minimum yield stress of beam material, MPa
$F_{ys}$ = specified minimum yield stress of stiffener material, MPa

The following width to thickness criterion should be satisfied in order to prevent local buckling of the stiffener plate:

$$L_{st} = \frac{h_{st}}{\tan 30^\circ} \tag{2.80}$$

$$\frac{h_{st}}{t_s} \geq 0.56\sqrt{\frac{E}{F_{ys}}} \tag{2.81}$$

where,
$h_{st}$ = height of the stiffener
$L_{st}$ = width of the stiffener

**Step 7) Check the rupture and bearing failure of bolts**
The design shear capacity ought to exceed the design shear strength as follows:

$\phi = 0.90$ for the non-ductile limit state (2005 AISC Seismic Provisions)

$$V_u \leq \phi B_{n,shear} = \phi N_b F_v A_b \tag{2.82}$$

where,

$$V_u = \frac{2M_{design}}{L'} \tag{2.83}$$

$A_b$ = nominal gross area of the bolts
$F_v$ = effective ultimate shear strength (See Eq. 2.14)
$L'$ = distance between plastic hinges

The tear-out failure at the end-plate and column flange due to bolt bearing is required to be checked next. The geometric parameters are shown in Fig. 2.17.

$\phi = 0.90$ for the non-ductile limit state (2005 AISC Seismic Provisions)

$$V_u \leq \phi R_{n,bearing} = \phi(N_i)R_{ni} + \phi(N_o)R_{no} \tag{2.84}$$

where,

$$R_{ni} = 1.2L_c tF_u \leq 2.4d_b tF_u \tag{2.85}$$

$$R_{no} = 1.2L_c tF_u \leq 2.4d_b tF_u \tag{2.86}$$

and where,
$N_i$ = number of inner bolts (2 for 4E and 4ES, and 4 for 8ES connections)
$N_o$ = number of outer bolts (2 for 4E and 4ES, and 4 for 8ES connections)
$R_{ni}$ = bearing strength at the inner bolts
$R_{no}$ = bearing strength at the outer bolts

### 2.4.3 T-stub connections

A practical design procedure for T-stub connections is presented in this section. The LRFD manual (AISC 2001) provides more information on the design and detailing of T-stub connections. Similar to the design procedure for end-plate connections, the governing mode should be ductile failure modes and the required design strength should exceed that given by the plastic moment of the beam flange. Some design steps from T-stubs are duplicate of those discussed in the design procedures for end-plate connection (See Section 2.4.2). Only new design steps pertinent only to T-stub connections will be discussed in detail in this section. Example 2 in Appendix B is a complete example of the application of the procedure described in the next subsections.

**Step 1) Determine the design force**
The design strength should be more than the factored plastic moment of the beam as shown in Eq. 2.71.

**Step 2) Select bolt type and size**
Components for the connecting elements in T-stub connections will be limited to high strength structural bars and super-elastic shape memory alloy (SMA) bars. A490 tension bars have a specific ultimate tensile strength for 1033.5 MPa and will be used exclusively in a 25.4 mm diameter size. They are fabricated from alloy steel and used in parallel with super-elastic SMA bars to transfer bending from beam to column panel zone. The SMA tension bars are also available in a 25.4 mm diameter and will be used in that size to maintain the geometric consistency. Bars above 25.4 mm diameter are difficult to pretension and bars smaller than 25.4 mm cannot provide the required force in a sufficiently small number of bars to make the connection economical.

As noted earlier, steel and SMA bars are used in parallel as the former provide energy dissipation and high strength and stiffness, while the later provide recentering capabilities.

Construction practice dictates that the tension bolts should share the same size and grade with the shear bolts. One of the most significant characteristics for the usage of shear bolts leads to the slip deformation. Slip is the preferred mode for increasing energy dissipation capacity and avoiding catastrophic failure. Therefore, the slip resistance should stay in the ductile region between the service load and the ultimate load. On the basis of this axiom, it is necessary for tension bars to satisfy the required diameter using the Step 3 procedure in the end-plate connection design. The general failure of bolts due to shear and tension needs to be checked. The failure modes will be illustrated with more details in the next chapter.

**Step 3) Determine the configuration of the bolts**
After Step 2, a preliminary configuration of the T-stub, including the spacing, gage length and arrangement for bolt holes can be determined. These design parameters have a significant influence on the effective width and strength of the T-stub members as shown by Eqs. 2.15 through 2.22. The shear bolt arrangements should guarantee an adequate edge distance for the beam flange so as to avoid stress concentrations in the stem.

**Step 4) Determine the required stem thickness**
The conventional net section failure calculation adopted in this T-stub component design (See Eqs. 2.26 and 2.27) can estimate the ultimate strength of the component element as the product of the material ultimate strength of the material used and the net cross section area defined as the gross section area minus punching area of the bolt holes. The required stem thickness can be determined after the configuration of bolts and the width of T-stub member are established as shown in below:

$\phi_f = 0.75$ for the resistant factor for the fracture

$$t_{stem,min} = \frac{M_{design}}{\phi_f F_u (W_{eff} - n_{sb} d_h) d} \tag{2.87}$$

The effective width ($W_{eff}$) as shown in Fig. 2.18 is not valid for the all stem tapering configuration and drives the simple approximate estimation.

**Step 5) Determine the T-stub flange width and thickness**
The ultimate strength for a T-stub flange subject to prying was described in Section 2.3.3.1. On the basis of this prying action phenomenon, an adequate T-stub width and thickness can be computed. The capacity of the existing T-stub can be determined by the failure modes based on the Eq. 2.15 to 2.17 which correspond to a pure flange mechanism, combined failure mode, and bolt fracture respectively. The balanced load ($T_o$) and critical thickness ($t_c$) for a T-stub flange can be derived by using a balanced failure approach in which the ultimate strength of the T-stub flange is reached at the same time as the bolt force including the prying action becomes critical (Astaneh, 1985). The balanced load and critical thickness for T-stub flange can be obtained by:

$\phi_f = 0.75$ for the bolt fracture

$$T_o = \frac{\phi_f B_{n,tension}}{1 + \left(\frac{\delta}{1+\delta}\right)\left(\frac{b'}{a'}\right)} \tag{2.88}$$

$$t_c = \sqrt{\frac{4B_{n,tension} b'}{pF_y}} \tag{2.89}$$

If the balanced load is equal to the design load ($T_{design}$), the value of a′ is equal to $2d_b$, and assuming that the tension bolt gage ($g_t$) will be relatively large compared to the flange width ($B_f$), the required value of b′ and flange width are (Swanson, 2001):

$$b' = \left(\frac{\phi B_{n,tension} - T_{design}}{T_{design}}\right)\left(\frac{1+\delta}{\delta}\right) a' \tag{2.90}$$

$$B_{f,design} = 2b' + 4d_b + 2t_{s,min} \tag{2.91}$$

**Step 6) Check the T-stub section**
After T-stub section has been selected, the capacity of the T-stub section should be checked by comparing the available failure modes which occur in either the flange or the stem. In order to ensure a ductile failure with significant deformation, the yielding capacity of the stem net section should be nearly equal to the ultimate capacity of the flange due to the prying mechanism.

$\phi_y = 0.90$ for the resistance factor against yielding

$$\phi_y F_y A_{net,stem} \leq \phi n_{tb} T \tag{2.92}$$

**Step 7) Check the yield and fracture for the component members**
Block shear failure is a combined yield fracture type of failure where the boundary of a block of tensile yielding in some area and tensile fracture in the remaining areas. (Eqs. 2.28 and 2.29). Block shear needs to be checked but it will only be critical is the bolt spacing and gages fall below the recommended value of $3d_b$.

**Step 8) Design the shear connection**
Finally, the shear connection should be designed. Failure modes for the shear connections and reliable design procedures are available in the current LRFD code (AISC 2001). In case of the partial strength moment connections, the usage of the short slotted holes on the shear tab connections in the loading direction can avoid a complete connection failure and reduce the torsion into the connection and beam (Swanson, 2001). Failure of the shear connection as the fracture of the net area, web bolts or welds can induce the catastrophic collapse of the connection and miss the opportunity to resist the gravity loads after the flange failure. Therefore, the shear connection has to endure large rotation before the yield failure.

## 2.4.4 Clip angle connection

A practical design procedure for the clip angle connections with heavy angles t (t = 25.4 mm) will be presented in this section. The design procedures are very similar

*Table 2.5* Summary and comparison of design procedures for T-stub and clip angle connection.

| *Design step* | *T-stub connection type* | *Thick clip angle connection type* |
|---|---|---|
| 1 | *Determine the design force*<br>$M_{design} = C_{pr}F_{ye}Z_x$ $\quad C_{pr} = \frac{F_y + F_u}{2F_y} \leq 1.2$ | $F_{ye} = R_yF_y$ |
| 2 | *Determine the adequate bar diameter*<br>Two rows of tension bar arrangement<br>$F_{nt} = \frac{\sum_i F_{nt,i}h_i}{\sum_i h_i}$ $\quad d_{b,req} = \sqrt{\frac{4M_{pr}}{2\pi\phi F_{nt}(h_1 + h_2)}}$ | *Determine the adequate bar diameter*<br>One row of tension bar arrangement<br>$F_{nt} = \frac{\sum_i F_{nt,i}h_i}{\sum_i h_i}$ $\quad d_{b,req} = \sqrt{\frac{4M_{pr}}{3\pi\phi F_{nt}(h_1)}}$ |
| 3 | *Determine the configuration of the bolts*<br>In general, four or five rows of the shear bolt arrangement<br>Check the gage and bolt spacing<br>(Typ. S = 76.2 mm) | *Determine the configuration of the bolts*<br>In general, two rows of the shear bolt arrangement<br>Check the gage and bolt spacing<br>(Typ. S = 76.2 mm) |
| 4 | *Determine the stem thickness*<br>$R_{n,net} = F_uA_{net,stem}$ $\quad A_{net,stem} = (W_{eff} - n_{sb}d_h)t$ | *Determine the clip angle thickness*<br>$R_{n,net} = F_uA_{net,angle}$ $\quad A_{net,angle} = (W_{eff} - n_{sb}d_h)t$ |
| 5 | *Determine the T-stub flange width and thickness*<br>Consider T-shape to find the effective width per bolt<br>$p = \frac{2W_{T\text{-}stub}}{n_{tb}}$ | *Determine the clip angle flange width and thickness*<br>Consider L-shape to find the effective width per bolt<br>$p = \frac{W_{Clip}}{n_{tb}}$ |
| 6 | Check the T-stub section<br>$\phi_yF_y A_{net,stem} \leq \phi n_{tb}T \quad \phi_y = 0.90$ | Check the clip angle section<br>$\phi_yF_y A_{net,angle} \leq \phi n_{tb}T \quad \phi_y = 0.90$ |
| 7 | *Check the yield-fracture failure (Block shear failure modes)*<br>If $F_u A_{nt} \geq 0.6F_u A_{nv}$ $\quad R_{n,block} = 0.6F_y A_{gv} + F_u A_{nt}$ | Otherwise $\quad R_{n,block} = 0.6F_u A_{nv} + F_y A_{gt}$ |
| 8 | *Design the shear connection*<br>Use shear tab or double web angle connection with short slotted bolt holes | |

to that for T-stub connections described in the previous section and for simplicity the procedure is summarized in Table 2.5. Example 3 in Appendix B is a complete example of the application of the procedure described in the next subsections.

## 2.5 CONNECTION DESIGN AND EXAMPLES

This chapter is describes the specimen design for the smart PR-CFT connections, with emphasis on connection materials, geometry, and constructability. Three connection models utilizing the components described in Chapter 2.3, 2.4 are developed: an end-plate, a T-stub, and a thick clip angle connection. Descriptions of the connections are given in Table 2.6. Detailed design examples for these connections, using Math-Cad worksheets, are provided in Appendix B.

*Table 2.6* Detailed specifications of the smart PR-CFT connections.

| *Unit: kN, mm* | *End-Plate Connection with RCTT* | *End-Plate Connection with CCFT* |
|---|---|---|
| *(a) A detailed specification of the smart PR-CFT connection with an end-plate component member* | | |
| Beam | W24 × 55 | W24 × 103 |
| Column | HSS16 × 16 × 500 | HSS18 × 500 |
| SMA Bar Material | Super-Elastic Nitinol | Super-Elastic Nitinol |
| Steel Bar Material | Corresponding to A490 Bolts | Corresponding to A490 Bolts |
| Bar Gauge Length | 406.4 mm | 457.2 mm |
| Bar Diameter | 25.4 mm | 25.4 mm |
| Bar Slenderness Ratio | 51.88 | 58.36 |
| Bolt Material | A 490 Bolts | A 490 Bolts |
| Shear Bolt Size | No Shear Bots | No Shear Bots |
| Web Bolt Size | No Web Bots | No Web Bots |
| Nut Type | A 563 Nuts Corresponding to 25.4 mm Dia. Bolts | A 563 Nuts Corresponding to 25.4 mm Dia. Bolts |
| Component Members | End-Plate 15 × 38.5 × 1 (8ES Type) | End-Plate 15 × 38.5 × 1 (8ES Type) |
| Shear Tab | No Shear Tab | No Shear Tab |
| Panel Zone Shape & Size | Rectangular 16 × 16 × 24.5 | Rectangular 18 × 18 × 24.5 |
| Inside Concrete Material | Rec. Confined Concrete ($F/F_u = 0.2$, $f'_c = 27.56$ MPa) | Cir. Confined Concrete ($F/F_u = 0.2$, $f'_c = 27.56$ MPa) |
| *(b) A detailed specification of the smart PR-CFT connection with a T-stub component member* | | |
| Beam | W24 × 55 | W24 × 103 |
| Column | HSS16 × 16 × 500 | HSS18 × 500 |
| SAIA Bar Material | Super-Elastic Nitinol | Super-Elastic Nitinol |
| Steel Bar Material | Corresponding to A490 Bolts | Corresponding to A490 Bolts |
| Bar Gauge Length | 406.4 mm | 457.2 mm |
| Bar Diameter | 25.4 mm | 25.4 mm |
| Bar Slenderness Ratio | 51.88 | 58.36 |
| Bolt Material | A 490 Bolts | A 490 Bolts |
| Shear Bolt Size | 1 × 4 (One Washer) | 1 × 4 (One Washer) |
| Web Bolt Size | 1 × 4 (One Washer) | 1 × 4 (One Washer) |
| Nut Type | A 563 Nuts Corresponding to 25.4 mm Dia. Bolts | A 563 Nuts Corresponding to 25.4 mm Dia. Bolts |
| Component Members | Cut from W16 × 100 (T-stub) | Cut from W16 × 100 (T-stub) |
| Shear Tab | Plate 4.5 × 9 × 0.56 (Fillet Welding 7.9375 mm) | Plate 4.5 × 9 × 0.56 (Fillet Welding 7.9375 mm) |
| Panel Zone Shape & Size | Rectangular 16 × 16 × 23.6 | Rectangular 18 × 18 × 23.6 |
| Inside Concrete Material | Rec. Confined Concrete ($F/F_u = 0.2$, $f'_c = 27.56$ MPa) | Cir. Confined Concrete ($F/F_u = 0.2$, $f'_c = 27.56$ MPa) |

*Table 2.6* Continued.

| *Unit: kN, mm* | *End-Plate Connection with KCTT* | *End-Plate Connection with CCFT* |
|---|---|---|
| *(c) A detailed specification of the smart PR-CFT connection with a clip angle component member* | | |
| Beam | W18 × 50 | W18 × 50 |
| Column | HSS12 × 12 × 500 | HSS14 × 500 |
| SMA Bar Material | Super-Elastic Nitinol | Super-Elastic Nitinol |
| Steel Bar Material | Corresponding to A490 Bolts | Corresponding to A490 Bolts |
| Bar Gauge Length | 304.8 mm | 355.6 mm |
| Bar Diameter | Steel Bar.25.4 mm, SMA Bar.26.9875 mm | Steel Bar.25.4 mm, SMA Bar.26.9875 mm |
| Bar Stenderness Ratio | Steel Bar.38.90, SMA Bar.32.43 | Steel Bar.45.39, SMA Bar.37.84 |
| Bolt Material | A 490 Bolts | A 490 Bolts |
| Shear Bolt Size | 1 × 4 (One Washer) | 1 × 4 (One Washer) |
| Web Bolt Size | 1 × 4 (One Washer) | 1 × 4 (One Washer) |
| Nut Type | A 563 Nuts Corresponding to 25.4 mm Dia. Bolts | A 563 Nuts Corresponding to 25.4 mm Dia. Bolts |
| Component Members | L6 × 8 × 1 (Thick Clip Angle) | L6 × 8 × 1 (Thick Clip Angle) |
| Shear Tab | Plate 4.5 × 9 × 0.56 (Fillet Welding 7.9375 mm) | Plate 4.5 × 9 × 0.56 (Fillet Welding 7.9375 mm) |
| Panel Zone Shape & Size | Rectangular 12 × 12 × 18.1 | Rectangular 14 × 14 × 18.1 |
| Inside Concrete Material | Rec. Confined Concrete ($F/F_u = 0.2$, $f'_c = 27.56$ MPa) | Cir. Confined Concrete ($F/F_u = 0.2$, $f'_c = 27.56$ MPa) |

This chapter will be organized as follows. First, the basic design principles are presented and the advantages of using these connections discussed (Section 2.5.1). Second, a description of the connection details, including the material properties, schematic drawings, and bolt/bar specifications, are presented (Section 2.5.2). Third, the governing failure modes for each connection are introduced, followed by the discussion of the structural characteristics of each component (Section 2.5.3). Fourth, the processing of the analytical data related to computing axial deformations and relative rotations is described (Section 2.5.4 and Appendix C). Finally, summary and conclusion are presented (Section 2.5.5).

## 2.5.1 Design principles

All connections were designed as full strength (FS) connections, meaning that they can transfer the full plastic beam moment ($M_{p,beam}$) from the beam to the column. They fulfill the requirements for connection design given in both the AISC LRFD Standard (*AISC 2001*) and the AISC Seismic Provisions (*2005 AISC Seismic Provisions*). However, the connections transfer the $M_{p,beam}$ at relatively large rotations and after significant yielding of the connection components. Because of results from the flexibility of the components, the connections will be considered as partial restraint (PR) ones. The primary purpose of this chapter is to develop design methods for the PR composite connections that will result in ductile connection behavior. Since some shear yielding and local buckling have been observed in the panel zone of connections to hollow steel columns before reaching $M_{p,beam}$, the columns' (and thus the panel zones') capacity was increased by filling the column with concrete. Thus all columns used were concrete-filled tubes or CFT columns.

The design approach used in this research explicitly considered the feasibility of integrating shape memory alloys (SMA) and regular steel bars into steel-concrete composite connections. Super-elastic Nitinol bars used as tension fasteners and subjected to large deformations can provide re-centering capabilities because their permanent strains remain small. When combined with A490 steel bars, they will result in connections with better energy dissipation and permanent deformation performance as compared with connections using either all super-elastic Nitinol bars or all conventional steel ones.

Finally, components such as shear/web bolts, shear tab plates, and clip angles or T-stubs were designed with the intent to avoid catastrophic losses of stiffness and strength due to brittle failure modes. All design strengths were checked against the demand from both code-based forces and those given by non-linear analyses. Details of those analyses are given in later chapters. The designs were based on weak beam-strong column conditions.

## 2.5.2 Specimen details

Three types of smart PR-CFT connections were designed and detailed as shown in Appendix B: an end-plate, a T-stub, and a clip angle one. Each connection type was designed to connect to both a rectangular concrete filled tube column (RCFT) and a circular concrete filled tube column (CCFT) to a wide flange shape. Conventional

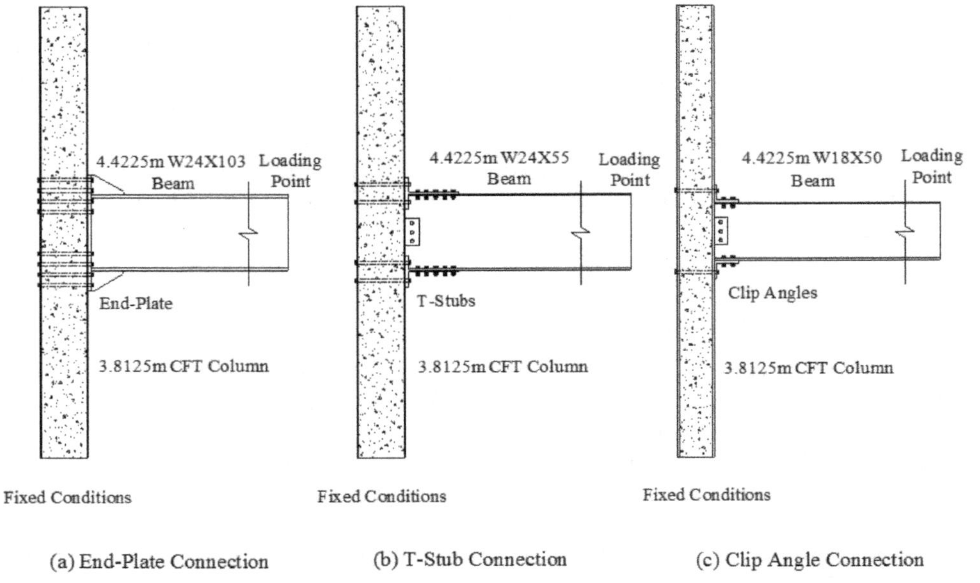


*Figure 2.26* Typical connection configurations.

A490 bolts were used to connect the flanges, and web bolts and a shear tab were used for the T-stub and clip angle connections.

### 2.5.3 Typical configurations

Typical connection sub-assemblage configurations are shown in Fig. 2.26. The column and beam lengths were selected for a building with 3.8 m floor heights and 8.8 m bays, and the models developed with the assumption of hinges forming at the mid-height and mid-span of the columns and beams, respectively, for an exterior bay. Based on a simplified model of a story subjected to lateral loads, column was taken as 3.8 m height and the beams as 4.4 m long for all the sub-assemblages. These models were used to study the behavior of the connection components and the adjacent areas of the beam and column.

The end-plate connection consisted of a concrete filled tube using either a HSS 16 × 16 × 500 (RCFT = rectangular CFT) or a HSS 18 × 500 (CCFT = circular CFT), and a W25 × 103 beam. Similar columns sections and lengths were used for the T-stub connections, but with a smaller beam (W24 × 55) size and a T-stub cut from a W16 × 100 section. The clip angle connection, which provided the smallest design capacity among the three types of connections, consisted of a either HSS12 × 12 × 500 (RCFT) or HSS 14 × 500 (CCFT), W18 × 50 beams and a clip angle member. The column sections were chosen such that the CCFT and RCFT columns had a similar equivalent area and capacity.

Monotonic and cyclic displacements were applied to the tip of beam. These analyses were used to investigate the deformation and strength performance of the different connections. These analyses or numerical simulations will be discussed exhaustively in the next two chapters.

### 2.5.4 Connection details

Detailed calculations for all three connection types are shown in Appendix B. The resulting configurations for typical interior joints are given in Figs. 2.27 to 2.32. The specimen identifications for the smart PR-CFT connections are composed of the shape of the column (i.e., CCFT or RCFT) and the connection type (i.e., End-plate, T-stub or Clip Angle).

572 Grade 50 steel was used for all members and joint components. A490 high strength bolt material was used for all bolts, with corresponding washers and nuts. Materials with equivalent properties to A490 bolts were sued for the steel tension bars, while the material properties for the super-elastic Nitinol (Shape Memory Alloy, SMA) were taken from work of other researchers (Davide, 2003). SMA tension bars were placed at the farthest practical locations from the centroid of the connection in order to take advantage of the re-centering effect during unloading (refer to Figs. 2.27 through 2.32 below).

The end-plate connection (Figs. 2.27 and 2.28), was composed of a 977.9 mm × 381 mm × 25.4 mm plate welded to the beam by with a 7.9375 mm fillet weld. The design required the use of extended stiffener plates welded between the beam flange and the end-plate. They were terminated at the beam flange and at the end of the end-plate with landings about 25.4 mm long.

The plate stiffeners had the same material strength (A572-Gr. 50l) as the beam and their thickness was equal to the beam web thickness. The tension fasteners that ran through the CFT column were sixteen 25.4 mm diameter, either 508mm long (RCFT column) or 558.8 mm long bars (CCFT column) with washers at either end.

The T-stub connection (Figs. 2.29 and 2.30) had tension fasteners that also ran through the CFT column. These were eight 25.4 mm diameter, either 508 mm long (RCFT column) or 558.8 mm long bars (CCFT column). Twelve 25.4 mm diameter, 101.6 mm long A490 bolts were used to fasten each T-stub stem to the beam flange and three 25.4 mm diameter, 101.6 mm long A490 bolts were used as web bolts.

The clip angle connection (Figs. 2.31 and 2.32), was composed of thick clip angles cut from a L6 × 8 × 1 and a 228.6 mm × 114.3 mm × 14.2875 mm shear tab. The tension fasteners through the CFT column were two 25.4 diameter (SMA bar) and one 26.9875 mm diameter (steel bar), either 406.4 mm long (RCFT column) or 457.2 mm long bars (CCFT column). Four 25.4 mm diameter, 101.6 mm long A490 bolts were used to fasten each clip angle leg to the beam flange and three 25.4 mm diameter, 101.6 mm long A490 bolts were used as web bolts. 3D configurations of all three smart PR-CFT connections are shown in Fig. 2.33.

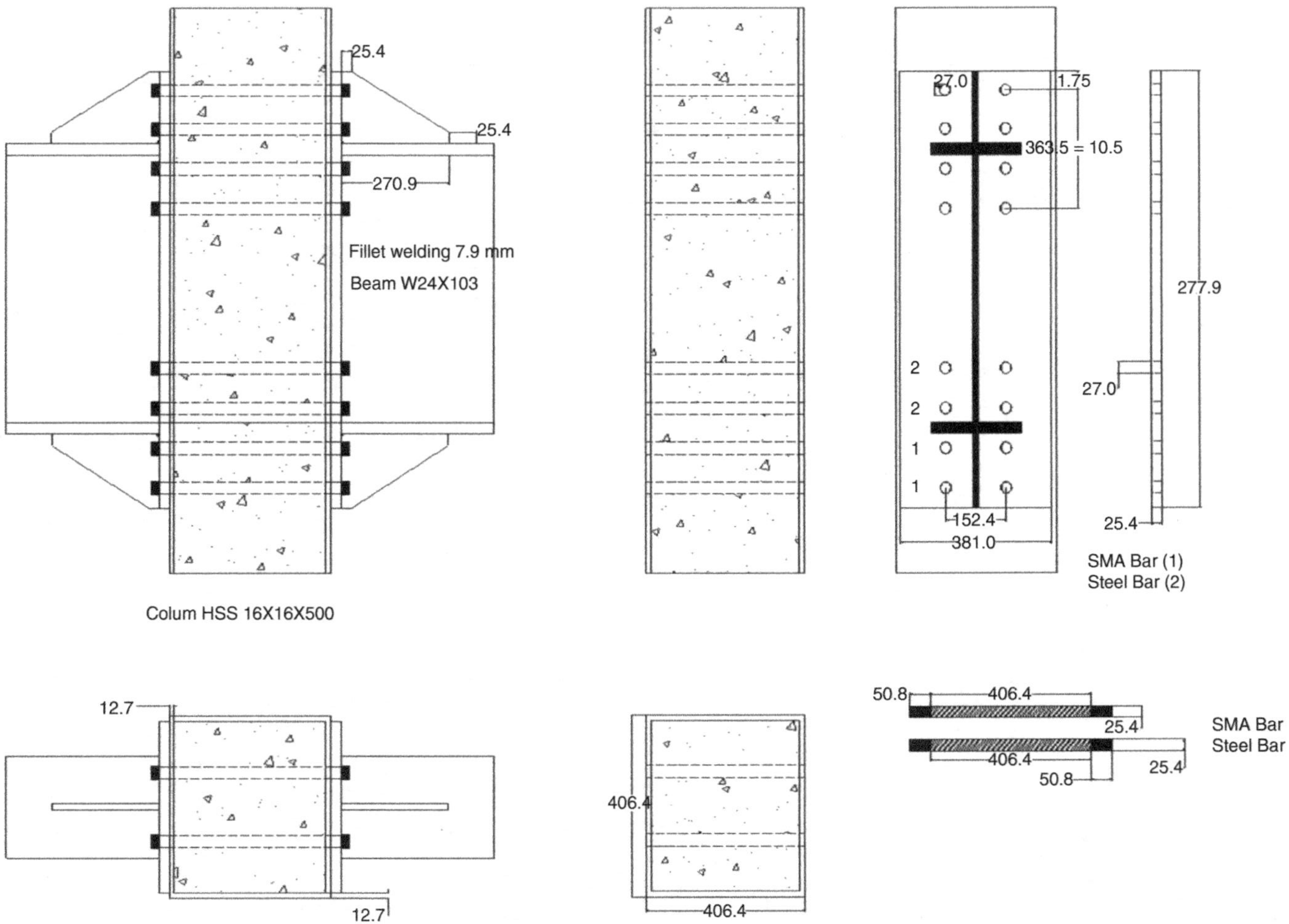


*Figure 2.27* Schematic drawing of RCFT end-plate connection detail (unit: mm).

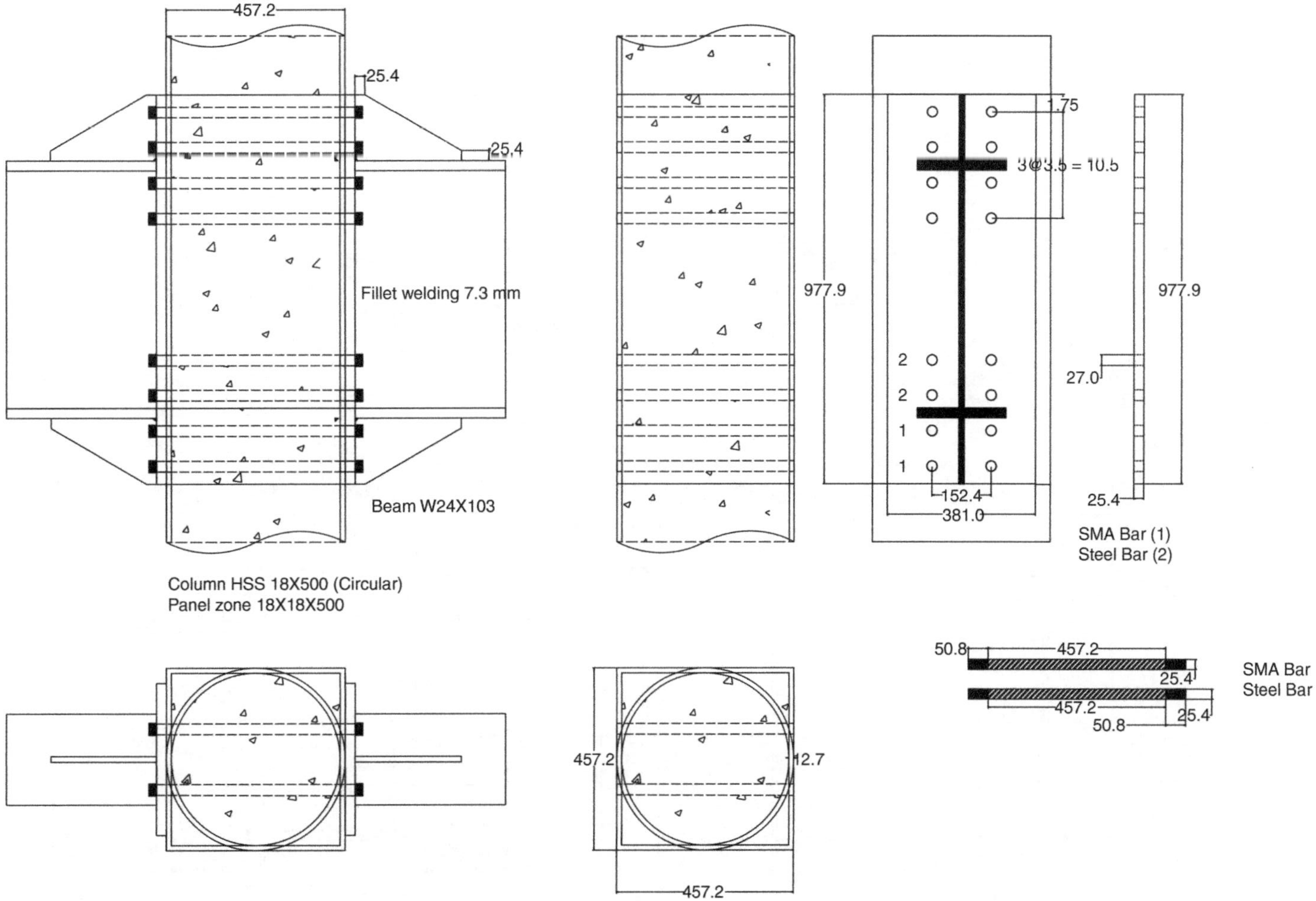


*Figure 2.28* Schematic drawing of CCFT end-plate connection detail (unit: mm).

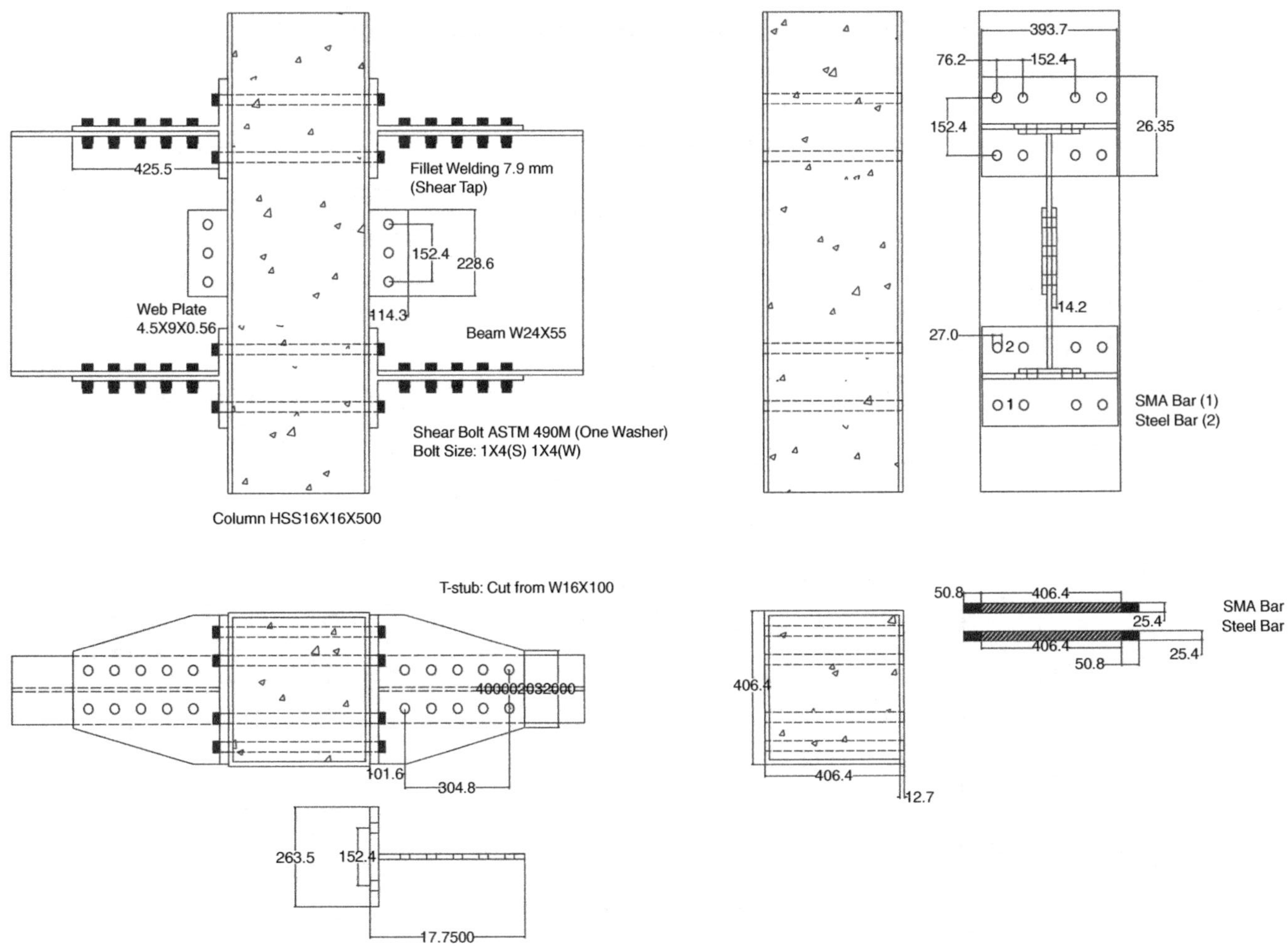


*Figure 2.29* Schematic drawing of RCFT T-stub connection detail (unit: mm).

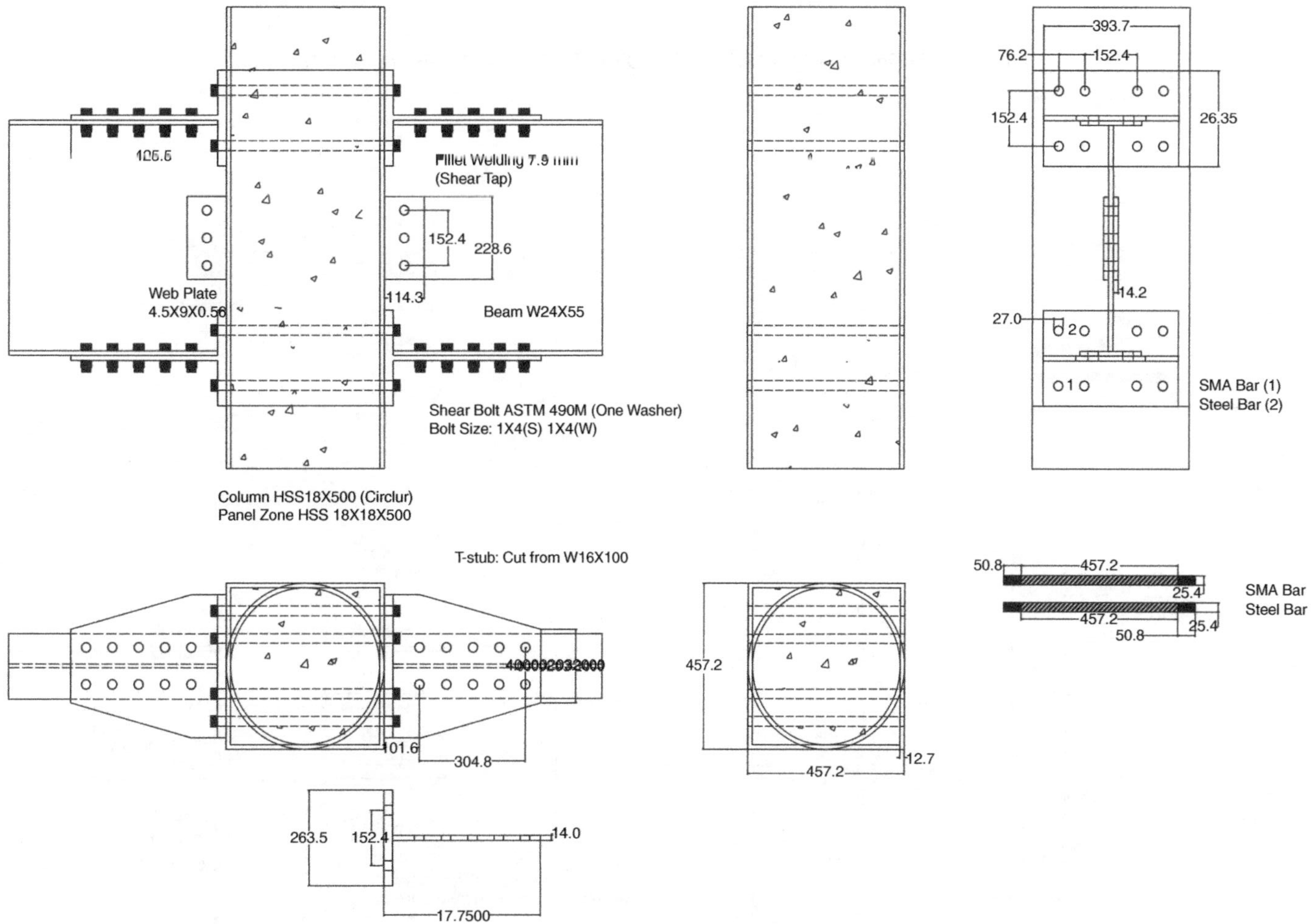


*Figure 2.30* Schematic drawing of CCFT T-stub connection detail (unit: mm).

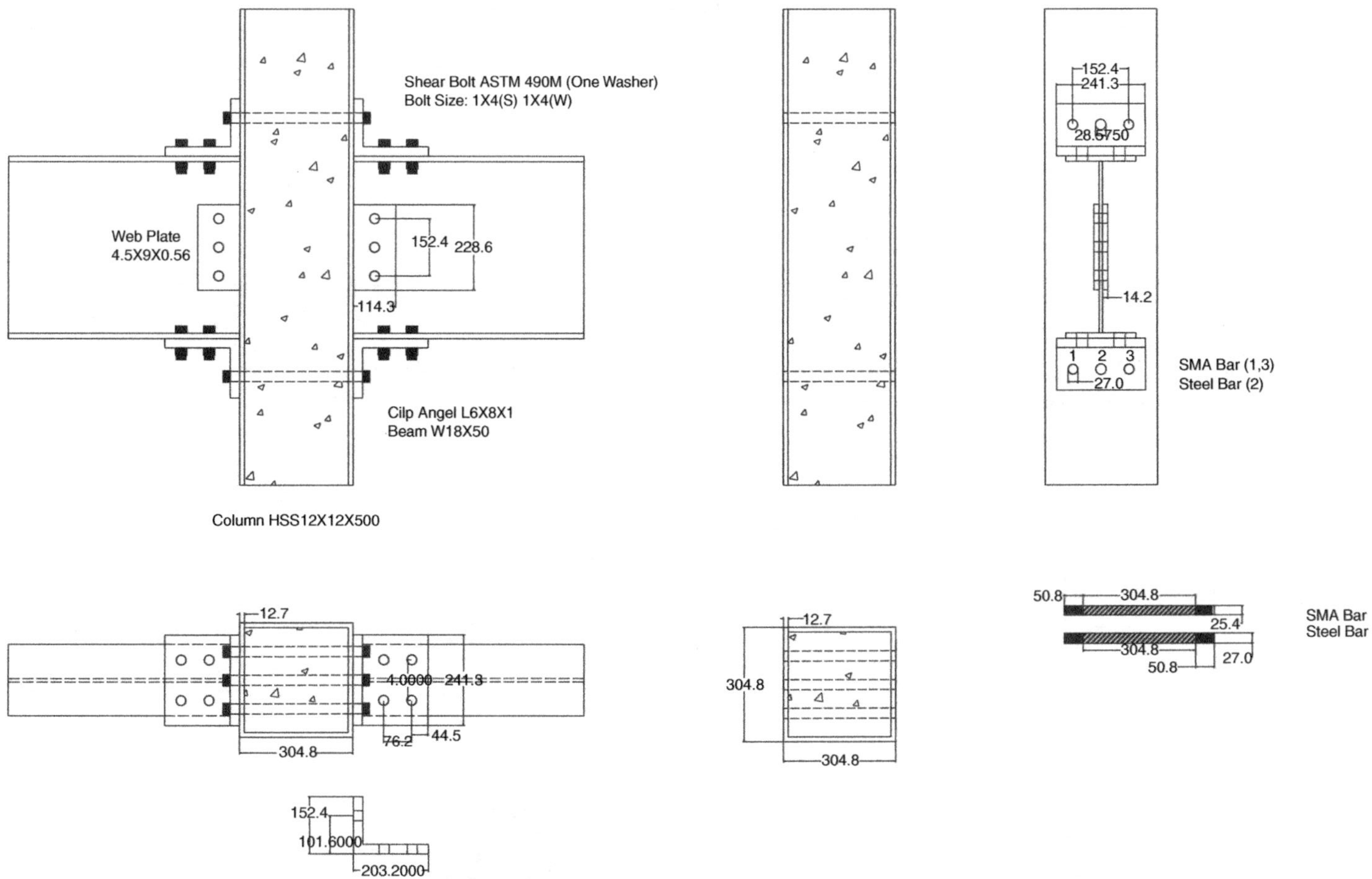


*Figure 2.31* Schematic drawing of RCFT clip angle connection detail (unit: mm).

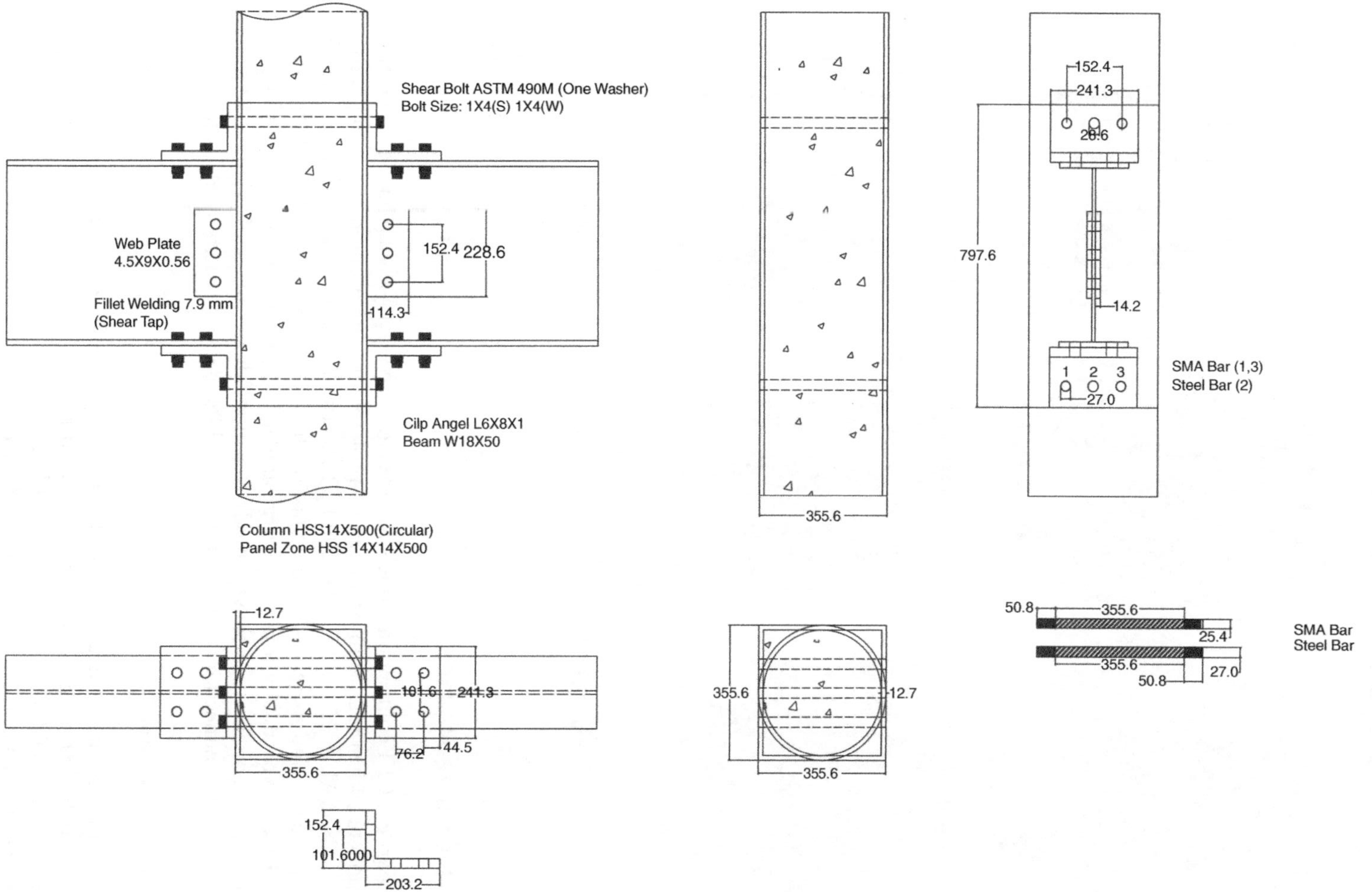


*Figure* 2.32 Schematic drawing of CCFT clip angle connection detail (unit: mm).

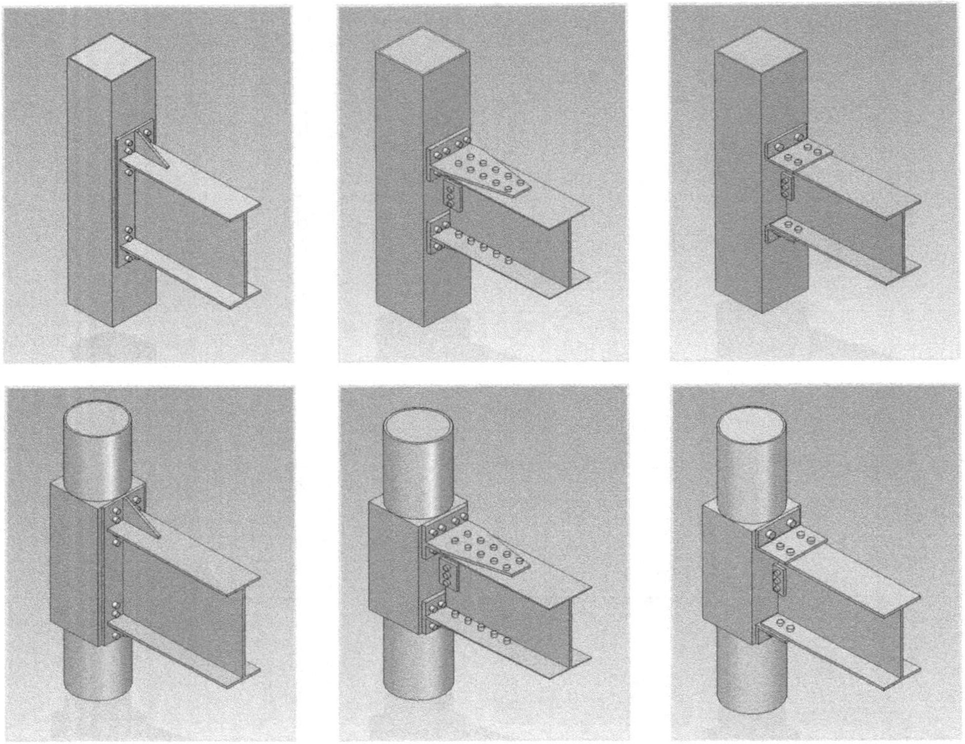

*Figure 2.33* 3D configurations of PR-CFT connection detail.

## 2.6 FAILURE MODES

All smart PR-CFT connection models were designed to reach yielding on the connection components when the beam produced its probable maximum moment at the plastic hinge (ANSI/AISC 358-05, FEMA 2000). This criterion satisfies a weak beam-strong column design and increases the ductility of the connection. Therefore, no fractures should occur even under the most severe ground motion and yielding of the components should be the dominating behavior mode. The possible yielding and failure modes for smart PR-CFT connections are given in Fig. 2.34 and can be categorized as follows:

**Ductile Failure Modes:**

- Slippage on the shear surface (T-stub and clip angle connections)
- Yielding of the gross area of component members
- Bearing yielding at the around bolt holes
- Plastic yielding of the gross area of the beam

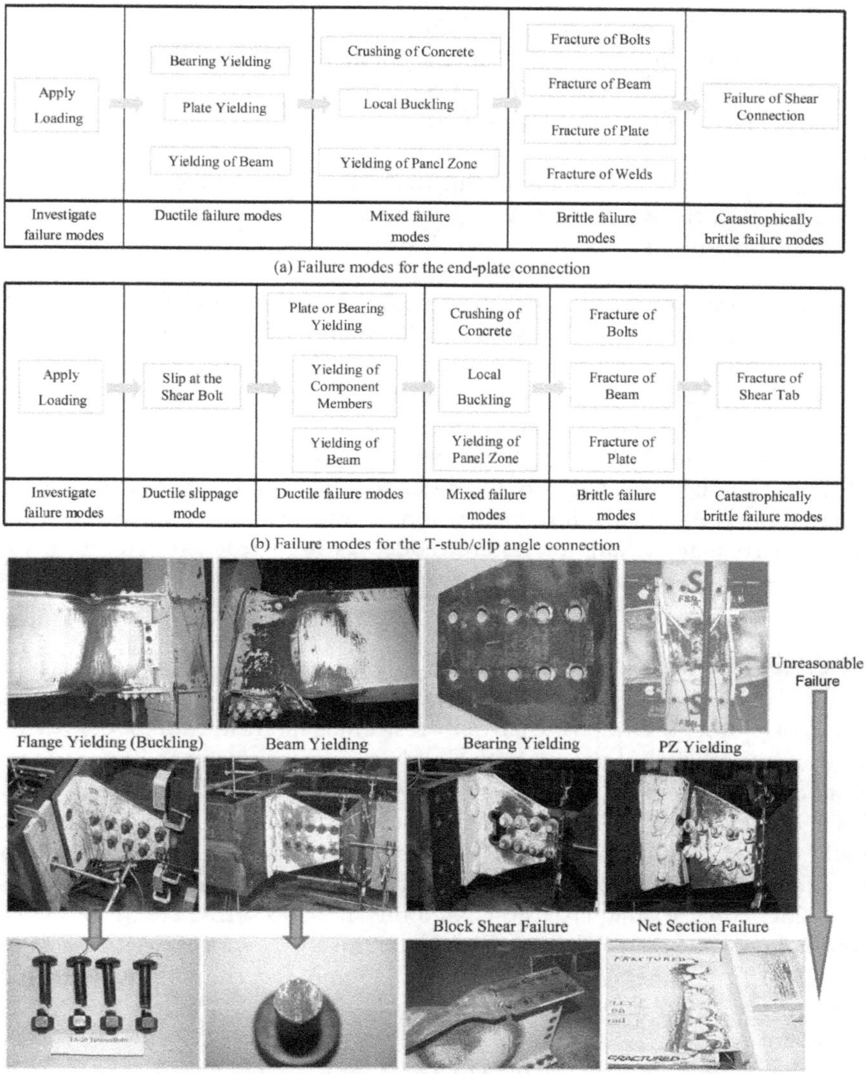


*Figure 2.34* Failure modes for PR-CFT connections.

## Failure Modes with Limited Ductility:

- Local buckling of component members
- Local buckling of the beam flange
- Shearing yielding of the composite panel zone
- Local buckling of CFT columns including crushing at the inside concrete

**Brittle Failure Modes:**

- Fracture of the edge distance or bolt spacing
- Block shear failure of T-stub/clip angle under direct shear force
- Fracture of the net section area of component members
- Block shear failure of the beam flange
- Shear fracture of the shear bolts
- Tension fracture of bars connecting the component member to the CFT column
- Fracture of the welds at the end-plate and plate stiffeners
- Shear fracture of web bolts and a shear tap

Slippage of the shear bolts is the most desirable failure mode while shear fracture of a shear connection is the least desirable failure mode. The smart PR-CFT connection with end-plates does not show slippage because of the lack of the shear surface. The feasible yielding and failure modes for smart PR-CFT connections are listed below in the order of desirability, as suggested by Astaneh-Asl (1995 and 1997). Fig. 2.34(c) shows photos of many failure modes achieved in previous experimental work at GT aimed at establishing the capacity of the different connection components. The state of stress based on Von-Mises failure theory under static loads will be used to determine whether a particular yielding or failure mode has been reached. The results of the numerical studies on these connections are shown in the next chapter.

## 2.7 COMPOSITE FRAME DESIGN

Three sets of prototype composite partially restrained moment frames (C-PRMF) were designed – one with end-plate, one with T-stub, and one with clip angle connections. In addition, companion composite special moment frames (C-SMF) with fully rigid (FR) welded connections were also designed in order to compare the behavior of both types (partially (PR) and fully restrained (FR)) composite frames. All composite moment frames presented here are designed in accordance with the *AISC 2005 Seismic Provisions* (AISC 2005 Seismic Provisions) and the *IBC 2003* (IBC 2003) for lateral and gravity loads, respectively. The gravity and lateral loads were determined following the ASCE 7-05 guidelines (ASCE 2005). Design limits, system requirements, and seismicity factors for these building located on a high seismicity area were determined by these guidelines.

This chapter will be structured as follows. Typical characteristics of these frames are described in Section 2.7.1. Descriptions of the PR building configurations used in this study are given in Section 2.7.2. Seismic design methods, design limits, and the equivalent lateral forces are described in Section 2.7.3. Finally, the design of the C-SMF specimens is given in Section 2.7.4.

### 2.7.1 Characteristics of composite moment frames

The frame designs are governed by the *AISC 2005 Seismic Provisions*. Four general classes of composite moment frame (C-MF) are identified in Part II of the *AISC 2005 Seismic Provisions* (AISC 2005 Seismic Provisions) as shown in Table 2.7. The buildings were designed to the loads prescribed by ASCE 7-05 (ASCE 2005). The primary

*Table 2.7* General classes of composite moment frame (C-MF).

| *The type of C-MF* | *Main deformation/yield shape* | *A total inter-story drift angle* | *SDC* | *Special system requirements* |
|---|---|---|---|---|
| C-PRMF | Limited yielding in column base<br>Main yielding in the ductile components | 0.04 radian | C or below | A nominal strength is at least equal to 50 percent of $M_p$ |
| C-SMF | Main yielding in the beams<br>Limited inelastic deformations in the columns and/or connections | 0.04 radian | D and above | The required strength shall be determined with the flexural strength (LRFD: $R_yM_n$) |
| C-IMF | Main yielding in the beams<br>Moderate inelastic deformations in the columns and/or connections | 0.03 radian | C and below | – |
| C-OMF | The imited inelastic action will occur in the beam, columns and/or connections | – | A and B | – |

purpose of the ASCE 7-05 standard is to provide information useful to determine the required strength, maximum inter-story drift, and seismic use groups for a given structure type and geographical location. The seismic design category (SDC) assigned to a building is a classification based upon the occupancy class (type of occupation and consequences to human life in case of collapse) and the seismicity of the site. SDC A, B and C generally correspond to structures in zones of low to moderate seismicity or low importance, while SDC D, E, and F require special seismic detailing as they address structures in areas of high seismic risk and/or critical structures (hospitals, fire stations, emergency response centers, for example). For this study, composite PR moment frames (C-PRMF), a moderately ductile system, and composite special moment frames (C-SMF), one of the most ductile systems, were selected for the trial design of several low-rise (4 to 6 stories) moment frames. The designs herein satisfy all the design requirements of C-PRMF or C-SMF for SDC D, E, or F.

Typical composite partially restrained frames (C-PRMF) are composed of I-shape steel columns and composite steel beams which are interconnected with PR composite connections (Leon and Kim 2004; Thermou, Elnashai, Plumier, and Doneaux 2004). However, composite PR frames with concrete filled tube columns and steel beams with PR composite connections have been recently proposed (Tsai *et al.*, 2004 and Wu *et al.*, 2006). PR composite connections use traditional shear and bottom flange connections, but take advantage of the floor slab to provide the top connection. Composite connections use shear studs to the beams and slab reinforcement in the negative moment regions to provide additional strength and stiffness as shown in Fig. 2.35. A PR composite connection has many beneficial characteristics including:

- The floor slab system results in a more efficient distribution of strength and stiffness between negative moment and positive moment regions of the beam. It also contributes to the redistribution of loads under inelastic state.
- In the design of PR composite connections, applied loads can be considered separately, with the bending moment assigned to the steel reinforcement in the slab and a clip angle or plate on the bottom flange, and the shear force assigned to the web angle or plate.

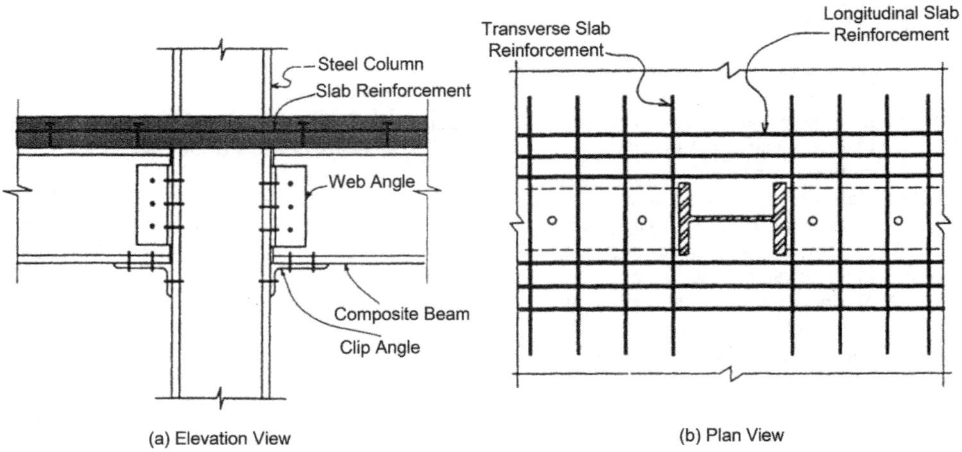


*Figure 2.35* Typical composite partially restrained moment connection (AISC 2005 Seismic Provisions).

- PR composite connections can undergo large deformations without fracture. The connections are generally designed for less than the full plastic strength of the beam. The ductility of the connection comes from deformations of its many components. The intent is to delay the occurrence of brittle failures such as web crippling, bolt or weld failures, and net section failures. If these brittle failures are avoided, the large available connection ductility can guarantee excellent frame performance under large inelastic deformations.
- PR moment frames are better at mitigating the effects of seismically induced loads as the lengthening of the natural period due to both the flexibility of the connection and its gradual yielding and stable hysteretic behavior of the connections.

C-PRMFs were originally conceived for areas of low to moderate seismicity in SDC C and below. However, C-PRMF can be used in areas of higher seismicity (Leon 1990) with appropriate detailing and analyses. In addition, the recently developed bidirectional bolted connections for CFT columns and I-beams provide superior earthquake performance in terms of in stiffness, strength, ductility, and energy dissipation. Recent studies demonstrate that the seismic resistance exceeds those requirements specified in the seismic design codes of Taiwan and the US (Wu *et al.*, 2007). The structural configurations of those connections are very similar to those of end-plate connections presented in this research (See Fig. 2.36). Therefore, C-PRMFs with those connection models have excellent seismic resistance, and this structural system can perform well and be put into practice.

Composite special moment frames (C-SMF) are composed of a variety of configurations where structural steel or composite beams are combined with either reinforced concrete or composite columns. Schematic connection drawings for C-SMF are shown in Fig. 2.37. In order to avoid the need for field welding of the beam flange adjacent to the critical beam-to-column junction, the steel beam can run continuously though the reinforcement concrete column as shown in Fig. 2.37(a). The steel band plates

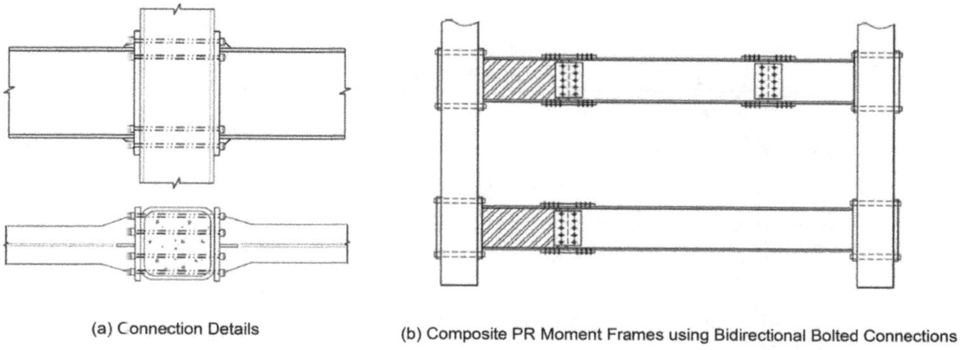


*Figure 2.36* Bidirectional bolted connections between CFT columns and H-beams (Wu *et al.*, 2007).

attached to the beam are one of possible ways to strengthen the joint by providing good confinement to the concrete. As shown in Fig. 2.37(b), connections between steel beams and encased composite columns have been used and tested extensively in Japan. One disadvantage of this connection detail is that it requires welding of the beam flange to the encased steel column. Composite filled tube column-to-steel beam connections as shown in Fig. 2.37(c) have been used less frequency but there has been recent research resulting in practical design recommendations (Azizinamini and Schneider 2004). Based on ASCE 7-05, C-SMF were originally designed for use in SDC D and above. C-SMF shall be designed with assumption that significant inelastic deformation will occur under the design earthquake, primarily in the beams, but with limited inelastic deformation in the columns and connections. Therefore, connections in C-SMF satisfy the story drift capacity of 0.04 radian as specified in the *AISC 2005 Seismic Provisions* (AISC 2005 Seismic Provisions) so that they are not susceptible to weld fracture.

### 2.7.2 Building configurations

This section describes the building configurations for C-PRMF and C-SMF used in this study. C-PRMF were designed with three types of moment connections: end-plate, T-stub and clip angle connections. The rest of the system comprises structural steel beams and CFT columns. On the other hand, C-SMF were designed with welded moment connections between composite filled tube columns and steel beams

Building configurations, materials, and modeling conditions were the same for both the C-PRMF and C-SMF in order to compare their inelastic behavior. Four and six story configurations with 3 by 5 long bays were used throughout this research. Perimeter moment resisting frame systems were used because the intent is to demonstrate the economy of this system for a market segment that constitutes about 90 percent of the steel frame construction in the USA. Most of all, these moment resisting frames have been very popular in many regions of high seismicity because of high ductility and excellent architectural versatility. Identical dead loads, live loads, seismic design, and

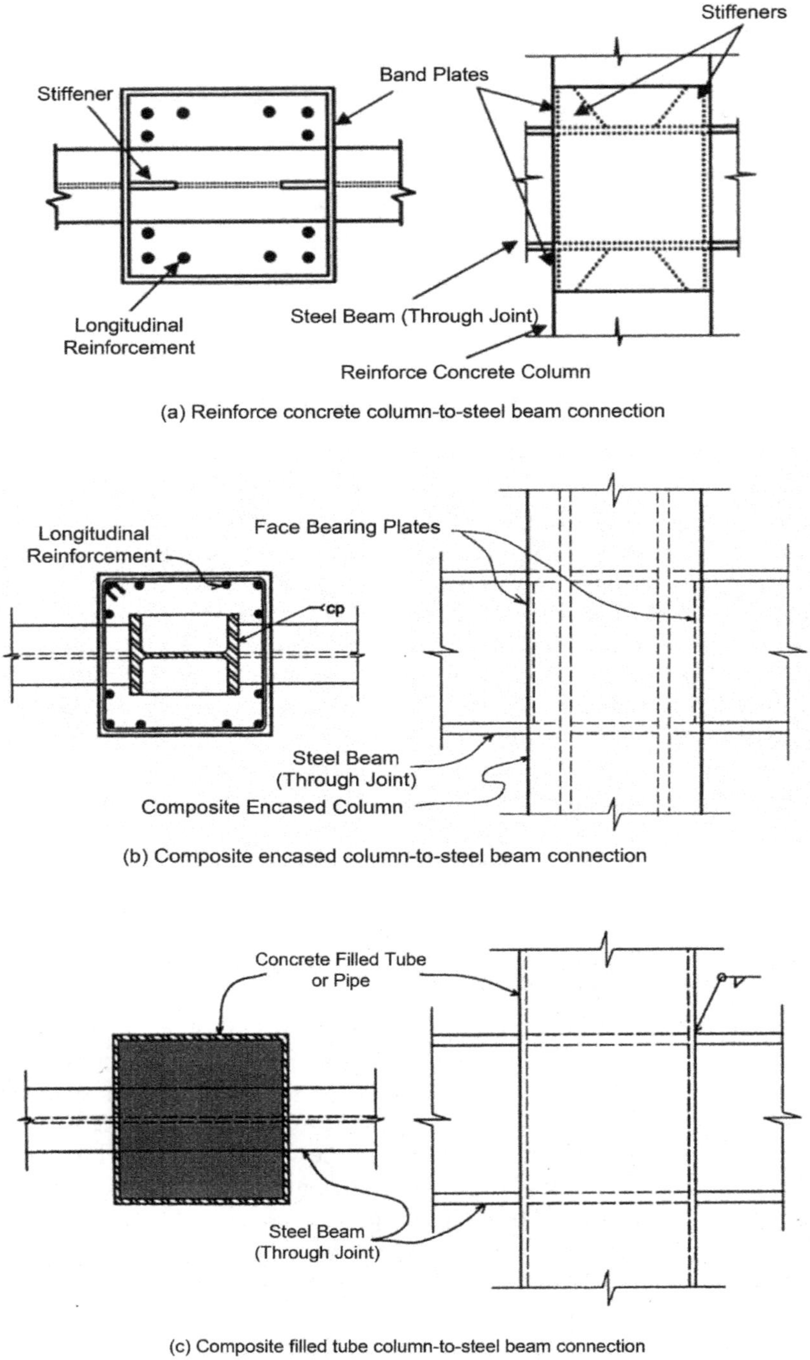


*Figure 2.37* Typical composite special moment connection (Azizinamini & Schneider, 2004).

*Table 2.8* Location, loads, and structural classifications common to all frames.

| *Located area* | *Gravity loads* | *SDC* | *Occupancy category* |
|---|---|---|---|
| LA Area | Dead: 4790 Pa Lice: 3832 Pa | D Class | Ordinary Structures |

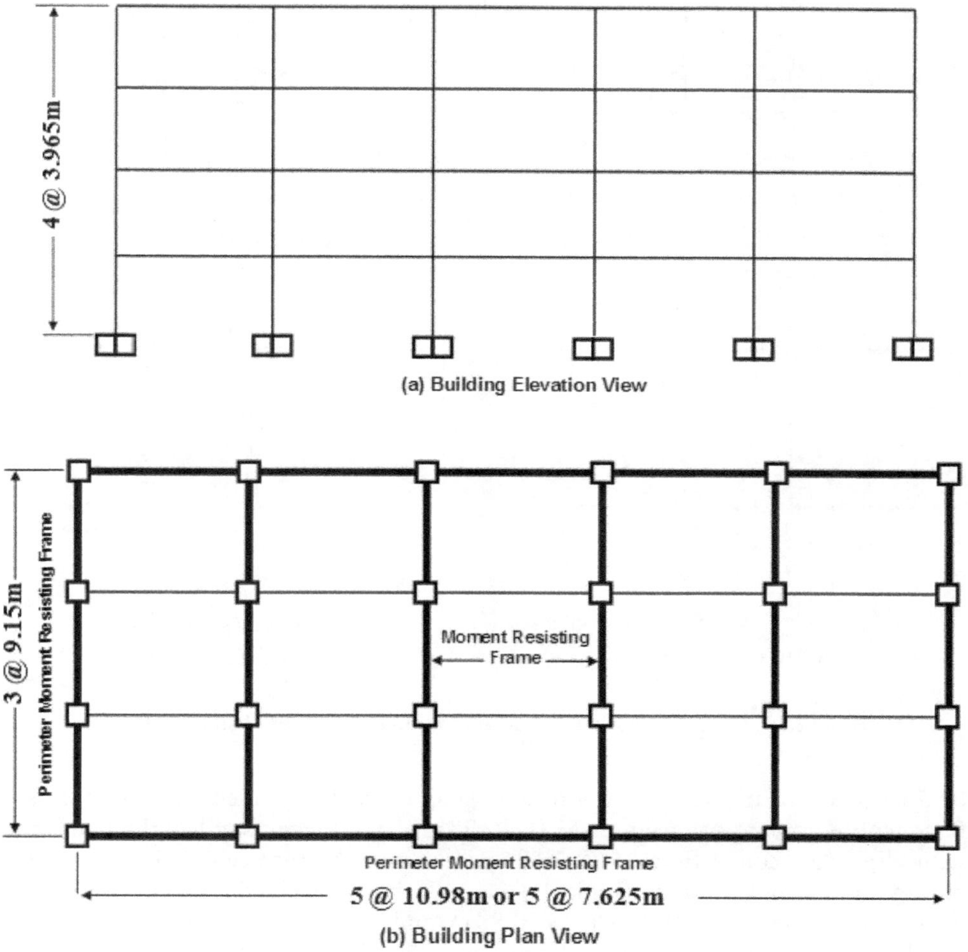


*Figure 2.38* Building elevation and plan view for the 4 story building.

occupancy category are used with all buildings, as given in Table 2.8. More detailed descriptions are given in the following sub-sections.

#### 2.7.2.1 Building description for 4 story building

The configuration for the 4 story buildings is shown in Fig. 2.38. The total height is 15.86 m, with a constant height 3.965 m for all stories. This building has 3 bays

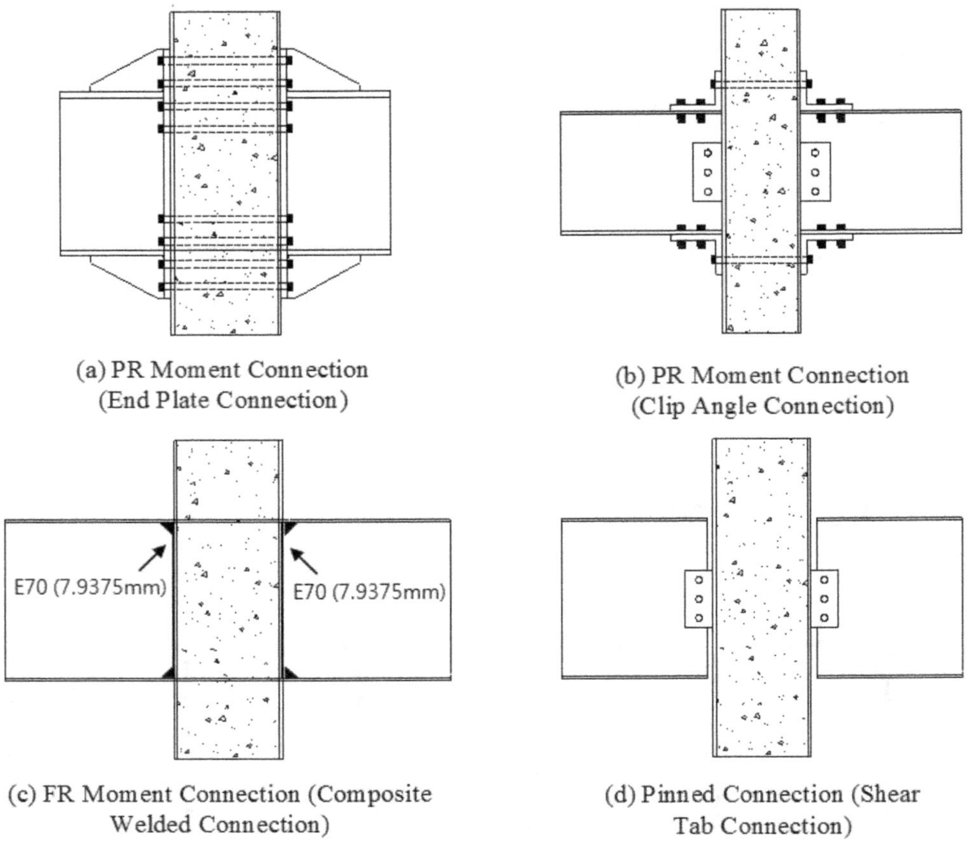


*Figure 2.39* Cruciform connection details for moment frames (4 story building).

by 5 bays. There are two bay lengths according to the PR moment connections used. Bays with end-plate connections (5 @ 10.98 m) have longer bay lengths than those with clip angle connections (5 @ 7.625 m) because the former connections consist of larger beam and columns (Table 2.8). C-SMF having welded moment connections were also designed with 10.98 m (5 @ 10.98 m) and 7.625 m (5 @ 7.625 m) bay lengths. In addition, the same member size for each C-PRMF was used in the corresponding C-SMF in order to compare the behavior of both composite frames.

Resistance against lateral forces is provided primarily by rigid frame action in the perimeter frames. These perimeter frames utilize composite PR connections between the CFT columns and beams while the interior CFT columns and beams aligned in one direction are interconnected by pinned connections. For the C-SMF, FR connections are used in the perimeter frames instead of PR connections. The moment resistant frames with either PR or FR connections are presented as thick lines in the building plans for each frame type shown in Fig. 2.38. The moment connections used in the moment resistant frames of the 4 story buildings are given in Fig. 2.39.

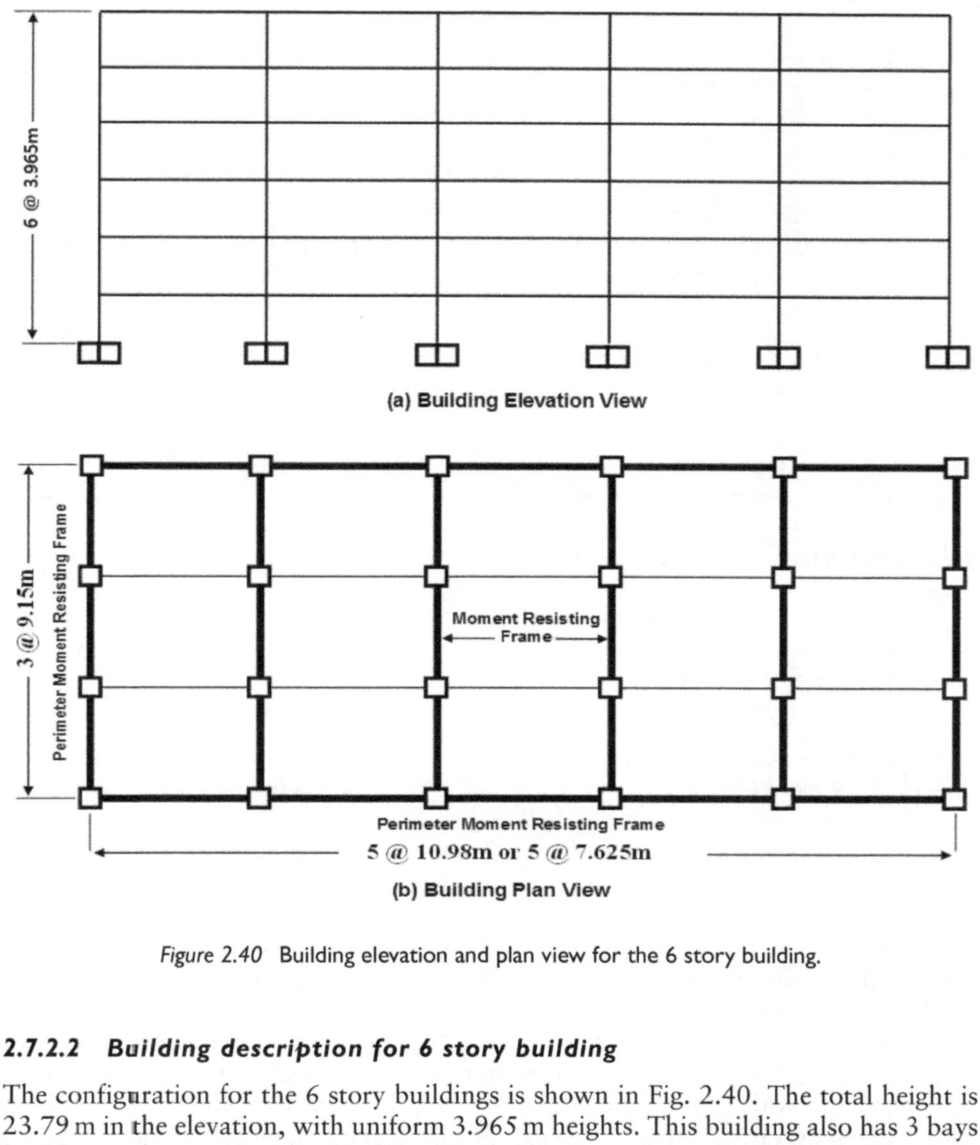


*Figure 2.40* Building elevation and plan view for the 6 story building.

#### 2.7.2.2 *Building description for 6 story building*

The configuration for the 6 story buildings is shown in Fig. 2.40. The total height is 23.79 m in the elevation, with uniform 3.965 m heights. This building also has 3 bays by 5 bays. There are two kinds of bay lengths determined according to the PR moment connections used. The bays with end-plate connections (5 @ 10.98 m) have longer bay lengths than those with T-stub connections (5 @ 7.625 m) because the former connections consist of larger beam and columns. The plan for the 6 story building is the same as that of 4 story building. Both buildings share the same conditions given in Table 2.8. C-SMF having welded moment connections were designed with five 10.98 m (5 @ 10.98 m) and five 7.625 m (5 @ 7.625 m) bay lengths. As for the 4 story buildings, the same member size within each C-PRMF was used in the corresponding C-SMF in order to compare the behavior of both composite frames.

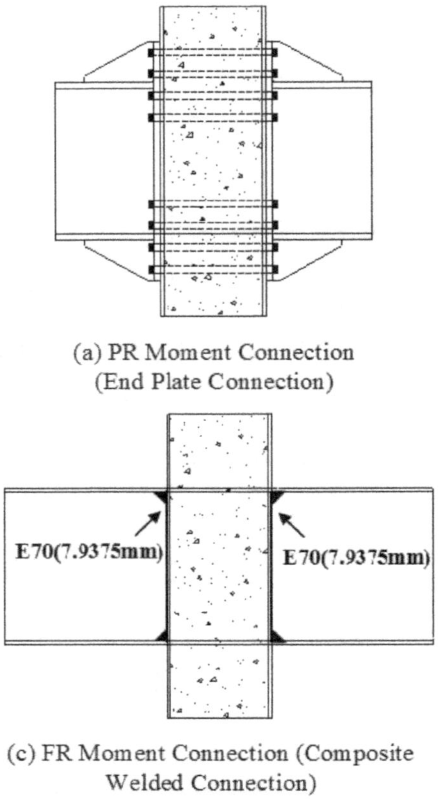


*Figure 2.41* The cruciform connection details for moment frames (6 story building).

Thick lines in the building plan Fig. 2.40 indicate the moment resisting frames. The moment resisting frames in 6 story building were located in the same positions as those in the 4 story building. However, T-stub connections were used as the PR connections in the five 7.625 m (5 @ 7.625 m) bay lengths. T-stub connections provide less inter-story drift as well as well as more resistance against to the lateral loads in comparison with clip angle connections. For this higher building system, T-stub connections are a better structural system than clip angle connections under the same modeling conditions for plan and component members. The moment connections use in the moment resistant frames for the 6 story buildings are given in Fig. 2.41. The detailed design procedures for all frame specimens will be described in the next section where the use of the SAP2000 programs (CSI, 1984–2004) is described.

### 2.7.3 Seismic design method

As stated above, the connections and the composite frames were designed as structures located in the L.A area in accordance with *AISC LRFD* (AISC 2001) and *AISC*

*2005 Seismic Provisions* (ANSI/AISC 341-05) respectively. The design of the prototype buildings was checked with the SAP 2000 design checking tool (SAP 2000, ver 11, 2007). Only Dead (D), live (L), and earthquake (E) loads are considered in this research as earthquakes dominate over wind in the L.A. area. The equivalent lateral loads (E) for these composite frames are calculated by using 2003 International Building Codes (IBC 2003).

#### 2.7.3.1 Load combinations

The design dead and live loads for the composite frames are assumed as 4790 Pa and 3832 Pa respectively. A572 Grade 50 steel was used for beams and steel section of CFT columns. ASCE 7-05 and LRFD design guidelines for load factors and combinations were used, as follows:

- Load Combination 1: 1.4D
- Load Combination 2: 1.2D + 1.6L + 0.5L
- Load Combination 3: 1.2D + 1.6S + (0.5L or 0.8W)
- Load Combination 4: 1.2D + 1.6W + L + 0.5S
- Load Combination 5: 1.2D + 1.0E + L
- Load Combination 6: 0.9D + 1.6W
- Load Combination 7: 0.9D + 1.0E

The terms related snow load (S) and wind load (W) were ignored in the above combinations. The earthquake effect includes the components from both vertical and horizontal accelerations. Based on the load combinations, load combination 5 dominated over other load combinations.

#### 2.7.3.2 Equivalent lateral loads

Composite moment frames should provide adequate strength, stiffness, and energy capacity so that they can withstand not only the lateral loads but also the gravity loads within specified limits of deformation and strength. The design utilized equivalent lateral load procedures as introduced in the ASCE 7-05 and the IBC2003 codes. The equivalent lateral load approach is based on a set of static lateral loads that corresponds to the 1st mode shape of deformation. For frame structures, the first mode generally contributes upwards of 90% of the effective seismic mass and dominates the behavior of the structure. Therefore, those procedures may not always be valid when higher mode shapes contribute more than 10% of the effective seismic mass. More details on the calculation procedures for equivalent lateral loads are given in Appendix D.

The design response spectra for these composite moment frames (CMF) in the LA area are summarized in Table 2.9 and Fig. 2.42. The site class for the area on which the building is located is one of the factors that determine the seismic response coefficients. A site class A for 4 story building or C for 6 story building, which corresponds to hard rock area or soft rock area respectively, was used in these designs. The fundamental time period of the building is computed by simplified analysis based on the code equations (ASCE 7-05 Section 9.5.4 and IBC2003 Section 1617.5). Therefore, the same value of fundamental time period can be applied to the buildings with same number of stories and heights regardless of the connection type. The final calculations

*Table 2.9* Design response spectra for CMF in LA area.

| Area<br>site class | *4 Story frame*<br>*90045 LA area*<br>*C* | *6 Story frame*<br>*90045 LA area*<br>*A* |
|---|---|---|
| $S_S$ | 1.5491 | 1.5491 |
| $S_1$ | 0.5897 | 0.5897 |
| $F_a$ | 1.0 | 0.8 |
| $F_v$ | 1.3 | 0.8 |
| $S_{ms}$ | 1.5491 | 1.2393 |
| $S_{ml}$ | 0.7666 | 0.4718 |
| $S_{Ds}$ | 1.033 | 0.8262 |
| $S_{Dl}$ | 0.511 | 0.3146 |
| T | 0.099 | 0.076 |
| $T_s$ | 0.495 | 0.381 |
| T | 0.661 | 0.914 |

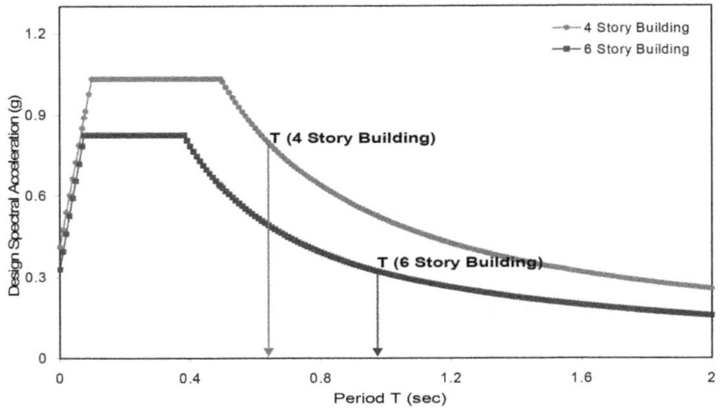


*Figure 2.42* Design response spectra for CMF in LA area.

for the equivalent lateral loads are summarized in Table 2.10. The vertical distribution of the equivalent lateral load is proportional to the area of the building plan which determines the weight of each story level.

#### *2.7.3.3 Regulations and limits*

The IBC 2003 code (IBC2003) is designed to protect public health and welfare in all communities through model code regulations. Minimum regulations for building systems using both prescriptive and performance based provisions are specified in this comprehensive building code. The IBC 2003 requires that the design story drift ($\Delta$) should not exceed the allowable story drift ($\Delta_a$) as obtained from Table 1617.3 of the IBC2003 for any story level. The design story drift can be calculated as the difference of the deflections ($\delta_x$) at the center of mass at the top and bottom of the

*Table 2.10* Design loads for all composite moment frames.

| Story | Dead load (D) | Live load (L) | Equivalent lateral load (E) | Story | Dead load (D) | Live load (L) | Equivalent lateral load (E) |
|---|---|---|---|---|---|---|---|
| (a) The design loads for 4 story composite frame with end-plate connections (5 @ 36′ bay length) | | | | (b) The design loads for 4 story composite frame with clip angle connections (5 @ 25′ bay length) | | | |
| 1 | 4790 Pa | 3832 Pa | 161.5 kN | 1 | 4790 Pa | 3832 Pa | 93.5 kN |
| 2 | 4790 Pa | 3832 Pa | 254.9 kN | 2 | 4790 Pa | 3832 Pa | 148.4 kN |
| 3 | 4790 Pa | 3832 Pa | 264.8 kN | 3 | 4790 Pa | 3832 Pa | 153.2 kN |
| 4 | 4790 Pa | 3832 Pa | 180.7 kN | 4 | 4790 Pa | 3832 Pa | 113.4 kN |
| (c) The design loads for 6 story composite frame with end-plate connections (5 @ 36′ bay length) | | | | (d) The design loads for 6 story composite frame with T-stub angle connections (5 @ 25′ bay length) | | | |
| 1 | 4790 Pa | 3832 Pa | 87.3 kN | 1 | 4790 Pa | 3832 Pa | 50.5 kN |
| 2 | 4790 Pa | 3832 Pa | 167.9 kN | 2 | 4790 Pa | 3832 Pa | 116.6 kN |
| 3 | 4790 Pa | 3832 Pa | 219.1 kN | 3 | 4790 Pa | 3832 Pa | 152.2 kN |
| 4 | 4790 Pa | 3832 Pa | 232.5 kN | 4 | 4790 Pa | 3832 Pa | 161.5 kN |
| 5 | 4790 Pa | 3832 Pa | 202.9 kN | 5 | 4790 Pa | 3832 Pa | 117.4 kN |
| 6 | 4790 Pa | 3832 Pa | 126.5 kN | 6 | 4790 Pa | 3832 Pa | 73.2 kN |

story. The deflections ($\delta_x$) of each story level $x$ and allowable story drift are determined as following equations:

$$\delta_x = \frac{C_d \delta_{xe}}{I_E} \tag{2.93}$$

$$\Delta_a = 0.02 h_{xe} \tag{2.94}$$

where,
$C_d$: the deflection amplification factor ($C_d = 5.5$ for C-SMF system)
$\delta_{xe}$: the deflections determined by an elastic analysis for C-SMF system
$I_E$: the occupancy importance factor ($I_E = 1.0$ for an ordinary occupancy)
$h_{xe}$: the story height at each story level $x$

The P-Δ effects on the story shears and moments, the resulting member forces and moments and the story drift caused by these effects need not be considered when the stability coefficient ($\theta$) is equal to or less than 0.1. The stability coefficient is determined by:

$$\theta = \frac{P_x \Delta}{V_x h_{sx} C_d} \tag{2.95}$$

where,
$P_x$ : total un-factored vertical design load at and above story level $x$
$V_x$: the seismic lateral force between story level $x$ and story level $x-1$
$\Delta$: the design story drift occurring simultaneously with $V_x$

The stability coefficient shall not exceed $\theta_{max}$ determined as below:

$$\theta_{max} = \frac{0.5}{\beta C_d} \leq 0.25 \tag{2.96}$$

where, $\beta$ is the ratio of shear demand to shear capacity for the story between story level $x$ and story level $x-1$, generally taken as 1.0.

Based on these limits, the design checks for deflection and drift limits for the composite frames can be conducted by comparing the factored deflections obtained by Eq. 2.93 and the stability coefficient obtained by Eq 2.95 with the allowable story drift and the stability coefficient limit (0.1 or $\theta_{max}$) respectively. The design checks of deflection and drift limit for C-SMF subjected to the dominant Load Combination 5 will be shown after the initial selection of member sizes using the SAP2000 programs (SAP2000).

### 2.7.4 Design of composite moment frame specimens

The design for all buildings that consist of structural composite columns and steel beams are summarized in Table 2.11. As mentioned above, material for all steel component members is assumed as A572 Gr. 50. The prototype building has nine variations according to the different combinations of connection types and column systems. A uniform size for all column members was selected because of fabrication and economy considerations. This means that the behavior of the lower columns is anticipated as the controlling factor, as the column sizes generally decrease with height. On the other hands, smaller beam sizes were selected for the higher stories in order to achieve an economical design. In addition, beam and column sizes presented in here are very close to those presented in the 3D FE models described in previous chapters in order to maintain the ideal failure modes. Other connection details follow those of the 3D FE connection models.

Structural models of both the 4 and 6 story buildings have symmetric configurations at all story levels. Because of the assumption that the composite floors behave as rigid diaphragms, the perimeter composite moment frames (C-MF) work together with the internal frames in resisting the lateral loads. Thus, analyses of a 2D perimeter composite moment frame can used to simulate the behavior of the buildings, avoiding the need for 3D analyses. Fig. 2.43 show plan views of the 3D building and perimeter moment frames of interest (C-SMF). Moment frames along the W-E direction deform more under the equivalent lateral loads because of the larger number of panel zones and members. The red dashed rectangular indicate the perimeter C-SMF modeled as 2D models on a SAP2000 program.

Analyses performed by OpenSEES and SAP2000 were used to estimate story drift, deflections, and P-Delta effects. After this initial analysis, a design check was run by using an AISC-LRFD 2001 code which is available on the design check menu of SAP2000. Fig. 2.44 shows the results of the design check for structural frame models with combined RCFT and CCFT columns. These design checks are described in terms of a strength ratio (capacity used/capacity available). The strength ratios are shown on the beam elements. The beam elements are deformed by the bending moments as well as the axial forces. Therefore, these strength ratios for the beams should include the

*Table 2.11* Design results for composite frame buildings.

| *Model ID* | *Connection type* | *Column system* | *Column size (All stories)* | *Beam size* | |
|---|---|---|---|---|---|
| | | | | *1st and 2nd Story* | *3rd and 4th Story* |
| (a) 4 story building with 5 @ 10.98 m bays and end-plate or welding connections | | | | | |
| 4END-C1 | End Plate | RCFT | HSS16 × 16 × 500 | W24 × 103 | W24 × 84 |
| 4END-C2 | End Plate | CCFT | HSS18 × 500 | W24 × 103 | W24 × 84 |
| 4END-C3 | End Plate | RCFT + CCFT | HSS16 × 16 × 500, HSS18 × 500 | W24 × 103 | W24 × 84 |
| 4END-C4 | End Plate | RCFT | HSS16 × 16 × 500 | W24 × 103 | W24 × 84 |
| 4END-C5 | End Plate | CCFT | HSS18 × 500 | W24 × 103 | W24 × 84 |
| 4END-C6 | End Plate | RCFT + CCFT | HSS16 × 16 × 500, HSS18 × 500 | W24 × 103 | W24 × 84 |
| 4END-C7 | Welding | RCFT | HSS16 × 16 × 500 | W24 × 103 | W24 × 84 |
| 4END-C8 | Welding | CCFT | HSS18 × 500 | W24 × 103 | W24 × 84 |
| 4END-C9 | Welding | RCFT + CCFT | HSS16 × 16 × 500, HSS18 × 500 | W24 × 103 | W24 × 84 |
| (b) 4 story building with 5 @ 7.625 m bays and clip angle or welding connections | | | | | |
| 4CLI-C1 | Clip Angle | RCFT | HSS14 × 14 × 500 | W18 × 50 | W18 × 50 |
| 4CLI-C2 | Clip Angle | CCFT | HSS16 × 500 | W18 × 50 | W18 × 50 |
| 4CLI-C3 | Clip Angle | RCFT + CCFT | HSS14 × 14 × 500, HSS16 × 500 | W18 × 50 | W18 × 50 |
| 4CLI-C4 | Clip Angle | RCFT | HSS14 × 14 × 500 | W18 × 50 | W18 × 50 |
| 4CLI-C5 | Clip Angle | CCFT | HSS16 × 500 | W18 × 50 | W18 × 50 |
| 4CLI-C6 | Clip Angle | RCFT + CCFT | HSS14 × 14 × 500, HSS16 × 500 | W18 × 50 | W18 × 50 |
| 4CLI-C7 | Welding | RCFT | HSS14 × 14 × 500 | W18 × 50 | W18 × 50 |
| 4CLI-C8 | Welding | CCFT | HSS16 × 500 | W18 × 50 | W18 × 50 |
| 4CLI-C9 | Welding | RCFT + CCFT | HSS14 × 14 × 500, HSS16 × 500 | W18 × 50 | W18 × 50 |

| *Model ID* | *Connection type* | *Column system* | *Column size (All stories)* | *Beam Size* | |
|---|---|---|---|---|---|
| | | | | *1st to 3rd Story* | *4th and 6th Story* |
| (c) 6 story building with 5 @ 36 bays and end-plate or welding connections | | | | | |
| 6END-C1 | End Plate | RCFT | HSS16 × 16 × 500 | W24 × 103 | W24 × 84 |
| 6END-C2 | End Plate | CCFT | HSS18 × 500 | W24 × 103 | W24 × 84 |
| 6END-C3 | End Plate | RCFT + CCFT | HSS16 × 16 × 500, HSS18 × 500 | W24 × 103 | W24 × 84 |
| 6END-C4 | End Plate | RCFT | HSS16 × 16 × 500 | W24 × 103 | W24 × 84 |
| 6END-C5 | End Plate | CCFT | HSS18 × 500 | W24 × 103 | W24 × 84 |
| 6END-C6 | End Plate | RCFT + CCFT | HSS16 × 16 × 500, HSS18 × 500 | W24 × 103 | W24 × 84 |
| 6END-C7 | Welding | RCFT | HSS16 × 16 × 500 | W24 × 103 | W24 × 84 |
| 6END-C8 | Welding | CCFT | HSS18 × 500 | W24 × 103 | W24 × 84 |
| 6END-C9 | Welding | RCFT + CCFT | HSS16 × 16 × 500, HSS18 × 500 | W24 × 103 | W24 × 84 |
| (d) 6 story building with 5 @ 7.625 m bays and T-stub or welding connections | | | | | |
| 6TSU-C1 | T-Stub | RCFT | HSS16 × 16 × 375 | W24 × 62 | W24 × 55 |
| 6TSU-C2 | T-Stub | CCFT | HSS18 × 375 | W24 × 62 | W24 × 55 |
| 6TSU-C3 | T-Stub | RCFT + CCFT | HSS16 × 16 × 375,HSS18 × 375 | W24 × 62 | W24 × 55 |
| 6TSU-C4 | T-Stub | RCFT | HSS16 × 16 × 375 | W24 × 62 | W24 × 55 |
| 6TSU-C5 | T-Stub | CCFT | HSS18 × 375 | W24 × 62 | W24 × 55 |
| 6TSU-C6 | T-Stub | RCFT + CCFT | HSS16 × 16 × 375,HSS18 × 375 | W24 × 62 | W24 × 55 |
| 6TSU-C7 | Welding | RCFT | HSS16 × 16 × 375 | W24 × 62 | W24 × 55 |
| 6TSU-C8 | Welding | CCFT | HSS18 × 375 | W24 × 62 | W24 × 55 |
| 6TSU-C9 | Welding | RCFT + CCFT | HSS16 × 16 × 375,HSS18 × 375 | W24 × 62 | W24 × 55 |

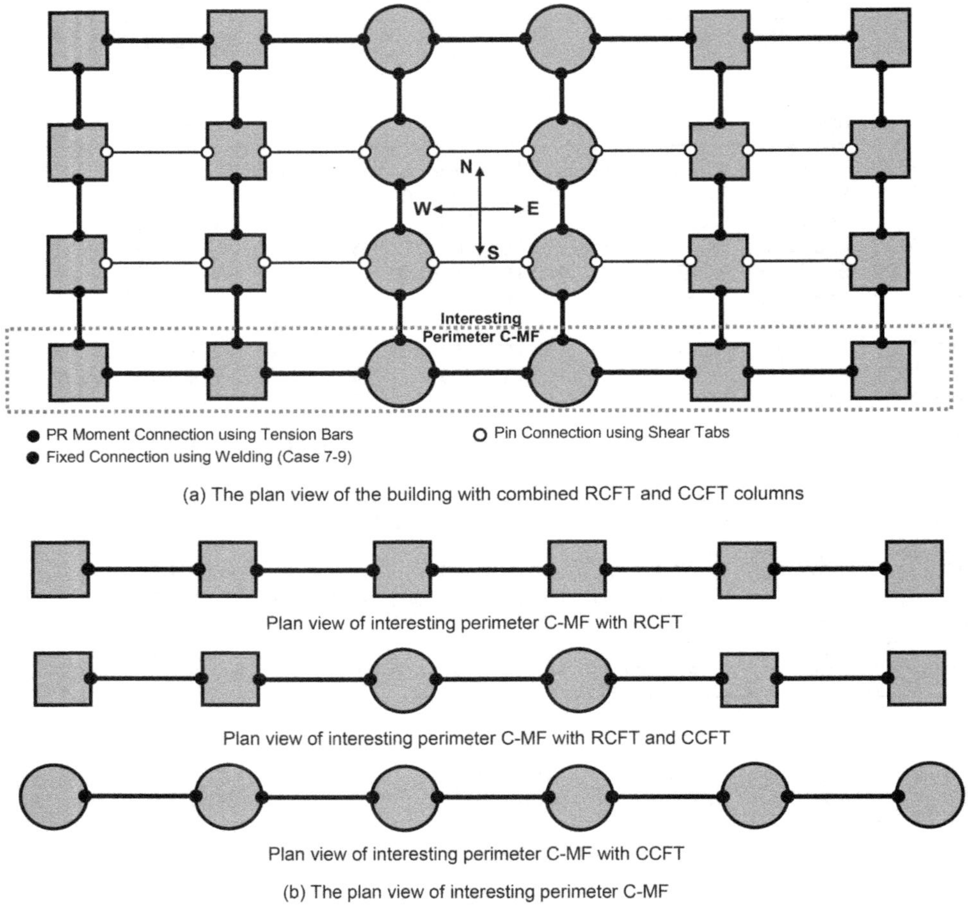


*Figure 2.43* Plan views of the 3D building and perimeter composite moment frames (C-MF).

available strength of the members as beam-columns Larger negative moments adjacent to the panel zone are the cause that the strength ratios acting on one end of beam elements are generally larger than those acting on the middle of beam elements as shown in Fig. 2.44(a-1). Other figures ranging from Fig. 2.44(a-2) to Fig. 2.44(d) show the largest strength ratio on each beam element. The design check function in SAP 2000 is only available for steel sections, so the design check for the composite columns will be calculated by other methods presented in Appendix A.

Design checks for deflection and drift ratios are given in Table 2.12. The units of story drifts and stable coefficients are cm and radian, respectively. The dominant load combination was applied to all frame models. All values from static analyses using the SAP2000 are less than the allowable limits.

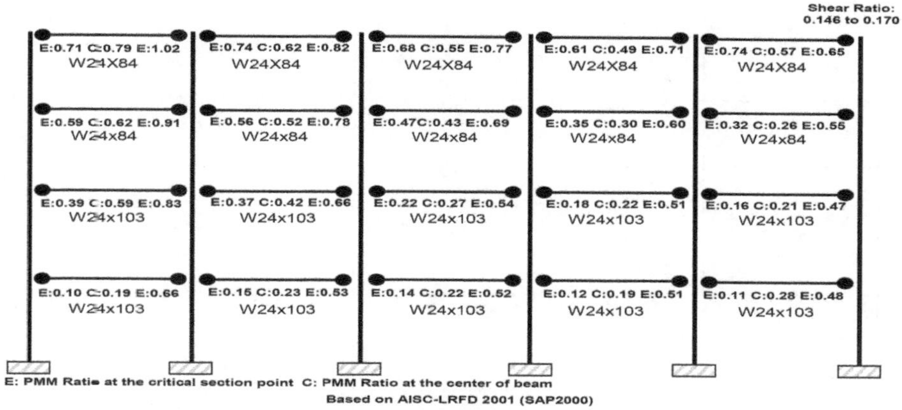


(a-1) Moment interaction ratio for beam members under dominant load combination 5 (SAP2000 PMM Ratio, 4END-C9)

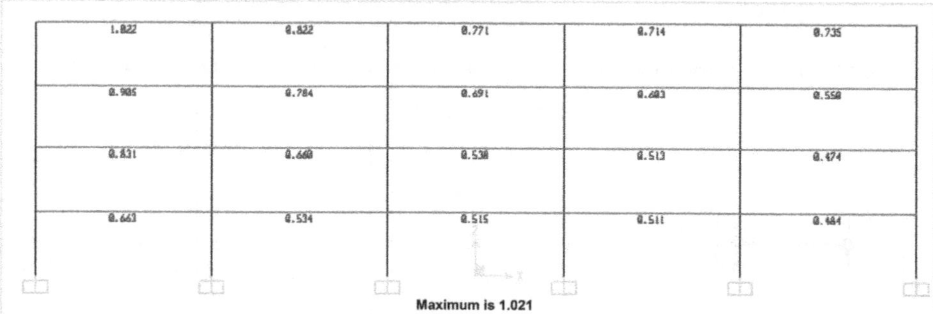


(a-2) Moment interaction ratio for beam members under dominant load combination 5 (SAP2000 PMM Ratio, 4END-C9)

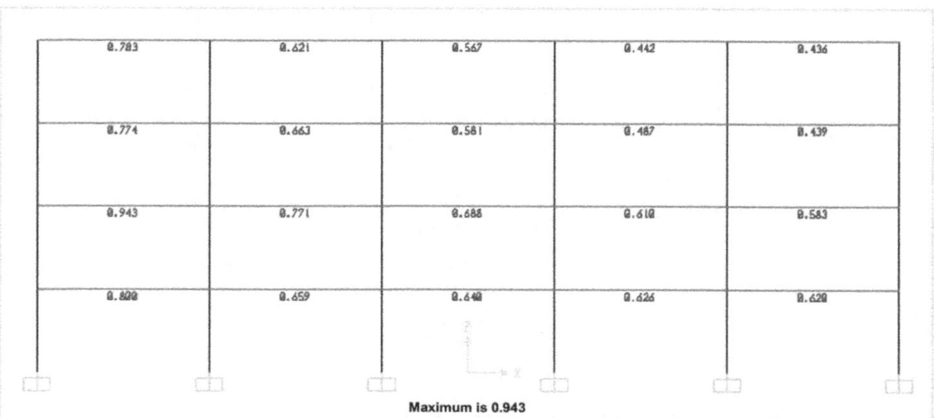


(b) Moment interaction ratio for beam members under dominant load combination 5 (SAP2000 PMM Ratio, 4CLI-C9)

*Figure 2.44* Moment interaction ratio for beam members under load combination 5.

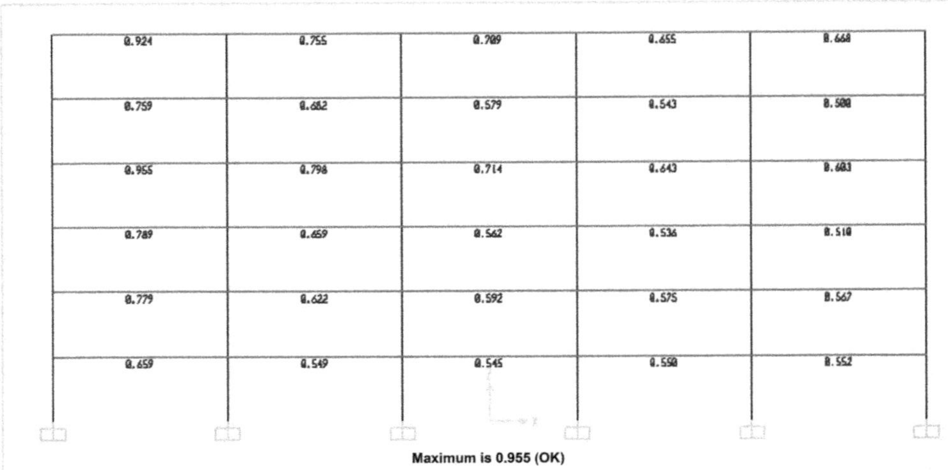


(c) Moment interaction ratio for beam members under dominant load combination 5 (SAP2000 PMM Ratio, 6END-C9)

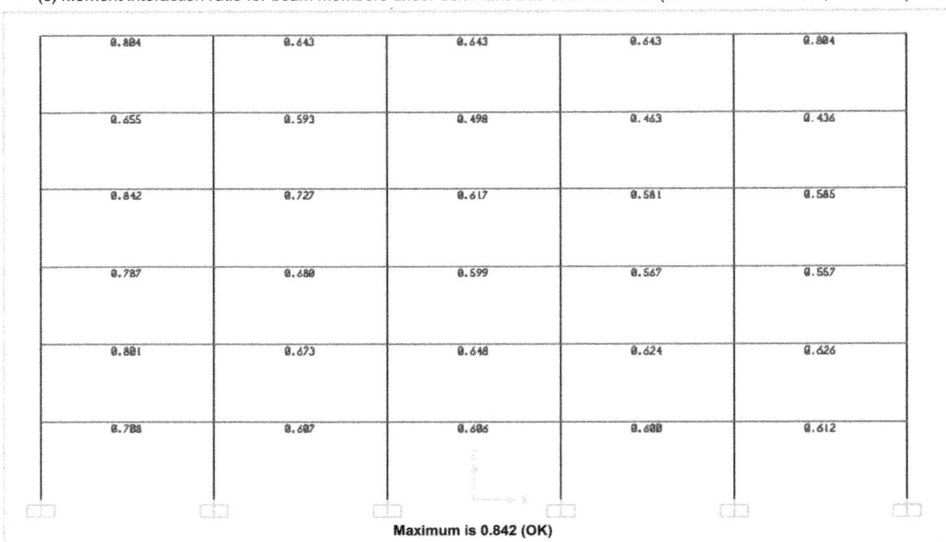


(d) Moment interaction ratio for beam members under dominant load combination 5 (SAP2000 PMM Ratio, 6TSU-C9)

*Figure 2.44* Continued.

## 2.8 FE MODELING FOR CONNECTION SYSTEMS

The ABAQUS (ABAQUS *Version* 6.6-1, 2006) finite element code was used to analyze the proposed PR-CFT connections. These numerical models consisted of a combination of elements, springs and constraints conditions. Amongst these were refined 3D

*Table 2.12* Design checks for deflection and drift ratio.

| Story | *Factored story drift* ($\Delta$) | *Allowable story drift* ($\Delta_a$) | *Stable coefficient* ($\theta$) | *Max. stable coefficient* ($\theta_{max}$) | Decision |
|---|---|---|---|---|---|
| *(a) Design checks for deflection and drift ratio (4END-C7)* | | | | | |
| 4 | 7.48 | 12.48 | 0.004 | 0.091 | OK |
| 3 | 6.49 | 9.36 | 0.008 | 0.091 | OK |
| 2 | 4.67 | 6.24 | 0.015 | 0.091 | OK |
| 1 | 1.99 | 3.12 | 0.026 | 0.091 | OK |
| *(b) Design checks for deflection and drift ratio (4END-C8)* | | | | | |
| 4 | 7.32 | 12.48 | 0.004 | 0.091 | OK |
| 3 | 6.38 | 9.36 | 0.008 | 0.091 | OK |
| 2 | 4.63 | 6.24 | 0.015 | 0.091 | OK |
| 1 | 2.10 | 3.12 | 0.027 | 0.091 | OK |
| *(c) Design checks for deflection and drift ratio (4CLI-C7)* | | | | | |
| 4 | 8.96 | 12.48 | 0.005 | 0.091 | OK |
| 3 | 7.81 | 9.36 | 0.010 | 0.091 | OK |
| 2 | 5.55 | 6.24 | 0.018 | 0.091 | OK |
| 1 | 2.26 | 3.12 | 0.028 | 0.091 | OK |
| *(d) Design checks for deflection and drift ratio (4CLI-C8)* | | | | | |
| 4 | 8.85 | 12.48 | 0.005 | 0.091 | OK |
| 3 | 7.71 | 9.36 | 0.010 | 0.091 | OK |
| 2 | 5.50 | 6.24 | 0.018 | 0.091 | OK |
| 1 | 2.31 | 3.12 | 0.029 | 0.091 | OK |
| *(e) Design checks for deflection and drift ratio (6END-C7)* | | | | | |
| 6 | 13.86 | 18.72 | 0.002 | 0.091 | OK |
| 5 | 13.14 | 15.6 | 0.006 | 0.091 | OK |
| 4 | 11.57 | 12.48 | 0.009 | 0.091 | OK |
| 3 | 9.22 | 9.36 | 0.015 | 0.091 | OK |
| 2 | 6.14 | 6.24 | 0.026 | 0.091 | OK |
| 1 | 2.56 | 3.12 | 0.042 | 0.091 | OK |
| *(f) Design checks for deflection and drift ratio (6END-C8)* | | | | | |
| 6 | 13.72 | 18.72 | 0.002 | 0.091 | OK |
| 5 | 13.02 | 15.6 | 0.006 | 0.091 | OK |
| 4 | 11.49 | 12.48 | 0.009 | 0.091 | OK |
| 3 | 9.20 | 9.36 | 0.015 | 0.091 | OK |
| 2 | 6.22 | 6.24 | 0.026 | 0.091 | OK |
| 1 | 2.71 | 3.12 | 0.044 | 0.091 | OK |
| *(g) Design checks for deflection and drift ratio (6TSU-C7)* | | | | | |
| 6 | 10.84 | 18.72 | 0.002 | 0.091 | OK |
| 5 | 10.33 | 15.6 | 0.004 | 0.091 | OK |
| 4 | 9.19 | 12.48 | 0.007 | 0.091 | OK |
| 3 | 7.36 | 9.36 | 0.012 | 0.091 | OK |
| 2 | 4.87 | 6.24 | 0.021 | 0.091 | OK |
| 1 | 1.99 | 3.12 | 0.032 | 0.091 | OK |
| *(h) Design checks for deflection and drift ratio (6TSU-C8)* | | | | | |
| 6 | 10.83 | 18.72 | 0.002 | 0.091 | OK |
| 5 | 10.34 | 15.6 | 0.004 | 0.091 | OK |
| 4 | 9.23 | 12.48 | 0.007 | 0.091 | OK |
| 3 | 7.45 | 9.36 | 0.012 | 0.091 | OK |
| 2 | 5.02 | 6.24 | 0.021 | 0.091 | OK |
| 1 | 2.17 | 3.12 | 0.035 | 0.091 | OK |

solid elements incorporating the full nonlinear material/ geometric properties, contact elements, surface interaction with friction, constraint conditions using equation points, concrete crack conditions and elastic foundation springs. These advanced modeling methods were intended to provide a detailed and accurate understanding of the overall behavior of the connections, including the stress distributions on the contact surfaces in spite of the high computational cost typically associated with this type of data.

This chapter will be structured as follows. First the detail modeling described in the previous paragraph is presented. The results of the analyses for the FE models, including the stress distribution, final deformation of each component model and comparison of ultimate strength are described next. Several behavior characteristics under monotonic load are then studied with these models, including the stiffness for all components. It is then shown that these characteristics have a significant influence on the total behavior of the PR-CFT connections. Failure modes for the PR-CFT connections are then described based on the response of the FE models at different load levels. The failure strengths obtained by FE model tests are then compared with current design methods. Finally, a summary and discussion about this chapter are presented. This chapter deals almost exclusively with monotonic behavior.

### 2.8.1 3D solid modeling method

FE models for the PR-CFT connections were constructed using the nonlinear FE program ABAQUS 6.6-1. In particular, ABAQUS/CAE was used to generate many of the models. ABAQUS/CAE is a dedicated FE preprocessor that offers powerful and flexible parametric modeling for users familiar with modern computer aided design (CAD) systems (ABAQUS 2006). In this research, most of the FE work, including the generation of parametric geometries and meshes was done using a version of ABAQUS/CAE incorporating file-based input to provide more advanced modeling options. A typical analysis of a model using a Pentium D 3.00 GHz computer with 1.0 GB of memory required between 12 and 96 hour running time.

### 2.8.2 Modeling parts and elements adopted

The FE models (i.e., T-stub connections) were subdivided into several independent bodies such as two T-stub members, three web bolts, 10 shear bolt-nuts, one beam, hollow steel column, and interior concrete that interacted with each other via contact definitions (Fig. 2.45). They were modeled as half symmetric models using symmetric boundary conditions (See Section 2.8.5). Shear bolts and nuts were modeled as one body in order to neglect the surface interaction between theses two surfaces without slippage. Merging two independent parts made a significant contribution to saving computational cost. The modeling parts of the typical connection with a RCFT column or a CCFT column are shown in Figs. 2.46 and 2.47 respectively.

All parts were made up of 3D solid elements. The six basic connection models studied are shown as assemblies of 3D solid elements in Fig. 2.47. Close-up views of the corresponding connection areas are shown in Fig. 2.48. An exploded view of Fig. 2.48A is shown in Fig. 2.49. In this figure, the meshes for the welded end-plate and stiffeners to the beam were made up of C3D8I elements, 8 node brick elements with the

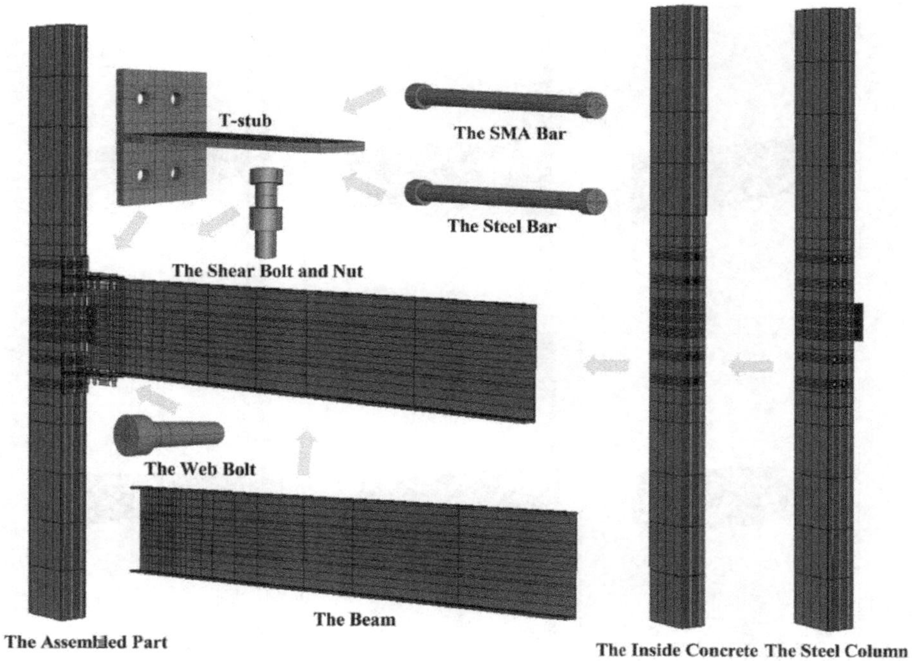


*Figure 2.45* Partitioned 3D solid models for the Smart SMA PR-CFT connection (RCFT case).

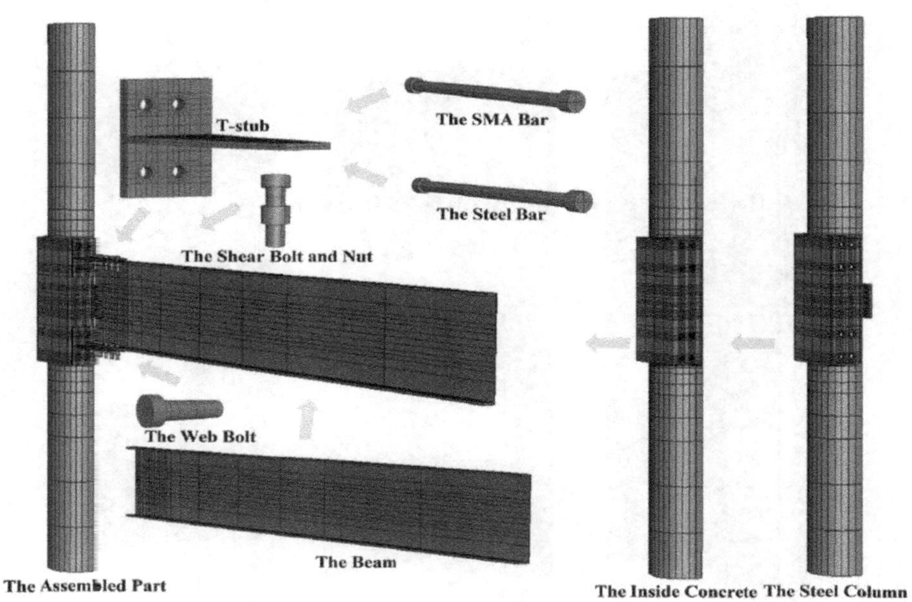


*Figure 2.46* Partitioned 3D solid models for the Smart SMA PR-CFT connection.

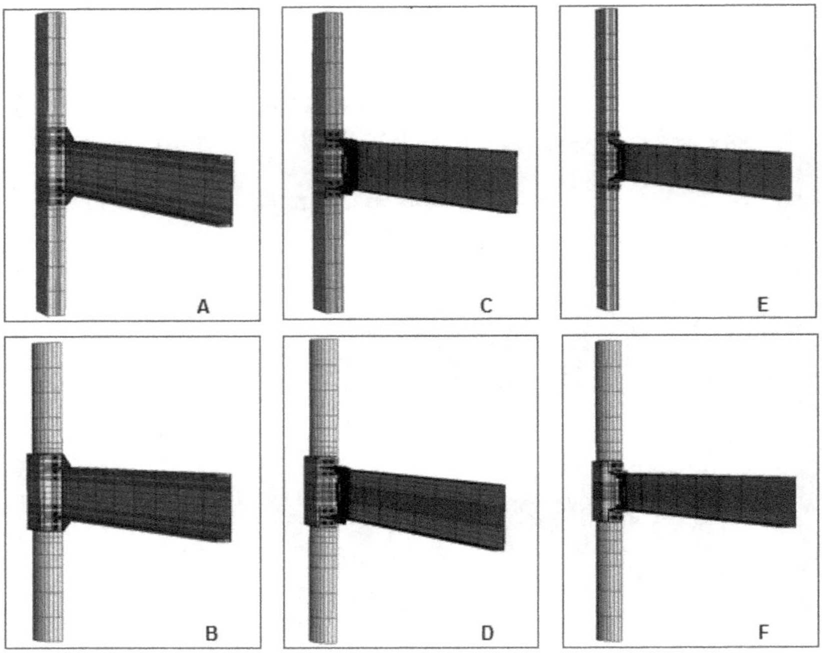


*Figure 2.47* 3D solid elements of the smart PR-CFT connections.

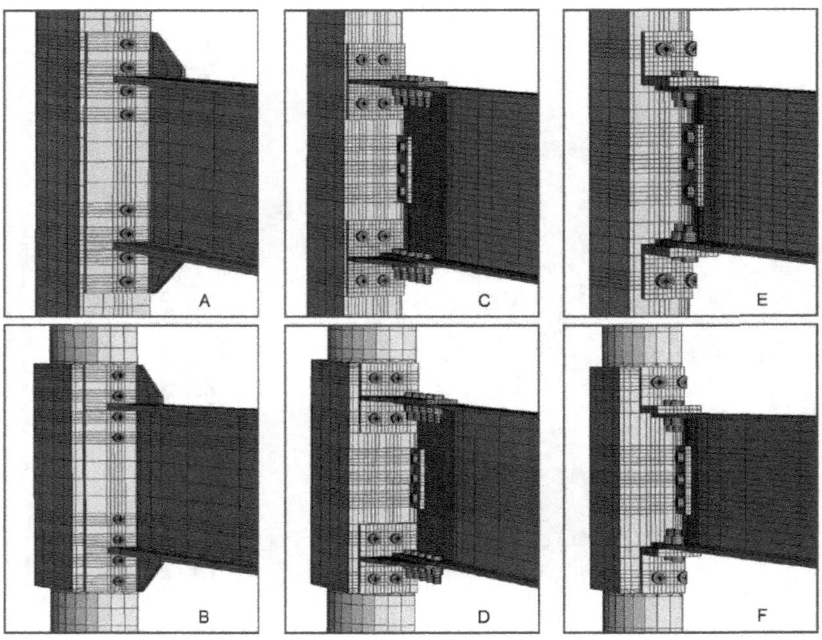


*Figure 2.48* 3D solid elements of the connection details.

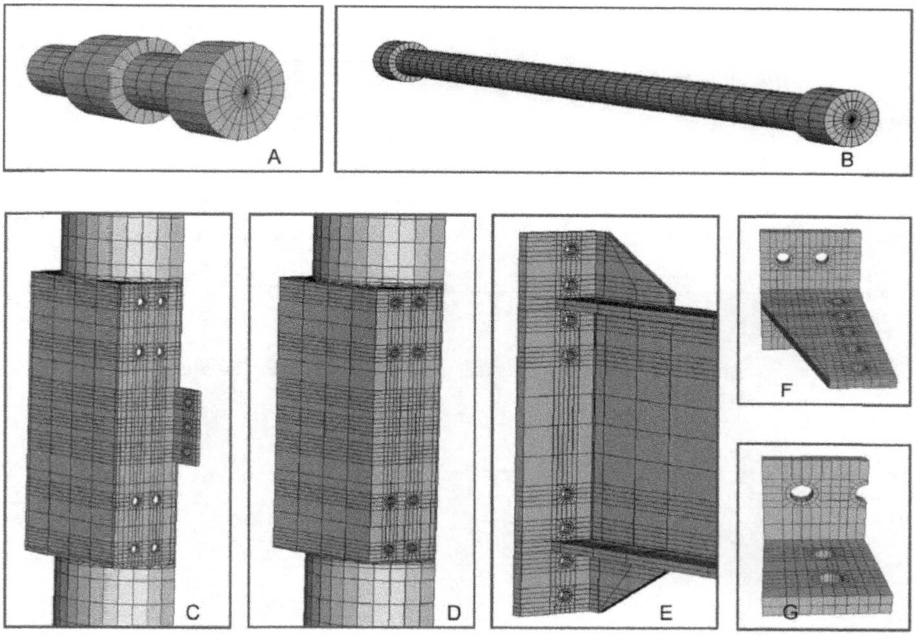


*Figure 2.49* Connection components: (A) bolt, (B) bar, (C) steel column, (D) inside concrete, (E) end-plate and steel beam, (F) T-stub & (G) clip angle.

full integration and incompatible modes. These elements provide robust modeling for meshes with elements having large aspect ratios, leading to a considerable reduction in the number of elements and associated computational efficiencies. For all models, beam members also consisted of C3D8I elements. Fig. 2.49 shows the component members made up of 3D solid elements. The bolts and bars were made up of C3D6 elements for the inner core and an outer layer of C3D8 elements as shown in Figs. 2.49A and 2.49B respectively. The two element nodes located on the contact surfaces between the steel column and the interior concrete part had the same initial coordinate positions. Contact interaction with an initial clearance and direction was generated by using a gap element which connects two element nodes. Therefore, the nodal points of all elements located on the inside steel column surfaces corresponded to those of all elements located on the inside concrete surfaces as shown in Figs. 2.49C and 2.49D. CCFT columns welded to a rectangular shaped panel zone were modeled using C3D4 elements, a 4 node tetrahedral element. Clip angles and T-stubs were made up of layered C3D8 elements, an 8 node brick element, with the leading edge of the T-stub stem made up of layered C3D6 elements, a 6 node wedge element.

### 2.8.3 Material properties

The steel material properties for the component members were modeled after A572-Gr.50 steel with fully nonlinear isotropic characteristics (Fig. 2.50), while the bolt

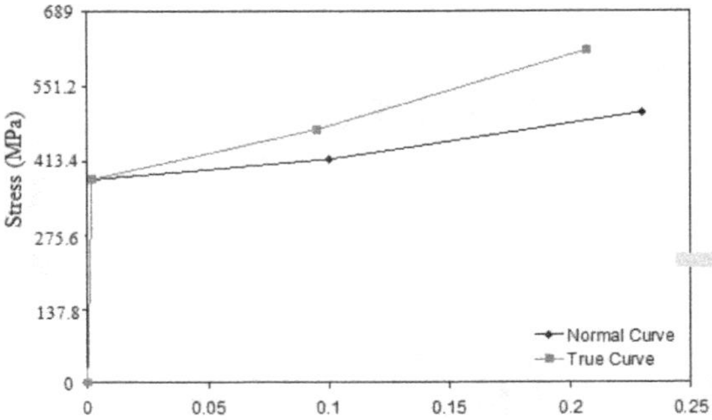


ABAQUS Input Value

| True Stress | True Plastic Strain |
|---|---|
| 379.639 MPa | 0 |
| 458.5295 MPa | 0.09302 |
| 617.2751 MPa | 0.204 |

Nominal Stress Value for Design

| Carbon Steel (A572-Gr.50) | Fy | Fu |
|---|---|---|
| E=199810 MPa | 378.95 MPa | 502.97 MPa |

*Figure 2.50* Tensile stress-strain curves for A572-Gr. 50 Steel.

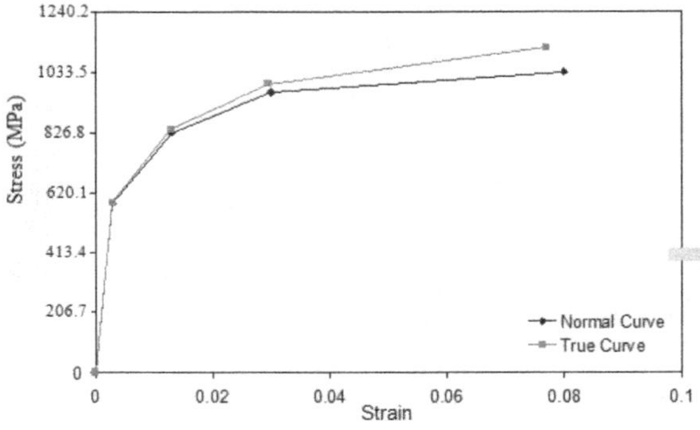


ABAQUS Input Value

| True Stress | True Plastic Strain |
|---|---|
| 587.3725 MPa | 0 |
| 837.5484 MPa | 0.00872 |
| 993.538 MPa | 0.02459 |
| 1116.18 MPa | 0.07137 |

Nominal Stress Value for Design

| High Strength Bolt (A490) | Fy | Fu |
|---|---|---|
| E=199810 MPa | 585.65 MPa | 1033.5 MPa |

*Figure 2.51* Tensile stress-strain curves for A490 bolt.

material properties for the bolts and nuts are modeled after A490 bolt material (Fig. 2.51). The true stress-logarithmic strain curve from a tensile test was used to specify the plastic part of the isotropic material model for elastic-plastic material model that uses a von Mises yield surface. When defining the plastic material data in ABAQUS, the true stress and true plastic strain should be used as shown in Figs. 2.50 and 2.51. The plastic strain is obtained by subtracting the elastic strain, defined as the value of true stress divided by Young's modulus, from the value of total strain (See Fig. 2.52).

The constitutive models for confined concrete contain different stress-strain curves for tension and compression response. These models incorporated a damaged concrete plasticity option, one of the material managers available in ABAQUS (ABAQUS 2006). The concrete damaged plasticity option takes advantage of concepts of isotropic damaged elasticity in combination with isotropic tensile and compressive plasticity to represent the inelastic behavior of concrete. The nonlinear constitutive models for the confined concrete columns were developed following the material models proposed by Hu (Hu *et al.*, 2005) and Torres (Torres *et al.*, 2004). The resulting typical stress-strain

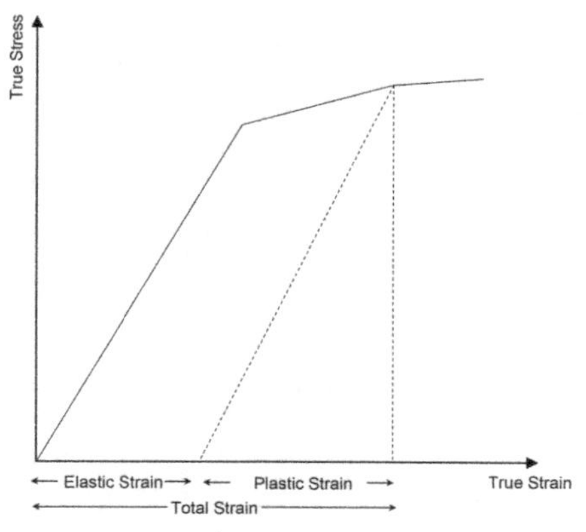


*Figure 2.52* Decomposition of the total strain into elastic and plastic strain.

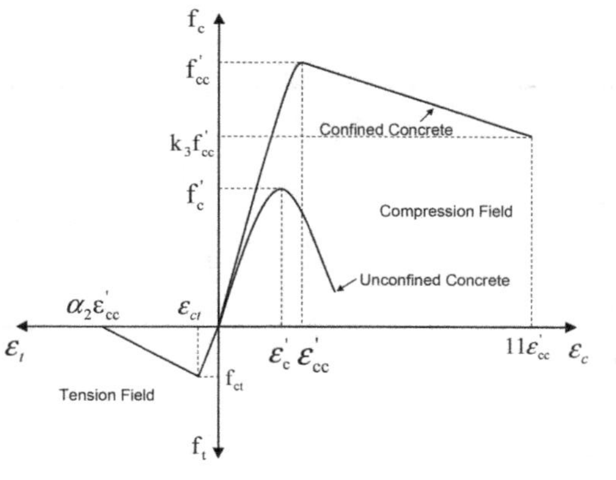


*Figure 2.53* Equivalent stress and strain curves for the concrete material (Hu *et al.*, 2005 and Torres *et al.*, 2004).

curves for the concrete material and the formulas involved with confined concrete are shown in Fig. 2.53. The axial load ratio ($F/F_u$) and the shape of the CFT columns are significant parameters to determine the material constitutive models. Material examples for RCFT and CCFT with $F/F_u = 0.2$ used in this research are illustrated in Figs. 2.54 and 2.55 respectively. Table 2.13 indicates the material input codes of the concrete damage plasticity for the ABAQUS program.

The last set of material properties, those for the SMA bars, was generated from the material properties of super-elastic Nitinol specimens (DesRoches *et al.*, 2004, McCormick, 2006). Fig. 2.56 shows a representative stress-strain curve for a 25.4 mm diameter SMA bar. These quasi-static tensile tests performed on Nitinol specimens provided the required information with respect to deformation under unequal cyclic

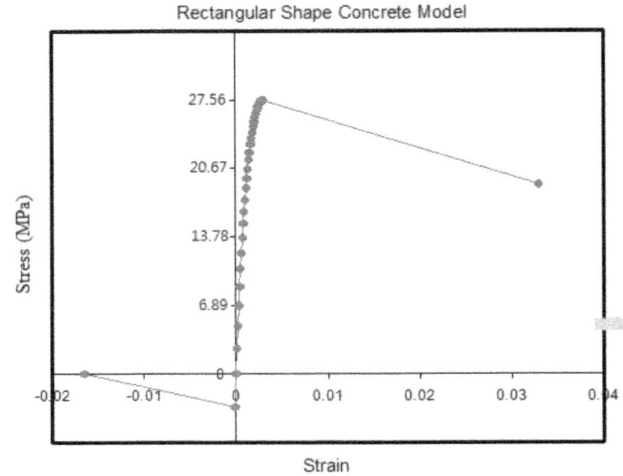


| Plastic Stress | Plastic Strain | Compression Damage |
|---|---|---|
| 15.0202 MPa | 0.000000 | 0 |
| 17.5006 MPa | 0.000067 | 0.01 |
| 20.6011 MPa | 0.000203 | 0.02 |
| 24.3906 MPa | 0.000504 | 0.05 |
| 26.2509 MPa | 0.000804 | 0.1 |
| 27.0777 MPa | 0.001060 | 0.15 |
| 27.56 MPa | 0.001533 | 0.2 |
| 23.2882 MPa | 0.016759 | 0.5 |
| 19.0853 MPa | 0.031985 | 0.9 |

| Plastic Stress | Plastic Strain | Tension Damage |
|---|---|---|
| 3.26586 MPa | 0 | 0 |
| 1.66738 MPa | 0.008312 | 0.5 |
| 0.0689 MPa | 0.016500 | 0.9 |

*Figure 2.54* The stress strain curve for confined concrete.

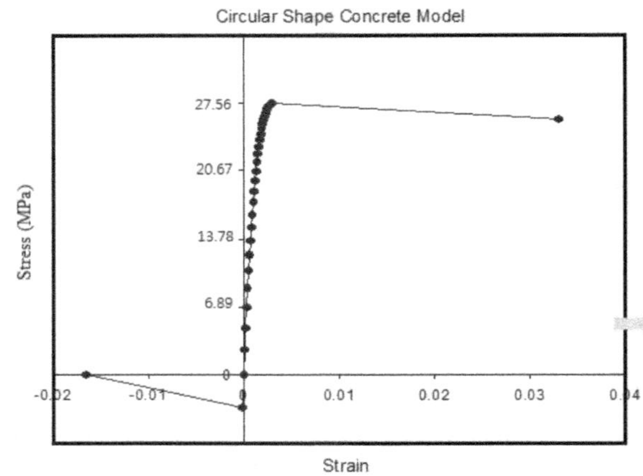


| Plastic Stress | Plastic Strain | Compression Damage |
|---|---|---|
| 15.0202 MPa | 0.000000 | 0 |
| 17.5006 MPa | 0.000067 | 0.01 |
| 20.6011 MPa | 0.000203 | 0.02 |
| 24.3217 MPa | 0.000504 | 0.05 |
| 26.2509 MPa | 0.000804 | 0.1 |
| 27.0777 MPa | 0.001060 | 0.15 |
| 27.56 MPa | 0.001533 | 0.2 |
| 26.7332 MPa | 0.016578 | 0.5 |
| 25.9064 MPa | 0.031622 | 0.9 |

| Plastic Stress | Plastic Strain | Tension Damage |
|---|---|---|
| 3.26586 MPa | 0 | 0 |
| 1.66738 MPa | 0.008311856 | 0.5 |
| 0.0689 MPa | 0.0165 | 0.9 |

*Figure 2.55* The stress strain curve for confined concrete.

loading. The complex non-linear behavior shown by SMA materials was idealized as a multi-linear stress-strain in ABAQUS.

### 2.8.4 Interface conditions

All interfaces between two contact surfaces were explicitly modeled. The general contact formulation used in ABAQUS involves a master-slave algorithm (ABAQUS 2006). This formulation considers the interactions for surfaces that are in contact, interpenetrate or slip and imposes a constraint on the nodes of the slave surface in order not

*Table 2.13* Summary of material constitutive models for confined concrete.

| Limit | Stress and Strain for Confined Concrete under Compression | | |
|---|---|---|---|
| Initaial Value: Unconfined Concrete Property $f'_c$ and $\varepsilon'_c$ | | | |
| $\varepsilon_c \le \varepsilon'_{cc}$ | $k_1 = 4.1 \quad k_2 = 20.5$ | | |
| | $f'_{cc} = f'_c + k_1 f_l \quad \varepsilon'_{cc} = \varepsilon'_c\left(1 + k_2 \frac{f_l}{f'_c}\right)$ | | |
| | $R_\sigma = 4 \quad R_\varepsilon = 4$ | | |
| | $E_c = 57000\sqrt{f'_{cc}}$(psi) | | |
| | $R_E = \frac{E_c\varepsilon'_{cc}}{f'_{cc}} \quad R = \frac{R_E(R_\sigma - 1)}{(R_\varepsilon - 1)^2} - \frac{1}{R_\varepsilon} \quad f_c = \frac{E_c\varepsilon_c}{1 + (R + R_E - 2)\left(\frac{\varepsilon_c}{\varepsilon'_{cc}}\right) - (2R - 1)\left(\frac{\varepsilon_c}{\varepsilon'_{cc}}\right)^2 + R\left(\frac{\varepsilon_c}{\varepsilon'_{cc}}\right)^3}$ | | |
| | Rectangular Shape | Circular Shape | |
| $\varepsilon_c > \varepsilon'_{cc}$ | $0 \le F/F_u \le 0.23$ | $0 \le F/F_u \le 0.34$ | $f_l/f_y = 0$ |
| | $f_l/f_y = 0 \quad k_3 = 1 - 0.304\ (F/F_u)$ | $0 \le F/F_u \le 0.45$ | $k_3 = 0.87 - 0.889\ (F/F_u)$ |
| | $0.23 \le F/F_u \le 0.56$ | $0.34 \le F/F_u \le 0.57$ | $f_l/f_y = -0.00517 + 0.0152(F/F_u)$ |
| | $f_l/f_y = -0.00859 + 0.0373(F/F_u)$ | $0.45 \le F/F_u \le 0.57$ | $k_3 = 0.508 - 0.083\ (F/F_u)$ |
| | $k_3 = 1 - 0.304\ (F/F_u)$ | | |
| | $0.56 \le F/F_u \le 0.74$ | | |
| | $f_l/f_y = 0.014 + 0.00333(F/F_u)\ k_3 = 0.55$ | | |
| Limit | Stress and Strain for Confined Concrete under Tension | | |
| $\varepsilon_t > \varepsilon_{ct}$ | $f_t = E_c\varepsilon_t\ f_{ct} = 7.5\sqrt{f'_{cc}}$ | | |
| $\varepsilon_t > \varepsilon_{ct}$ | $\alpha_2\varepsilon'_{cc} = 5.5\varepsilon'_{cc}$ | | |

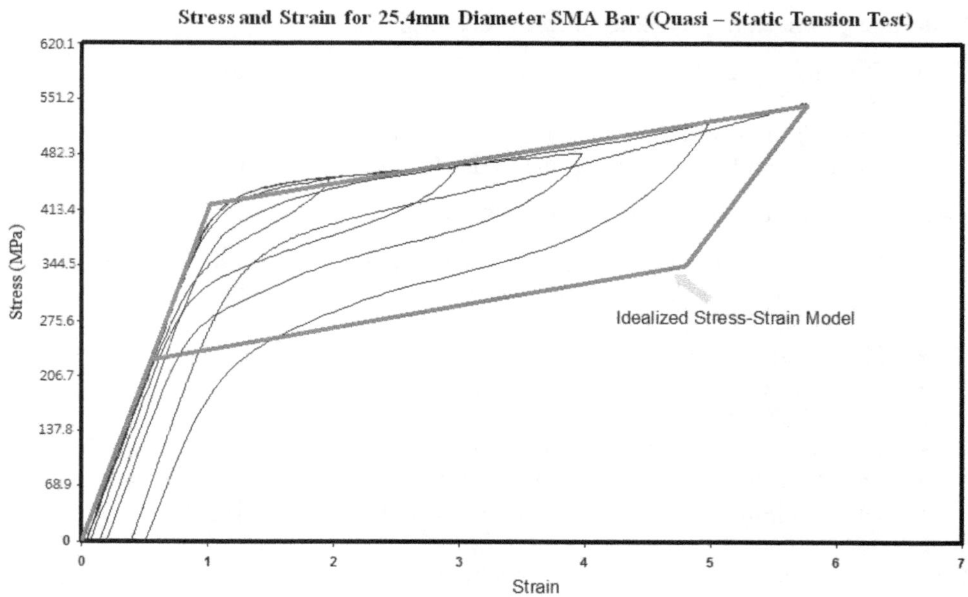


*Figure 2.56* The tensile stress strain curve for the super-elastic SMA bar (DesRoches *et al.*, 2004).

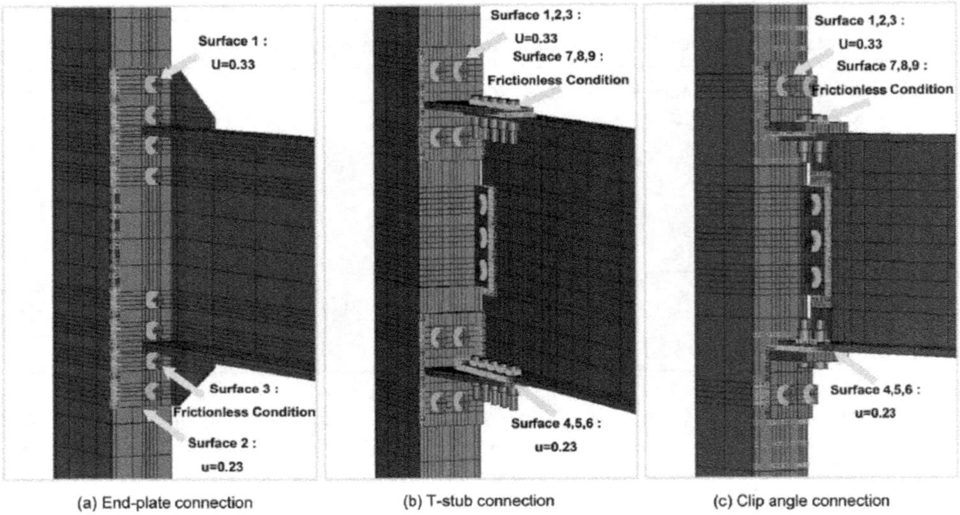


*Figure 2.57* Surface interactions with the friction coefficient.

to penetrate the master surface (Citipitioglu *et al.*, 2002). Surface interactions with a friction coefficient were defined as shown in Fig. 2.57 and as below:

**End-Plate Connection** – Contact surfaces between:

- The underside of bar heads and end-plate surface surrounding the bolt holes (Surface 1)
- The end-plate and the CFT column (Surface 2)
- The bar shank and the hole surface in the CFT column wall (Surface 3)

**T-sub Connection** – Contact surfaces between:

- The underside of fastener heads (i.e. bar heads, shear bolt heads, and web bolt heads ) and the T-stub/shear tab surface surrounding the bolt holes (Surfaces 1, 2, and 3)
- The T-stub flange and the CFT column (Surface 4)
- The shear tab and the beam (Surface 5)
- The T-stub stem and the beam (Surface 6)
- The bar shank and the hole surface in the CFT column wall (Surface 7)
- The shear bolt shank and the hole surface in the T-stub stem (Surface 8)
- The web bolt shank and the hole surface in the beam web (Surface 9)

**Clip Angle Connection**– Contact surfaces between:

- The underside of fastener heads (i.e. bar heads, shear bolt heads, and web bolt heads ) and the clip angle/shear tab surface surrounding the bolt holes (Surface 1, 2, and 3)

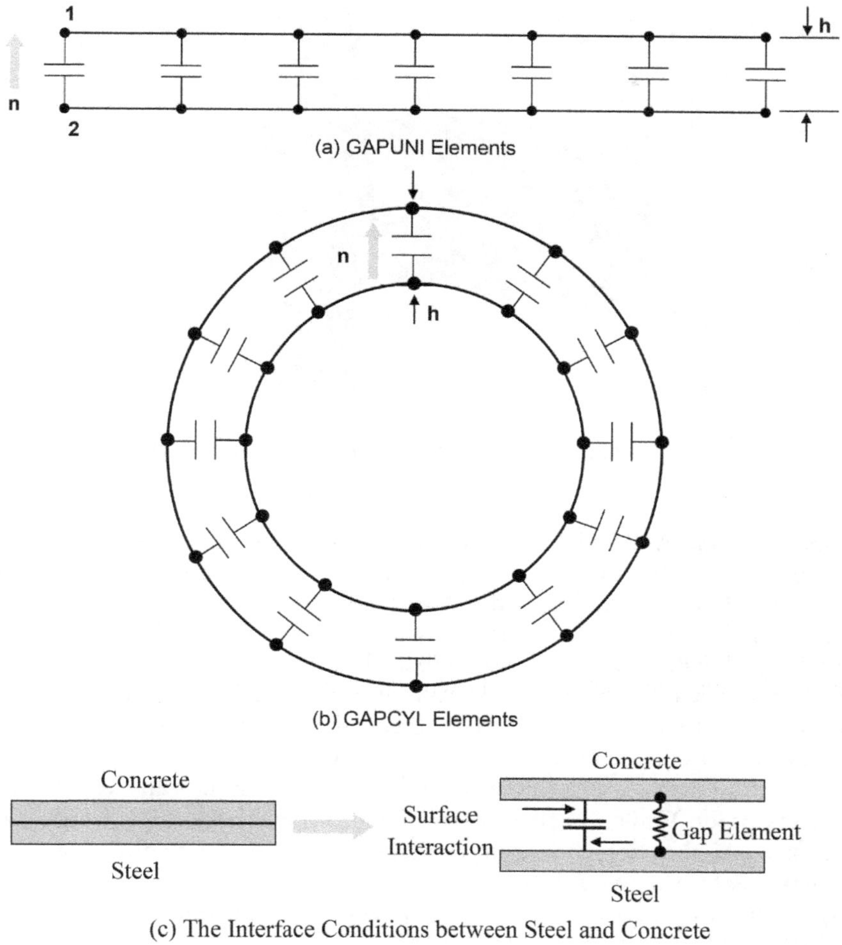


*Figure 2.58* The contact interactions between steel and concrete.

- The clip angle flange and the CFT column (Surface 4)
- The shear tap and the beam (Surface 5)
- The clip angle leg and the beam (Surface 6)
- The bar shank and the hole surface in the CFT column wall (Surface 7)
- The shear bolt shank and the hole surface in the clip angle leg (Surface 8)
- The web bolt shank and the hole surface in the beam web (Surface 9)

The contact behavior at the interface between the steel column surface and the interior concrete core was modeled with gap elements. They were defined by specifying two nodes with an initial separation clearance (h) and a contact normal direction (n) as shown in Fig. 2.58. The generation of the gap elements benefited from the input

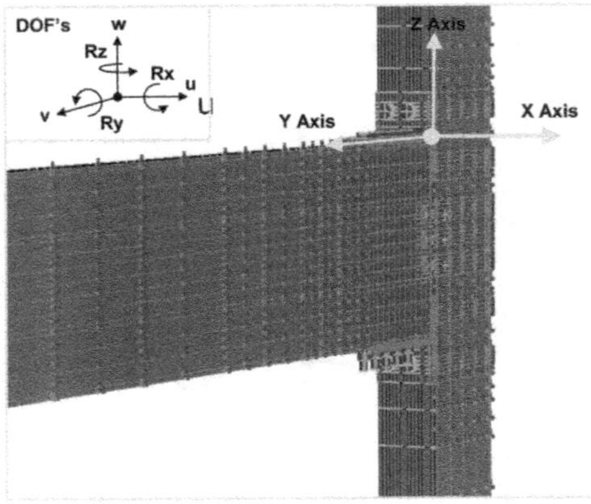


*Figure 2.59* Symmetric BCs for the half modeling (Left).

file based option in ABAQUS. Microsoft Excel worksheets were used to generate the connections between two nodes having the same coordinate position on the interface. Two different kinds of gap elements, classified according to the contact situation, were used. GAPUNI elements modeled contact between two nodes when the contact direction was fixed in space. For instance, GAPUNI elements were used to model the contact between two flat planes as shown in Fig. 2.58(a). GAPCYL elements modeled contact between two nodes when the contact direction was orthogonal to an axis. For instance, GAPCYL elements were used to model the contact between two circular tubes. As a consequence, the separation and force between the gap elements was provided as an output. While gap elements are defined along the normal direction, surface interactions with a friction coefficient are defined along the shear direction in order to generate the real surface condition between steel and concrete interface. The values of the coefficient of friction were taken as 0.3. The interface conditions are depicted in Fig. 2.58(c).

## 2.8.5 Initial conditions

Half of PR-CFT connection was modeled by using symmetry about the centerline of the web, CFT column, and other components. The half models satisfy the precondition that both geometric configurations and loads were perfectly symmetric about the boundary plane. The "type" boundary condition available in ABAQUS/CAE was specified instead of constraining individual degrees of freedom. An example of symmetric boundary conditions for the half model is shown in Fig. 2.59. A prescribing boundary condition of type XSYMM to the symmetric plane represents the surface on a plane of symmetry normal to the X Axis. This boundary condition is identical to prescribing a boundary condition using the direct format to degrees of freedom 1, 5, and 6 in the symmetric plane since symmetry about a plane X = constant indicates $u = 0$, $R_y = 0$ , and $R_z = 0$.

The models were loaded in two steps. The first step was used to pretension the bolts while the second step was used to apply the main load with the propagation of the

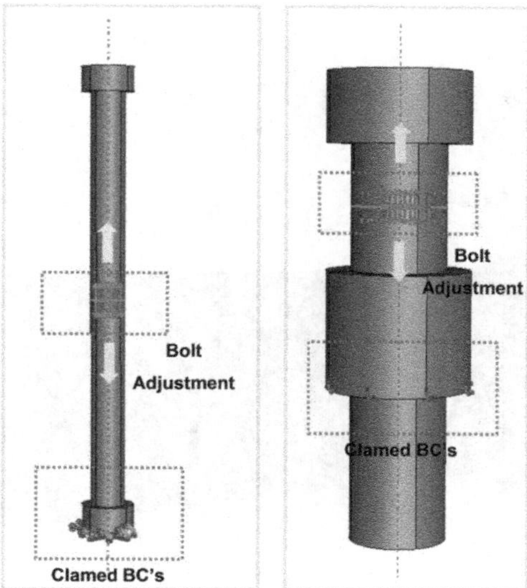


*Figure 2.60* Initial pretension force of a bar and bolt (Right).

bolt pretension. Great care was taken to attempt model the bolt behavior correctly, including the oversized bolt holes. The all bolts were pre-tensioned by applying an adjustment length/displacement to the center of the bolt shank as shown in Fig. 2.60. The direction of the pretension force was taken as the normal axis to the loading surface. The prescribed bolt displacements were calculated by assuming that bolts remain elastic with the axial pretension. In ABAQUS, the prescribed bolt displacement can be converted into the pretension force using the 'History Output' tool.

### 2.8.6 Loading

The second step was used to apply the external displacement. For static loading cases, the load was generated by imposing a support displacement to the tip of the beam as shown in Fig. 2.61. A displacement-type boundary condition was used to apply a prescribed displacement magnitude of −254 mm in the Z direction to the middle of the beam tip. The postprocessor in ABAQUS automatically calculated the equivalent forces foe each displacement step. The force-displacement response of the connections was changed into a corresponding moment-rotation response using the equations for the "instruments" described in the previous chapter.

### 2.8.7 Steps and solution

A sequence of one or more analysis steps had to be defined for each FE model. As noted above, two time steps were required to analyze the bolted connection models with the time increments. This approach was generally used for stable problems and

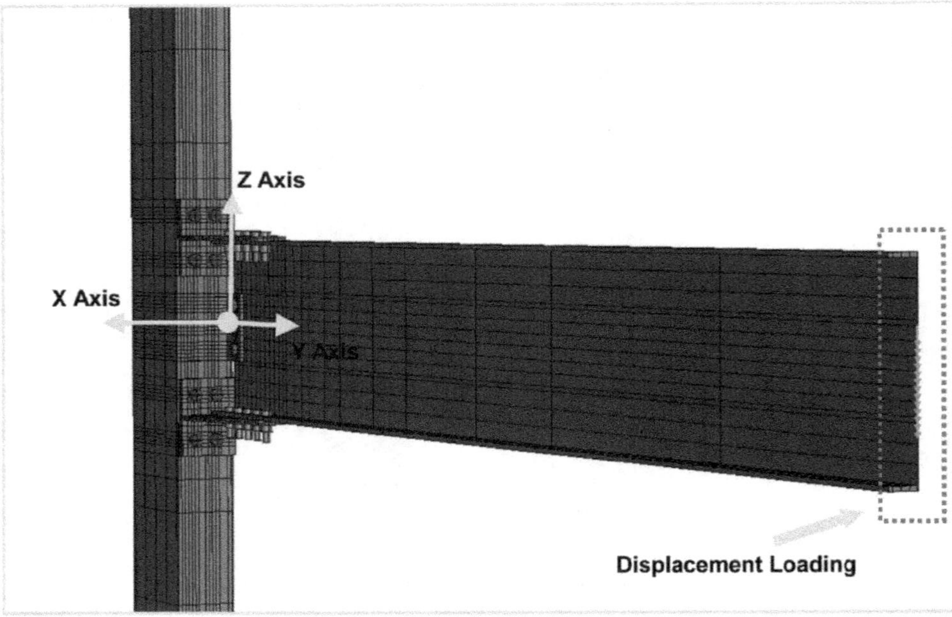


*Figure 2.61* The main loading condition.

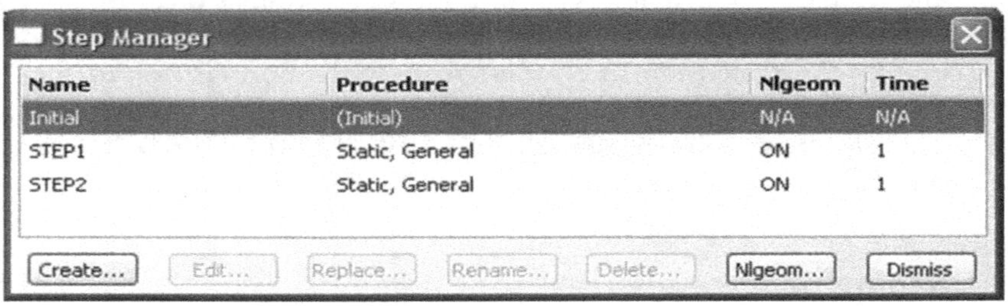


*Figure 2.62* The basic step manager.

can include linear or nonlinear response but without inertia effects. Multiple analysis steps can be assigned during the analysis as shown in Fig. 2.62. For each step in the analysis, the step manager indicates whether the FE model will account for geometric nonlinear effects due to large deformation with the setting of Nlgeom parameter. The initial step was used to define boundary conditions, predefined fields, and interactions which are applicable at the beginning of the analysis.

After the initial step, several steps were lumped into step 1 to introduce the pretension in the bolts (Fig. 2.63). The contact interaction calculations generally converged successfully within the maximum number of allowed iterations, typically taken as 12. Fig. 2.64 shows the analysis solution and control for the computations associated with

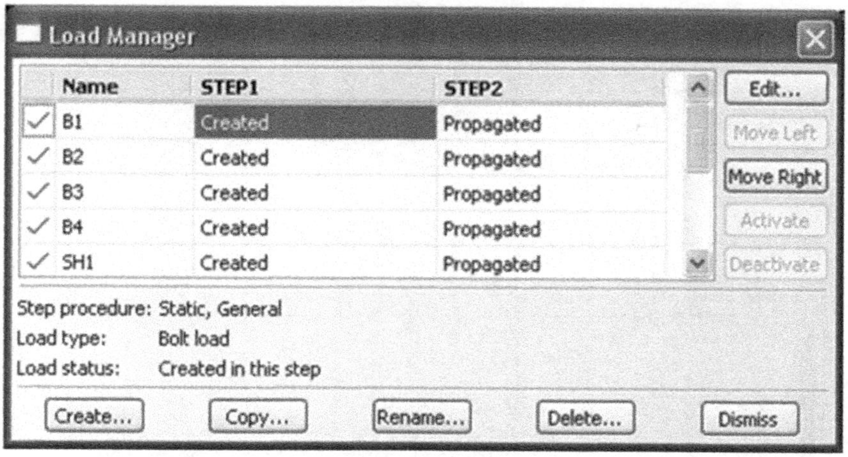


*Figure 2.63* Load manager for bolt pretensions incorporated with time steps.

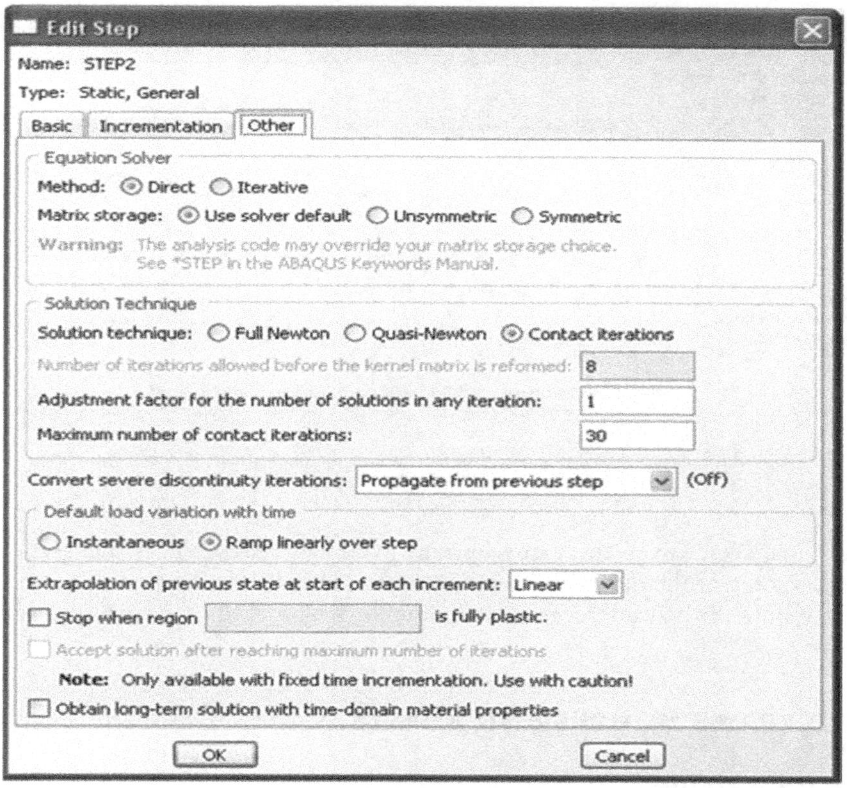


*Figure 2.64* Solution and control associated with the ABAQUS step manager.

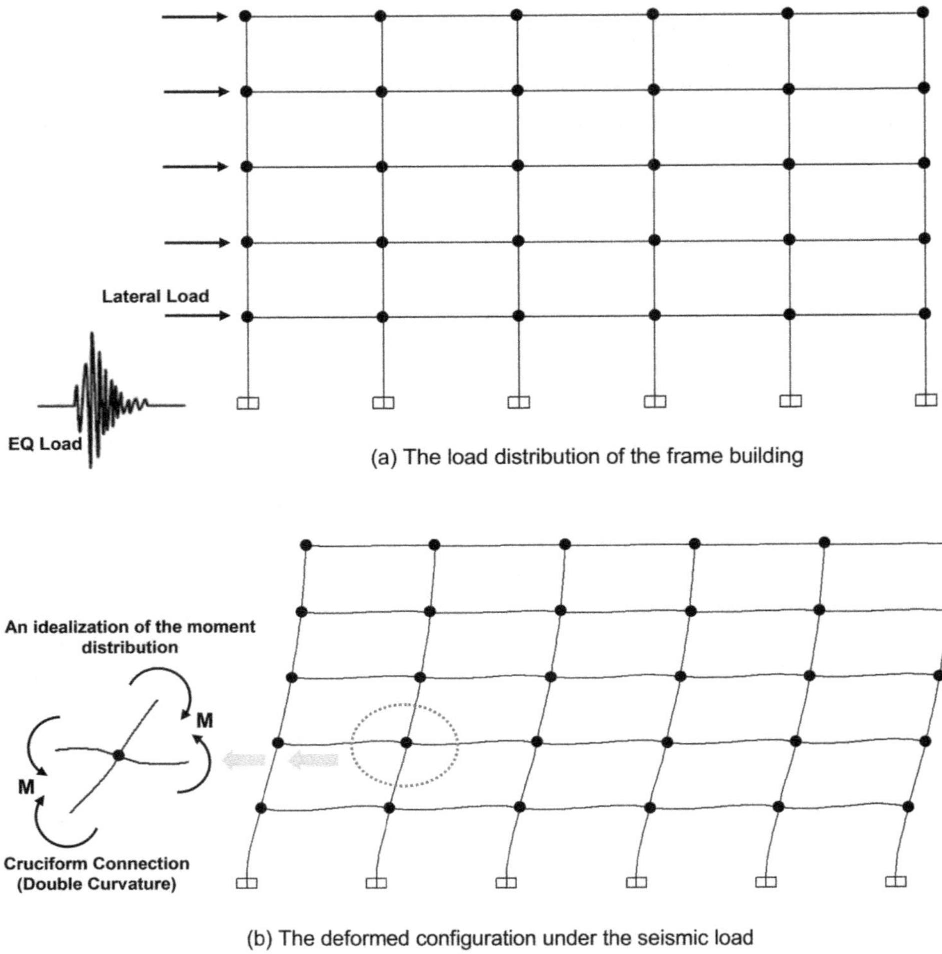


*Figure 2.65* Load distribution and deformed configuration of the frame building.

the step manager. Given the many options available in ABAQUS, this section only highlights some of the most important parameters used. It is felt that this is sufficient to clearly state the procedures followed and allow reproduction of the analyses.

## 2.9 2D FRAME MODELS AND ANALYSES

### 2.9.1 Joint model

A frame structure should provide adequate stiffness, strength, and energy dissipation capacity to withstand the lateral as well as gravity loads. This must be achieved within

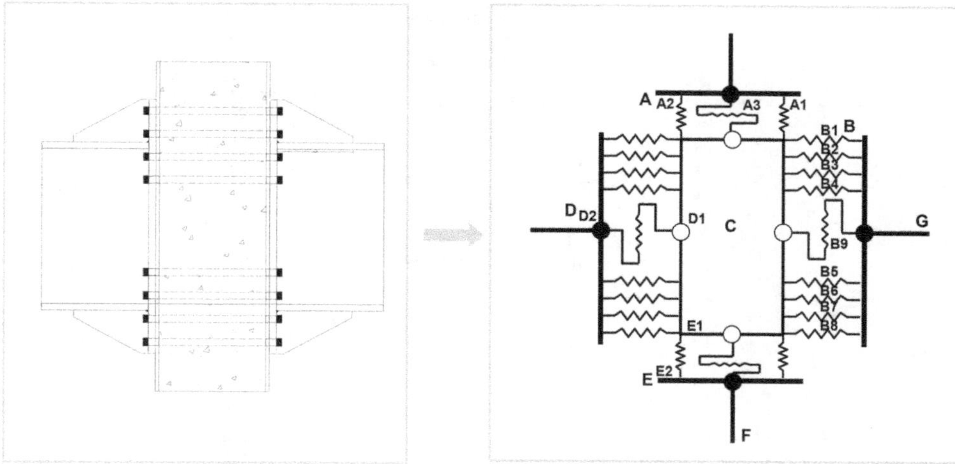


(a) The end-plate connection detail

(b) The idealized simple joint model

*Figure 2.66* Idealized joint model of the end-plate connection.

allowable limits of deformation and strength. The dynamic random seismic inertial forces introduced into a building by an earthquake can be converted into the equivalent lateral loads for design. This approach accounts for seismic zoning, site characteristics, and structural system and configuration at a level suitable for the preliminary sizing of structural members. Fig. 2.65 shows the deformed shape for a typical moment frame building and corresponding beam-column sub-assemblage under moderate to severe earthquake loading. Typically, the members are deformed in double curvature and the joints by shear as lateral loads dominate. In addition, the first mode is assumed to dominate behavior; higher modes are ignored in design. The modeling described in this chapter is meant to develop connection models that will speed up the non-linear analyses of frames subjected to this type of deformation pattern.

## 2.9.2 Joint model of the end-plate connection

Fig. 2.66 shows the proposed 2 dimensional idealization for a simplified joint model for an end-plate connection. The connection components designed to yield, such as tension bars and the CFT panel zone, are converted into equivalent spring elements. The end-plate, which is designed as a rigid plate, is shown as rigid elements. Detail explanations on the assemblage of the joint models will be given in this section.

Fig. 2.67 show the idealized force distribution at the perimeter of the joint for an end-plate connection subjected to seismic loads. Generally, the beam develops its flexural strength (i.e., plastic hinging) while the column carries the axial gravity loads elastically. These internal member forces are shown as equivalent concentrated forces acting on the joint (blue arrows in Fig. 2.67(a)). The internal reactions in the connection

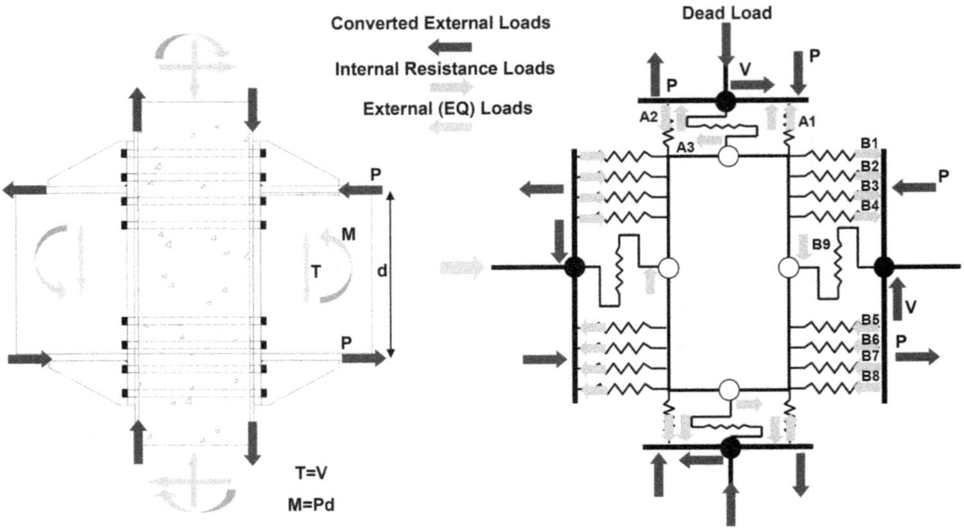


(a) Idealized loading distribution at the PZ (b) Idealized loading distribution at the joint model

*Figure 2.67* External and internal forces at the joint for the end-plate connection.

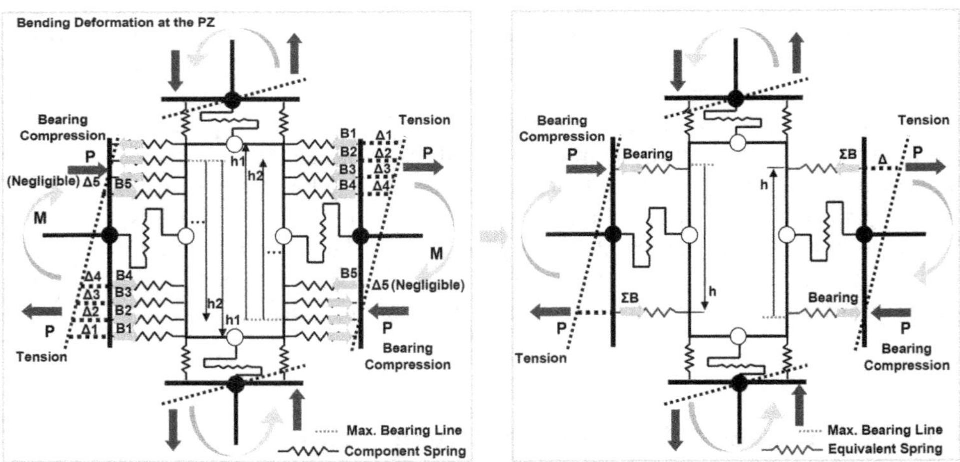


(a) Bending deformation at the joint model with component spring elements (b) Bending deformation at the simplified joint model with equivalent spring elements

*Figure 2.68* Response mechanism of the joint element under bending deformation.

components act against these external forces (green arrows in Fig. 2.67(b)) in order to satisfy equilibrium.

The response of the joint element under the shear deformations resulting from the bending forces in the framing members is shown in Fig. 2.68. It is deformed in a scissors-line manner. The internal tension loads are carried by tension bars, which correspond to

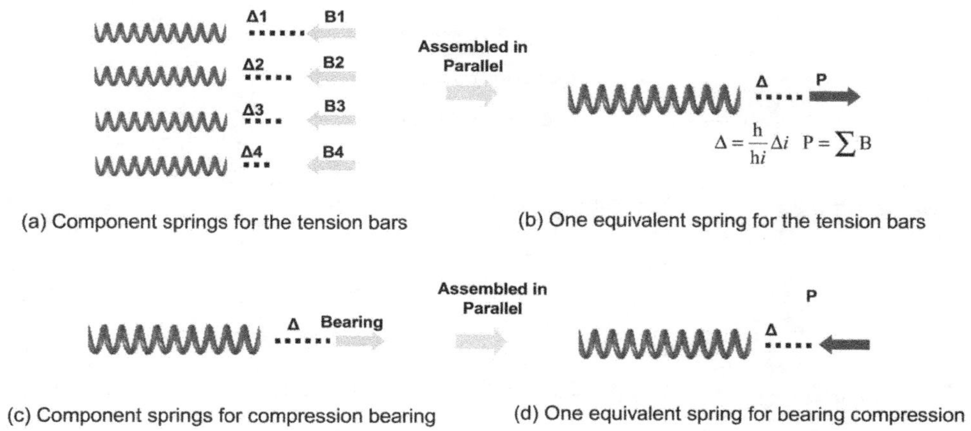


*Figure 2.69* Assemblage procedures for spring elements.

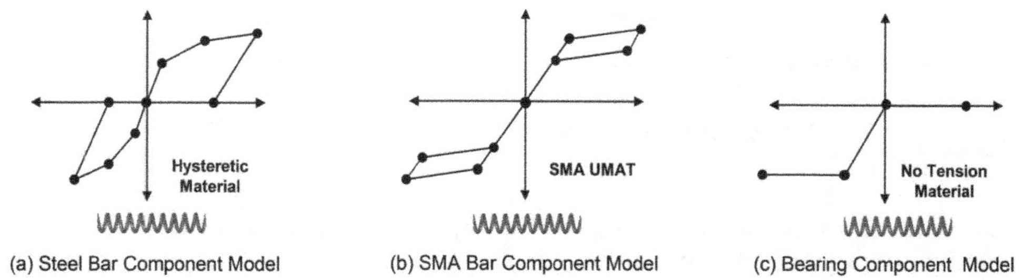


*Figure 2.70* Properties of the individual component.

the top springs on the right side of the connection as shown in Fig. 2.68(a). On the other hand, the internal compression loads resulting from the bearing forces between the beam flange and the CFT column surface are transferred to the springs on the bottom right. The bars inside the compression zone do not make a significant contribution to the response mechanism of the joint model due to bearing effects. The end-plate is assumed to behave as a rigid plate, resulting in a linear strain pivoting about the center of bearing. The latter is determined from the advanced 3D analyses. For end plates, it tends to be lower than shown in the sketch.

The assumptions about strain distribution, basic material stress-strain characteristics, and basic statics (equilibrium of the force resultants) provide the theoretical basis to condense the numerous springs in Fig. 2.68(a) into single springs as shown in Fig. 2.68(b) and Fig. 2.68. This parallel system is constrained to maintain force equilibrium between the summation of bar reaction force and the converted axial force ($P = \Sigma B$). The behavior under tension loading is determined by the theoretical equations shown in Fig. 2.69(b), while those under compression (bearing) are determined from observations of the 3D FE model described in Chapter 2.8. The behavior of the components is depicted in Fig. 2.70. The behavior of the steel tension bars and

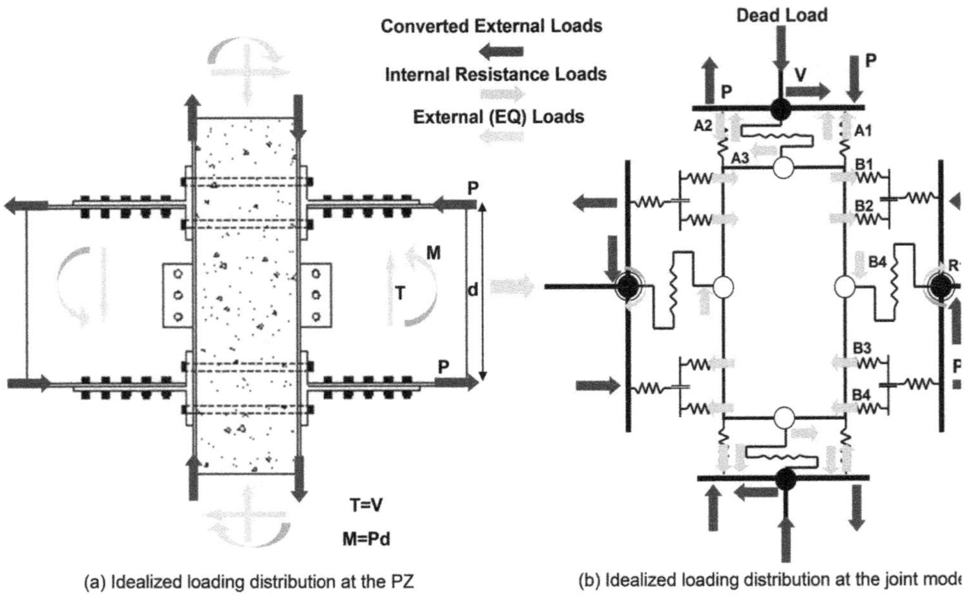


*Figure 2.71* External and internal forces in the joint model for the T-stub connection.

that of the bearing component are generated by using the default material code in the program, with the SMA materials utilizing a user defined material code.

An algorithm based on incremental displacement control is appropriate for the theoretical verification of the equivalent spring element formulation For a given displacement of the equivalent spring element, the displacement of each component can be computed by using the simple geometric ratios.

### 2.9.3 Joint model of the T-stub connection

The component model of the T-stub component was developed following similar approaches and procedures to those used for the end plate connection (Section 2.9.2). The force distribution and deformed configuration of the joint model are depicted in Figs. 2.71 and 2.72, respectively. Unlike the joint model for the end-plate connection, it contains a rotational spring for the shear tab and includes a sliding component to model slip. The individual spring elements are also assembled into one equivalent spring element.

The component model for the OpenSEES program is shown in Fig. 2.73. The data obtained from comprehensive experiments on T-stub components (Swanson 1999), such as that for fracture of the T-stem shown in Fig. 2.74, was used to develop the component springs.

The total force vs. deformation behavior is given in Fig. 2.76(d). The total displacement is due primarily to three basic mechanisms: (a) bar yielding/uplift and flexural

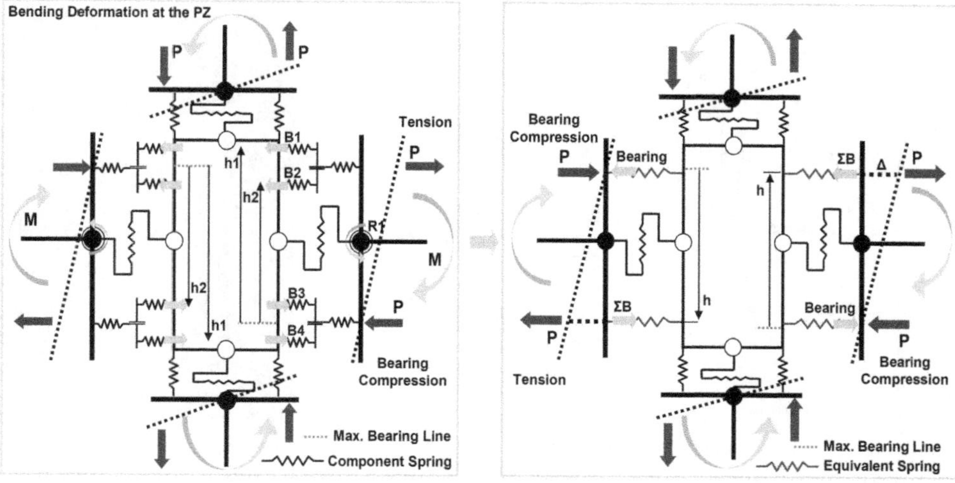


(a) Bending deformation at the joint model with component spring elements

(b) Bending deformation at the simplified joint model with equivalent spring elements

*Figure 2.72* Response mechanism of the joint element under bending deformation.

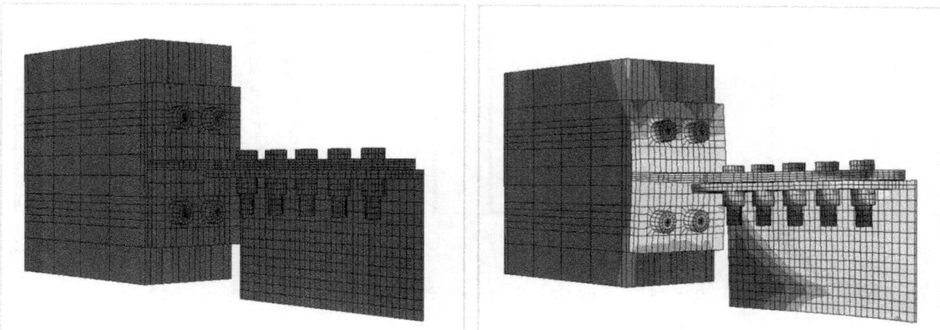

(a) A FE component model for the T-Stub connection

(b) Deformed shape of a component model under the axial force (P)

*Figure 2.73* Component model for the 3D FE T-stub and its deformed Configuration under axial force.

deformation of the T-flange (Fig. 2.75(a)), (b) T-stem deformation (Fig. 2.75(b)), and (c) slip (Fig. 2.75(c)). The bar uplift resultant is produced by assembling all spring elements modeling bar components in the parallel system. The cyclic analytical and experimental data performed by Swanson (1999) result in a good visual match of the shape of the hysteresis loops.

The behavior of each component is shown in Fig. 2.76. The stiffness for each component was developed by observations of experimental results. The bar uplift model considers the prying action and yield lines in the flange and results in a tri-linear backbone curve: initial elastic behavior, followed by the formation of two yield lines

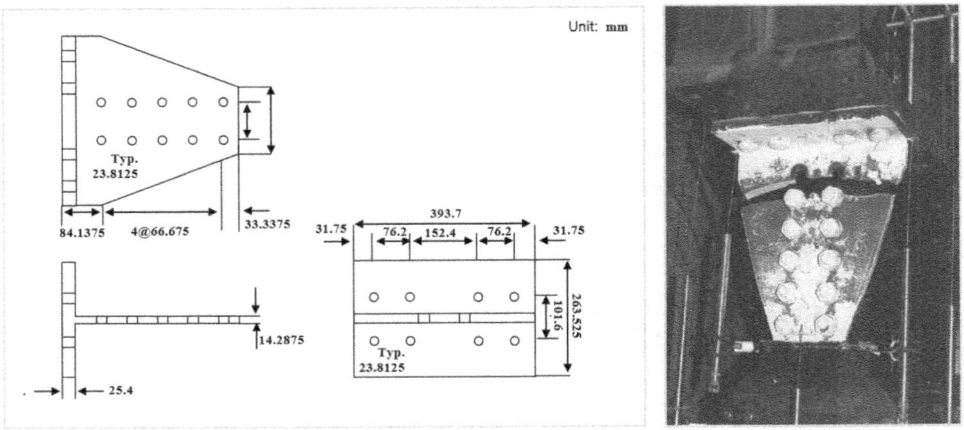


(a) T-Stub component detail (TA01 Model)

(b) Component test (Swanson 1999)

*Figure 2.74* Specimen details of T-stub component model.

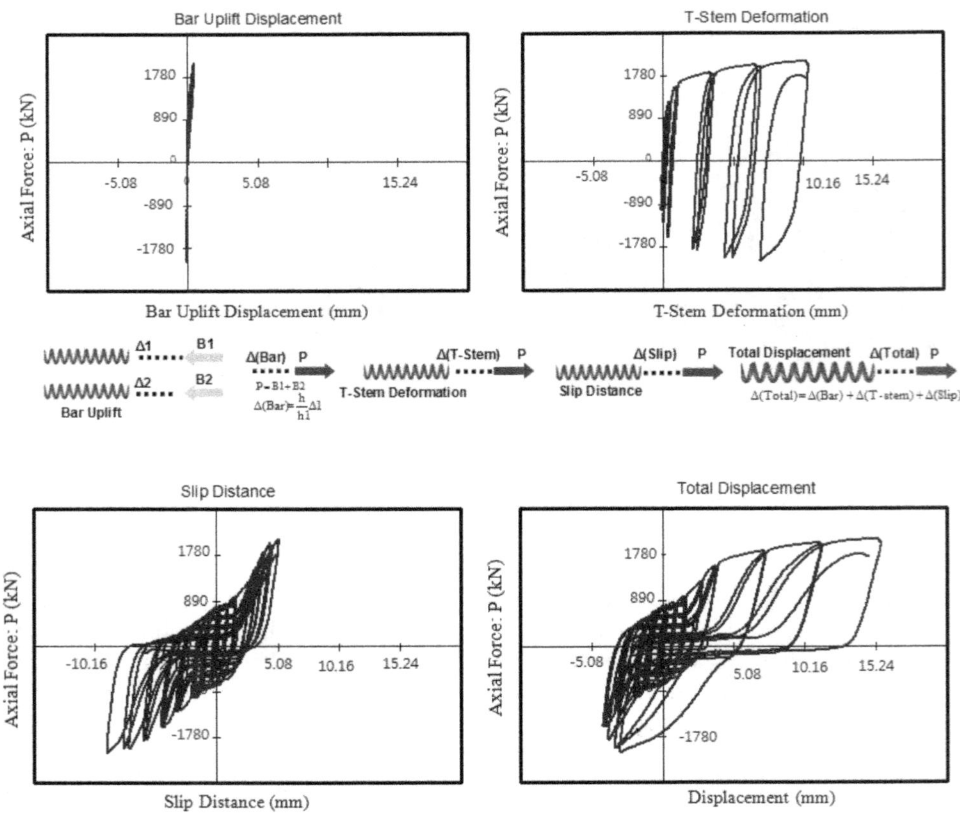


*Figure 2.75* Force vs. deformation of T-stub component model.

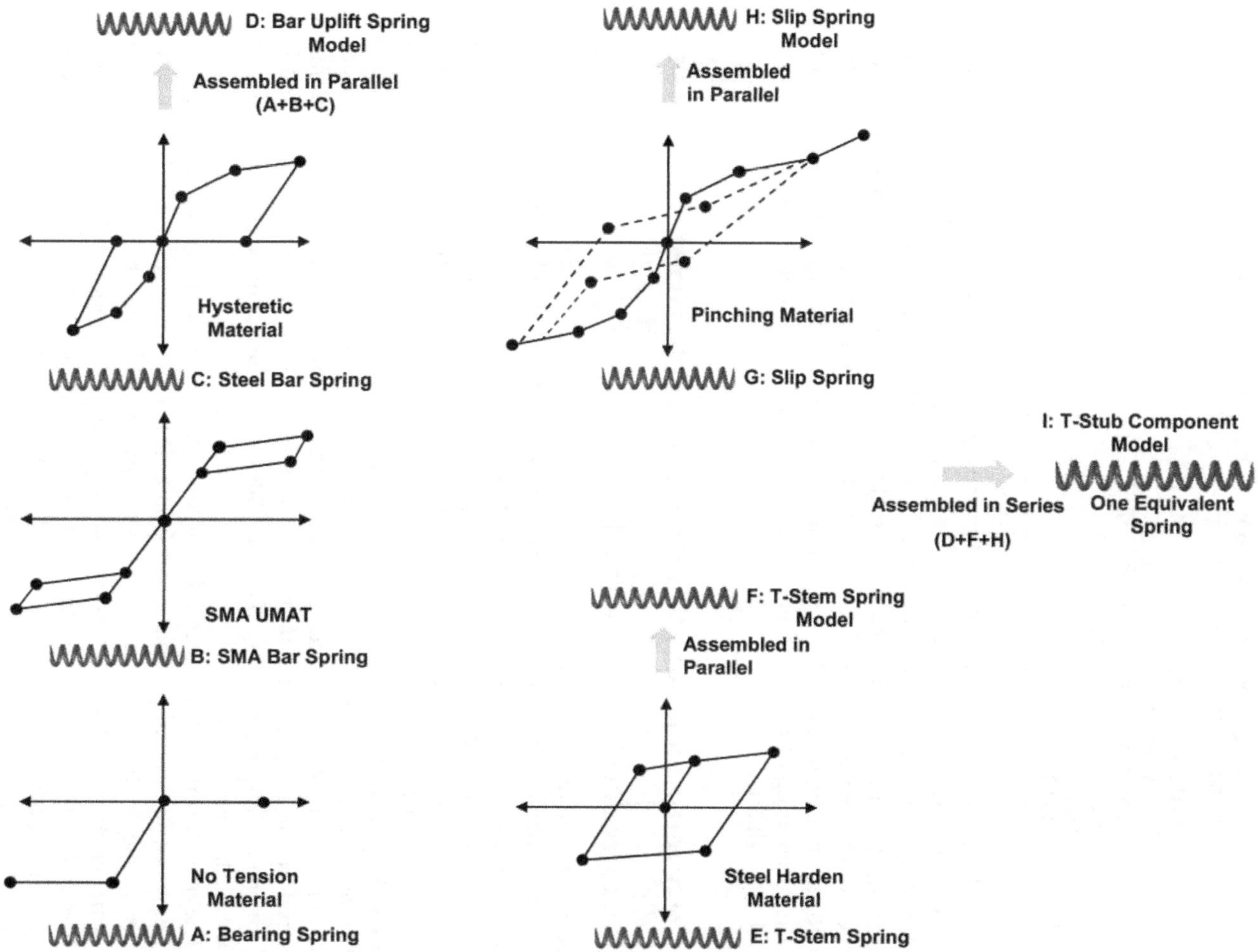


*Figure 2.76* Properties of the individual component model.

*Table 2.14* Overall frame analyses and data measurements.

| |
|---|
| (a) Frame Analysis (Second Order Inelastic Analysis) |
| The nonlinear pushover analysis with the equivalent lateral loads |
| The nonlinear dynamic analysis with the ground motion |
| (b) Measurement |
| Inter story drift ratio (ISDR) |
| Panel zone rotation angle (PZRA) |
| Member force and deformation |
| Fiber stress and strain |
| Nodal displacement |

and/or yielding of the tension bolts. The stem component has an elastic-hardening model for the base material coupled with a slip one to mimic the sliding due to the oversize of the holes. The bearing on the column flange is modeled with a compression-only spring. The model is not capable of tracking the softening behavior shown in the last cycle of the test as this was due to the propagation of the fracture in the stem; this behavior cannot be modeled by the simple springs used here. Attempts to model this by some summation of total strains (rain flow counting) and similar simple techniques proved unworkable.

## 2.10 NONLINEAR ANALYSES OF COMPOSITE MOMENT FRAMES

### 2.10.1 Introduction to nonlinear analyses

Numerical frame models which include the 2D joint models described in previous section are assumed to be accurate and able to replicate the real behavior of PR frames. The programs could not model fracture of the connections and connection rotational ductility was assumed as infinite; however, peak rotations were checked to ensure that they did not reach unusual levels (0.07 radians). The expected strength and deformation demands for composite frame systems can be estimated by these nonlinear frame analyses. The performance data selected for comparisons are the inter story drift ratios (ISDR), panel zone rotation angles (PZRA), forces, deformations and fiber stresses for key members and nodal displacement. In particular, member forces and deformations measured at the integration points of the nonlinear beam-column elements will be used for calculating elastic strength ratios (ESR) and inelastic curvature ductility ratios (ICDR) in the next chapter.

The overall frame analyses and data measurements are summarized in Table 2.13. The dominant load combination, Load Combination 5 (see Section 2.7.3), was used to perform the nonlinear pushover analyses. The detailed load profiles for each frame model are shown in Table 2.10. Equivalent point loads, simulating the uniform dead and live loads, were applied to the beam elements in the gravity direction using the Constant Time Series function associated with the load pattern in OpenSEES. The equivalent lateral loads on the joints were simulated by using the Linear Time Series function, so these loads can be applied in a linearly incremental fashion associated with a predefined time step. For each time step, a static analysis was performed through a displacement control algorithm.

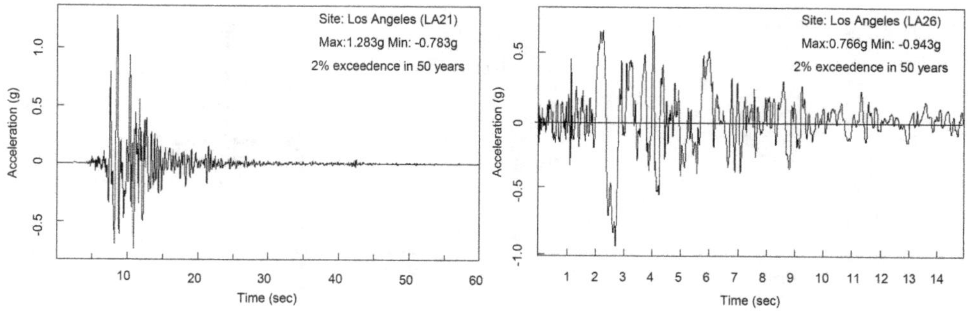


*Figure 2.77* Ground motions used in nonlinear dynamic analysis.

Ground motions selected from the SAC suite of ground motions (Somerville *et al.*, 1997) were used to perform the nonlinear dynamic analyses. Some selected ground motions are shown in Fig. 2.77. More information on the ground motions used in this research is given in Appendix F. For each time step, a transient analysis was performed using the Newmark method (Newmark 1959). A value of 2.5 percent was used for the structural damping as defined by the Rayleigh command in OpenSEES. In order to include second order effects (P-Delta effect) due to dead and live loads along the gravity direction, these loads were also applied to the beam elements. In addition to the gravity loads to model the P-Delta effects, lumped masses were assigned to nodes so as to generate the story shear force due to the ground acceleration. Lumped masses consisted of 1.0 times dead loads plus 0.2 times live loads. The calculation of the lumped mass is described in Appendix D.

From these nonlinear analyses, the primary response data were collected by utilizing recorder commands in the OpenSEES program. A schematic view of the data collected for the composite frame performance is depicted in Fig. 2.78. Two types of data recorder commands, Node Recorder and Element Recorder, were used to collect the data of interest. The Node Recorder function was used to record the response of the global nodes, while the Element Recorder function collected data on the local response of members and fiber sections. For example, displacement, velocity, acceleration, and reaction force at the Record 1 or Record 2 position as shown in Fig. 2.78 was monitored by the Node Recorder. On the other hand, member forces, deformations, fiber stresses, and strains at Record 3 to Record 5 positions were monitored by the Element Recorder. The ISDA shown in Fig. 2.78(a) was calculated by observations of the global response data. The base shear force ($V_{base}$) is the summation of the reaction forces at the column bases, and is another example of a global measurement. On the other hand, both the PZRA and stress hinge sequences are derived from monitoring local response data.

## 2.10.2 Failure mechanism for composite frame models

From the pushover results, it can be shown that hinges occurs at the bottom of the composite columns as well as other column locations. The occurrence of a hinge, and

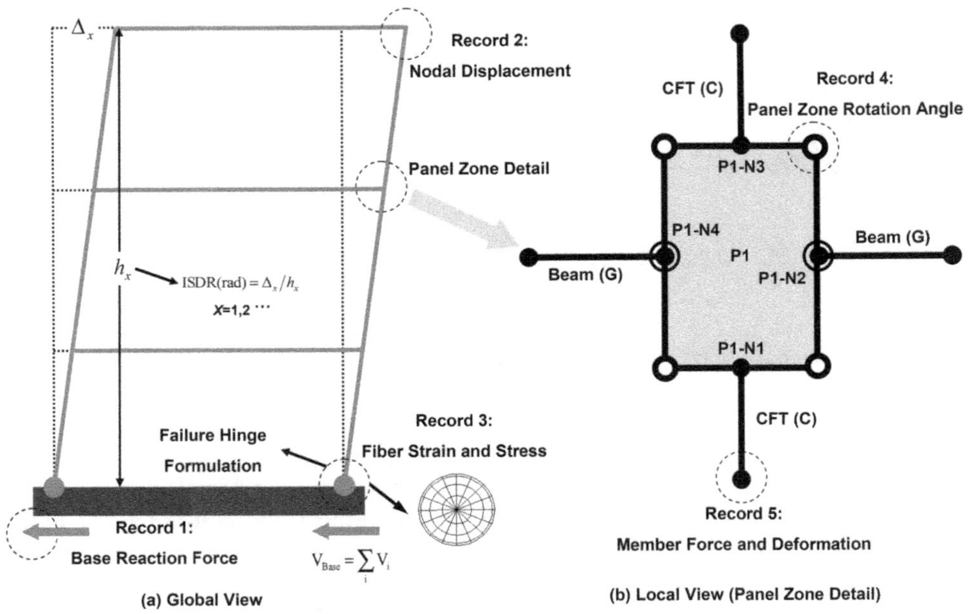


*Figure 2.78* The schematic view of data collections.

its increasing rotations, can be detected by measuring the stresses and strains in the sections during the pushover analysis. The composite columns subjected to lateral loads accommodate the imposed frame deformations through a double curvature deflection mode.

The composite columns subjected to lateral loads accommodate the imposed frame deformations through a double curvature deflection mode. The member forces acting on the composite frames are shown in Fig. 2.79(a). As the applied lateral loads are increased, the bending moments are also increasing at the ends of composite columns. The bending stress due to the bending moments contributes to creating the hinges as shown in Fig. 2.79(b). Generally, the shear stresses in the members are negligible in a frame analysis. Both hinge levels are defined in Fig. 2.79(c). The yield stress hinge is determined when the design yield stress (378.95 MPa). The ultimate failure hinge is determined when the post design ultimate stress (509.86 MPa) is reached. The design stress level was based on the steel materials used in the composite columns.

Only two performance levels (the yield point and the ultimate points) are shown because no yielding occurred at the design point. Hinges start to occur at the column bases and extend to the upper stories. The hinges occur due to the combination of axial and bending stresses. At the yield base shear force, the number of hinge points due to compression is more than that of hinge points due to tension because of the axial stresses generated by the gravity loads. However, as the applied lateral loads are increased, the bending stresses generated by the bending moments become dominant.

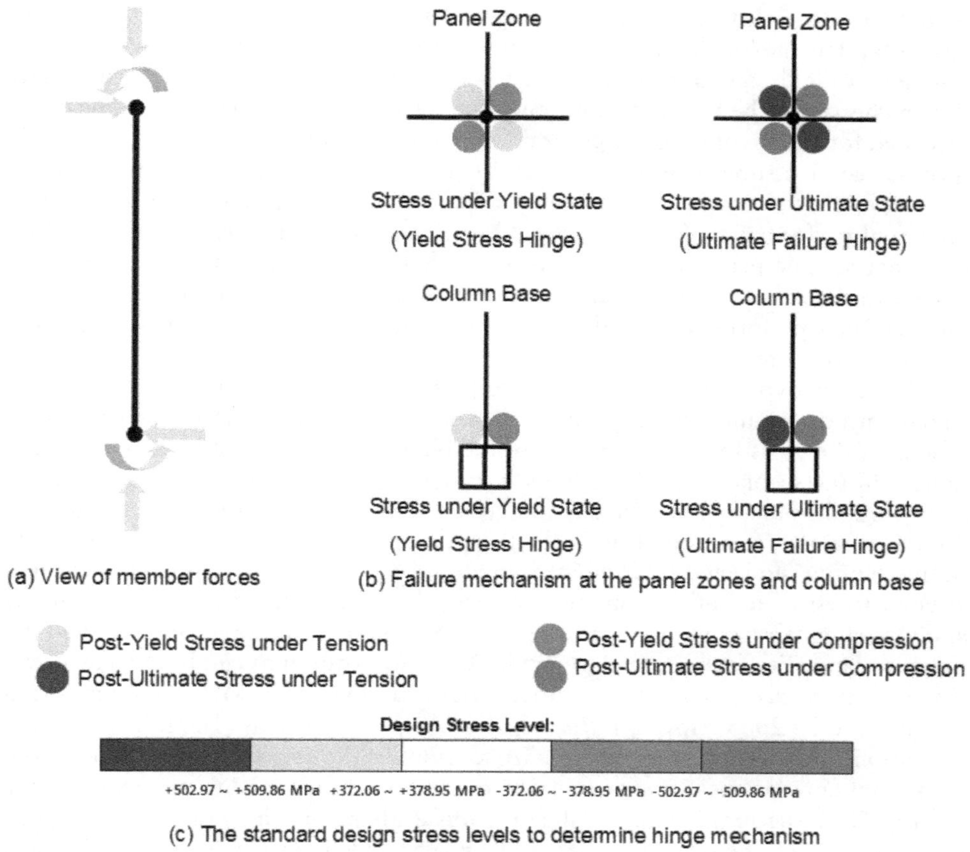


*Figure 2.79* Determination of the failure mechanism using the failure hinge.

Moreover, while many hinges are shown in the columns, no complete story mechanisms form, so the structures are still stable.

Overall, the interior composite columns are more susceptible to failure than the external composite columns due to the larger P-Delta effect. The positions of hinges are symmetric to the center of the moment frame. Composite moment frames with welded connections show more hinges than those with PR connections at the yield point because of the sudden brittle failure.

### 2.10.3 Performance evaluation (ESR)

The basic design objective for the frames designed in this research was to enforce a strong column-weak beam mechanism, i.e., reaching the full plastic moment capacity of the steel beam before any other failure mode was reached. In addition to strength, this requires large hinge rotation capacity in the critical sections. Exceedance of any ultimate limit state in the columns indicates the most severe type of damage for the building as it can lead to complete collapse. Therefore, a careful investigation of the structural damage for the composite columns is emphasized in this chapter. The more popular available design-oriented programs do not provide, in general, the correct design checks for the beam-columns and composite sections in particular. The checks provided for beam-columns are generally very conservative and strength based as this process requires subjective judgment and is thus impossible to automate. Even if it were possible to automate them, these checks would not provide any information on actual performance. Finally, it should be noted that there were no investigations on either seismic performance or damage evaluation for composite moment frames similar to those studied here found in the technical literature. This chapter will focus on the seismic performance and the damage evaluation for the composite CFT columns because of these reasons.

There are two major steps in order to perform the damage evaluation, one associated with determining the capacity and the other with assessing the demand. In the first step, the cross-sectional capacity of the hinging regions must be carefully determined. In the second step, the demand at these critical sections must be established from careful numerical and detailed analyses of entire structures subjected to large ground motions. To accomplish the first step, monotonic and cyclic behavior of CFT beam-columns subjected to combined axial and moment loading was studied in an attempt to estimate both the maximum strength and ductility for doubly-symmetric and axis-symmetric composite cross sections. From these studies it can be shown that ultimate capacities for rectangular/circular CFT beam-columns can be estimated with reasonable accuracy using the simplified axial and moment (P-M) interaction formulas provided by *2005 AISC Specification* for composite systems. The P-M interaction formations are given in Appendix A. To accomplish the second step, advanced computational simulations were carried out on a series of composite moment resisting frames. The structural damage evaluation was evaluated in this research through the comparison of elastic strength ratios (ESR) ESRs are defined as the ratios of the member response to the strength capacity for the member cross section. These concepts are taken from the work by Hajjar *et al.* (Hajjar 1998).

The P-M interaction diagram for a composite section based on a full plastic stress distribution can be generated as a linear interpolation between five points. The theoretical descriptions for determining these five points were given in Section 2.3.2 (See Fig. 2.9 and Table 2.2). For the damage evaluation, a simplified bilinear interpolation may be used between three points as shown in Fig. 2.80. The simplified expressions shown as Eqs. 2.97 and 2.98 can be used for determining the member capacity to use in the calculation of the elastic strength ratios (ESR) for composite columns. For combinations at axial loads and bending moments, the ESR can be defined as:

If $P_r \leq P_D$

$$\mathrm{ESR} = \frac{M}{M_B} \tag{2.97}$$

otherwise,

$$\mathrm{ESR} = \frac{P - P_D}{P_A - P_D} + \frac{M}{M_B}. \tag{2.98}$$

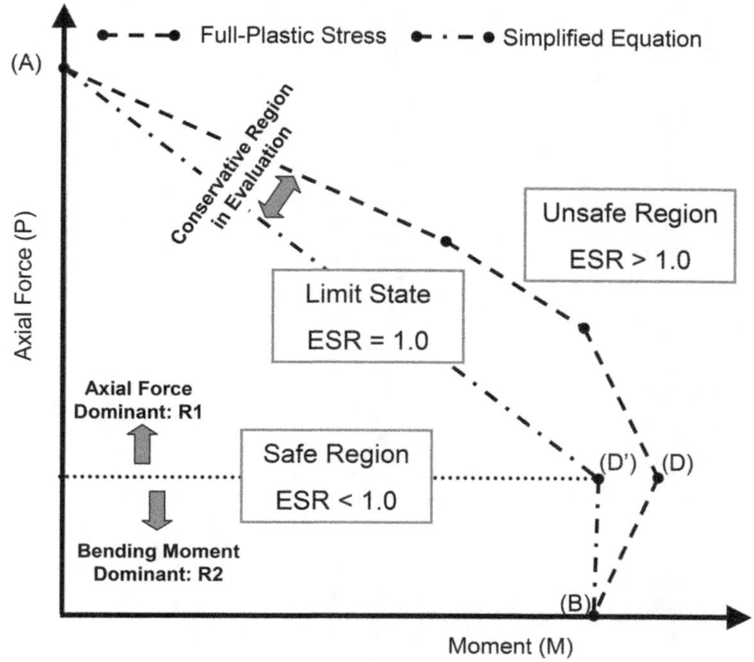


*Figure 2.80* Basic concept for the elastic strength ratio (ESR).

This approach is reasonably accurate for steel columns and should provide a conservative estimation for composite structures. Because the simplified equation line is taken conservatively as being bilinear instead of multi-linear, the value of ESR from the analysis can exceed 1.0 as the design level is exceeded.

Chapter 3

# Analyses for smart PR-CFT connections (end plate connection type)

## 3.1 GENERAL OVERVIEW

In recent years, concrete filled steel tube (CFT) columns have become widely accepted and used in multistory buildings as well as bridges. These elements provide the synergetic advantages of ductility and toughness associated with steel structures and high compressive strength associated with confined concrete components. The advantages of CFT columns over other so-called mixed or hybrid systems (fully encased or partially encased systems) include the fact that the concrete prevents local buckling of the steel tube wall and the confinement action of the steel tube extends the usable strain of the concrete (Wu *et al.*, 2005; Wu *et al.*, 2007). In addition, CFT columns have improved fire resistance and significant cost reductions in comparison with traditional steel construction. Composite CFT columns are especially efficient as the vertical elements in moment resisting frames in high seismic areas because they have a high strength to weight ratio, provide excellent monotonic and dynamic resistance under biaxial bending plus axial force, and the concrete provides additional damping (Tsai *et al.*, 2004; Tsai *et al.*, 2008; Tsai & Hsiao, 2008; Hu and Leon, 2010). A typical moment connection that was part of a composite braced frame consisting of steel I shape girders and either circular or rectangular (CCFT or RCFT) columns tested by Tsai *et al.* (2004) is illustrated in Fig. 3.1.

To evaluate the performance of a moment frame subjected to lateral-loads, the accurate modeling of both flexural and shear deformation of the framing members and connections is the most critical issue. In this study, the shear deformations were deemed to play a minor role because the concrete inside the tubular columns considerably reduces the contribution of shear deformations to the overall joint displacements. In addition, because the behavior of both beams and columns as isolated members is well understood, this study focuses on connection behavior only. This behavior can be represented in its simplest form by a moment-rotation curve (Fig. 3.2). As shown in Fig. 3.2, connections are classified by three main parameters: stiffness, strength, and ductility (Leon, 1997; Green *et al.*, 2004; Rassati *et al.*, 2004). For stiffness, connections are classified as fully restrained (FR), partially restrained (PR) or simple connections. An ideal pinned connection only transmits shear forces from the beam to columns and is considered as a simple connection. At the other extreme, connections where no change in angle occurs between the beam and column are classified as FR.

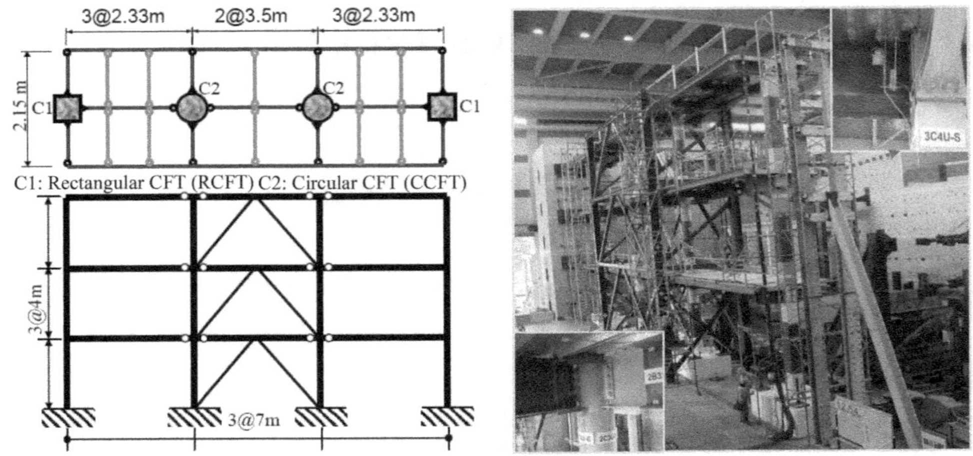


*Figure 3.1* 3 story by 3 bay CFT composite frame with buckling restrained brace (Tsai *et al.*, 2004; Tsai *et al.*, 2008).

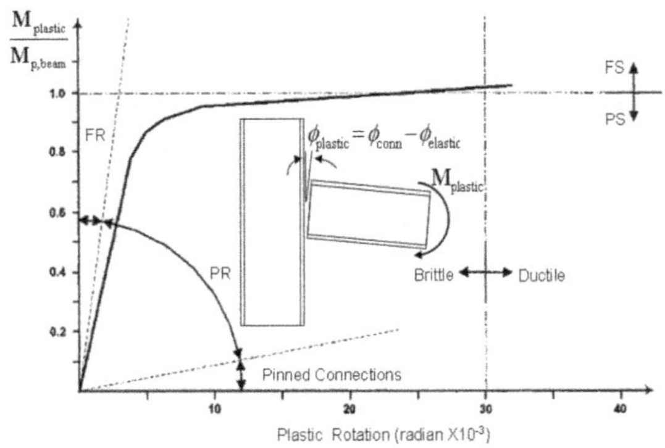


*Figure 3.2* Typical moment-rotation curves for PR connections.

Most real connections in steel buildings fall somewhere between these two extremes and are PR connections. For strength, connections are classified as either full strength (FS) or partial strength (PS) depending on whether they can transmit the full plastic moment of the beam ($M_{p,beam}$). Finally, connections are classified as brittle or ductile connections based on their ability to achieve a certain plastic rotational demand. For example, in the aftermath of Northridge earthquake, an elastic rotation of 0.01 radians and a plastic rotation of 0.03 radians under cyclic loading have been accepted as the rotational limit to distinguish between ductile and brittle connections in special moment resisting frames. More recently, a total interstory drift of 0.04 has been accepted as the requirements for ductile systems. The major failures of fully welded moment connections during the 1994 Northridge and 1995 Kobe earthquakes have led

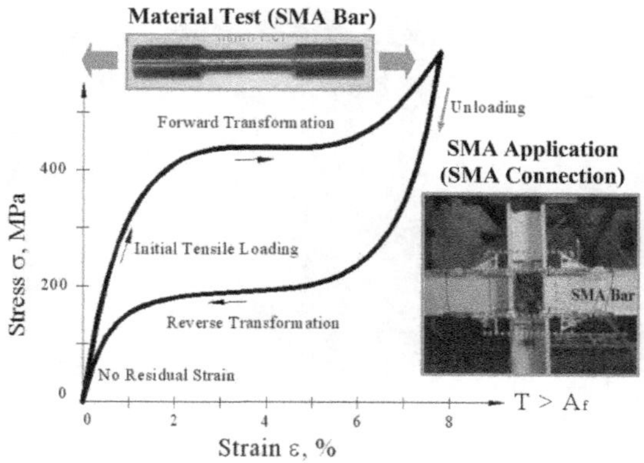


*Figure 3.3* Stress and strain behavior for super-elastic (SE) SMA materials.

to the conclusion that the traditional fully welded moment connections (FR/FS) require expensive and careful quality control and quality assurance measures during construction. In contrast, bolted connections or combinations of field bolted-shop welded connections pose an attractive solution to this brittle failure dilemma (Swanson & Leon, 2000; Swanson & Leon, 2001). It also has been demonstrated that well-detailed PR structures can provide similar or superior seismic behavior to their FR counterparts (Rassati *et al.*, 2004)

More recently, work at Georgia Tech on nickel-titanium (Nitinol) shape memory alloys (SMA) has explored the applications of this material to the design of connections in steel structures subjected to large cyclic loads. SMA materials can undergo large deformations with little permanent residual strain through either the shape memory effect or the super-elastic effect. The deformations, generally in the range of 6% to 8% strain (Fig. 3.3), can be recovered with changes in either temperature or stress. A connection incorporating SMA components not only contains all of the merits of bolted PR connections mentioned above, but also adds a desirable recentering capacity due to the SMA material characteristics (Ocel *et al.*, 2004; Penar, 2005).

This research intends to explore a mixture of steel bars and super-elastic Nitinol bars as connecting elements to composite CFT columns (Fig. 3.4). It is hypothesized that such combinations of CFT columns and SMA connections will achieve excellent ductility, high strength, and recentering capability. Based on these premises, a number of original connection models were developed to investigate the optimal distribution of steel and SMA components. Because of space limitations, this paper deals with only one of those designs (i.e., end plate connections).

Several numerical studies on the connection models were performed using refined 3D finite element (FE) analyses. These initial detailed FE analyses were used to develop an understanding of the monotonic moment-rotation behavior of the connections, as well as the contribution and influence of different deformation mechanisms. In addition, these studies were intended to identify and improve detailing in any areas of stress concentration that would be susceptible to brittle failure under large amplitude cyclic

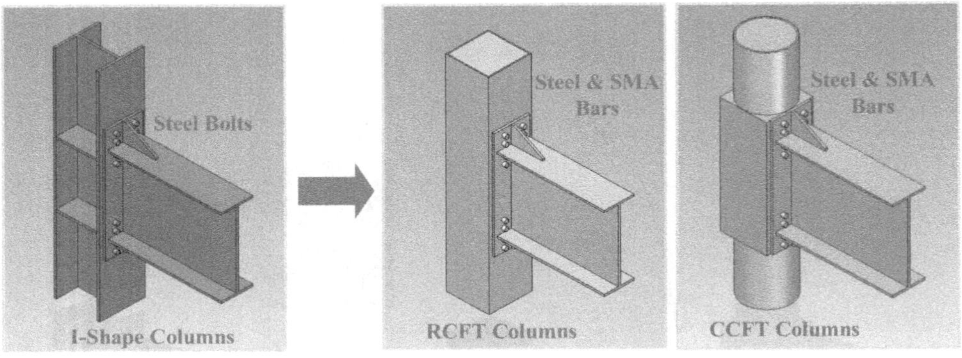


*Figure 3.4* From traditional steel to a new smart composite connection.

displacement reversals. The results from FE analyses were applied to the development of less complex joint models based on equivalent non-linear springs whose behavior would be simpler to extend to the case of cyclic deformations. The numerical cyclic simulations were conducted with these joint models using the OpenSEES program, a research computer code widely used in the USA for this type of analyses. The structural advantage of new connection designs in terms of strength, ductility, and rehabilitation is validated through these numerical simulations which are consistent with results of refined 3D FE analyses.

## 3.2 DESIGN FOR NEW CONNECTIONS

### 3.2.1 Design philosophy

All connections in this study were designed as full strength (FS), meaning that they can transfer the full plastic beam moment ($M_{p,beam}$) calculated according to the AISC-LRFD Standard (AISC, 2001). The connection design, however, did not aim to achieve full restraint (FR or full end rigidity); it intended to utilize PR behavior to obtain ductile connection behavior. The connection selected for discussion in this paper is an end plate one (Fig. 3.5). Since some shear yielding and local buckling have been observed in the panel zone of I-shape steel columns before reaching the full moment capacity of the connected beams, it was decided that the panel zone capacity should be upgraded by using a concrete-filled tube section to minimize these effects. Component members such as shear/web bolts, shear tab plates, and clip angle or T-stubs were designed with the intent of preventing or reducing loss of stiffness and strength due to brittle failure modes. Therefore, the dominant behavioral modes for both the steel and composite components should be ductile yield modes such as slip of the bolts, yielding of steel, and minor local bucking. This will avoid the potential for local connection failures leading to overall collapse of the frame (Astaneh-Asl, 1995). All designs were carried out based on demand capacity principles, evaluated using the typical assumptions of weak beam and strong column conditions.

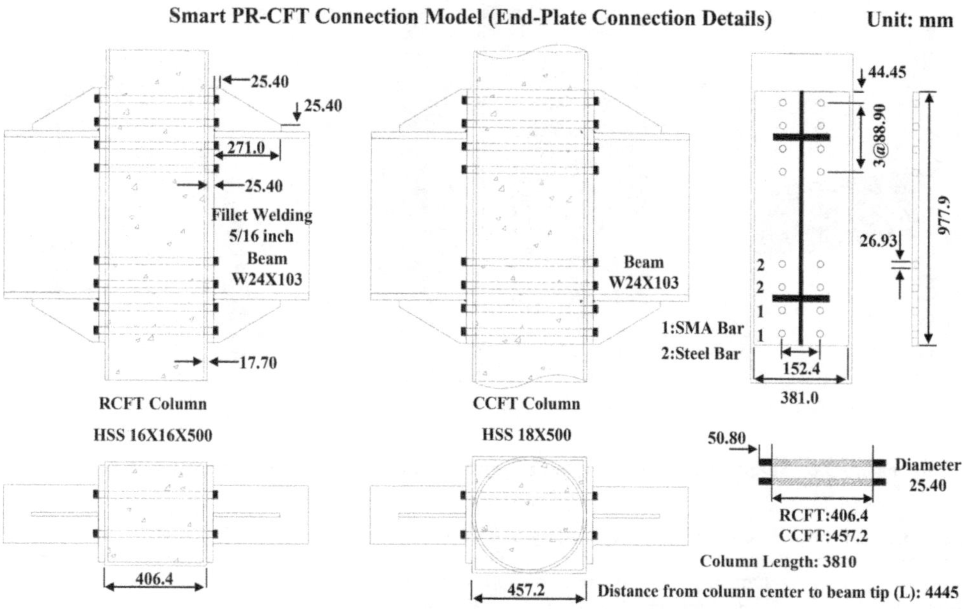


*Figure 3.5* Connection details.

### 3.2.2 Connection details

The connections are to be fabricated as an assembly of various steel members cut from standard shapes available in the current design specification (AISC, 2001). A572 Grade 50 and A500 Grade C steel were used for all members and joint components. A490 high strength bolt material was used for steel bars, washers, and nuts. Super-elastic (SE) Nitinol bars, with the characteristics shown in Fig. 3.3, were located where the largest deformations were likely to occur. Extended stiffener plates welded between the connected beam flange and the end-plate were required to maintain stiffness (Fig. 3.5). The plate stiffeners (A572 Grade 50) had the same thickness as the beam web. Panel zones were designed as rectangular, even in the case of circular CFT columns, in order to allow for bidirectional end-plate connections and steel and SMA rods penetrating through the connections to tie the end plates together.

## 3.3 THREE-DIMENSIONAL FINITE ELEMENT ANALYSES

### 3.3.1 Modeling procedures

The ABAQUS (ABAQUS, 2006) finite element code was used to analyze these PR-CFT connection subassemblies. The subassemblies used were meant to simulate typical connections as tested in the laboratory to satisfy prequalification requirements (AISC, 2005). The numerical models consisted of refined 3D solid elements (8 node

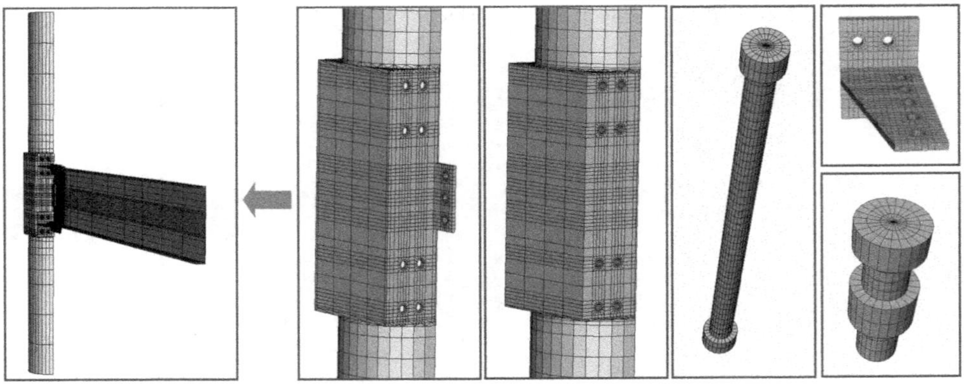


*Figure 3.6* 3D solid elements for the assembled connection and component members.

*Table 3.1* Number of elements, nodes, DOFs, and contact interactions for FE models.

| *FE Model* | *Number of Solid Elements* | *Number of Gap Elements* | *Number of Nodes* | *Total Number of Degree of Freedom* | *Number of Surface Interactions* |
|---|---|---|---|---|---|
| Connection with RCFT Columns | 22499 | 1637 | 26439 | 162966 | 41 |
| Connection with CCFT Columns | 29209 | 1783 | 29961 | 194562 | 41 |

brick element, C3D8) incorporating the full nonlinear material properties, geometric nonlinearity, contact elements, surface interaction with friction, constraint conditions using equation points, concrete cracked conditions, and elastic foundation modeling. These modeling methods were useful to provide detailed and accurate understanding of the overall behavior of the connection, including the stress distributions on the contact surfaces. Of course, these advantages come both at a high computational cost during the runs and substantial time needed to overcome numerical instabilities in the solution algorithms.

In this research, most of the FE work, including the generation of parametric geometries and meshes, was done by using ABAQUS/CAE with an input file-based edition to add more advanced modeling flexibility. Two FE models replicated the connection details including the size of beam and columns used, plus the details of the connecting components, as shown in Fig. 3.5. End-plate connections plus beam and column stubs (Fig. 3.6(a)) were subdivided into several independent bodies: 8 tension bars, a welded end-plate component, an I beam, an outside hollow steel column, and an interior concrete material (Fig. 3.6(b) to (e)). Each of these independent parts has its own material properties and interacts with each other via contact definitions. Some details of the number of elements, nodes, DOFs, and contact interactions are given in Table 3.1. The connections were modeled as half symmetric models to save

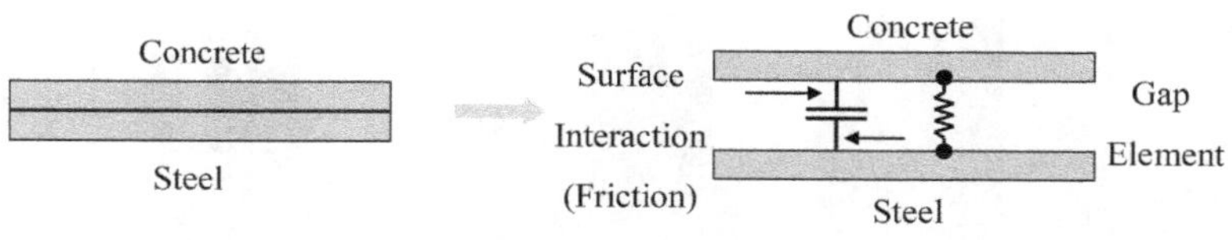


*Figure 3.7* Modeling of the contact between steel and inside concrete.

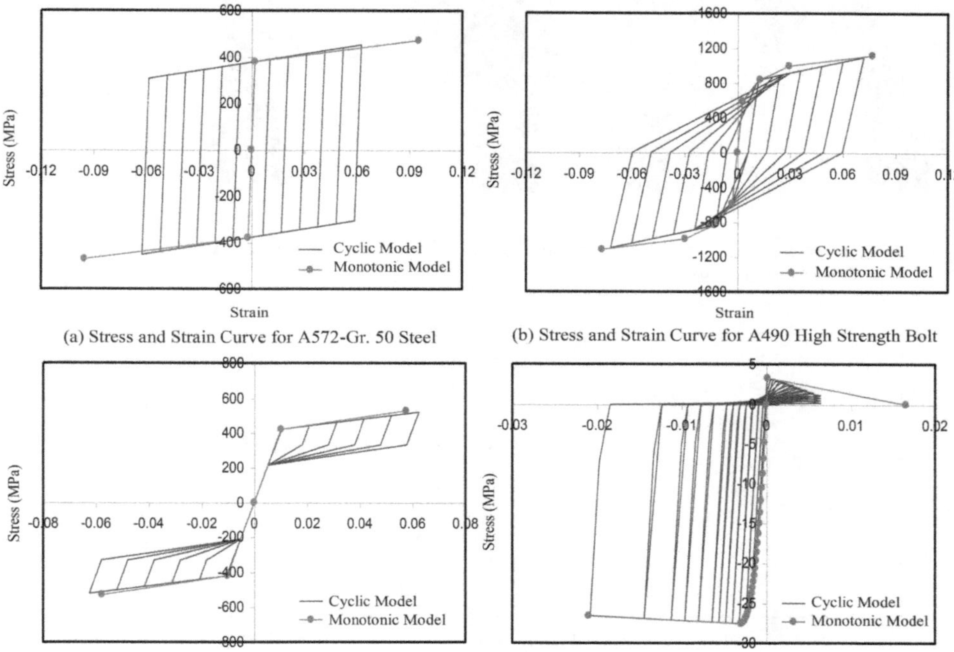


(a) Stress and Strain Curve for A572-Gr. 50 Steel
(b) Stress and Strain Curve for A490 High Strength Bolt
(c) Stress and Strain Curve for Super-elastic SMA
(d) Stress and Strain Curve for Confined Concrete

*Figure 3.8* Material properties for the analyses.

computation time. The assembled connection and component members are shown in Fig. 3.6(a).

Surface interactions with friction coefficients were defined for all component interfaces. In addition, gap elements were used to generate the contact behavior along the normal direction to the surface between the steel column surface and inside concrete core as shown in Fig. 3.7. The 3D solid elements incorporated nonlinear material properties as shown in Fig. 3.8. The cyclic material behavior for the SMA materials was generated by a user-defined code (Davide, 2003) in OpenSEES because ABAQUS lacks such model. Cyclic modeling will be described in the next section.

The models were loaded in two steps. The first step was used to pretension the bolts while the second step was used to apply the actual load. The bolts were pre-tensioned by applying an adjustment length/displacement to the surface of the bolt shank.

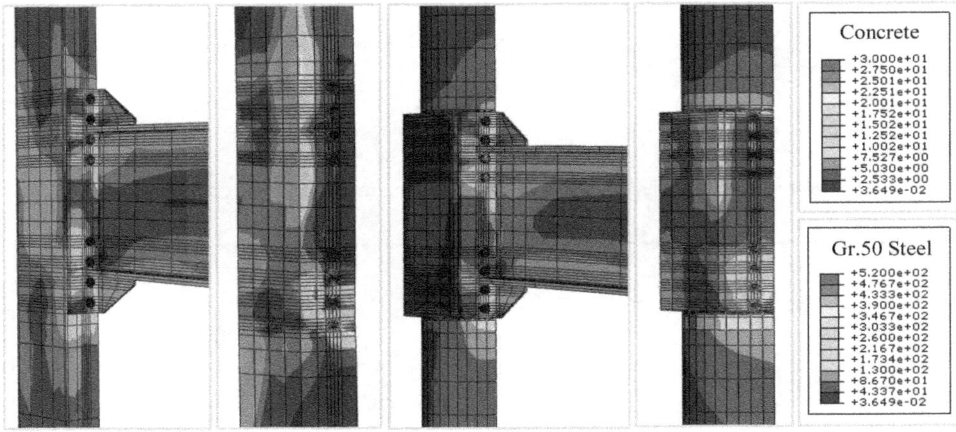


*Figure 3.9* Stress distribution at ultimate state for RCFT and CCFT columns (Left: Stress in steel, Right: Stress in concrete).

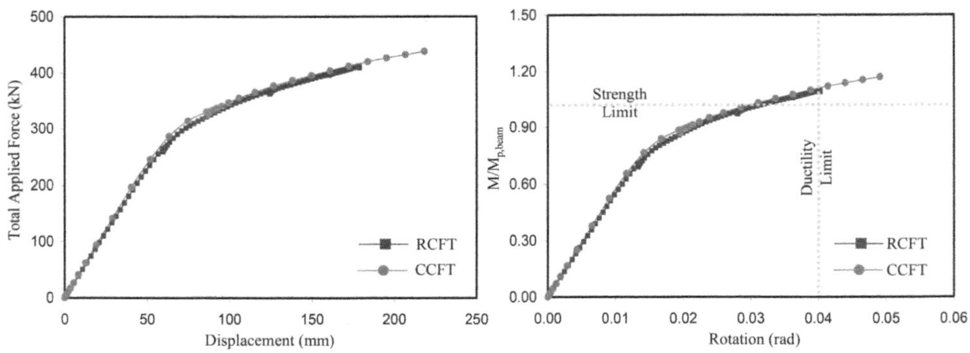


*Figure 3.10* FE test results (Monotonic loading).

### 3.3.2 FE analysis results

Static monotonic loads were generated by imposing a support displacement to the tip of the beam. The force-displacement response of the connections ($T - \Delta$) was changed into a moment-rotation response using simple relationships ($M = TL$, $\theta = \Delta/L$).

For the end-plate connection, the stress distributions at the ultimate state are shown in Fig. 3.9.

Due to the contact interaction definition, bearing forces acting on the steel tube could be transferred into the concrete core. High bearing pressures and crushing were observed on the concrete underneath the bar heads.

The nonlinear moment-rotation behavior curve is shown in Fig. 3.10. These connections exceeded both the full plastic strength of the beam ($M_{p,beam}$) and the required rotational limit for ductility (refer to Fig. 3.2). This implies that this type of connection

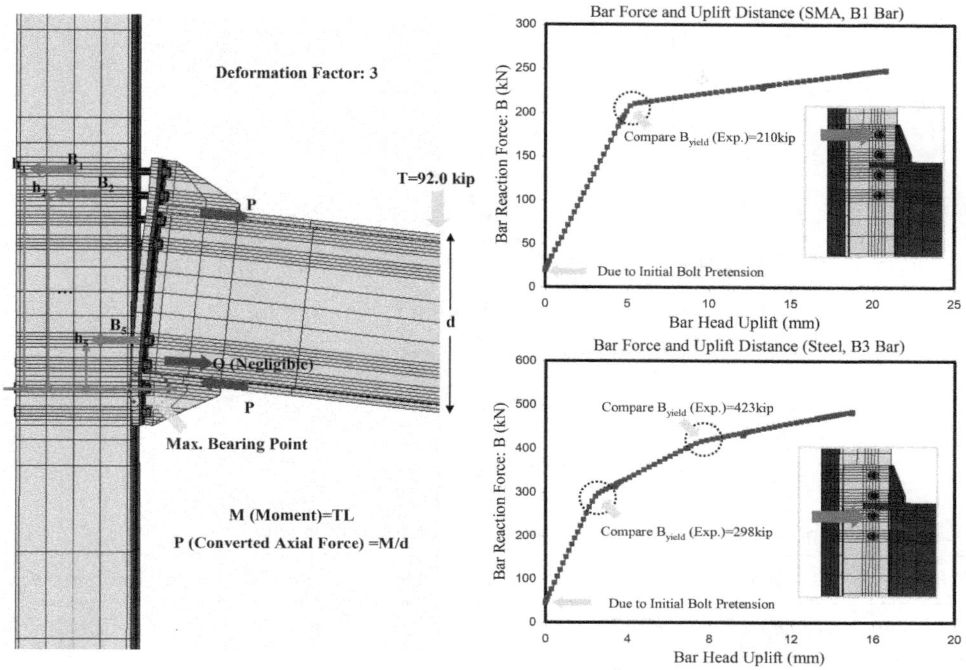


*Figure 3.11* Reaction forces and deformations under the total applied force (T).

can potentially satisfy current design criteria with respect to strength, stiffness, and deformation capacity for use in special moment frames.

### 3.3.3 Observations

The FE test results, which were obtained from two FE test models (i.e. one connection with a RCFT column and one with a CCFT column), provide detailed insight into the monotonic behavior of these SMA PR-CFT connections. They also provide the foundation for the development of simpler models, as the 3D FE connection models shown above are unsuitable for the analysis of entire buildings. The later models will be based on the global behavior or moment-rotation curve for the entire connection. The conversion from these detailed results to a global model is carried out in a straightforward fashion.

The total applied force (T) at the tip of the beam is first converted into a bending moment ($M = TL$). This bending moment is transmitted to the connection as a set of concentrated axial forces ($P = M/d$), as shown in the left side of Fig. 3.11. The right side of Fig. 3.11 shows the key points for the behavior of both the steel and SMA bars as the monotonic load increases. In order to fulfill equilibrium constraints, the summation of bar reaction forces ($\Sigma B_i$) must equal the summation of the converted axial force and prying force ($\Sigma B_i = P + Q$).

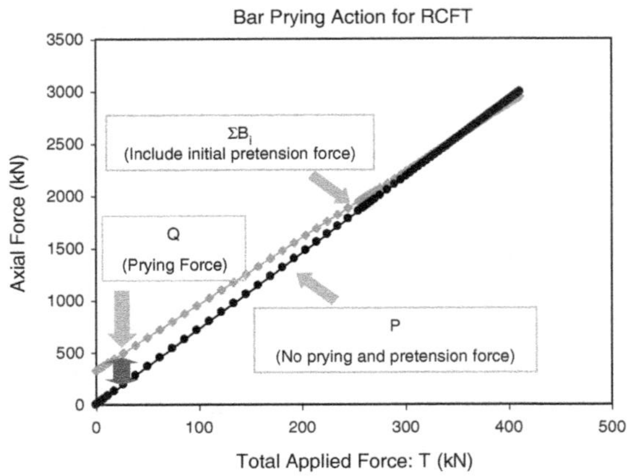


*Figure 3.12* Bar prying response mechanism.

*Table 3.2* Comparisons between external moment and internal moment.

| *T* | *P = M/d* | *ΣBi (i = 1 to 5)* | *M = TL** | *M = ΣBihi* | *Moment Difference*** |
|---|---|---|---|---|---|
| 247.1 kN | 1783.8 kN | 1872.7 kN | 1111.9 kN-m | 1136.8 kN-m | 2.18% |
| 335.8 kN | 2451.0 kN | 2464.3 kN | 1525.5 kN-m | 1508.6 kN-m | −1.12% |
| 409.7 kN | 2971.4 kN | 2935.8 kN | 1853.2 kN-m | 1801.2 kN-m | −2.88% |

L*: Distance from the tip of the beam to the centerline of the CFT column.
Moment Difference** = $(\Sigma B_i h_i - TL^{**})/(\Sigma B_i h_i) * 100$.

The prying response of the end-plate connections is given in Fig. 3.12. At the beginning of the loading history, the prying force (Q) corresponds to the summation of the bolt reaction forces due to the initial pretensions. However, the effect of prying force becomes negligible after the considerable axial force (P) is applied to tension bars. Thus the difference between P (total force) and $\Sigma B_i$ (total force including prying) decreases as the load increases and vanishes at the ultimate deformation.

This bar prying response has an influence on both the external moment (M = TL) and the internal moment ($M = \Sigma B_i h_i$). Comparisons between the external moment and the internal reaction moment are given in Table 3.2. The external moments show good agreement with the internal moments, indicating that the simplifications made do not affect the prediction of response appreciably.

The uplift of the end-plate, depicted in Fig. 3.13, shows the deformations at the bar locations. These results indicate that the FE models can capture the local behavior well. The components subjected to tension force show a different behavior from those subjected to compression force. SMA and steel bars subjected to tension follow their material property paths in terms of the relative deformation and the reaction force. Those subjected to compression, however, were affected by the bearing because of their penetration into the surface of the CFT columns.

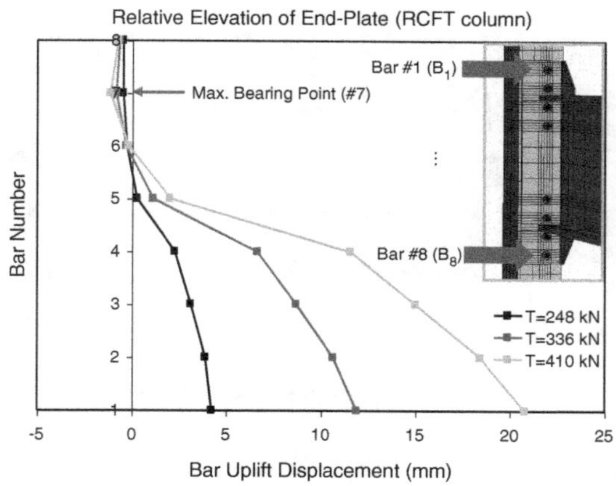


*Figure 3.13* Measurement of end-plate separation at tension bar heads.

## 3.4 CONNECTION MODELS UNDER CYCLIC LOADS

Simple beam-to-column joint models for the PR-CFT connections were constructed using the nonlinear finite element (FE) program OpenSEES (Mazzoni *et al.*, 2006). A 2D joint model is appropriate for use in frame analyses to estimate the inelastic response of moment frames under seismic loads (Hu, 2008; Leon & Hu, 2007; Leon & Hu, 2008) if the model is carefully developed and robust. The primary objective of this study is the development of such a simplified joint model, one which both reflects the real connection behavior and is computationally efficient to allow its use in the study of large frames.

### 3.4.1 Simplified 2D joint models

The idealization of the moment distribution at the perimeter of a typical panel zone under cyclic loads is shown in Fig. 3.14(a). Figs 3.14(b) and (c) show the idealization of the force distribution at the perimeter of the joint model for the end-plate connection. As shown in these figures, the connection components, such as the tension bars, are converted into equivalent spring elements. In addition, the end-plate is modeled as a rigid plate, as the end plate contributes little to the deformation of the overall connection when designed by current US seismic design requirements. Note that this assumption means that these studies are not applicable to connections with thin end plates, where the plate deformations will play an important role in the overall deformation characteristics. Such deformations will negate the benefits of the SMA recentering ability as they will lead to large plastic deformations elsewhere in the system.

The joint model in Fig. 3.14(c) includes spring elements for the CFT column and panel zone as well as those for the connection bars. The external moments acting on the beams and columns are assumed as equivalent concentrated force resultants (P forces

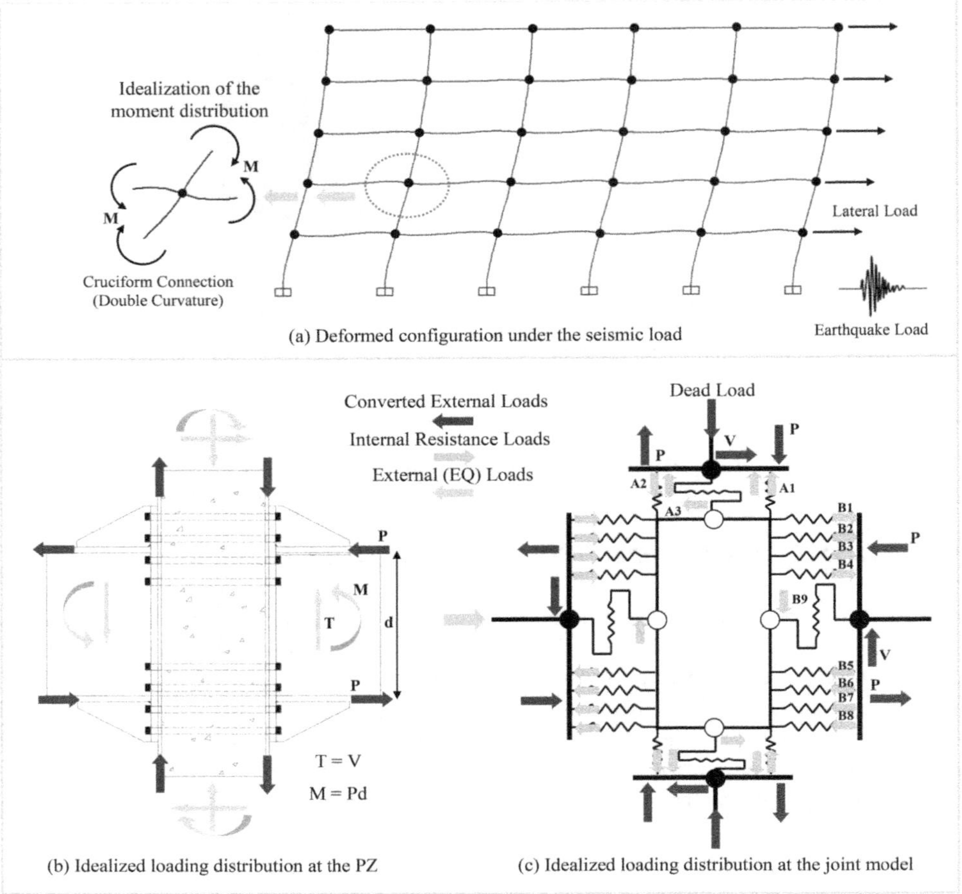


*Figure 3.14* External and internal forces in the joint model.

in Fig. 3.14(b)). The internal resistance of the connection components acts against these converted external forces as shown in Fig. 3.14(c) in order to satisfy equilibrium.

The response of the joint element under bending deformation is shown in Fig. 3.15. The joint model under seismic loading is deformed in a scissors-like (shear) mode (Lee & Foutch, 2002). The internal tension loads corresponding to the external forces are carried directly by the tension bars as shown in Fig. 3.15(a). Fig. 3.16 shows the reduction (condensing) of the large number of springs in Fig. 3.16(a) to just a few equivalent springs shown in Fig. 3.16(b). The behavior (i.e. force vs. deformation) of the equivalent spring elements under tension loading was assumed as shown in Fig. 3.17. The behavior under compression bearing was determined by observations from the 3D FE models. The equivalent element shown in Fig. 3.16(b) reduces the number of

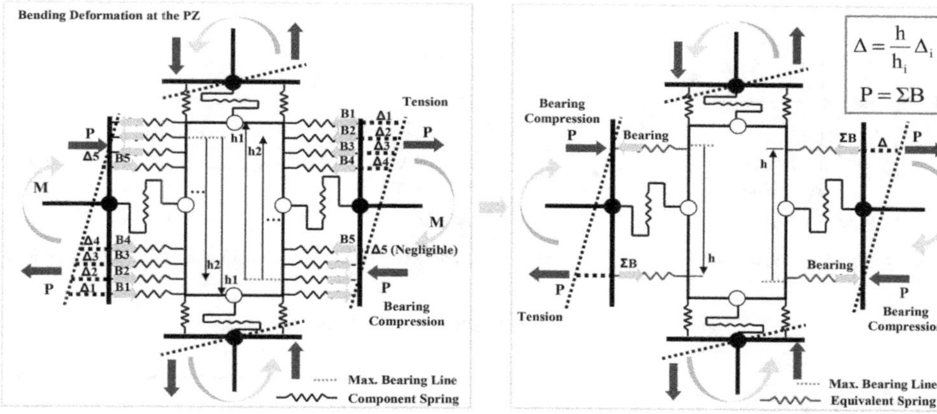


(a) Bending deformation at the joint model with component spring elements

(b) Bending deformation at the simplified joint model with equivalent spring elements

*Figure 3.15* Response mechanism of the joint element under bending deformation.

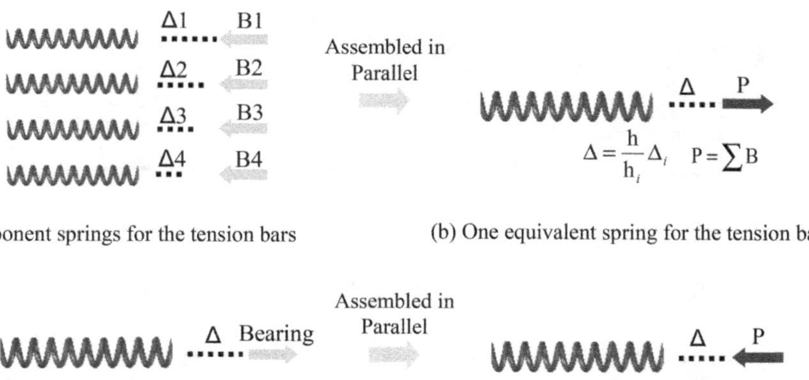


(a) Component springs for the tension bars

(b) One equivalent spring for the tension bars

(c) Component springs for compression bearing

(d) One equivalent spring for bearing compression

*Figure 3.16* The behavioral properties of the individual component model.

variables during the analysis, both saving running time and avoiding numerical convergence problems. The penalty paid is that the contribution of individual mechanisms to the overall joint deformation becomes difficult or impossible to extract from the results.

The cyclic behavior of the equivalent spring elements is compared with the converted axial force vs. corresponding deformation obtained from static 3D FE results in Fig. 3.18. During the loading path, the results from the 3D FE models show good agreement with that of the equivalent spring element.

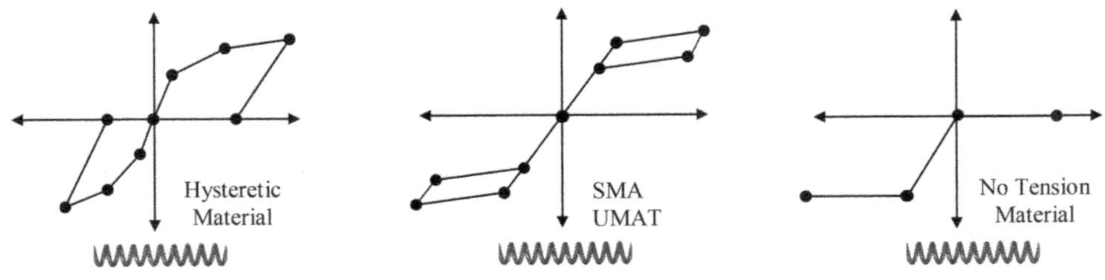


(a) Steel Bar Component Model (b) SMA Bar Component Model (c) Bearing Component Model

*Figure 3.17* Behavioral of the individual components.

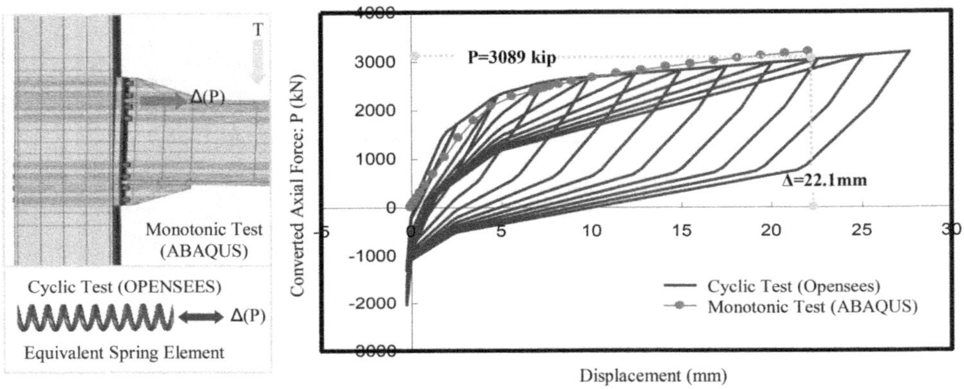


(a) Measuring Position (b) Component Test under Axial Loading

*Figure 3.18* Comparisons of results between two analytical runs.

*Table 3.3* Comparisons of the bar responses for two analytical cases (Figure 18(b)).

| *Basic* | *OPENSEES (Δ, Displacement)* | *ABAQUS (Δ, Displacement)* | *OPENSEES (B, Bolt Force)* | *ABAQUS (B, Bolt Force)* |
|---|---|---|---|---|
| Δ = 22.1 mm | Δ1 = 26.67 mm | Δ1 = 27.18 mm | B1 = 261.5 kN | B1 = 256.2 kN |
| | Δ2 = 23.62 mm | Δ2 = 24.38 mm | B2 = 254.9 kN | B2 = 250.0 kN |
| | Δ3 = 20.57 mm | Δ3 = 20.10 mm | B3 = 516.9 kN | B3 = 495.2 kN |
| | Δ4 = 17.27 mm | Δ4 = 16.20 mm | B4 = 509.3 kN | B4 = 521.3 kN |
| | | | P = 2 × ΣB (i = 1 to 4) = 3083 kN | 2 × ΣB (i = 1 to 4) = 3035 kN |

*2 × ΣB (i = 1 to 5) obtained by ABAQUS analysis was 3158 kN.

The solution for the case of cyclic loads algorithm was based on incremental displacement control. Comparisons between results of the algorithm and those of 3D FE model test are given in Table 3.3. Both the displacements and forces results show very good agreement.

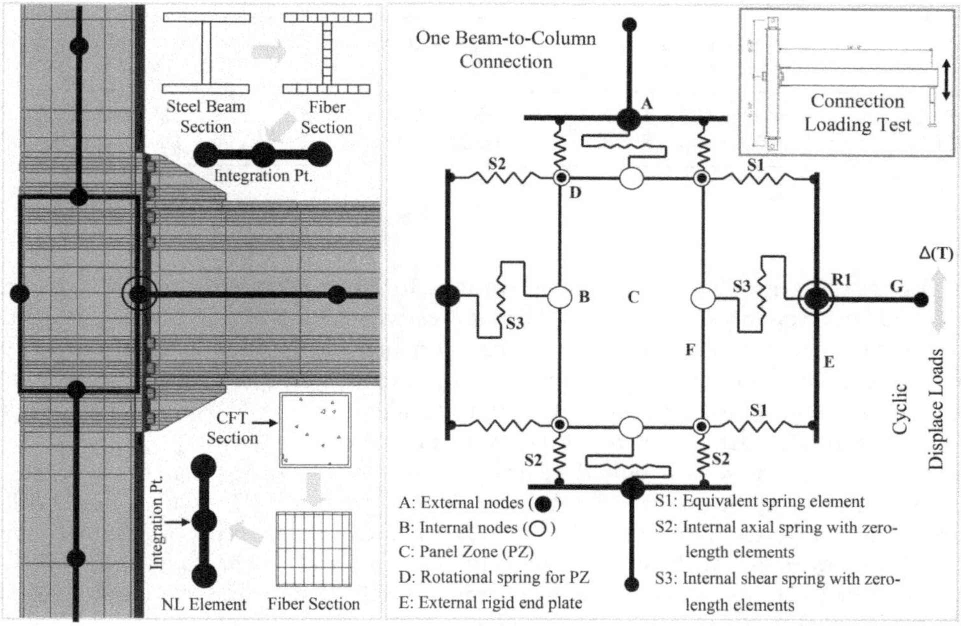


(a) The element formulation on the connection (b) The joint element of one beam-to-column connection

*Figure 3.19* Element formulation and joint element for SMA PR-CFT connection.

The joint models presented above were modeled as 2D joint elements in the OpenSEES program, which allows the development of user-defined elements such as the one described here (Fig. 3.19). This element includes two equivalent spring elements (S1) to reproduce the behavior of the component model, four internal spring elements (S2) to reproduce the axial deformation of the CFT column, four internal shear springs (S3) to reproduce the shear deformation of the CFT column and the beam, and one shear panel element (C) which is intended to reproduce the failure of the panel zone under severe loading. The joint model includes one element of the beam and column. In OpenSEES, the beam and CFT column are modeled as a nonlinear beam-column element with 2D fiber sections (Mazzoni *et al.*, 2006).

### 3.4.2 Model validation

The cyclic test results obtained from the numerical tests, which were performed on the joint elements, were compared with those obtained from the experimental tests performed by Wu *et al.* (Wu *et al.*, 2008). They studied the seismic behavior of bidirectional bolted beam-to-column connections with CFT columns. For the connection details, the ends of the beam were welded with end-plates. The end-plates were tied to the rectangular CFT columns in both directions (i.e. east-west or north-south direction). The

*Table 3.4* Details of the specimens (Wu *et al.*, 2007; Wu *et al.*, 2005).

| *Specimen* | *Column Sections (A572-Gr. 50)* | *Beam Sections (A36)* | *Tension Fasteners (A490)* |
|---|---|---|---|
| FSBE6 | 400 × 400 × 6 × 6 | H500 × 200 × 10 × 16 | 16–30 mm Dia. Bars |
| FSBE8 | 400 × 400 × 8 × 8 | H500 × 200 × 10 × 16 | 16–30 mm Dia. Bars |

Unit: mm.

tied bars were installed at different elevations for the two directions. Furthermore, the tied bars were pre-stressed after the compressive strength of the concrete was fully developed. The adequacy of the numerical modeling for the joint element developed in this study was validated by comparison to these experimental results. In addition, individual deformation components were backcalibrated to the experimental results by Swanson *et al.* (Rassati *et al.*, 2004; Swanson & Leon, 2000).

There are two sets of bolted beam-to-column connection specimens compared in this study, as summarized in Table 3.4 (Wu *et al.*, 2007; Wu *et al.*, 2008). The columns were composed of 400 × 400 mm square cross-section steel tubes with the thickness of 6 and 8 mm, respectively. The material of the steel tube is A572 Gr. 50. The steel tube was filled with concrete with the design compressive strength of 27.95 MPa. The beams of the specimens were made up of an H shaped cross-section of H500 × 200 × 10 × 16 mm. The material of the beam is A36 steel. The dimension of the end-plate is 25 mm (thickness) by 400 mm (width) by 720 mm (height). This end-plate was tied to the rectangular CFT columns with bolts. There are 8 bolts located above and below beam flanges, respectively, and in all 16 bars on the bidirectional end-plates. The bolts were A490 having a 30 mm diameter. The total length of CFT columns and the clear length of steel beams are 3750 mm and 1300 mm, respectively. The cyclic displacement loads were applied to the tip of the steel beam while the column tops and bottom were pinned.

The shear force and shear deformation of the panel zone obtained from the full connection tests (Wu *et al.*, 2008) were used to determine the stiffness of the shear panel zone in the joint element. Other modeling attributes for the material properties and analysis procedures were the same as those mentioned in Section 3.4.1. Two joint elements corresponding to two test specimens, FSBE 6 and FSBE 8 (Wu *et al.*, 2008), were generated for the numerical studies.

Total applied force vs. displacement curves are shown in Fig. 3.20. The cyclic curves of the connection tests for the test specimens are plotted as dotted lines, while those of the numerical studies are plotted as solid lines. According to the studies (Wu *et al.*, 2005; Wu *et al.*, 2007; Wu *et al.*, 2008), the welded beam ends cracked and then caused brittle failure shortly after the maximum load was achieved. The rapid strength degradation due to this failure type is evident from the experiment results. Cracking of any connection component was not taken into consideration in the numerical simulations, as the only welds in the innovative connections are the fillet welds in the end plate. There is no evidence that cracking will occur at this location when the fillet welds are properly fabricated. The simulations and experiments matched very well until the weld cracking began. As the cracking progressed and the strength degraded

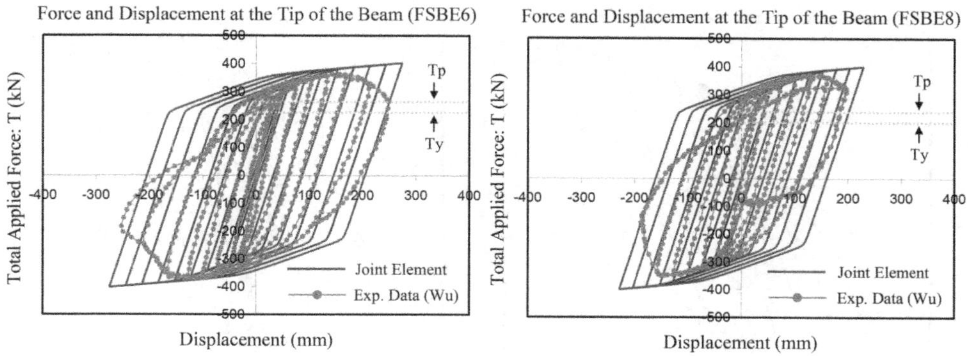


*Figure 3.20* Total applied force and displacement curves for the bidirectional beam-to-column connection.

in the experiment, the two curves diverged. Before that occurred, the energy absorption capacity is somewhat overestimated (about 15%) by the model but otherwise the results are in good agreement with respect to the initial slope, loading envelope, and the unloading/reloading slope. The initial yield load ($T_y$) and the plastic load ($T_p$) were determined by reference to the yield and plastic strength of the steel beam. Both curves exceed 250 kN, the limit of the plastic load. This indicates that connections performed as designed, i.e. as full strength connections. Further, the good fit between the simulation and experiment suggests that the joint element is adequate to predict the trend of the connection behavior.

### 3.4.3 Cyclic test results and observations

The 2D joint models for the connections were subjected to loads applied to the tip of the beam corresponding to the position of a loading actuator (see force T in Fig. 3.11). The size and length of the joint element models are the same as those of component members for test models or 3D FE models.

Comparisons between the monotonic results for the 3D FE models and those for the 2D joint element model are given in Fig. 3.21. Both show a good agreement in terms of initial stiffness, ultimate strength, and envelopes for both the force-displacement and moment-rotation behavior curves. For the cyclic behavior, the recentering effect can be observed during unloading procedures due to the tensile recovering forces in the SMA bars.

More results for the recentering effect, according to different percentages of the two bar materials, are given in Fig. 3.22. The model equipped only with SE-SMA tension bars shows excellent recentering capabilities. However, plastic deformation of the beam causes a permanent displacement in the moment vs. rotation curve. On the other hand, the behavior of the model with steel tension bars only shows much fuller hysteresis loops than that of the SE-SMA tension bars only model. As expected, the steel tension bars increase the energy dissipation capacity and provide improved resistance. The joint equipped with both steel tension bars and SE-SMA tension bars

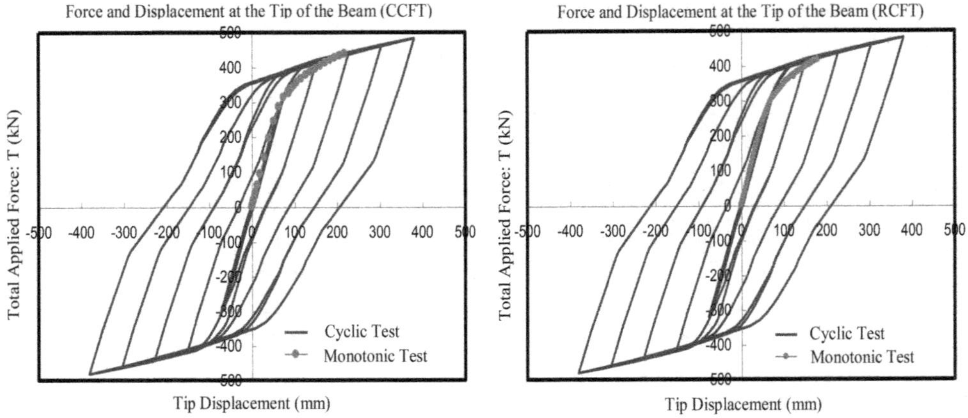


*Figure 3.21* Comparisons between cyclic tests and monotonic tests.

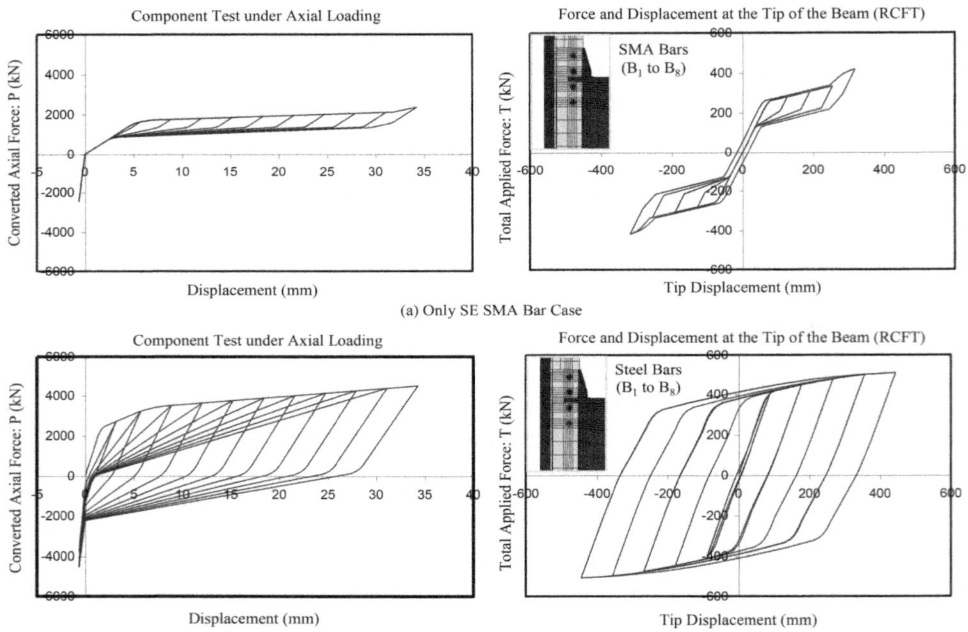


(a) Only SE SMA Bar Case

(b) Only Steel Bar Case

*Figure 3.22* Connection behavior results for different bar arrangements (SMA vs. Steel Bars).

takes advantage of both effects, i.e., shape memory effect for recentering and upgraded energy dissipation capacity.

Finally, comparisons of the connection behavior for two different connection types (a fully welded (FR/FS) connection and one with steel and SMA tension bars) are given in Fig. 3.23. For the static monotonic curves, the initial slope of the welded connection

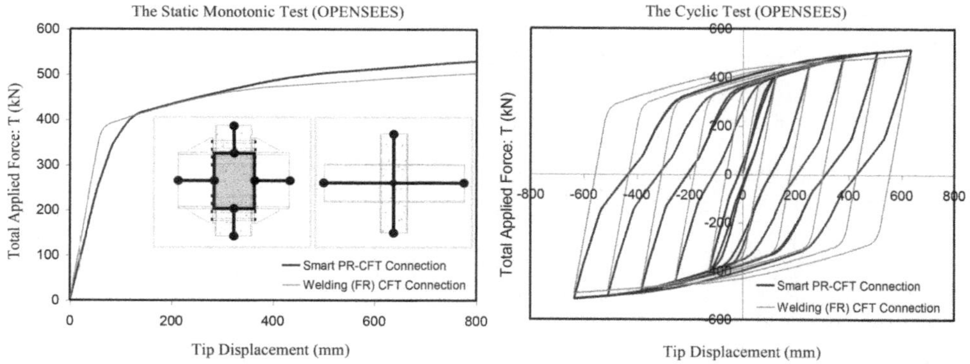


*Figure 3.23* Comparisons of connection behavior for different connection types (PR vs. FR).

is steeper than that of the smart PR-CFT connection. However, the welded connection evidences shallower hardening after yielding. The welded connection also shows much more permanent deformation during unloading.

## 3.5 COMPOSITE MOMENT FRAMES

Composite special moment frames (C-SMF) located on the perimeter of the building are economical and efficient seismic force resisting systems (SFRS) as recognized in Section 9/Part II of the current AISC seismic provisions (AISC, 2005). In these systems, deformations due to seismic actions are accommodated by the formation of hinges in the beams and stability is assured through the enforcement of strong column-weak beam mechanisms in design (Lee & Foutch, 2002; Tsai *et al.*, 2008). Typical connection configurations for these frames include fully restrained (FR) concrete filled tube (CFT) column-to-steel beam welded connections and bolted partially restrained (PR) connections as shown in Fig. 3.24. In these studies, the emphasis is on the design of low-rise (4 and 6 story) C-SMF using composite special bolted full strength, partially restrained (FS/PR) connections that incorporate steel and shape memory alloy (SMA) through rods as the main flexural connection elements. Companion frames with welded connections (FR/FS) were also designed and analyzed to provide comparison data.

Typical C-PRMFs (Section 8/Part II of [1]) are composed of I-shape steel columns and composite steel beams which are interconnected with PR composite connections (Leon & Kim, 2004; Thermou *et al.*, 2004). In these systems, the connections are generally flexible and all the yielding is concentrated in the connection elements themselves. In this paper we will refer to C-PRMFs which are composed of CFT columns and steel beams with innovative SMA PR composite connections (Lee & Foutch, 2002; Wu *et al.*, 2005; Wu *et al.*, 2007; Braconi *et al.*, 2007) as shown in Fig. 3.24(c). In the original investigation (Hu, 2008), three variations of the connection shown in Fig. 3.24(b) were studied: (1) top-and-seat angles, (2) T-stubs, and (3) end plates. These three connections encompass the range of PR behavior, from very flexible to very stiff, respectively. The end plate connections are basically rigid connections and

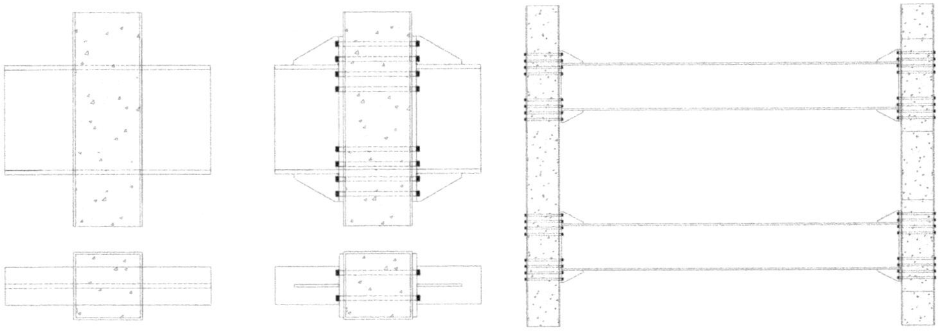

(a) The detail of FR-CFT column-to-steel beam connection

(b) The detail of the bolted PR-CFT connection

(c) Composite PR moment frames using bolted PR-CFT connections

*Figure 3.24* Connection details and composite PR moment frames (C-PRMF).

recognized as such by design specifications (AISC, 2005). Only end plate connections, for which the development of the connection model is described in a companion paper (Hu & Leon, 2008), are treated in this paper.

C-PRMFs were originally conceived for areas of low to moderate seismicity in seismic design category (SDC) C and below. The SDC assigned to a building is a classification based upon the occupancy class and the seismicity of the site (ASCE, 2002). However, C-PRMFs can be used in areas of higher seismicity (Leon, 1990) with appropriate detailing and treated as C-SMF if the connection capacity at an interstory drift of 0.04 radians exceeds 80% of the flexural capacity of the beam and the behavior of the connection is modeled explicitly. Recent research on similar bidirectional bolted connections for CFT columns and H-beams has shown superior stiffness, strength, ductility, and energy dissipation characteristics (Wu *et al.*, 2007). These results demonstrate that the seismic resistance exceeds the requirements specified in the seismic design codes of Taiwan and USA (Lee & Foutch, 2002; Tdai & Hsiao, 2008; Azizinamini & Schneider, 2001). Therefore, C-PRMFs systems can perform well and can become a practical option in design.

A companion paper (Hu & Leon, 2008) describes the behavior and development of simple models for the proposed new connections. The studies included initial detailed 3D advanced FE models of the connection subjected to pushover loads. These detailed models were then used to develop simplified 2D component models for connections subjected to cyclic loads. These models were implemented in the Open System for Earthquake Engineering Simulation (OpenSEES), an open source program widely used in the USA for this type of study (Mazzoni *et al.*, 2006) and calibrated to tests data. These studies indicated that good recentering capabilities and overall excellent hysteretic behavior can be achieved by these connections.

This paper describes the design of two prototype C-MF models designed in accordance with a new proposed procedure (AISC, 2005; ASCE, 2002; AISC, 2005; ICC, 2003). Both pushover and nonlinear dynamic analyses are then conducted and the dynamic response of these new frames is quantified.

*Table 3.5* The basic conditions applied to the composite moment frames (C-MF).

| *Located Area* | *Gravity Loads* | *SDC* | *Building* | *Occupancy Category* |
|---|---|---|---|---|
| LA Area | Dead: 4.79 kPa Live: 3.83 kPa | D Class | 4 or 6 Composite Building | Ordinary Structures |

## 3.6 DESIGN METHODOLOGY FOR COMPOSITE MOMENT FRAMES

### 3.6.1 Frame design

Two sets of four and six story frames were designed. For each of the frame heights, two connection types were used between steel beams and circular or square concrete-filled columns: (1) welded connections used for benchmark frames (henceforth C-SMF), and (2) connections with a mix of steel and SMA through tension bars (henceforth C-PRMF) (Hu & Leon, 2008). Building configurations, materials, and modeling conditions for the C-PRMFs are the same as those for C-SMF in order to compare both frame models in terms of inelastic behavior, strength, stability limits, and rehabilitation. The dead loads, live loads, seismic design, and occupancy category used for all prototype building are given in Table 3.5. The SDC is assumed to be a high seismicity area as defined for class D in the 2003 International Building Codes (IBC 2003). The design was based on the mapped maximum considered earthquake spectral acceleration for the Los Angeles (LA) area.

### 3.6.2 Building configuration

Four and six story buildings with three by five bays were utilized in this research (Fig. 3.25(a)). Typical elevations of a six-story building and a four-story prototype building are shown in Fig. 3.25(b) and (c). The story height and bay length were kept constant at 3.96 m and 10.97 m, respectively. Moment resisting frames in the six-story building were installed in the same perimeter positions as those in the four-story building. Overall, structural models of both 4 and 6 story buildings have symmetric configurations in all story levels.

Moment resisting frames are shown as thick lines in the building plan shown in Fig. 3.25(a). All connections are assumed as moment resisting, except for those in interior floor beams in the EW directions. A 2D perimeter moment resisting frame enclosed by the dotted line in Figure 2(a) was selected for the numerical model. The plan views of the three variations of perimeter frames studied (labeled C1 through C3 in Table 3.6) are illustrated in Fig. 3.26. C3 represents a mix of column types for which the center most columns could be used as part of a biaxial system.

Initial member sizes and design combinations are summarized in Table 3.6. The first number of the acronym shown in the model ID indicates the total numbers of stories (i.e. 4 or 6). The letter of the acronym following this number represents the connection type (END: end-plate connections, WED: welded connections). The last letter indicates the column combination as illustrated in Fig. 3.26. For instance, the

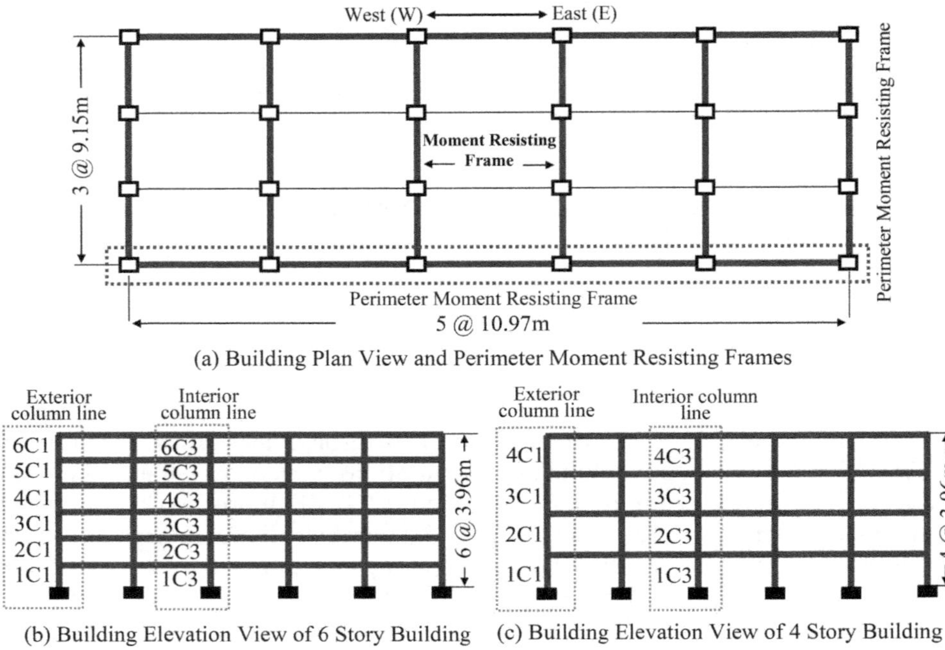


*Figure 3.25* Elevation and plan view for prototype building.

model ID for the 6 story composite moment with end-plate connections having RCFT column systems is denoted as the 6END-C1.

Material for all steel component members is assumed as A572 Gr. 50 with 1.5 percent strain hardening. The concrete in the column sections was assumed as the normal weight (NW) concrete with a 28-day 27.6 MPa design compressive strength. For the end-plate connections, A 490 high strength bolt and super-elastic (SE) Nitinol material (DesRoches *et al.*, 2004) were used for the steel tension bars and SMA tension bars, respectively. Smaller beam sizes were designed for the higher stories in order to achieve an economical design. In addition, beam and column sizes presented here are very close to those presented in the 3D FE models described in the companion paper (Hu & Leon, 2008).

### 3.6.3 Seismic design loads

The structures were designed as special moment frames for the area in accordance with both the AISC-LRFD manual (AISC, 2001) and the AISC 2005 Seismic Provisions (Tsai *et al.*, 2008). Load combinations including only dead (DL), live (LL), and earthquake (E) loads were selected for the nonlinear pushover analyses because earthquake loads dominate over wind loads in the LA area. Load combination 5 ($LC5 = 1.2DL + 1.0LL + 1.0E$) stipulated in the ASCE 7-02 (ASCE, 2002) dominated over all other load combinations in this study.

*Table 3.6* The initial design selections for composite frame buildings.

| *Model ID* | *Connection Type* | *Column System* | *Column Size (All Stories)* | *Beam Size 1st to 3rd Story* | *4th and 6th Story* |
|---|---|---|---|---|---|
| (a) 6 Story Composite Moment Frames | | | | | |
| 6END-C1 | End Plate | RCFT | HSS16 × 16 × 500 | W24 × 103 | W24 × 84 |
| 6END-C2 | End Plate | CCFT | HSS18 × 500 | W24 × 103 | W24 × 84 |
| 6END-C3 | End Plate | RCFT + CCFT | HSS16 × 16 × 500, HSS18 × 500 | W24 × 103 | W24 × 84 |
| 6WED-C1 | Welding | RCFT | HSS16 × 16 × 500 | W24 × 103 | W24 × 84 |
| 6WED-C2 | Welding | CCFT | HSS18 × 500 | W24 × 103 | W24 × 84 |
| 6WED-C3 | Welding | RCFT + CCFT | HSS16 × 16 × 500, HSS18 × 500 | W24 × 103 | W24 × 84 |
| *Model ID* | *Connection Type* | *Column System* | *Column Size (All Stories)* | *Beam Size 1st and 2nd Story* | *3rd and 4th Story* |
| (b) 4 Story Composite Moment Frames | | | | | |
| 4END-C1 | End Plate | RCFT | HSS16 × 16 × 500 | W24 × 103 | W24 × 84 |
| 4END-C2 | End Plate | CCFT | HSS18 × 500 | W24 × 103 | W24 × 84 |
| 4END-C3 | End Plate | RCFT + CCFT | HSS16 × 16 × 500, HSS18 × 500 | W24 × 103 | W24 × 84 |
| 4WED-C1 | Welding | RCFT | HSS16 × 16 × 500 | W24 × 103 | W24 × 84 |
| 4WED-C2 | Welding | CCFT | HSS18 × 500 | W24 × 103 | W24 × 84 |
| 4WED-C3 | Welding | RCFT + CCFT | HSS16 × 16 × 500, HSS18 × 500 | W24 × 103 | W24 × 84 |

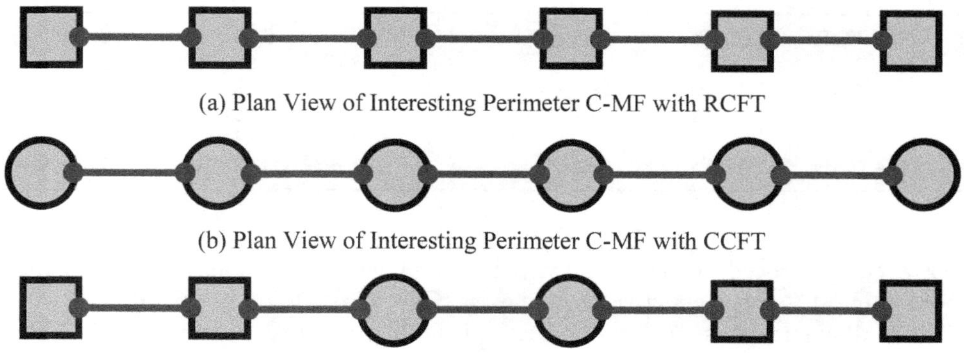

(a) Plan View of Interesting Perimeter C-MF with RCFT

(b) Plan View of Interesting Perimeter C-MF with CCFT

(c) Plan View of Interesting Perimeter C-MF with RCFT and CCFT

*Figure 3.26* The plan views of interesting perimeter C-MF with CFT column combinations.

The seismic performance on the frame structures was first estimated through pushover analyses. The distributions of equivalent lateral loads for the pushover analyses were determined by a set of static lateral loads that corresponded to the 1st mode shape of deformation as introduced in the ASCE 7-02 (ASCE, 2002) and the IBC 2003 codes (ICC, 2003). For frame structures, the first mode generally contributes upwards of 90% of the effective seismic mass and dominates the behavior of the structure.

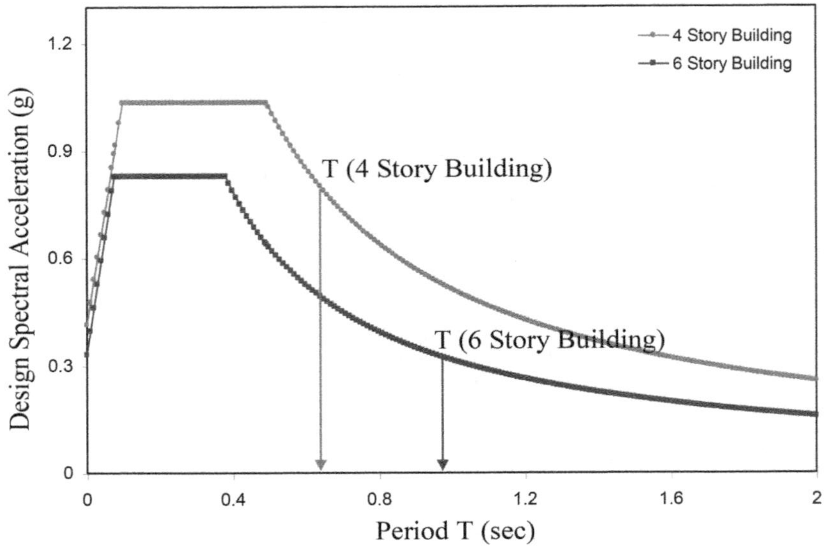


*Figure 3.27* Design response spectra for C-MF in LA area.

*Table 3.7* Design loads (Dead, Live, and Equivalent Lateral Load).

| *Story* | *Dead Load (DL)* | *Live Load (LL)* | *Equivalent Lateral Load (E)* | *Story* | *Dead Load (DL)* | *Live Load (LL)* | *Equivalent Lateral Load (E)* |
|---|---|---|---|---|---|---|---|
| 1 | 4.79 kPa | 3.83 kPa | 161.5 kN | 1 | 4.79 kPa | 3.83 kPa | 87.3 kN |
| | | | | 2 | 4.79 kPa | 3.83 kPa | 167.8 kN |
| 2 | 4.79 kPa | 3.83 kPa | 256.2 kN | 3 | 4.79 kPa | 3.83 kPa | 219.2 kN |
| 3 | 4.79 kPa | 3.83 kPa | 264.7 kN | 4 | 4.79 kPa | 3.83 kPa | 232.5 kN |
| | | | | 5 | 4.79 kPa | 3.83 kPa | 202.9 kN |
| 4 | 4.79 kPa | 3.83 kPa | 180.6 kN | 6 | 4.79 kPa | 3.83 kPa | 126.4 kN |

However, this procedure may not always be valid when higher mode shapes contribute more than 10% of the effective seismic mass (Chopra & Desroches, 2008).

The design response spectra for area code 90045 (LA) is shown in Fig. 3.27. A soil site class A (hard rock) was used for the 4 story building, while a soil site class C (soft rock) was used for the 6 story building. The fundamental time period of the building was computed by simplified analysis based on the code equations (i.e. ASCE 7-02 Section 9.5.5.3 (ASCE, 2002)). Therefore, the same fundamental period (T) was used for the design of all buildings with the same number of stories regardless of the connection types ($T = 0.66$ sec for 4 story building and $T = 0.91$ sec for 6 story building).

The seismic base shear force ($V_{Base}$) was determined by the simplified lateral force procedure (ASCE 7-02, Section 9.5.4 (ASCE, 2002)). The equivalent story forces were calculated based on the portion of the seismic dead load located at each story level. The values for both the equivalent lateral and gravity loads are summarized in Table 3.7. Finally, design checks for the initial selections of the member sizes were performed to

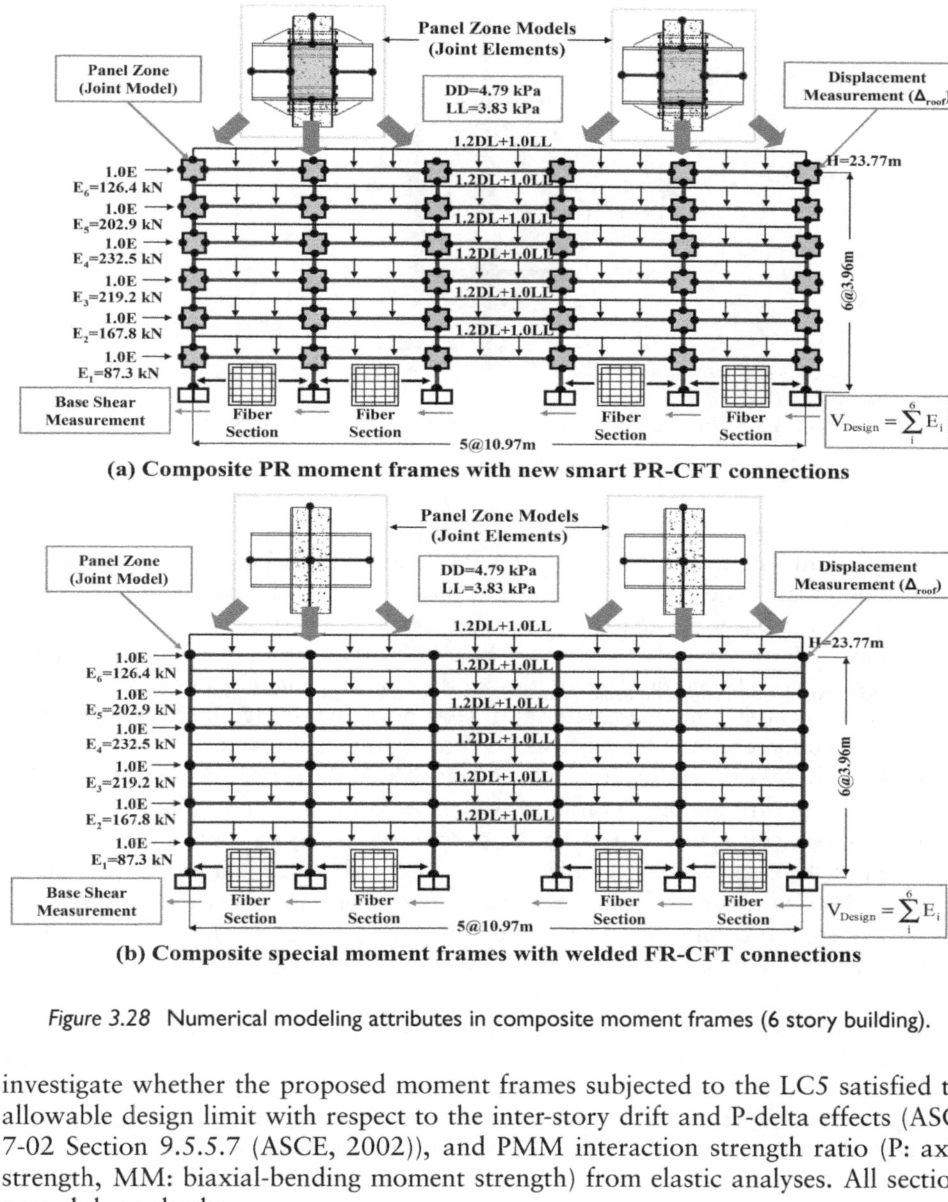


*Figure 3.28* Numerical modeling attributes in composite moment frames (6 story building).

investigate whether the proposed moment frames subjected to the LC5 satisfied the allowable design limit with respect to the inter-story drift and P-delta effects (ASCE 7-02 Section 9.5.5.7 (ASCE, 2002)), and PMM interaction strength ratio (P: axial strength, MM: biaxial-bending moment strength) from elastic analyses. All sections passed these checks.

### 3.6.4 Numerical modeling attributes

Because of symmetry and the assumption of rigid floor diaphragms, the 2D perimeter frames shown inside the dotted box in Fig. 3.25(a) can be considered as representative of the typical frame behavior. Fig. 3.28 shows the modeling attributes for the applied

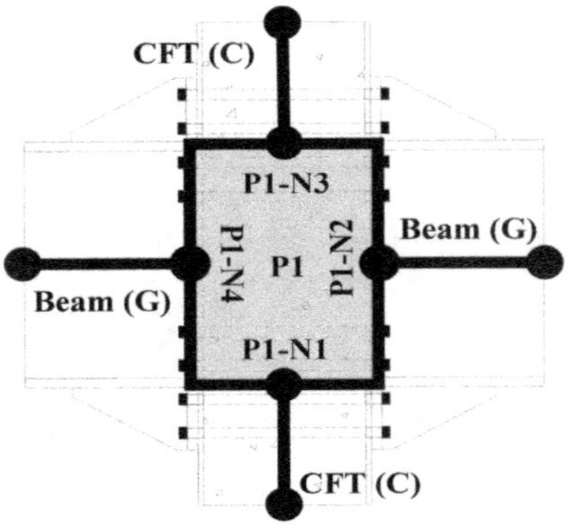


*Figure 3.29* The details of the joint element for SMA PR-CFT connections.

load combination (LC 5), elements, cross-sections, panel zones, numerical analyses, and response measurements using the OpenSEES program (Mazzoni *et al.*, 2006).

The dominant load combination was applied to the steel beams and panel zones for the nonlinear pushover analyses. The CFT columns and steel beams were modeled as nonlinear 2D discrete fiber sections (Hu, 2008; Mazzoni *et al.*, 2006; Padgett & Desroches, 2008). The material properties for the steel and concrete members were simulated by the default nonlinear material models provided in the OpenSEES program. A user defined material model was used for the SMA tension bars.

One of the most significant characteristic of the 2D frame models proposed in this study is the careful consideration of the panel zone modeling. The real behavior of the composite PR-CFT connections was replicated using a robust joint element as described in the original investigation (Hu, 2008) and the companion paper (Hu & Leon, 2008). The structural details of the joint element are given in Fig. 3.29. The nonlinear beam-to-column elements were attached to 4 external nodal points of the joint element. The behavior of the panel zones was defined by the tri-linear model shown in Fig. 3.30. The required information for this tri-linear model, such as initial stiffness ($K_{pro}$), yield shear strength ($V_{ypro}$), and ultimate shear strength ($V_u$), was estimated by using the equations proposed by Wu *et al.* (Wu *et al.*, 2007). These equations take into account both (a) the stiffness loss due to the bolt holes and (b) the superposition of the strength between two materials due to the composite effect into consideration so as to simulate the shear force-deformation of the panel zone accurately. This tri-linear model can be simulated by using the hysteretic material in the OpenSEES program. The properties of panel zones for all C-PRMF models are summarized in Table 3.8.

The frame models for the C-SMFs with welded connections were only composed of nonlinear beam-column elements. The beams and columns extended from centerline to centerline and met together at a node as shown in Fig. 3.28(b), but rigid end-offset were used in the beam elements. In addition, shear distortions at the panel zones were neglected in these frames.

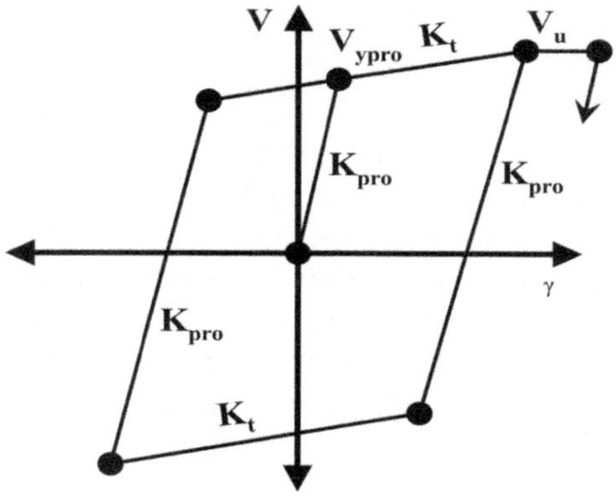


*Figure 3.30* The tri-linear model to simulate the behavior of the panel zone.

*Table 3.8* The shear strength and stiffness information of the panel zone models (PR Connections).

| *Connection Type* | *Beam Size* | *Column Size* | *PZ Size** | $V_{ypro}$** | $K_{ypro}$** | $V_u$** | $K_t$ |
|---|---|---|---|---|---|---|---|
| End-Plate | W24 × 103 | HSS16 × 16 × 500 | 41 × 41 (1.3)* | $7.814 \times 10^3$ | $3.031 \times 10^6$ | $8.291 \times 10^3$ | $0.01K_{ypro}$ |
| | | HSS18 × 500 | 46 × 46 (1.3)* | $9.370 \times 10^3$ | $3.580 \times 10^6$ | $9.877 \times 10^3$ | $0.01K_{ypro}$ |
| | | HSS16 × 16 × 500 | 41 × 41(1.3)* | $7.743 \times 10^3$ | $3.002 \times 10^6$ | $8.228 \times 10^3$ | $0.01K_{ypro}$ |
| | W24 × 84 | HSS18 × 500 | 46 × 46 (1.3)* | $9.508 \times 10^3$ | $3.586 \times 10^6$ | $1.012 \times 10^4$ | $0.01K_{ypro}$ |

*The unit is cm.
**The unit is kN. ( )* indicates the thickness of steel tube section.

The general modeling methods follow those introduced in the guidelines given in FEMA 355C (FEMA, 2000), and include the following assumptions:

- All frame models were designed in accordance with a strong column-weak beam capacity philosophy so that the plastic yielding of the beam is the dominant deformation mechanism.
- A mass corresponding to 1.0DL (Dead Load) + 0.2 LL (Live Load) was applied for the nonlinear dynamic analyses.
- All steel members included 1.5% strain hardening.
- 2.5% Rayleigh damping was used in the first mode.
- Soil-structure interaction at the ground support was neglected.
- The uniform dead and live loads on the beams were converted into equivalent point loads.
- The plastic zone in the CFT columns was determined by measuring the fiber stresses at all cross-sections along the member length.

The primary response data obtained from the non-linear analyses included the nodal displacements, member reaction forces, base shear forces, and fiber stresses. These were collected by utilizing the recorder command in OpenSEES.

## 3.7 NONLINEAR ANALYSES

### 3.7.1 Introduction

The frame models described above were used to examine the seismic behavior and performance for both new C-PRCF and C-SMF structures by means of both nonlinear pushover analyses and nonlinear dynamic analyses. A total of 24 nonlinear pushover analyses (12 static and 12 cyclic pushover analyses) and 120 nonlinear dynamic analyses were carried out. The main interest in nonlinear analyses is to show in detail the ability of the C-PRMF connections using SMA tension bars to improve frame performance. The benefits of C-PRMF were verified by comparing their behavior under a performance-based framework with their corresponding welded counterpart (e.g. 6END-C1 vs. 6WED-C1).

The dominant load combination, LC5, was used to perform the nonlinear pushover analyses. The detailed load profiles for each frame model are summarized in Table 3.7. Equivalent point loads, simulating the uniform dead and live loads, were applied to the beam elements in the gravity directions using the Constant Time Series function associated with one of the load patterns in the OpenSEES program. The equivalent lateral loads on the joints were simulated by using the Linear Time Series function, so these loads can be applied in a linearly static- or cyclic-incremental fashion associated with a predefined time step. For each time step, a static or cyclic pushover analysis was conducted through a displacement control algorithm (Mazzoni *et al.*, 2006).

Two SAC suites of ground motions (LA21 to LA30 and SE21 to SE30) corresponding to a seismic hazard level of 2% probability of exceedence in 50 years for the western US area (Los Angeles (LA) and Seattle area (SE)) were used to conduct the nonlinear dynamic analyses. These 20 ground motions were developed from both historical records and simulations as part of the FEMA/SAC project on steel moment frames (Somerville *et al.*, 1997). The peak ground acceleration (PGA) ranged from 0.42 g for LA21 to 1.75 g for SE27. To solve the time dependent-dynamic problem, a transient equilibrium analysis was performed using the Newmark method (Newmark, 1959). A value of 2.5 percent was used for the damping as defined by the Rayleigh command in the OpenSEES program (Mazzoni *et al.*, 2006). In order to include second order effects (P-Delta effect) due to dead and live loads along the gravity direction, equivalent point loads converted from these gravity loads were also applied to the beam elements. In addition to the gravity loads used to model the P-Delta effects, lumped masses were assigned to nodes so as to generate the story shear forces due to the ground acceleration. Lumped masses consisted of 1.0 times dead loads plus 0.2 times live loads.

A schematic view of the data collection for the composite frame performance is depicted in Fig. 3.31. The important performance parameters selected for comparisons are the inter-story drift ratios (ISDR), forces, deformations, and fiber stresses for key members. The ISDR was calculated from the results of the global deformation data. The ISDR were defined as the ratio of the story displacement to its height as shown in Fig. 3.31. The base shear force ($V_{Base}$) is the summation of the reaction forces at the nodes corresponding to the column bases. Member reaction forces and fiber stresses measured at the integration points of the nonlinear beam-column elements

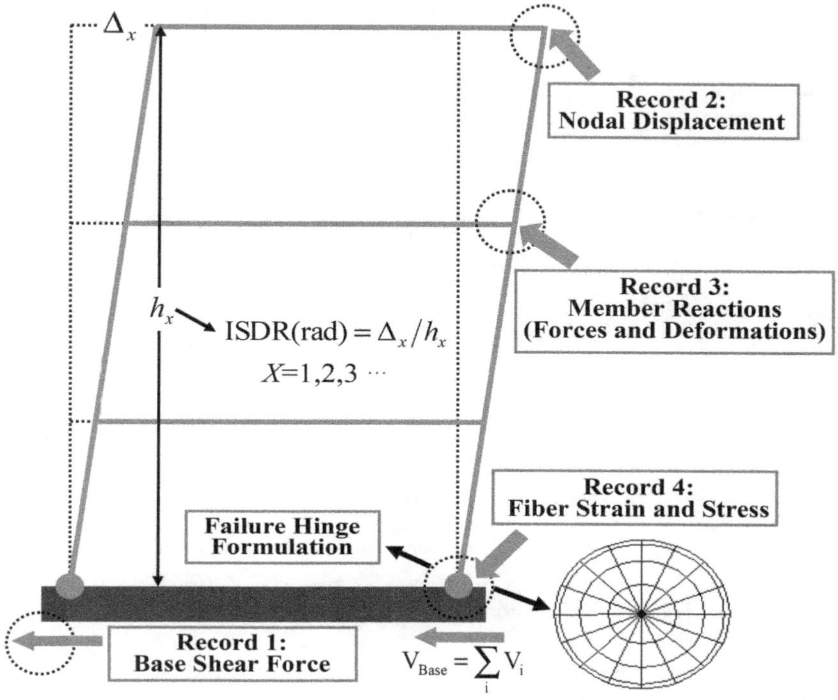


*Figure 3.31* Schematic view of primary indices.

were used to calculate elastic strength ratios (ESR) and for tracing the sequence of hinge formation.

### 3.7.2 Nonlinear pushover analysis

The resulting static pushover curves plotted in terms of the ISDR at the roof level vs. the base shear force normalized by the design base shear force ($V_{Base}/V_{Design}$) are shown in Fig. 3.32. The design base shear force ($V_{Design}$) is defined as the summation of the equivalent lateral loads (E) shown in Table 3.7. The design base shear forces for the 4- and 6-story frames are 859 kN and 1041 kN, respectively. The figures show comparisons of static pushover curves for composite moment frames with the same composite column systems, but different connection types.

There are some important transition points in the static pushover curves, which can be related to the ISDR. The limits to determine these transition points consist of the elastic range (proportional limit), initial yielding, initiation of strength hardening, ultimate strength, and strength degradation or the stability limit. For example, the 6 END-C1 model indicates that an ISDR of approximately 0.01 radians is the limit for the elastic range; that an ISDR, of about 0.02 corresponds to the yielding points

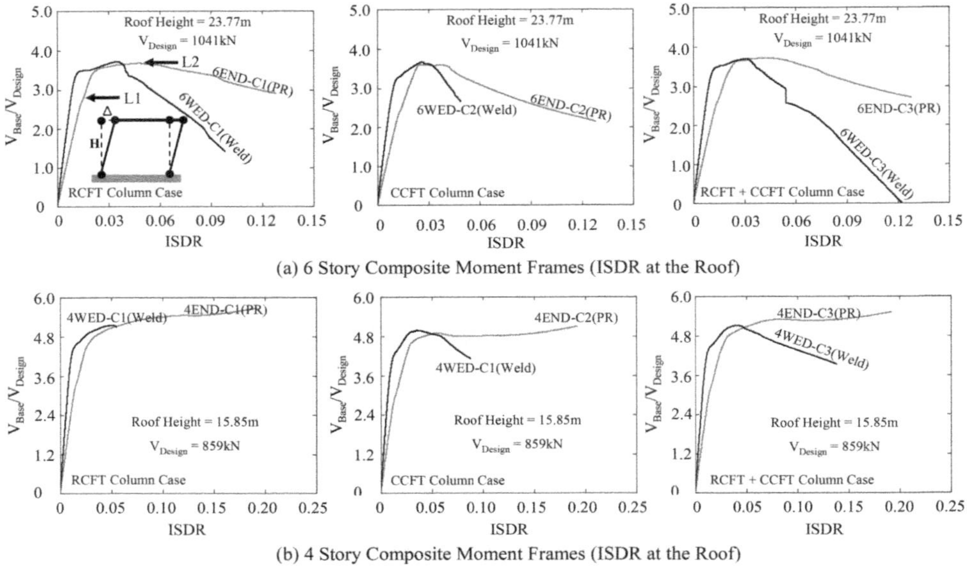


*Figure 3.32* The nonlinear monotonic pushover curves.

(L1 at the 6END-C1 model); that an ISDR of 0.03 corresponds to initiation of strength hardening; that an ISDR of 0.05 corresponds to the ultimate strength (L2 at the 6END-C1 model): and that an ISDR of approximately 0.09 corresponds to the stability limit.

From all pushover curves, the initial slope of the composite moment frames with welded connections is steeper than that of the composite moment frames with PR connections (e.g. 6WED-C1 vs. 6END-C1). The stiffness loss due to the oversize bolt holes in the panel zone and the structural characteristics of PR connections cause the composite moment frames with PR connections (i.e. 6END-C1 to 6END-C3) to have lower initial stiffness. An abrupt instability of the whole system due to the significant increasing P-Delta effect occurs in the welded frames immediately following this point. This is due to the assumed maximum rotational capacity of these connections. In contrast, connections with the more flexible tension bars provide more ductility and additional resistance. Therefore, the composite frames with PR connections show more gradual strength degradation after reaching their ultimate strength.

The strength of the composite columns, as represented by the axial force and bending moment interaction (P-M interaction) diagram, has great influence on the performance of the moment frames. The larger column sizes obviously increase the resistance against lateral loads. CCFT columns were designed with slightly smaller steel area and reinforcement (roughly 5% less) than the RCFT columns. As a result, the strength of composite frames with CCFT columns is identical or slightly smaller than that of composite frames with RCFT columns (e.g. 6END-C1 vs. 6END-C2). Composite moment frames with CCFT columns are susceptible to stress concentration at the panel zone because the rectangular shape panel zone was welded to the circular

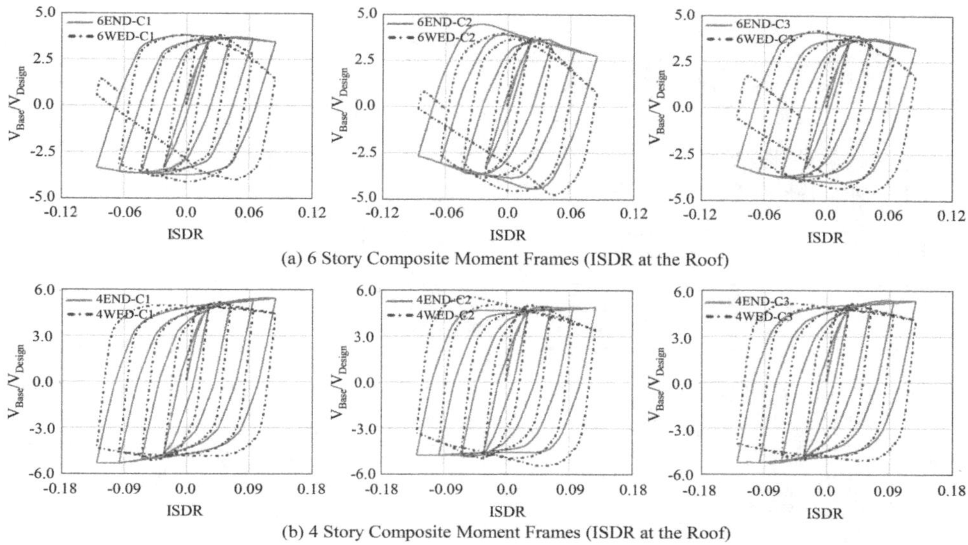


*Figure 3.33* The nonlinear cyclic pushover curves.

columns. This characteristic of panel zones causes the strength to deteriorate more rapidly. Finally, the strength of the taller frames (6 story moment frames) deteriorates more rapidly than that of the shorter frame (4 story moment frames) due to the larger P-Delta effect by larger gravity loads (roughly 33% larger in total). Comparisons between the pushover curves for 6END-C3 and those for 4END-C3 provide good examples to support this argument.

The same numerical models that were analyzed for monotonic loads were also used for cyclic pushover analyses. The modeling decisions and data collected, where possible, were identical to those for the static pushover tests. The energy dissipation capability as well as stiffness, strength, and permanent deformation for the moment frames can be estimated via cyclic pushover curves.

The cyclic pushover curves for all composite moment frames are shown in Fig. 3.33. Overall, the envelope of the static curve corresponds to that of the cyclic curve when the same models are tested. As expected, all transition points and limits obtained by the static pushover test are equal to those obtained by the cyclic pushover test. This illustrates an important limitation of these pushover analyses, which cannot capture substantial strength degradation unless the monotonic ultimate strength is reached. From all cyclic curves, the unloading slopes were taken as equal to the initial slope.

Composite moment frames with PR connections show smaller residual displacement than those with welded connections after unloading. This benefit results from the recentering effect due to the shape memory rods (Ocel *et al.*, 2004; Leon & Hu, 2008). Similarly, composite frames with PR connections show gentler strength degradation after reaching their ultimate strength than those with welded connections

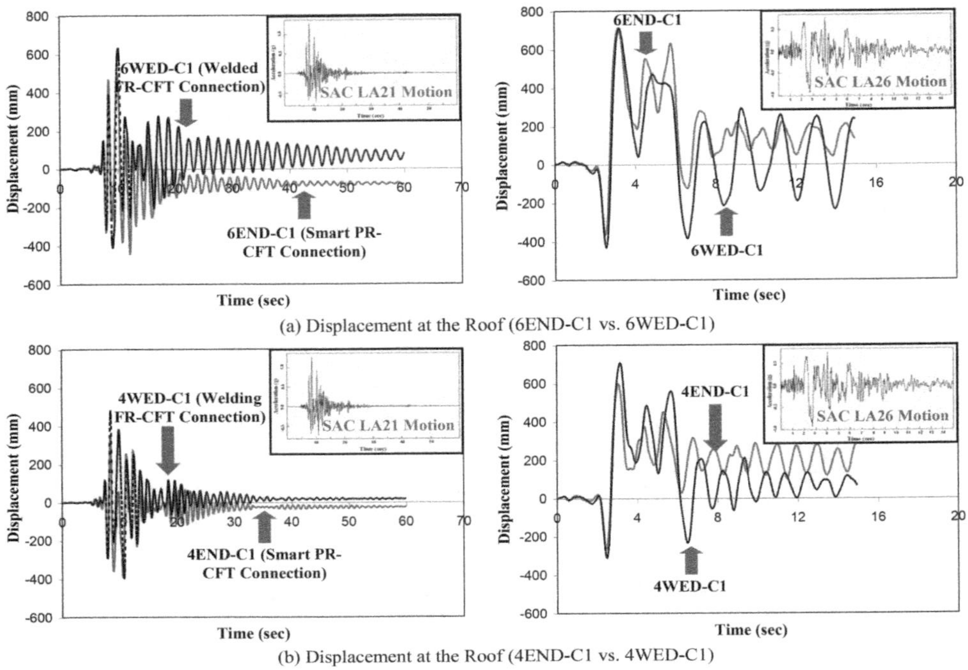


*Figure 3.34* Results of nonlinear dynamic analyses (Displacement time histories).

(e.g. 6END-C1 vs. 6WED-C1). Larger energy dissipation capability can be expected at the cyclic behavior of PR connection frames.

### 3.7.3 Nonlinear dynamic analysis

12 composite moment frame models were also used for the nonlinear dynamic analyses. For a comparison between the behavior of the PR connection frames and that of the welded FR frames, two ground motions, LA 21 and LA26, were selected. The LA 21 ground motion has a relatively long duration (60 second) with a PGA value of 1.28 g, while the LA26 ground motion has a relatively short duration (15 second) with a PGA of 0.944 g. Displacement time histories of the roof level are shown in Fig. 3.34. Generally, composite moment frames with PR connections show smaller peak roof displacements than those with welded connections under strong ground motions. Under the LA21 ground motion, the peak roof displacements for the 6END-C1 and 6WED-C1 model are approximately 480 and 600 mm, respectively, indicating roof ISDR of 0.02 and 0.025 radians. The occurrence time for each peak value was slightly different for the different frames and lagged behind the PGA (roughly 0.2 second time lag). The smaller peak roof displacements associated with PR connection frames are attributed to the gentle strength degradation and the recentering capability of the used SMA material. The ability of the new PR connections to restore the structure

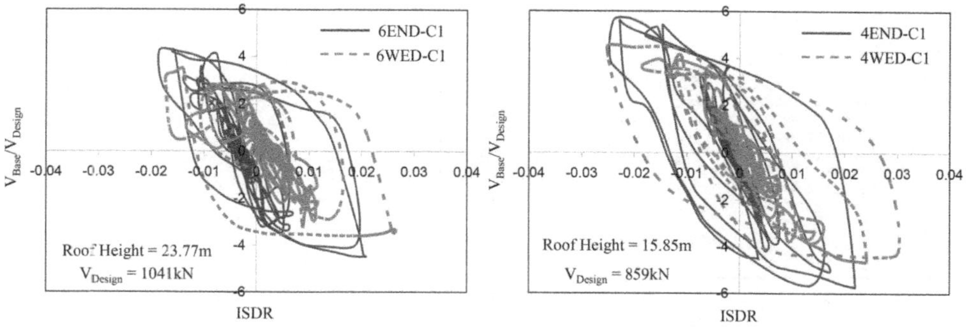


*Figure 3.35* Results of nonlinear dynamic analyses (Normalized base shear vs. ISDR).

to its initial conditions also reduces the peak displacement and amplitude during the ground motion.

Fig. 3.35 shows the influence of the applied ground motion and P-Delta effects. Larger maximum ISDRs occur in the frames with welded connections after the maximum strength ($V_{Base}/V_{Design} = 3.90$) is reached. The ISDR is 0.025 radians for the 6WED-C1 model, whereas it is 0.02 radians for the 6END-C1 model. This is, due to the primarily to the recentering capability, and gentle strength degradation guaranteed by the use of new PR connections.

## 3.8 DAMAGE EVALUATIONS

### 3.8.1 Performance-based evaluation

Based on the results of the non-linear analyses, a more comprehensive study for PR and welded connection frames was conducted to assess their seismic performance and the extent of structural damage suffered. The performance levels resulting from the transition points on the static pushover curves (Fig. 3.32) were used as the drift limits. The ISDRs at the yield and ultimate level are shown in Fig. 3.36. The applied performance levels refer to the difference in the base shear forces (V). For example, V = 2820 kN and V = 3434 kN were the yield strength level of 6END-C1 and that of 6WED-C1, respectively. All frames were stable up to the yield strength level. After reaching the ultimate strength level, plastic deformations began to migrate to the CFT column sections and concentrate on the lower stories as illustrated in Fig. 3.36(b). The maximum ISDR for the 6 story moment frames occurs at the 3rd story at the yield strength level, but it gradually moves to the lowest level as the lateral loads increase. The increased plastic deformations at the lower story cause this shift.

The findings of the nonlinear pushover tests are consistence with the ISDR investigations. The composite moment frames with welded connections show a stiffer initial slope than those with bolted PR connections in the nonlinear pushover curves (see Fig. 3.32). As a result, the composite moment frames with welded connections include

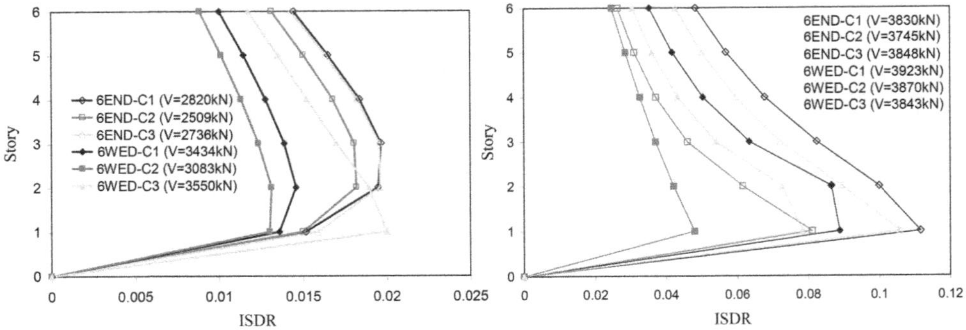


(a) ISDR at the yield base shear point (b) ISDR at the ultimate base shear point

*Figure 3.36* Inter story drift ratios at the measurement points (6 story frames).

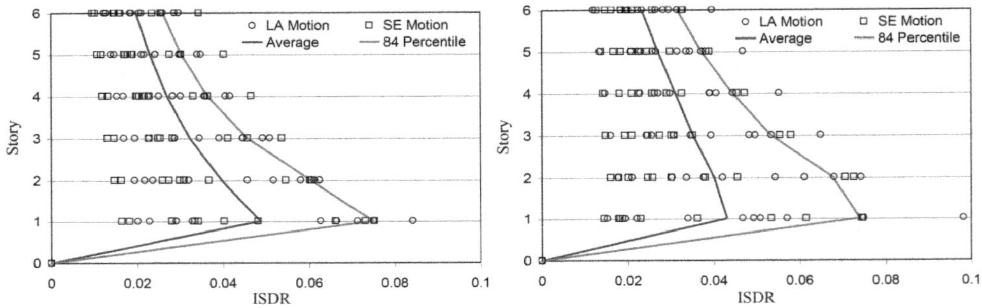


(a) 6END-C1 Model (b) 6WED-C1 Model

*Figure 3.37* The peak ISDR under seismic loads with ground motions.

smaller inter-story drift than those with PR connections within the yield strength level which belongs to the elastic range of the nonlinear pushover curves.

The composite moment frames with welded connections had a lower average ISDR (approximately 0.015 radians) as compared to those with PR connections (approximately 0.019 radians.) The composite moment frames with welded connections had lower plastic deformation capacity, resulting in relatively smaller ISDR values (roughly 40% less at the 1st story), as compared to those with PR connections under the ultimate strength level (see Fig. 3.36(b)). Similar to the nonlinear pushover test results (Fig. 3.32), the investigation of the ISDR under the ultimate strength level indicates that the frames with new PR connections perform well due to the excellent ductility and energy dissipation of the new connection system.

In order to examine the dynamic performance, the peak responses were investigated. The graphs of the scatter data for peak ISDR are provided in Fig. 3.37. These graphs show the average and 84 percentile ISDR together with individual peak data points obtained from all dynamic analysis results (LA21 to LA30 and SE21 to SE30). The values of the 84 percentile are employed from here on to indicate the statistical

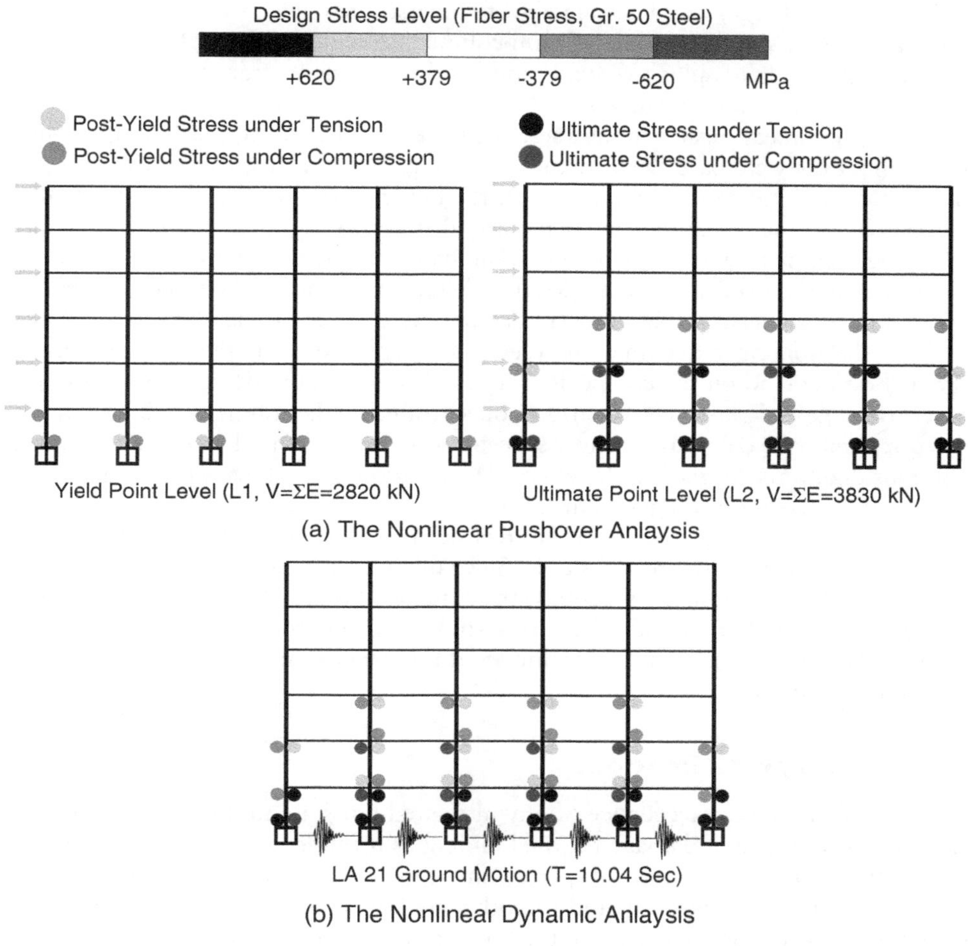


*Figure 3.38* Investigation of the failure hinges during the nonlinear analyses (6END-C1).

values of the peak ISDR as defined in FEMA 355C (FEMA, 2000). Similarly to the ISDR at the ultimate point obtained by the nonlinear pushover analysis, the peak ISDR obtained from the dynamic analysis show the largest values at the first story level and diminished with height. This implies that the composite columns located in the lower story levels are susceptible to severe plastic deformations under these ground motions. As expected, the composite moment frame with welded connections (6WED-C1) shows slightly higher statistical values of the peak ISDR (approximately 10%) than that with new PR connections (6END-C1).

A typical sequence of hinge formation during nonlinear analyses is shown in Fig. 3.38. The stress level was determined from the base steel materials in the composite columns. The expected strength values (Swanson *et al.*, 2002), rather than the nominal design strength ones, were used (379 MPa and 620 MPa for the yield and ultimate stress, respectively). The state of the hinge (i.e. the yield stress hinge or the ultimate failure hinge) was determined form fiber stresses, which were measured at the integration points of the nonlinear beam-column elements. Both axial forces transmitted from gravity loads plus live loads and bending moments transmitted from lateral loads act on the column members.

In the nonlinear pushover analyses, yield stress hinges start to occur at the column bases under the yield base shear force (L1, see Fig. 3.32(a)). At this performance level, only yield stress hinges occur and only in the lower story level. As the applied lateral loads are increased, bending moments are also increasing at the column members. Thus, bending moments contribute predominantly to creating hinges at the ultimate performance level (L2). These hinges then begin to extend to the upper stories. The column bases are susceptible to severe damages due to the ultimate failure.

For the nonlinear dynamic analyses, the hinges were computed at the time of the highest demand on the composite frames. The maximum demand, which implies severe damage and plastic deformation, concentrated on the column bases. Overall, the interior composite columns are more susceptible to failure than the external composite columns due to the larger P-Delta effect. The location of the hinges is symmetric with respect to the center of the moment frame.

Though many hinges occur in the steel beams before the yield performance level is reached, the structures are still stable. Since the frames were designed with a strong-column weak-beam design philosophy, the complete failure of the columns causes the entire frame to collapse. Therefore, this study also focused on studying composite columns and panel zones, and many hinges at the steel beam were not investigated, as shown in Fig. 3.38.

### 3.8.2 Damage estimations

As mentioned above, exceedance of any ultimate limit state in the columns indicates the most severe type of damage for the building as it can lead to complete collapse.

There are two major steps in performing the damage evaluation, one associated with determining the capacity and the other with assessing the demand. In the first step, the cross-sectional capacity of the hinging regions must be carefully determined. Monotonic and cyclic behavior of beam-columns subjected to combined axial and moment loads was studied in an attempt to estimate both the maximum strength and ductility for doubly-symmetric and axis-symmetric composite cross sections. From these studies, it can be shown that ultimate capacities for rectangular/circular CFT beam-columns can be estimated with reasonable accuracy using the simplified axial and moment (P-M) interaction formulas provided by 2005 AISC Specification (AISC, 2005) for composite systems. Figure 16 shows P-M interaction diagrams. The P-M interaction diagram for the composite sections based on a full plastic yield stress distribution can be generated as a linear interpolation between five points, as shown in the line AECDB. For the damage evaluation, a simplified-bilinear interpolation may be used between three points, as shown in the line segment AD'B. This approach is reasonably accurate for the steel columns and should provide a conservative estimation for the composite structures. To accomplish the second step, the demand at these critical sections must be established from advanced computational simulations, which were carried out on a series of entire composite moment resisting frames.

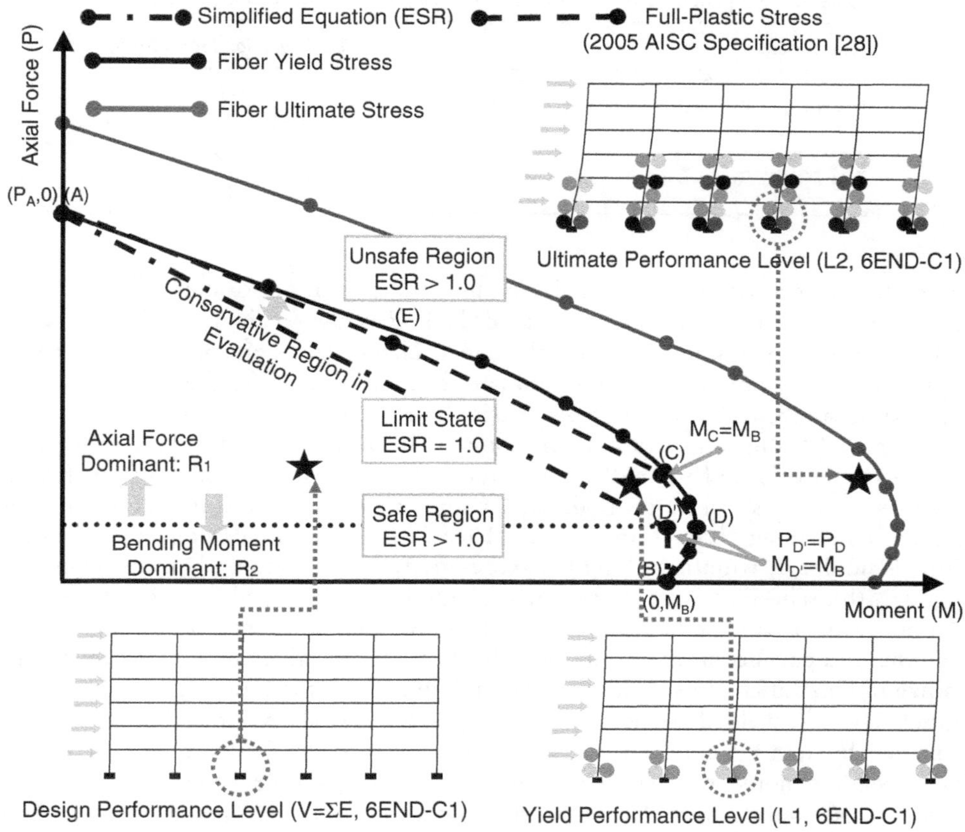


*Figure 3.39* ESR Calculations using P-M interaction diagrams for CFT Columns.

The structural damage was estimated in this study through the comparisons of the elastic strength ratios (ESR), which were defined as the ratios of the demand to the strength capacity for the member cross section. Thus, the value of the ESR can be determined by the position of the required strength on the domain of a simplified P-M interaction diagram. Two regions can be identified insofar as axial strength is concerned (Fig. 3.39). If the required axial strength is low ($R_1$), the behavior of CFT columns is controlled by yielding in tension due to the bending moments. The behavior in this region will be very ductile. It the required axial load is high ($R_2$), the behavior of CFT columns will be dominated by compression due to the axial loads. This will result in only limited ductile behavior. The ESR are defined with the combination of the required axial load ($P_r$) and bending moment ($M_r$) as summarized in Table 3.9.

The star signs shown in Fig. 3.39 indicate specific examples for the required strength anchored to the domain of the P-M interaction diagram. As the pushover loads increase, the local moments acting on the composite column increase significantly as compared to the axial loads because of the story drift and second order effects. P-M

*Table 3.9* Simplified equations for ESR (The simplified-bilinear interpolation).

| *Prior condition* | *Simplified equation (Safe Limit)* |
|---|---|
| $P_r < P_D$ ($R_2$ Region) | $\frac{M_r}{M_B} \leq 1.0$ |
| $P_r \leq P_D$ ($R_1$ Region) | $\frac{P_r - P_D}{P_A - P_D} + \frac{M_r}{M_B} \leq 1.0$ |

interaction forces acting on the CFT columns under both performance levels exceed the conservative limit state, so the value of the ESR is greater than 1.0. This indicates that the possibility of failure is relatively higher. The red star sign corresponding to the required member interaction at the ultimate state is very close to the P-M interaction limit based on the full ultimate stress distribution for the composite section.

Finally, PR connection frames were compared with welded connection frames with respect to the values of the ESR under either ISDR = 0.015 or 0.030 radians for the 20 ground motions. All frame models are symmetric in plan and subjected to uniform gravity loads along all bays and stories. Therefore, the value for damage evaluation show a similar distribution along the representative CFT column lines as defined in Fig. 3.25(b). The ESR were investigated at the bottom (B) and top (T) of the CFT columns. Calculation and comparison of the ESR values are summarized in Fig. 3.40.

The dashed lines in these figures indicate the ESR values of 1.0. Severe plastic damage occurs at the lower story of the building where the larger ESRs are concentrated. Overall, welded connection frames (6WED-C1) show larger ESRs along the story height than PR connection frames (6END-C1) at the 0.015 radian ISDR. The larger bending moments due to the stiffer pushover behavior for the welded connection frames causes more severe damage to the composite columns. The difference in ESR between two frame decreases at the 0.03 radian ISDR, with larger ESR along the story height still found for the welded connection frames. The peak ESR along the column lines were evaluated statistically as shown in Fig. 3.40(c). Rather than using a certain level of the ISDR as done for the static pushover cases, the ESR were calculated at the time when the maximum base shear force occurred. Similarly to the results for the ESR evaluations performed for the pushover analyses, the most severe damage concentrated on the lower story level. The column bases fail as their ESR exceeds 1.50. As expected, the larger ESRs are distributed to the interior columns having larger masses as compared with the exterior columns.

## 3.9 SUMMARY AND CONCLUSION

The smart PR-CFT connection developed in this study is an innovative structural element that takes advantage of the synergistic characteristics of the composite system, flexible PR connections, and use of new materials. The primary purposes of this study were to develop design methodologies for robust low-rise composite frames with bolted PR connections incorporating new smart materials and to assess the seismic performance of these frames though nonlinear analyses as well. Composite moment frames

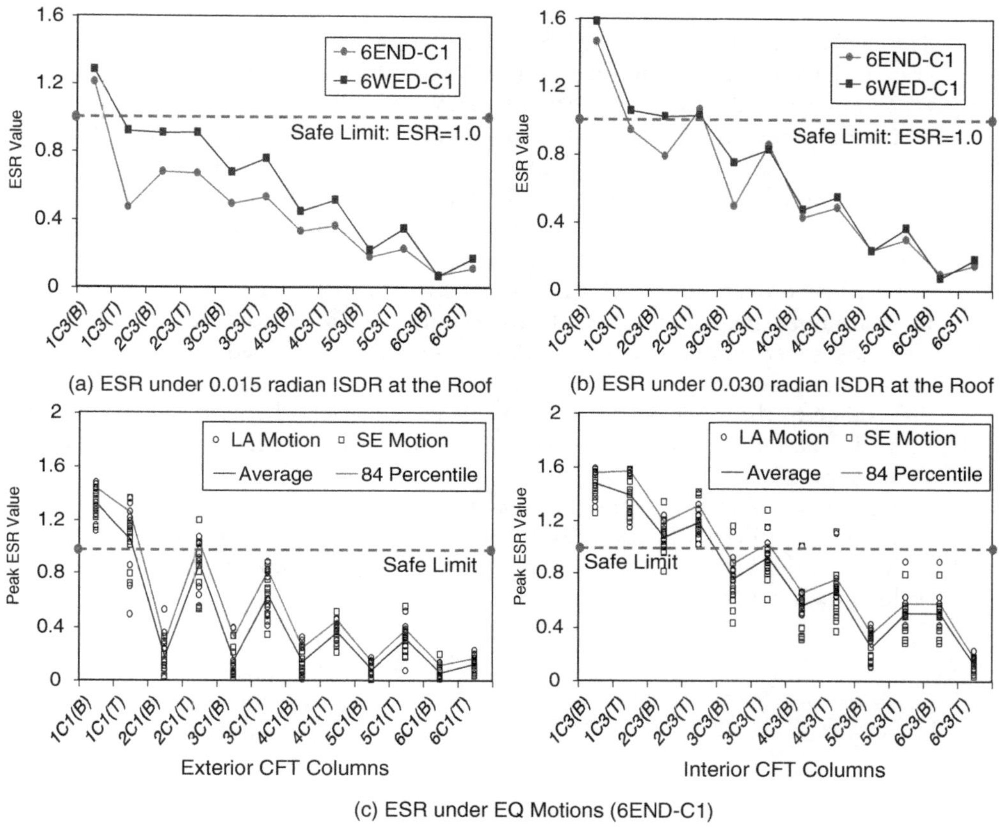


*Figure 3.40* Comparisons of ESR under static pushover loads and various ground motions.

were designed as either PR frames (C-PRMF) or special moment frames (C-SMF). They were modeled as 2D numerical frames used for nonlinear analyses. The exact behavior for the composite panel zone was simulated using 2D joint elements with a tri-linear backbone curve. The major conclusions from this study are summarized below:

a. Connection prototypes designed in accordance with the current AISC LRFD and seismic design standards performed very well. The connection was able to reach the full plastic strength of the beam after large deformation had occurred in the joint. Thus, deformations were distributed between the beam and joint, leading to a balanced design.
b. In these connections, tension bars running through the panel zone provide excellent ductility and considerable recentering due to their length and material properties. These characteristics for new connection models were validated by careful modeling.
c. Stiffness models were developed from observations of the FE analysis results and calibrated to full-scale component and joint tests. These monotonic models were

used as the foundation to formulate simpler spring models for use with cyclic loads.

d. The cyclic behaviors acting on the beam flanges were simulated by equivalent spring elements. The simplified component and joint model for the smart PR-CFT connection developed as part of this investigation provided very good results when compared to refined 3D FE analyses results. They served as a good platform to explore the optimization of the amount and location of the steel and SMA bars.
e. The pushover behavior of composite PR frames was significantly improved by the use of these new PR connections with respect to stiffness, strength degradation, permanent deformation, and energy dissipation capacity. Likewise, the advantage of new PR connections in achieving great structural efficiency was demonstrated by examining the peak response from nonlinear dynamic analyses.
f. The most severe damage occurred at the column bases. Welded connection frames were more susceptible to severe damage than PR connection frames at a reasonable range of the story drift (i.e. 0.015 to 0.03 radians) because higher bending moments acting on the welded connections caused a significant increase of the damage ratios.
g. These results obtained from the analytical study propose promise for implementing new smart PR connections in composite moment frames in an effort to address the lack of energy capacity and ductility in conventional composite welded moment frames.

Chapter 4

# Analyses for smart PR-CFT connections (T-stub connection type)

## 4.1 GENERAL OVERVIEW

As the application of performance-based seismic design begins to take root, it is clear that new structural concepts need to be developed to meet the demand performance for low and moderate levels of seismic loading. These demands can be met through the development of entirely new structural systems or the combination of existing structural elements that exploit their synergistic behavior in conventional systems. This paper describes an example of the latter, where the robustness and structural efficiency of concrete-filled tube (CFT) columns are combined with the ductility and recentering capabilities of a partially restrained (PR) connection utilizing a mix of materials (Fig. 4.1). This type of system is applicable in moment-resisting buildings from 3 to 10 stories, and combines a number of the advantages associated with composite systems (Mao & Xiao, 2006; Wu *et al.*, 2006). The advantages for CFT columns include the fact that the concrete prevents local buckling of the steel tube wall and that the confinement action of the steel tube extends the usable strain of the concrete after ultimate compressive strength. These advantages result in stronger and stiffer columns with superior cyclic performance (Choi *et al.*, 2010; Hu *et al.*, 2010). In particular, concrete-filled tubes show enhanced ductility and reduced rates of degradation under cycling at large drifts when compared to conventional steel or RC columns (Mao & Xiao, 2006; Wu *et al.*, 2006; Choi *et al.*, 2010).

In this proposed structural system, the efficiency of the CFT columns is matched by the use of an innovative connection that utilizes through bolted rods and T-stubs. The rods are manufactured from two different materials: (a) conventional mild low-carbon steel to provide hysteretic energy dissipation and (b) shape-memory alloys (SMA) bars that provide recentering forces to reduce permanent drifts after a major earthquake and to almost eliminate them under low to moderate seismic events. This connection provides dependable strength with predictable hardening at large deformations. The drift is accommodated through recoverable joint deformation, inelastic T-stub deformation, and beam hinging. These new composite PR connections should present a viable and attractive alternative to fully welded connections that have proven to be vulnerable in recent earthquakes (Hu & Leon, 2010; Hu *et al.*, 2011; Leon *et al.*, 1998; Leon, 1997).

The behavior of these connections is modeled based on the various response mechanisms of individual connection components (see Fig. 4.1). The behavior of each component under monotonic loads is fairly simple to model with bi-linear or tri-linear

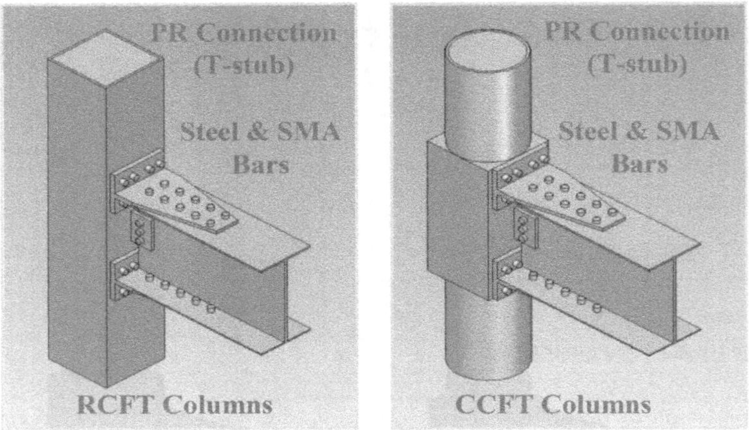


*Figure 4.1* New smart PR-CFT connections.

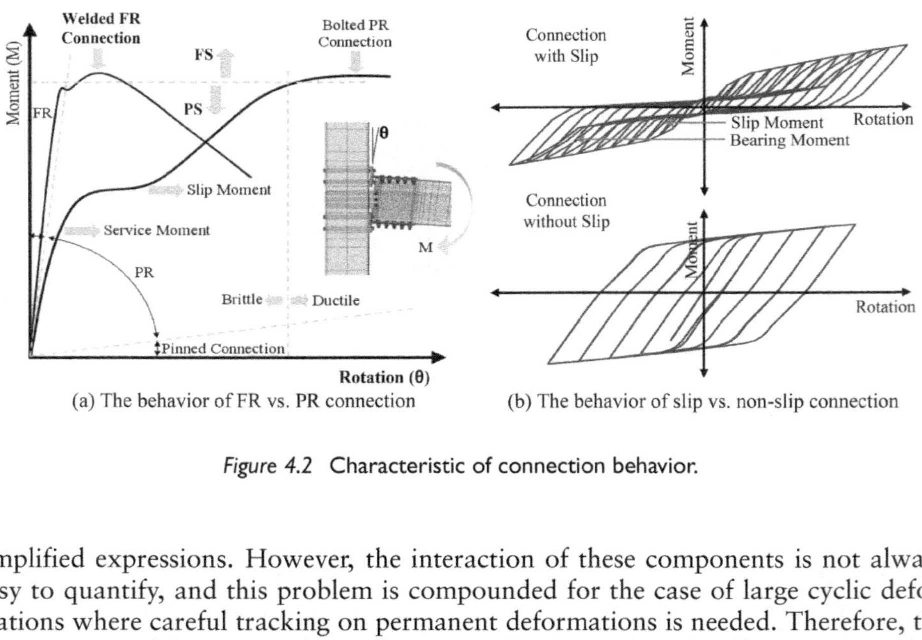


*Figure 4.2* Characteristic of connection behavior.

simplified expressions. However, the interaction of these components is not always easy to quantify, and this problem is compounded for the case of large cyclic deformations where careful tracking on permanent deformations is needed. Therefore, the connection models are complex and require a large number of stiffness components. For computational convenience, traditional approaches to frame design overlook the actual moment-rotational behavior of connections and adopt extreme behavioral models, i.e., the hinge model in simple pinned connections and the fully restrained (FR) or ideally rigid model in welded connections (Leon *et al.*, 1998; Leon, 1997). In reality, most connections, including bolted ones, exhibit semi-rigid or partially restrained (PR) behavior, which results in the intermediate behavior between two extremes, as shown in Fig. 4.2(a). The research described in this paper attempts to develop the simple

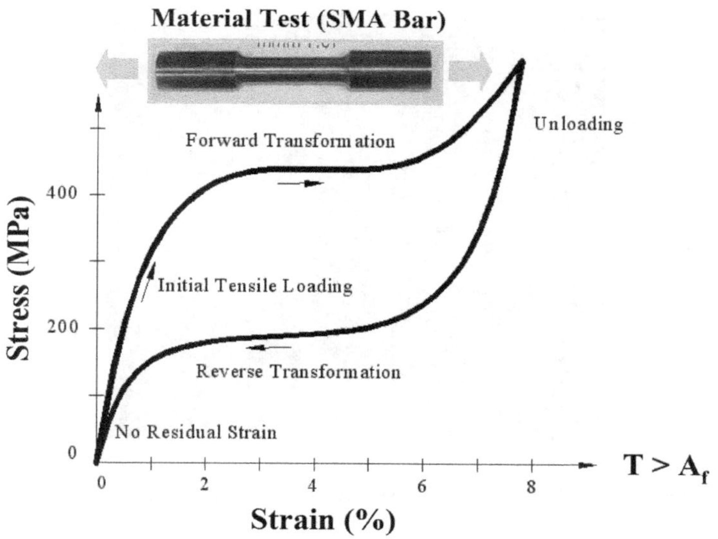


*Figure 4.3* Stress and strain curve for the super-elastic (SE) Nitinol material.

analytical model in order to predict the exact behavior of bolted connections. As shown in Fig. 4.2(a), connections are also classified by two other parameters: strength and ductility (Green *et al.*, 2004; Rassati *et al.*, 2004). As far as strength is concerned, connections are classified as full strength (FS) or partial strength (PS) based on whether or not the connection can sustain the full plastic moment of the framing beams. The stiffness of the connection is classified as fully restrained (FR) or partially restrained (PR) based on whether the rotational stiffness of the connection limits the reduction of the buckling load for the frame to a small value (typically 5%) of that for the theoretical case with fully rigid connections. Finally, connections are classified as brittle or ductile based on their ability to achieve certain plastic rotational demands (e.g., 0.03 radians of rotation or 4% interstory drift for special moment connection frames) (Green *et al.*, 2004; Rassati *et al.*, 2004; Swanson & Leon, 2000). In the bolted connections, slip between surfaces of bolted components occurs as the shear force on the shear faying surface exceeds that provided by the clamping force from the pre-tensioned shear bolts. Slip gives rise to temporary loss of stiffness that acts as a fuse during dynamic or seismic loading (Leon, 1997), and increases the energy capacity for the connection behavior (see Fig. 4.2(b)).

Shape memory alloys (SMAs) are used to enable large deformations with little permanent residual strain through either the shape memory effect or the super-elastic effect common in these materials. SMAs also provide high damping, durability, and fatigue resistance. In spite of the high cost, super-elastic SMA rods are applied to new PR connection models for these reasons. A stress-strain curve for a super-elastic SMA is shown in Fig. 4.3. Contrary to the accumulated permanent strain of ordinary steel materials, SMA materials can recover their original shape with changes in either temperature or stress (DesRoches *et al.*, 2004; McCormick *et al.*, 2005). Ocel *et al.*

*Figure 4.4* Application of SMA materials (SMA connection) (Penar, 2005).

*Table 4.1* Reported works on the topic of SMA connections.

| *Researchers* | *Institution* | *Year* | *Connection type* | *Modeling approach* | *SMA application* |
|---|---|---|---|---|---|
| Ocel *et al.* | Georgia Tech | 2004 | Tendon Connection | Experimental Connection Model | Shape Memory Tendon (Martensite SMA) |
| Penar | Georgia Tech | 2005 | Tendon Connection | Experimental Connection Model | Super-elastic SMA Tendon (Nitinol SMA) |
| Abolmaali *et al.* | UT Arlington | 2006 | T-stub Connection | Experimental Connection Model | Super-elastic SMA Bolt (Nitinol SMA) |
| MA *et al.* | Inha University | 2007 | End-plate Connection | FE Connection Model | Super-elastic SMA Bolt (Nitinol SMA) |
| Sep'ulveda *et al.* | University of Chile | 2008 | End-plate Connection | Experimental Connection Model | Super-elastic SMA Rod (Nitinol SMA) |
| Hu | Georgia Tech | 2008 | Composite Clip-Angle, T-stub, and End-Plate Connection | FE Connection Model | Super-elastic SMA Bar (Nitinol SMA) |

(Ocel *et al.*, 2004) and Penar (Penar, 2005) used SMA tendons or bars as the load transferring medium in beam-to-column connections as shown in Fig. 4.4. Based on loading tests on full-scale SMA connections (Ocel *et al.*, 2004; Penar, 2005; Abolmaali *et al.*, 2006), they concluded that connections fastened by SMA tendons can exhibit a recentering moment-rotational behavior due to the super-elastic characteristics of the SMA. The reported works on the topic of SMA connections are summarized in Table 4.1 (Ocel *et al.*, 2004; Penar, 2005; Abolmaali *et al.*, 2006 Ma *et al.*, 2007; Sep'ulveda *et al.*, 2008; Hu, 2008).

This research intends to explore new smart PR connections that incorporate the advantages of bolted PR connections, new materials, and composite construction. An interesting aspect of this research is an attempt at determining the optimal combinations of SMA and steel rods in the connections based on the entire system performance. Because of limited space, this study deals only with PR connections utilizing T-stubs. A logical variation using end-plates has been explored and described in detail elsewhere (Hu & Leon, 2010). This paper consists of five parts: (1) connection design; (2) development of simplified 2D cyclic joint model; (3) model verification; (4) composite-moment frame design; and (5) nonlinear analyses. The composite-moment frames with new smart PR connections were designed to highlight the benefit of new connection design in the whole frame structures.

## 4.2 NEW CONNECTION DESIGNS

All connection models considered were designed as full strength (FS) connections, meaning that they can transfer the full plastic beam moment ($M_P$) from the beam to the column. The steel beams were connected to the CFT columns using T-stub components and through tension bars. Because of the flexibility of the components, the connections are partially restrained (PR). To ensure ductile connection behavior, the connection design strength was based on the flexural plastic capacity of the beam or other components to avoid catastrophic losses of stiffness and strength. The design carefully avoided brittle failure modes, such as net section failure of the T-stub, bolt shear, and block shear fracture. The connections satisfied the design requirements given in both the AISC LRFD Standard (AISC, 2001) and the AISC Seismic Provisions (AISC, 2005).

The connection details are shown in Fig. 4.5. These connection models were used to study the behavior of the connection components and the adjacent areas of the beam and composite column. The frames were designed as composite-moment frames intended to form hinges at the ends of beams or the connection area.

The steel members were fabricated with standard shapes provided by the current design code (AISC, 2001). Composite CFT columns consisted of steel tubes (A500 Grade C steel) using either an HSS 16 × 16 × 500 for rectangular CFT (RCFT) columns or an HSS 18 × 500 for circular CFT (CCFT) columns. The T-stub connection was composed of thick T-stubs cut from a W16 × 100 section and 114 × 228 × 14 mm plate for the shear tab (A572 Grade 50 steel).

Super-elastic Nitinol bars and rod material equivalent to that of A490 bolts were used for the through rods. These tension fasteners included sixteen 25 mm (1″) diameter rods, either 510 mm long bar for RCFT columns or 560 mm long bar for CCFT columns. The super-elastic Nitinol bars were located where the larger deformations were likely to occur (see Fig. 4.5) with a view to use the recentering effect to the maximum. Forty 25 mm diameter, 102 mm long A490 bolts were used to fasten each T-stem to the beam flanges and six 25 mm diameter, 102 mm long A490 bolts were used as web bolts.

## 4.3 JOINT MODELS AND CYCLIC TESTS

The primary objective of this study was the development and numerical implementation of a joint model appropriate for use in refined 2D frame analyses, including

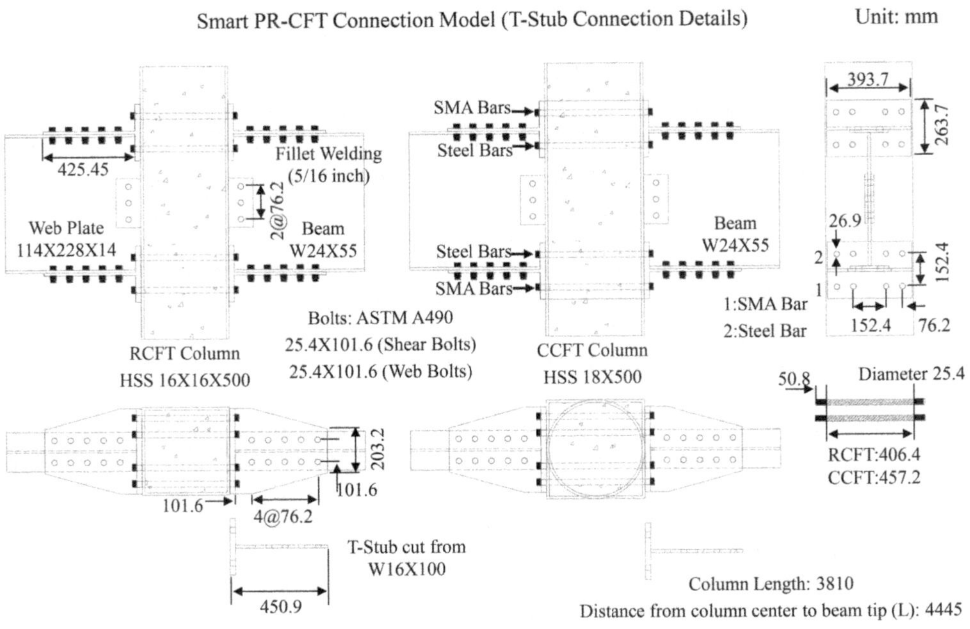


*Figure 4.5* Design details of the new connection models (T-stub, CFT column, and tension bars).

panel zones (PZ) effects. Simple joint models for new PR-CFT connections presented in Fig. 4.5 were constructed with the OpenSEES program (Mazzoni *et al.*, 2006), a nonlinear FE open source platform widely used in the U.S. for this type of studies.

## 4.3.1 Joint models

Fig. 4.6 shows the proposed 2D connection idealization. The connection components design to yield, such as tension bars, T-stubs, and composite panel zones, are modeled as component springs. The model also contains a rotational spring for the shear tab and a sliding component to model slip. The idealized force distributions at the perimeter of the joint for a T-stub connection subjected to seismic loads are also shown in this figure.

The response of the joint model resulting from seismic moments in the framing members is shown in Fig. 4.7. The edge of the steel beam is assumed to behave as a rigid plane, resulting in a linear displacement pivoting about the center of bearing. The different component springs in the joint model are assembled using a combination of parallel and series arrangements. These are then condensed to reduce the large number of component springs (see Fig. 4.7(b)), reducing the number of variables, saving computation time cost and avoiding numerical convergence problems.

## 4.3.2 Component springs

The complex connection behavior was decomposed into the load-deformation characteristics of three primary mechanisms such as bar deformations, T-stem elongation,

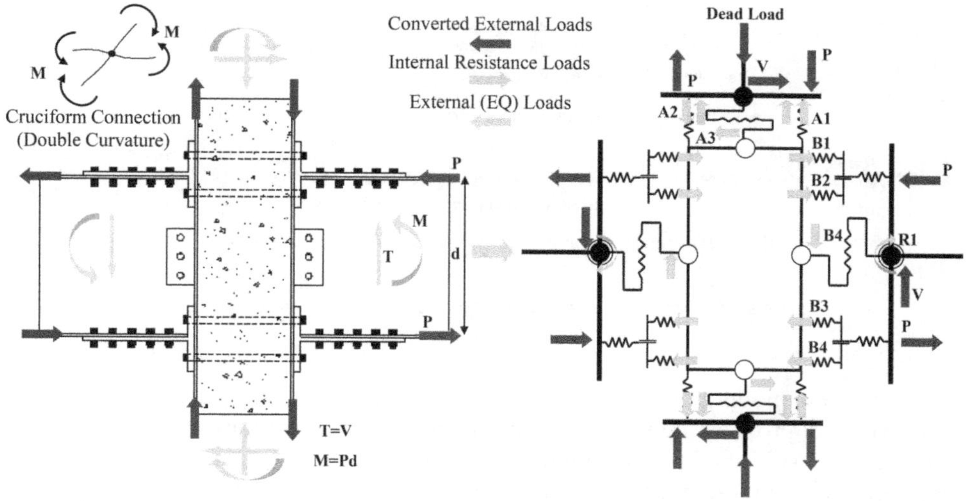


(a) Idealized loading distribution at the PZ (b) Idealized loading distribution at the joint model

*Figure 4.6* Force distribution at the joint model for the composite PR-CFT connection.

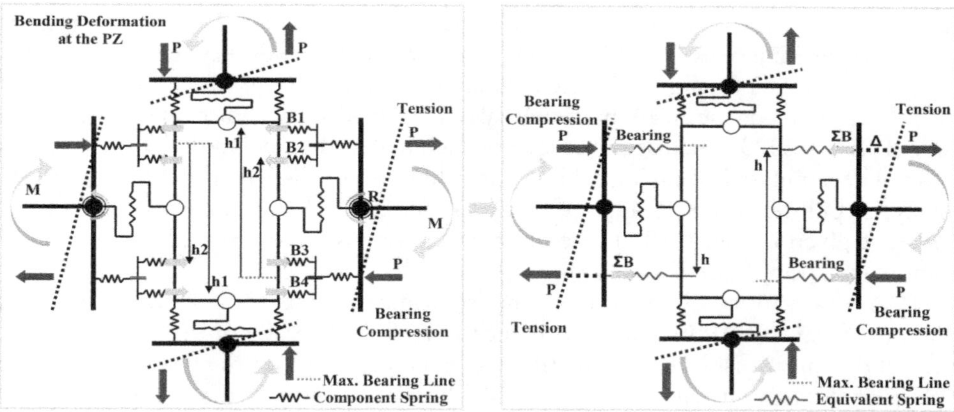


(a) Bending deformation at the joint model with component spring elements

(b) Bending deformation at the simplified joint model with equivalent spring elements

*Figure 4.7* Force and deformation response mechanism of the joint model.

and slippage (Swanson & Leon, 2000). Force-deformation response mechanisms and interactions for idealized component springs are illustrated in Fig. 4.8. Because of the different responses of the components in tension and bearing compression, two systems for load transfer were used for the spring models under cyclic loading (see Fig. 4.8(a) and (b)).

The force-deformation response of component springs was transformed into the multi-linear stiffness models shown in Fig. 4.9. There were three methods used for

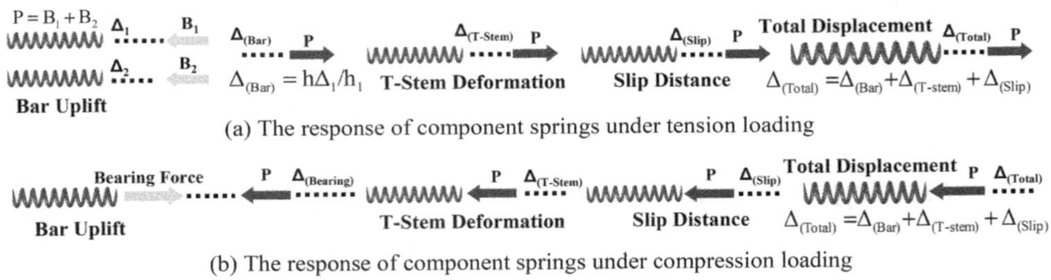


*Figure 4.8* Force-deformation responses of component springs.

obtaining these models: (1) material experimentation, (2) existing stiffness model, and (3) curve fitting to existing test data. The stiffness models for tension bars, which are assumed to be subjected to only axial tensile forces, were based on measured material properties. The force-deformation curves for other three components were generated either by fitting to the curves obtained from experimental data or established stiffness model (Swanson & Leon, 2000; Hu, 2008).

The cyclic behaviors of T-stub components in the new PR connection models proposed herein are shown in Fig. 4.9(b). These curves based on the stiffness models were simulated using the default uni-axial material commend in the OpenSEES program. Available pinching material and hardening material commands were used to construct the stiffness model for slip and T-stem elongation, respectively. However, the stiffness model for super-elastic SMA materials was simulated using a user defined material code (Davide, 2003) because of the lack of such material model in the OpenSEES program. The simulated behavior, which was calibrated to the experimental result from the cyclic pull-test on a 25.4 mm diameter super-elastic SMA bar (DesRoches *et al.*, 2004), is given to Fig. 4.10. These uni-axial materials were assigned into individual component springs (see Fig. 4.9(a)).

The observations obtained from the experimental tests for full-scale T-sub connections (Swanson, 1999), such as (1) location and magnitude of the force resultants and (2) the linear displacement distribution from the center of the bearing to the centerline of the bars in tension, provided the bases to condense the numerous springs into single equivalent spring (Spring I in Fig. 4.9(a)). The simplified joint model based on equivalent component springs shown in Fig. 4.7(b) performs well and leads to a computationally efficient implementation into a conventional frame analysis program.

### 4.3.3 Joint elements

As shown in Fig. 4.11(a), the joint element includes (1) four equivalent springs (S1) which are intended to reproduce the major deformations of component members (i.e., tension bars and T-stub) due to bending forces from the framing members; (2) four internal springs (S2) which are intended to reproduce the axial deformations of the CFT columns; (3) four internal shear springs (S3) which are intended to reproduce the shear deformations of the steel beams; (4) one shear panel element (C) which

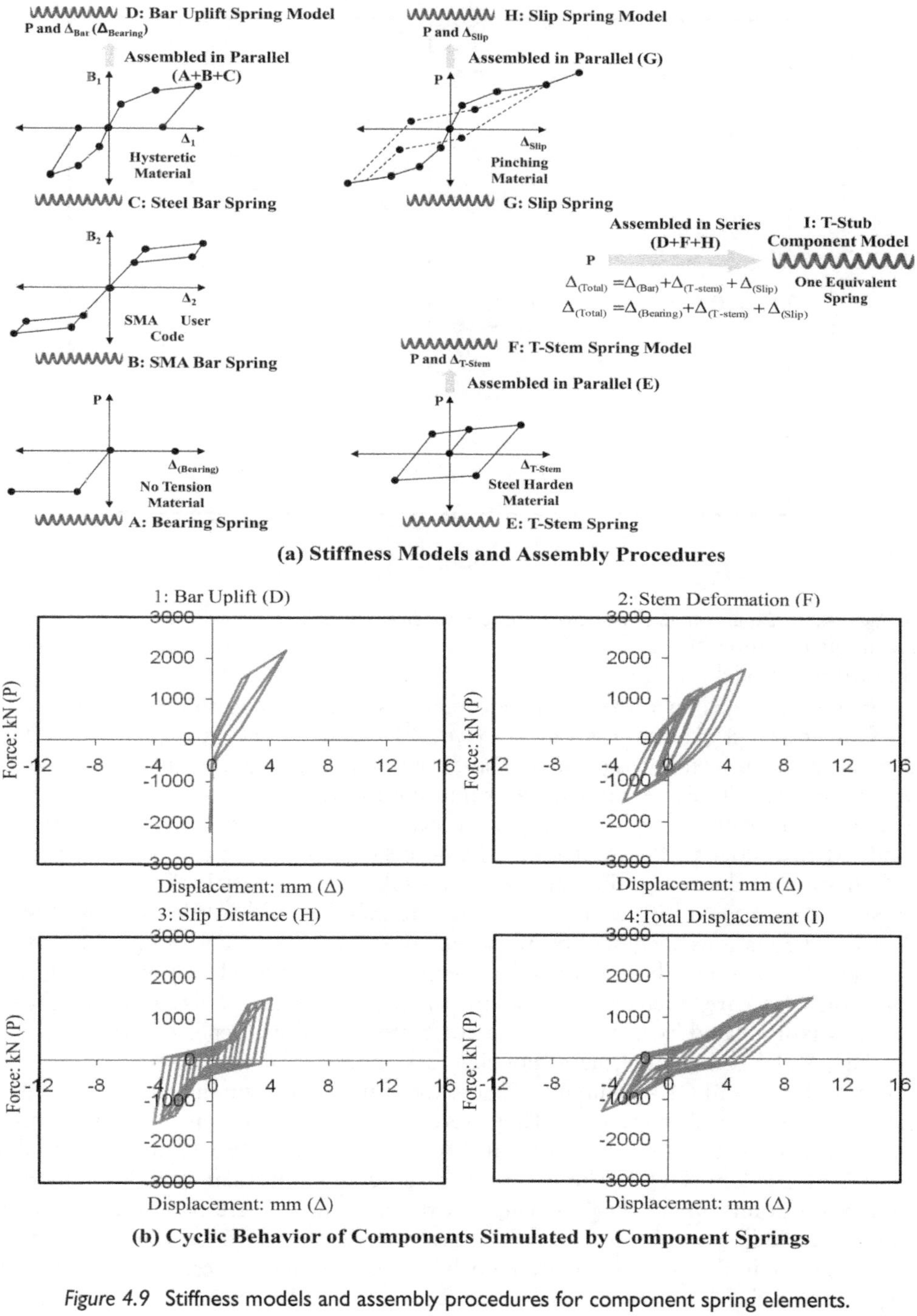


*Figure 4.9* Stiffness models and assembly procedures for component spring elements.

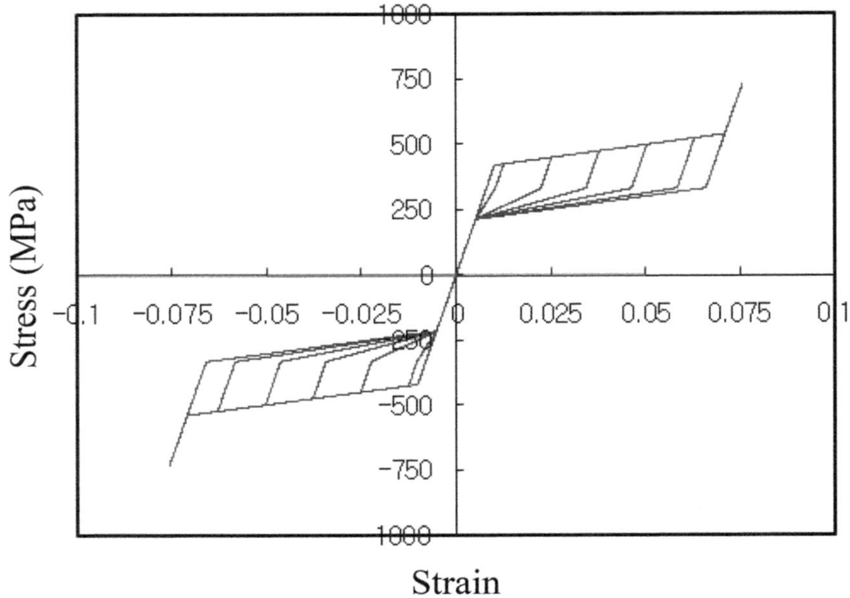


*Figure 4.10* Simulated material behavior of the super-elastic SMA bar (DesRoches *et al.*, 2004).

is intended to reproduce the shear deformations of the composite panel zone; and (5) four nodes on external rigid planes (Hu, 2008). Rotational spring elements (R1) were used if a shear tab was present. These component springs were modeled as zero-length and were implemented with four interior and exterior rigid planes coincident. The behavior of the composite panel zone was characterized by a tri-linear stiffness model (Krawinkler & Popov, 1982). The necessary parameters to include the shear deformation of the composite panel zone are taken as the initial stiffness ($K_i$), yield shear ($V_y$), post-yield stiffness ($K_t$), and ultimate shear ($V_u$), and were calculated using theoretical equations suggested by Wu *et al.* (Wu *et al.*, 2005; Wu *et al.*, 2007). Both the strength loss due to bolt holes and a strength superposition algorithm to model the composite effect are taken into consideration in these equations. This tri-linear stiffness model was simulated by using the hysteretic material in the OpenSEES program.

Fig. 4.11(b) shows the main modeling assumptions for the beam and columns. The steel beam and CFT column members were modeled as nonlinear beam-column elements with 2D fiber sections. They were connected to the joint element at four nodes, which include two translational displacements and one rotation as available degrees of freedom (DOF). The nonlinear beam-column element includes numerical integration points where displacement and resultant force were measured. 2D fiber sections, including nonlinear material behavior, were also assigned to these points.

As shown in Fig. 4.12, the material behavior of confined concrete at the CFT column contains different stress-strain curves for tension and compression response. The resulting stress-strain curves are based on HSS $16 \times 16 \times 500$ section for rectangular

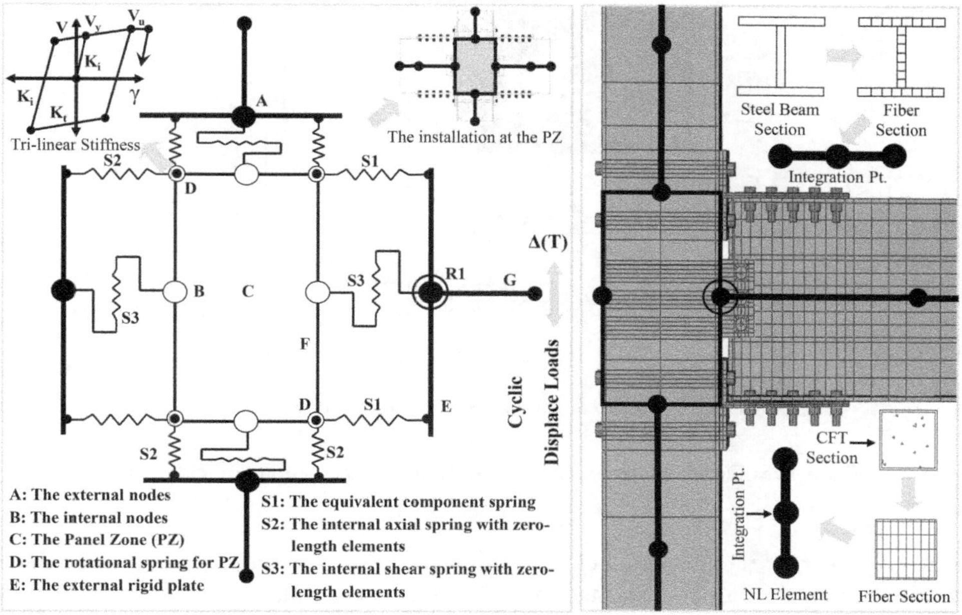


**(a) The joint element of one beam-to-column connection** **(b) The element formulation on the connection**

*Figure 4.11* Composition of the joint element.

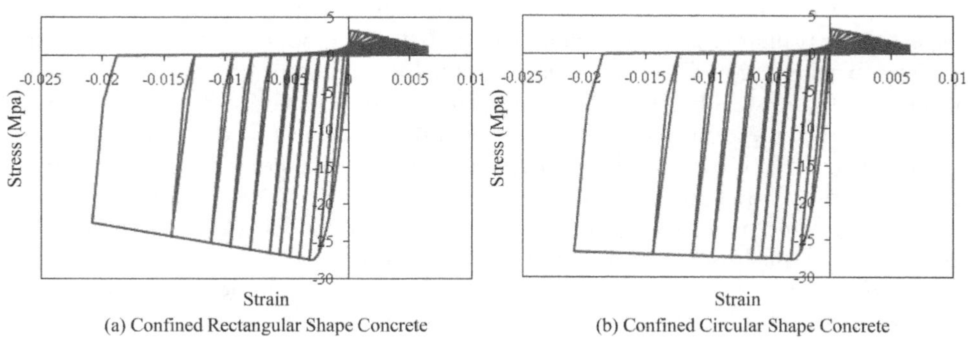


(a) Confined Rectangular Shape Concrete (b) Confined Circular Shape Concrete

*Figure 4.12* Cyclic stress-strain curves for confined concrete at the CFT column.

CFT (RCFT) columns or HSS 18 × 500 section for circular CFT (CCFT) columns. An equivalent uni-axial stiffness model suggested by Torres *et al.* (Torres *et al.*, 2004) was used for concrete in tension while that suggested by Hu *et al.* (Hu *et al.*, 2005) was used for confined concrete subjected to axial compressive forces. The degradation after peak compressive strength was affected by the shape of the CFT column section. The uni-axial concrete material commend also available in the OpenSEES program was

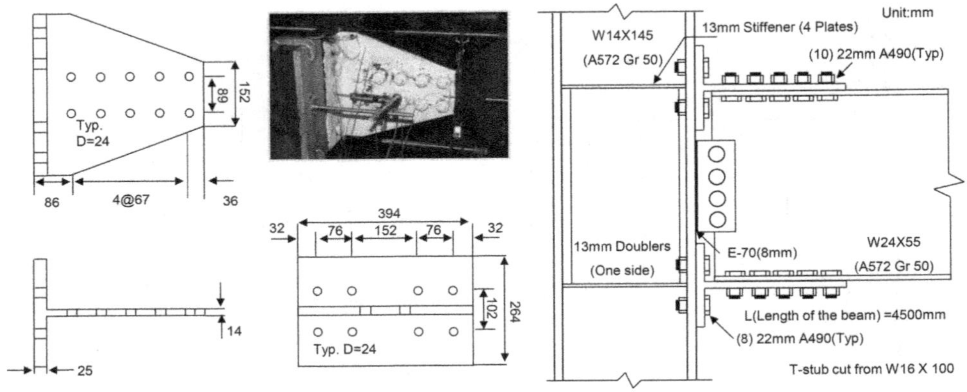


(a) T-Stub Component Detail (TA01 Model) (b) Full Connection Detail (FS05-TA01 Model)

*Figure 4.13* Details of experimental models for the calibration (Swanson, 1999).

used to construct the inelastic behavior of the confined concrete material with linear tension softening, tensile cracking, and compressive crushing.

### 4.3.4 Model verification and validation

Results obtained from existing T-stub connection test data obtained from SAC test programs (Swanson, 1999) were used for the calibration of 2D joint element models with respect to the primary response mechanisms of T-stub components (e.g., bar/bolt deformations, prying action, T-stem elongation, slippage and so on), because the detailing of the T-stub and steel beam is identical to that of the connection component used in the proposed connection models. The details of the specimens used for calibrations are given in Fig. 4.13. However, these specimens do not completely match the proposed connection models in that two crucial components such as SMA bars and CFT columns are missing. There is no way to perfectly calibrate the models that were used in this study owing to the absence of relevant physical test data.

Comparisons between experimental test results and numerical test results, with respect to force-displacement responses for both the T-stub component and the full-scale connection tested cyclically, are shown in Fig. 4.14. As shown in Fig. 4.14(a), significant mechanisms which occur at the axial force-deformation curve from the T-stub component test performed on the TA01 model, such as slippage, bearing, hardening, and the Bauschinger behavior, were well reproduced by the equivalent component spring. The major discrepancies are attributable to the overestimation of the slip behavior in the stiffness model. The length of the slip plateau was determined by the clearance between the shank of the shear bolt and the bolt hole (i.e., approximately 1.6 mm). After the shank of the shear bolt came into contact with the inner surface of the bolt hole, the strength and stiffness increase due to bearing.

Fig. 4.14(b) shows a comparison of numerical and experimental test values with respect to the force-displacement curve for the full-scale connection (FS05-TA01

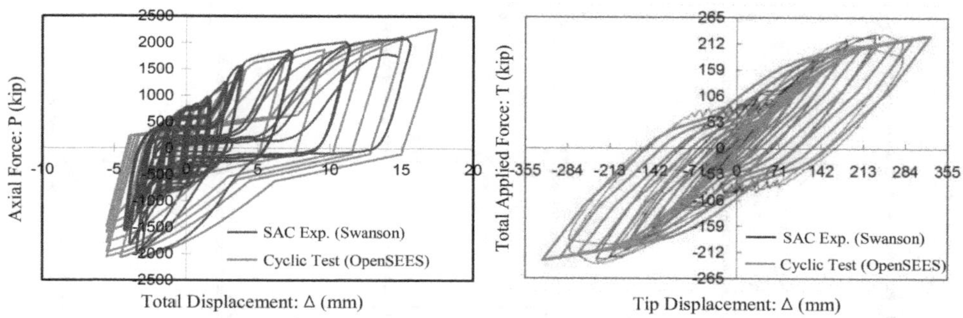

(a) Force and Total Displacement at the T-stub
(b) Force and Displacement at the Tip of the Beam

*Figure 4.14* Comparisons between experimental test results and numerical test results.

model). As the joint element was subjected to displacement loads applied to the tip of the beam corresponding to the position of a loading actuator, the behavior of the full-scale connection loaded cyclically was simply simulated through the numerical test. These curves show that the connection behaved in a ductile manner (e.g., 0.04 radian rotational limit; 0.04 L = 180 mm) and exceeded the full plastic strength of the beam (i.e., $T_p = M_p/L = 187$ kN). The cyclic curve simulated was not capable of modeling the strength degradation because component springs in the joint element did not include the ability to track the propagation of fracture. However, this simulated curve generally fits well the experimental result in terms of the initial slope, the level of slip plateau, trends in the nonlinear envelope, and even the unloading slope.

## 4.4 FRAME MODELS

Two prototype composite partially restraint moment frames (C-PRMFs) were designed, one with RCFT columns and one with CCFT columns. In addition, companion composite-special moment frames (C-SMFs) with fully rigid (FR) welded connections were also designed in order to compare the inelastic behavior of both types (partially restrained (PR) and fully restrained (FR)) composite frames. Building configurations, materials, and modeling conditions were the same for both the C-PRMF and C-SMF prototypes. The frames were designed in accordance with the AISC 2005 Seismic Provisions (AISC, 2005) and the IBC 2003 [30] for lateral and gravity loads, respectively. The gravity and lateral loads were determined following the ASCE 7-02 guidelines [31]. Design limits, system requirements, and seismicity factors for buildings located in a high seismicity area (i.e., LA area) were determined by these guidelines (AISC, 2005; ICC, 2003; ASCE, 2002). The plan view of the building is shown in Fig. 4.15. Six story configurations with 3 by 5 bays were used throughout this study. The total height (H) is 23.77 m (78′) in elevation, with uniformly 3.97 m (13′) heights (see Fig. 4.16). Buildings were designed with 7.63 m (25′) bay lengths. Uniform dead loads (479 kPa; 100 psf), live loads (383 kPa; 80 psf), seismic design category (SDC D class), and occupancy category (ordinary structures) were used for all building designs. The buildings

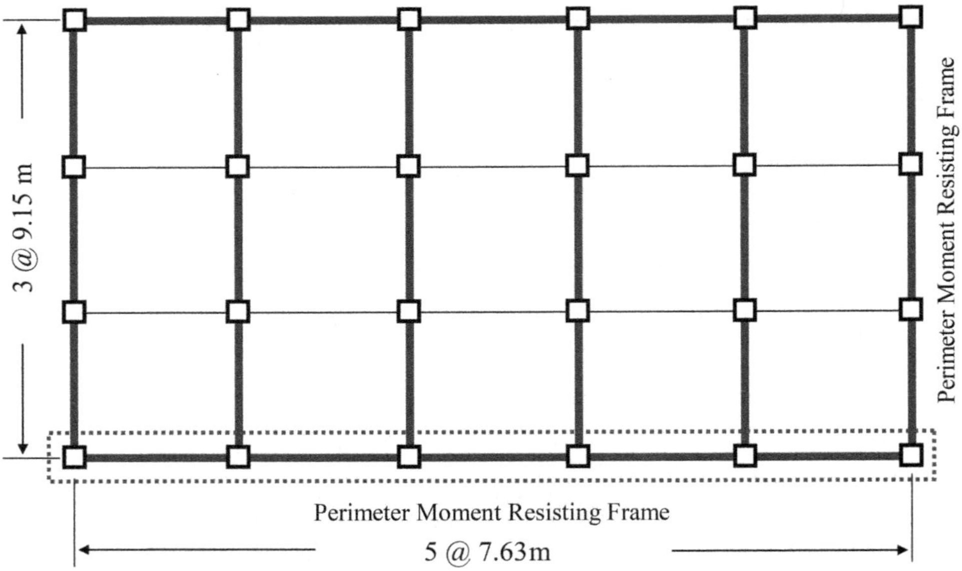


*Figure 4.15* Building plan view and perimeter moment resisting frames.

were assumed to have perimeter moment frames and only these were analyzed in a 2D frame model. The moment resistant frames with either PR or FR connections are presented as thick lines in the building plans for each frame type shown in Fig. 4.15. The sections for the composite columns and steel beams are summarized in Table 4.2. The prototype building has four variations shown by the model ID. This model ID varies according to the different combinations of connection types (i.e., T-stub (TSU) vs. welded (WED)) and column systems (i.e., rectangular (C1) vs. circular (C2)).

The elevation views of the 2D perimeter moment resisting frames are shown in Fig. 4.16. Numerical modeling attributes associated with the panel zone system, the dominant load combination, element, and data measurement points are also illustrated in this figure. The shear deformations in the panel zone were accounted for only in the PR frames. Load combination 5, consisting of factored dead (DL), live (LL), and earthquake (E) loads (LC 5; 1.2DL + 1.0LL + 1.0E) (ASCE, 2002), was used for the nonlinear pushover analyses. The equivalent lateral loads were based on a set of static lateral loads that corresponds to the 1st mode shape of deformation. Finally, the designs satisfy all requirements of the composite-moment frame for SDC D in terms of the allowable inter-story drift and the stability limit (ICC, 2003; ASCE, 2002).

## 4.5 NONLINEAR ANALYSES

The performance and efficiency of C-PRMF with new smart PR connections were investigated through nonlinear analyses performed on four frame models shown in

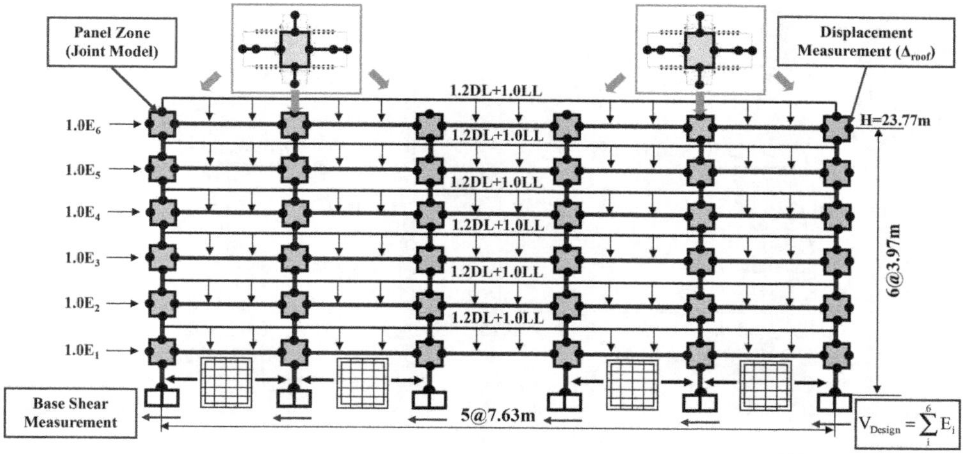


(a) Composite Moment Frame with Smart PR-CFT Connections

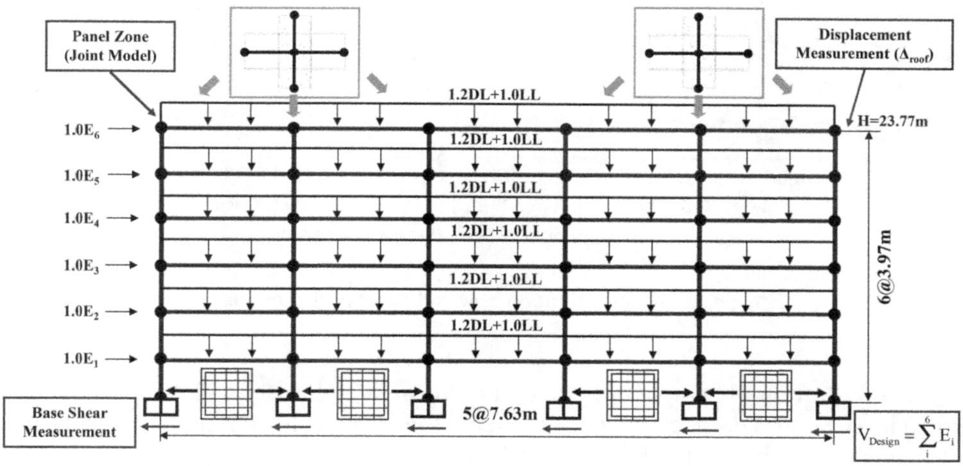


(b) Composite Moment Frame with Welded FR-CFT Connections

*Figure 4.16* Elevation views and modeling attributes for 2D moment resisting frames.

*Table 4.2* Design results for composite frame buildings.

| *Model ID* | *Connection type* | *Column system* | *Column size (all stories)* | *Beam size* | |
|---|---|---|---|---|---|
| | | | | *1st to 3rd story* | *4th and 6th story* |
| 6TSU-C1 | T-Stub | RCFT | HSS16 × 16 × 375 | W24 × 62 | W24 × 55 |
| 6TSU-C2 | T-Stub | CCFT | HSS18 × 375 | W24 × 62 | W24 × 55 |
| 6WED-C1 | Welded | RCFT | HSS16 × 16 × 375 | W24 × 62 | W24 × 55 |
| 6WED-C2 | Welded | CCFT | HSS18 × 375 | W24 × 62 | W24 × 55 |

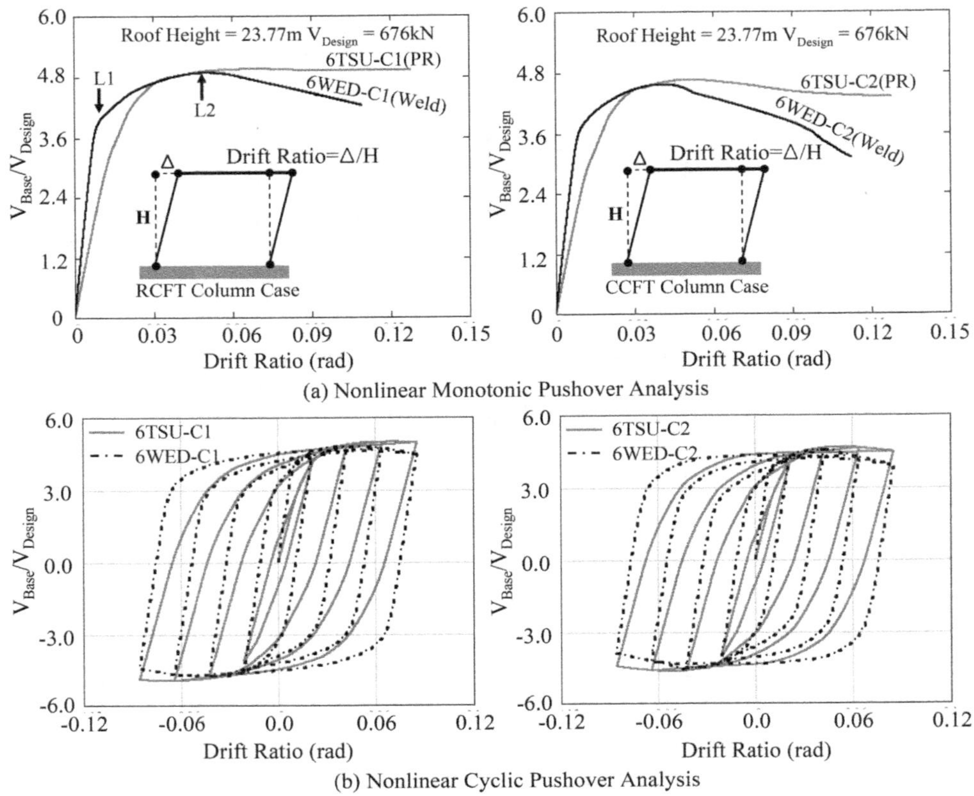


*Figure 4.17* Results of monotonic and cyclic pushover analyses.

Table 4.2. Both nonlinear pushover and nonlinear time-history analyses were carried out. The structural damage evaluations were based primarily on inter-story drift ratios (ISDRs) as described in the following sub-sections.

### 4.5.1 Nonlinear pushover analysis results

The maximum strength and deformation capacity of the composite-moment frames are demonstrated by the monotonic pushover analyses (Fig. 4.17(a)). The recentering behavior and energy dissipation capacity are demonstrated by the cyclic pushover analyses (Fig. 4.17(b)). The base shear forces are normalized by the design base shear force for the C-SMF ($V_{Design} = 676$ kN). The deformations are given as the drift ratios, defined as the roof displacements divided by the total frame height ($\Delta_{roof}/H$).

Fig. 4.17 also presents comparisons of the pushover curves between composite-moment frames with the same composite column systems but different connection types (i.e., 6TSU-C1 vs. 6WED-C1). As expected, the composite-moment frames with new PR connections (6TSU-C1) have a lower (more flexible) initial stiffness than those with welded connections (6WED-C1) because of the nature of PR connections and the influence of the bolt holes in the composite panel zones. The average initial stiffness of PR frames is approximately 5670 kN/m, whereas that of the welded frames is approximately 11750 kN/m. However, both types of frames reach a similar ultimate base shear of approximately 3250 kN representing a $V_{Base}/V_{Design} = 4.8$. The welded frames (i.e., 6WED-C1 and 6WED-C2) show more rapid strength degradation after reaching their ultimate strength as shown in Fig. 4.17(a). This is due to P-Δ effects, which become dominant after the welded frames exceeded an ISDR of 0.06 radians. A number of the welded connections, which are susceptible to brittle fracture, begin to fail at this stage. In contrast, the composite frames with PR connections (i.e., 6TSU-C1 and 6TSU-C2) show a nearly constant strength level or slight strength degradation after reaching their ultimate strength in the monotonic pushover curve. The flexible tension bars installed in the PR connections and the other large number of deformation mechanisms can prevent rapid strength degradation.

Comparisons of the cyclic curves of the two types of composite-moment frames (see Fig. 4.17(b)) suggest that new PR connections, as envisioned in their design, can provide both recentering and supplemental energy dissipation. Though the proposed PR connection frames show a greater global deformability due to the flexible joint installed on the panel zone, this may prevent the formation of plastic hinges in the beam elements leading to plastic deformations. The ability of PR frames to redistribute deformations, not just within the joint area but also throughout the structure, results in superior performance.

### 4.5.2 Nonlinear dynamic analysis results

Twenty ground motion records from the SAC suites (Somerville *et al.*, 1997) were used to conduct nonlinear dynamic analyses. These ground motions were developed from historical records for the western U.S. area, and consisted of 10 ground motions for the Los Angeles area (LA21 to LA30) and 10 ground motions for the Seattle area (SE21 to SE30). They correspond to a seismic hazard level of 2% probability of exceedence in 50 years.

In order to investigate the advantages of new PR connection design using SMA fastener systems, a detailed comparative study of a single representative ground motion will be described. The 6.9 magnitude 1995 Kobe ground motion (LA21) with a PGA value of 1.28 g and duration of 60 seconds was selected for this case study. The time history for the roof displacement and the normalized base shear force vs. total ISDR is shown in Fig. 4.18. The average peak roof displacement for the welded frames (6WED-C1 and 6WED-C2) and those with new PR connections (6TSU-C1 and 6TSU-C2) is approximately 600 mm and 460 mm (approximately 0.025 and 0.020 radians if interstory drift). The larger peak displacements for the welded frames result from the instability due to P-Δ effects that follows immediately after reaching the maximum strength. The excellent damping and recentering capability obtained from SMA materials as well as PR connection characteristics decreased the peak roof displacement by approximately 25% for the 6 story moment frame structures. The ability of SMA materials to recover from the large displacements reduced the residual displacement by about 75%.

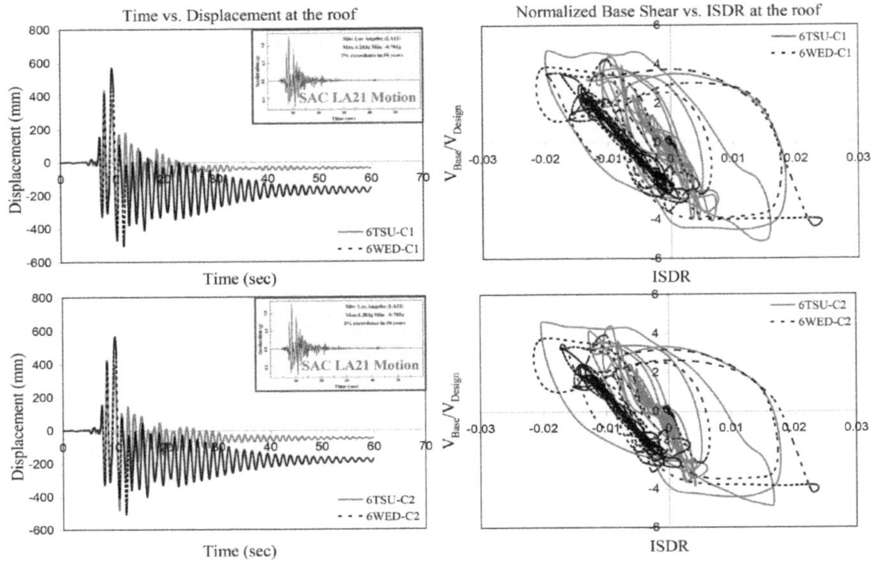


*Figure 4.18* Results of nonlinear dynamic analyses (LA 21 ground motion).

### 4.5.3 Performance evaluations

The inter-story drift ratios (ISDRs) were used to evaluate the seismic performance of the prototype frame models. These ratios are computed by dividing the difference of the deflections at the top and bottom of each story level into each story height. Performance levels defined from the monotonic pushover curves were used to conduct the performance evaluations. The performance levels result from two transition points on the pushover curves: Yield Point (L1) and Ultimate Point (L2) (see Fig. 4.17(a)).

The ISDRs for all frame models at each story level are shown in Fig. 4.19. They were measured at the yield base shear point (L1) corresponding to $V_{Base}/V_{Design} = 4.1$ ($V_{Base} = 2802$ kN) and the ultimate base shear point (L2) corresponding to $V_{Base}/V_{Design} = 4.8$ ($V_{Base} = 3261$ kN) for the 6WED-C1 model. Base shear forces at each performance level for other frame models are given in an inset of Fig. 4.19. The ISDR at the design base shear point corresponding to $V_{Base}/V_{Design} = 1.0$ ($V_{Base} = V_{Design} = 676$ kN) satisfies the allowable story drift limit (0.02 radian) defined in the seismic design codes (AISC, 2005; ICC, 2003; ASCE, 2002). All frame models are stable up to the yield point level. The maximum ISDRs generally occur at 2nd or 3rd story level until the yield point is reached. Relatively lower ISDRs are shown for the upper stories. Overall, welded frames show smaller ISDRs than those with new PR connections. This results from the stiffer initial behavior of the welded connection as shown in the elastic range of the pushover curves. As the lateral loads increase, the maximum ISDR gradually moves to the lowest story level, with the maximum ISDR found at the 1st story level when the base shear force reaches its maximum level. A considerable increase (localization) of plastic deformations occurs at the lower story.

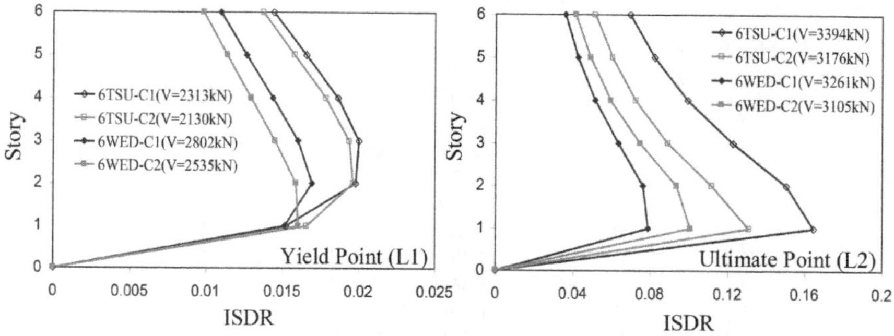


(a) ISDR at the Yield Base Shear Point (b) ISDR at the Ultimate Base Shear Point

*Figure 4.19* ISDR at each story level.

*Table 4.3* Peak responses of 6TSU-C1 model under various ground motions.

| *EQ motion* | *Max. PGA (g)* | *Max. base shear (kN)* | *Max. displ. (mm)* | *Max. vel. (m/sec)* | *Max. pseudo-accel. (g)* |
|---|---|---|---|---|---|
| LA21 | 1.283 | 3560 | 473 | **2.971** | 2.05 |
| LA22 | 0.921 | 3343 | 392 | 2.412 | 1.86 |
| LA23 | 0.418 | 3014 | 216 | 1.350 | 1.08 |
| LA24 | 0.473 | 3618 | 515 | 1.853 | 1.28 |
| LA25 | 0.869 | 3427 | 513 | 2.694 | 1.68 |
| LA26 | 0.944 | 3583 | **676** | 2.759 | 1.68 |
| LA27 | 0.927 | 3649 | 644 | 1.761 | 1.26 |
| LA28 | 1.330 | 3640 | 497 | 2.010 | 1.54 |
| LA29 | 0.809 | 3316 | 276 | 1.522 | 1.47 |
| LA30 | 0.992 | **3680** | 440 | 1.565 | 2.16 |
| SE21 | 0.755 | 3578 | 559 | 2.085 | 1.95 |
| SE22 | 0.485 | 2476 | 209 | 1.446 | 1.28 |
| SE23 | 0.605 | 3485 | 341 | 1.839 | 1.22 |
| SE24 | 0.539 | 3663 | 516 | 1.731 | 1.25 |
| SE25 | 0.895 | 3338 | 259 | 1.965 | 1.27 |
| SE26 | 0.821 | 3174 | 263 | 1.520 | 1.52 |
| SE27 | **1.755** | 3343 | 505 | 2.092 | **2.44** |
| SE28 | 1.391 | 3303 | 347 | 1.938 | 2.05 |
| SE29 | 1.636 | 3276 | 353 | 2.507 | 2.24 |
| SE30 | 1.573 | 3600 | 395 | 2.022 | 2.06 |
| **Average** | **0.971** | **3403** | **419** | **2.002** | **1.67** |

This implies that the most severe structural damage occurs at the bottom of the CFT columns. The new PR connections and the use of CFT columns provide superior ductility both in terms of local rotations and ability to globally redistribute forces and limit local damage.

The peak responses, such as maximum displacement, velocity, and pseudo-acceleration, for all 20 ground motions for frame model 6TSU-C1 are provided in Table 4.3. In general, peak responses occur at the roof story level. The occurrence

*Table 4.4* Peak ISDRs of 6 TSU-C1 model under various ground motions.

| *EQ motion* | *Peak time* | *1* | *2* | *3* | *4* | *5* | *6* |
|---|---|---|---|---|---|---|---|
| LA21 | 10.76 | 0.0314 | 0.0309 | 0.0284 | 0.0255 | 0.0226 | 0.0198 |
| LA22 | 8.88 | 0.0182 | 0.0210 | 0.0219 | 0.0210 | 0.0188 | 0.0165 |
| LA23 | 11.49 | 0.0157 | 0.0152 | 0.0134 | 0.0117 | 0.0103 | 0.0091 |
| LA24 | 8.92 | 0.0464 | 0.0437 | 0.0361 | 0.0297 | 0.0250 | 0.0214 |
| LA25 | 3.09 | 0.0354 | 0.0352 | 0.0327 | 0.0288 | 0.0248 | 0.0214 |
| LA26 | 3.18 | **0.0531** | **0.0523** | **0.0464** | **0.0394** | **0.0333** | **0.0284** |
| LA27 | 7.22 | 0.0520 | 0.0496 | 0.0436 | 0.0371 | 0.0314 | 0.0267 |
| LA28 | 4.82 | 0.0384 | 0.0367 | 0.0317 | 0.0275 | 0.0236 | 0.0202 |
| LA29 | 11.34 | 0.0209 | 0.0196 | 0.0171 | 0.0149 | 0.0131 | 0.0116 |
| LA30 | 11.48 | 0.0371 | 0.0346 | 0.0288 | 0.0245 | 0.0211 | 0.0183 |
| SE21 | 3.26 | 0.0472 | 0.0458 | 0.0392 | 0.0322 | 0.0269 | 0.0229 |
| SE22 | 3.52 | 0.0070 | 0.0088 | 0.0097 | 0.0098 | 0.0095 | 0.0087 |
| SE23 | 3.63 | 0.0262 | 0.0247 | 0.0219 | 0.0188 | 0.0162 | 0.0141 |
| SE24 | 3.57 | 0.0463 | 0.0433 | 0.0361 | 0.0300 | 0.0253 | 0.0215 |
| SE25 | 4.88 | 0.0181 | 0.0175 | 0.0157 | 0.0138 | 0.0121 | 0.0106 |
| SE26 | 11.18 | 0.0216 | 0.0203 | 0.0172 | 0.0144 | 0.0122 | 0.0105 |
| SE27 | 8.72 | 0.0323 | 0.0332 | 0.0315 | 0.0284 | 0.0247 | 0.0212 |
| SE28 | 10.66 | 0.0298 | 0.0283 | 0.0243 | 0.0205 | 0.0173 | 0.0146 |
| SE29 | 35.13 | 0.0165 | 0.0183 | 0.0188 | 0.0179 | 0.0162 | 0.0143 |
| SE30 | 29.93 | 0.0333 | 0.0310 | 0.0267 | 0.0226 | 0.0193 | 0.0166 |
| **Average** | | **0.0313** | **0.0305** | **0.0271** | **0.0234** | **0.0202** | **0.0174** |
| **84 Percentile** | | **0.0464** | **0.0437** | **0.0361** | **0.0300** | **0.0252** | **0.0215** |

time for each peak value lagged slightly behind the peak ground acceleration (PGA). The average maximum base shear force obtained by the nonlinear dynamic tests (i.e., $V_{Base} = 3403$ kN) is slightly larger than the ultimate base shear force obtained from the nonlinear pushover tests (i.e., $V_{Base} = 3394$ kN). The largest maximum pseudo-acceleration occurs for the record with the strongest PGA (SE27). The bold letters in the table indicate the largest value. The peak values of ISDR were also investigated, as given in Table 4.4. Characteristic values, (i.e., average and 84th percentile) as described in FEMA 355C (FEMA, 2000), are also shown in this table. The peak ISDRs obtained from the nonlinear dynamic analyses are concentrated in the 1st story level, and decrease as one moves up the frame.

## 4.6 SUMMARY AND CONCLUSIONS

This study investigated the performance of composite-moment frames with smart partially-restraint (PR) concrete filled tube (CFT) column connections through simplified but refined 2D frame analyses. It provides practical design and numerical modeling methodologies for innovative PR connections utilizing both the recentering properties of super-elastic SMA tension bars and the energy dissipation capacity of steel tension bars. Through the nonlinear frame analyses, the behavior of new PR connections was evaluated and compared to the performance of conventional C-SMF with welded connections in terms of time history responses, residual roof drifts, and inter-story drift ratios (ISDRs). The major results from this study are the following:

1. The stiffness values for individual mechanisms should be generated from either material behavior characteristics, if the behavior is primarily one-dimensional (i.e., tension bars), or the observation of experimental test results, if the loading is more complex (i.e., compression bearing).
2. Whenever possible, these models then need to be calibrated against test data that reflect both the joint and frame behavior. Ideally, the joint and frame behavior needs to be separated in the experiments, as was done for the T-stub work in the SAC project.
3. The behavior of the connections, including slippage, Bauschinger effect, and recentering effect, can be extrapolated to cyclic loads, provided the number of cycles is small enough to prevent low-cycle fatigue failures, and localization of damage is limited (i.e., no kinking of braces as is common in braced frames).
4. For the joints and frames investigated, comparisons between test results and numerical simulations showed good agreement. Of course, great care must be taken not to extrapolate beyond reasonable material limits; the latter can only be obtained from realistic test setups at large scale.
5. From the cyclic pushover tests, as well as the monotonic pushover curves, it was shown that the connection types play a significant role on the performance of the moment frame. Composite-moment frames with new PR connections showed more flexible behavior at the design and yield level. More gradual strength degradation was evident for these composite-moment frames, when compared to fully welded ones as the maximum strength was exceeded.
6. From the nonlinear dynamic analyses, it was shown that considerable plastic deformations occurred in the lower story composite columns immediately after the PGA of the ground motions had been reached. CFT columns provide much greater toughness and rotation capacity than conventional steel compared.
7. When composite-moment frames with welded connections were compared to those with new PR connections, welded connection frames show stiffer behavior and larger resistance than PR ones at low to moderate base shears, but they begin to fail rapidly after reaching their ultimate resistance due to weld failures.

Part 2

# Design and analyses for bolted connections

Chapter 5

# Bolted connections (FE models)

## 5.1 GENERAL OVERVIEW

In the past two decades, the incorporation of bolted connections into steel moment frame design has attracted attention since fully welded moment connections conventionally utilized in the practical construction field had inherent drawbacks (Kim & Han, 2007; Kim *et al.*, 2008; Leon, 1988; Hu & Leon, 2005; Hu *et al.*, 2010; Yang & Jeon, 2009). The unexpected failure often occurring at the welded moment connections is related to the connection geometry, which accelerates large strain demand in critical sections. These strain demands will cause the fracture of the base metal around the weld access hole. It results in low rotational ductility and poor behavioral performance under cyclic loads (Leon, 1988; Hu & Leon, 2005). Therefore, other connection types should be introduced to moment frame design with a view to remove the brittle failure of corresponding welded connections.

Bolted steel connections are one alternative on the ground that they can avoid the strain concentration and localization problems due to the connection geometry. (Swanson & Leon, 2000; Swanson & Leon, 2001; Fan & Wu, 2004; Leon, 1997) It is the cause that these bolted connections are generally fabricated by the assemblage of various connection components (e.g., plates, connectors, and bolts). Accordingly, the bolted connection system throughout the moment frame provides a high level of redundancy and the intermediate level of stiffness in comparison with fully welded connections (Leon, 1997; Azizinamini & Schneider, 2004; Rassati *et al.*, 2004).

The global behavior of bolted connections commonly represented by the moment-rotation curve becomes complex because the various response mechanisms of individual components interact with each other. These response mechanisms are associated with the large variety of connection configurations, geometrical discontinuities, frictional forces that lead to slip, and bolt pretensions that lead to prying action (Gantes & Lemonis, 2003; Al-Khatab & Bouchair, 2007). In this study, the types of interesting bolted connections are the full-scale T-stub connections which exhibit complex behavior.

Though the large number of variables related to the connection geometry, the complicate interaction of connection components, and constitutive relationships related to their material nonlinearities are taken into consideration, this research attempts to fully understand the response mechanism for the complete behavior of T-stub connections through analytical studies provided by finite element (FE) methods. The FE models are adequate to simulate complex and nonlinear behavior with considerable accuracy

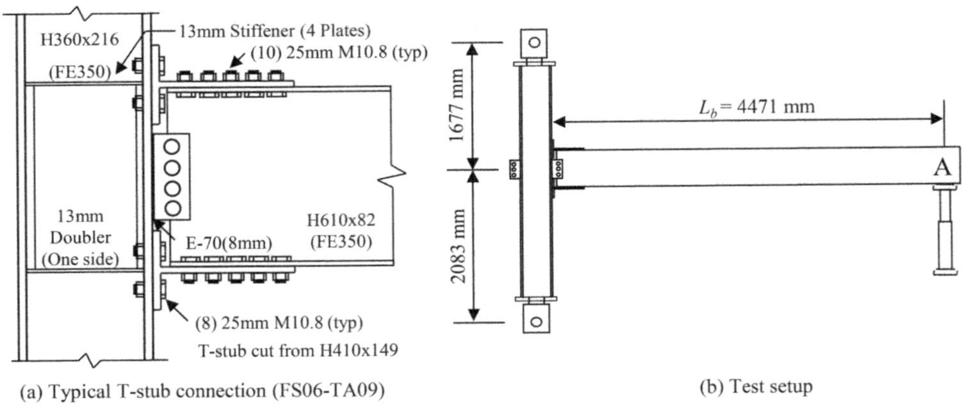


*Figure 5.1* Typical T-stub connection and its test setup.

because they require the sophisticated preparation process (Lemonis & Hantes, 2009). Nevertheless, in spite of this profit, the previous FE studies conducted on T-stub connections have been limited to only local T-stub components (Swanson & Leon, 2000; Swanson & Leon, 2001; Al-Khatab & Bouchair, 2007; Lemonis & Hantes, 2009; Piluso & Rizzano, 2008; Hu *et al.*, 2011) instead of entire beam-to-column subassemblages. For component tests, the axial loads assumed to be beam-flange forces in the actual connection are only applied to the T-stub components. A large number of parameters were investigated economically and rapidly through the component tests. However, they produce a restricted prediction to estimate strength, ductility, and stiffness of full-scale connections, since both localized bending moments and shear forces presented in the actual condition are missing in these component tests. Therefore, this study intends to develop full-scale FE connection models in order to generate complete-nonlinear moment-rotation curves, which are involved with modern structural analysis programs and up-to-date computational equipment.

This paper firstly presents a short illustration of the experimental programs which are usually utilized as the basic calibration for the numerical studies, and then discusses the application of FE models in the parametric analyses of bolted T-stub connections, with highlight on the behavior of the T-stub component. The stresses distributed into the beam-to-column connection are also investigated to verify that the connection capacity is directly associated with the strength models based on the failure mode of individual connection components. Finally, the results of parametric investigation on T-stub components and their bolts are presented to demonstrate how the geometric variations in those components can affect the behavior and strength capacity of the connection models.

## 5.2 EXPERIMENTAL PROGRAM

An experimental program was performed as a part of the SAC Subtask 7.03 research project at Georgia Tech. Smallidge (1999) and Schrauben (1999) individually tested eight full-scale connection specimens with T-stub components. The specimens were built with beams and columns connected by T-stubs, according to the standard AISC-LRFD manual (AISC, 2001).

A typical specimen and its test set-up are illustrated in Fig. 5.1. The details of all specimens such as section and material properties are summarized in Table 5.1, along with the ultimate moments obtained in the experiment. The typical test configuration was made up of a H360 × 216 column with pinned end supports, a 4.5 m length beam which varied from a H530 × 66 to H690 × 125 section, and three connecting members which are two T-stubs and a shear tab. Except for the beam in the FS08 test, which was FE250 (A36) steel, all members and joint components were fabricated from FE350 (A572-Gr.50) carbon steel. The steel strengths were obtained from the mill certificates. The cyclic displacement load history, which consists of stepwise increasing deformation cycles, was applied to the tip of the steel beam (see A in Fig. 5.1(b)).

The specimen identifications (IDs) consist of the test number and the T-stub classification. For instance, the FS06-TA09 model consists of "FS" which represents full-scale, "06" which represents the test number, and "TA09" which indicates the T-stub classification. The test series of T-stub components, which are TA, TB, TC, and TD listed in Table 5.1, were classified by the T-stub flange thickness ($t_f$) ranged from 14 mm to 32 mm. As listed in the table, 22 mm and 25 mm diameter tension bolts were used in a pair of test series (e.g., TA01 vs. TA09), along with the tensile capacity of used steel bolts based on the M10.8 (ASTM A490) high-strength bolt material. The dimensions and instrumentation scheme of T-stub components are given in Fig. 5.2. The T-stubs were cut from for standard wide flange sections so as to install more than 8 shear bolts on the tapered T-stem. As summarized in Table 5.2, the size, grade, number gauge, and spacing of the bolts were varied to investigate the effect of prying action on the tension flange, slip on the T-stem, and bearing compression around the bolt holes. The oversized bolt holes (i.e., bolt clearance of 1.6 mm) were applied to all specimens for construction.

The experimental tests were extensively instrumented with linear variable displacement transducers (LVDTs) which were mounted on the top and bottom of the beam flanges (see Fig. 5.2). In addition to connection rotations, each pair of LVDTs measured individual T-stub displacements which result in the different components of the overall T-stub deformation. Thus, LVDTs B measured the slip, LVDTs C measured the uplift of the T-stub flange from the face of the column flange, LVDTs D measured the elevation of the T-stub flange at the bolt line, LVDTs E measured the elongation of the T-stem, and LVDTs F monitored the overall T-stub deformation.

## 5.3 THEORETICAL BACKGROUND

### 5.3.1 Full connection behavior

The total applied force measured at the tip of the beam ($T$) should be changed into the bending moment by applying a first order approximation (see Eq. 5.2) in order to obtain the characteristics of connections from the experimental results. The resisting response mechanism for connection behavior is illustrated in Fig. 5.3. The reaction responses transformed from external forces are computed as follows:

$$M_r = M_x + Vx = TL_b \tag{5.1}$$

*Table 5.1* Details of full scale T-stub connection models.

| | *T-Stub* | | | | *Beam* | | | *Column* | | | *Bolt* | | *Shear Tab* | *Peak Moment* |
|---|---|---|---|---|---|---|---|---|---|---|---|---|---|---|
| *Model ID* | *Model ID* | *Size** | *Yield Stress*** | *Ultimate Stress*** | *Size** | *Yield Stress*** | *Ultimate Stress*** | *Size** | *Yield Stress*** | *Ultimate Stress*** | *Diameter**** | *Tensile Capacity***** | *Size**** | *Exp. Model* |
| FS03 | TD04 | H410 × 67 | 396 (F)<br>427 (W) | 557 (F)<br>569 (W) | H530 × 66 | 400 | 489 | H360 × 216 | 386 | 516 | 22 | 334 (T)<br>336 (S) | 10 × 127 × 229 | 650 kN-m |
| FS04 | TD08 | H410 × 67 | 396 (F)<br>427 (W) | 557 (F)<br>569 (W) | H530 × 66 | 400 | 489 | H360 × 216 | 386 | 516 | 25 | 438 (T)<br>435 (S) | 10 × 127 × 229 | 681kN-m |
| FS05 | TA01 | H410 × 149 | 318 (F)<br>352 (W) | 460 (F)<br>469 (W) | H610 × 82 | 420 | 523 | H360 × 216 | 386 | 516 | 22 | 334 (T)<br>336 (S) | 10 × 127 × 305 | 1010kN-m |
| FS06 | TA09 | H410 × 149 | 318 (F)<br>352 (W) | 460 (F)<br>469 (W) | H610 × 82 | 420 | 523 | H360 × 216 | 386 | 516 | 25 | 438 (T)<br>435 (S) | 10 × 127 × 305 | 1010kN-m |
| FS07 | TB01 | H530 × 138 | 362 (F)<br>378 (W) | 498 (F)<br>501 (W) | H610 × 82 | 420 | 523 | H360 × 216 | 386 | 516 | 22 | 334 (T)<br>336 (S) | 10 × 127 × 305 | 1060kN-m |
| FS08 | TB05 | H530 × 138 | 362 (F)<br>378 (W) | 498 (F)<br>501 (W) | H610 × 82 | 372 | 487 | H360 × 216 | 386 | 516 | 25 | 438 (T)<br>435 (S) | 10 × 127 × 305 | 985kN-m |
| FS09 | TC01 | H840 × 251 | 390 (F)<br>415 (W) | 523 (F)<br>530 (W) | H690 × 125 | 415 | 510 | H360 × 216 | 386 | 516 | 22 | 334 (T)<br>336 (S) | 10 × 127 × 381 | 1890kN-m |
| FS10 | TC09 | H840 × 251 | 390 (F)<br>415 (W) | 523 (F)<br>530 (W) | H690 × 125 | 415 | 510 | H360 × 216 | 386 | 516 | 25 | 438 (T)<br>435 (S) | 10 × 127 × 381 | 1990kN-m |

*: Metric Size  **: MPa (Coup on Tests)  ***: mm  ****: kN
F: Flange  W: Web  T: Tension Bolt  S: Shear Bolt

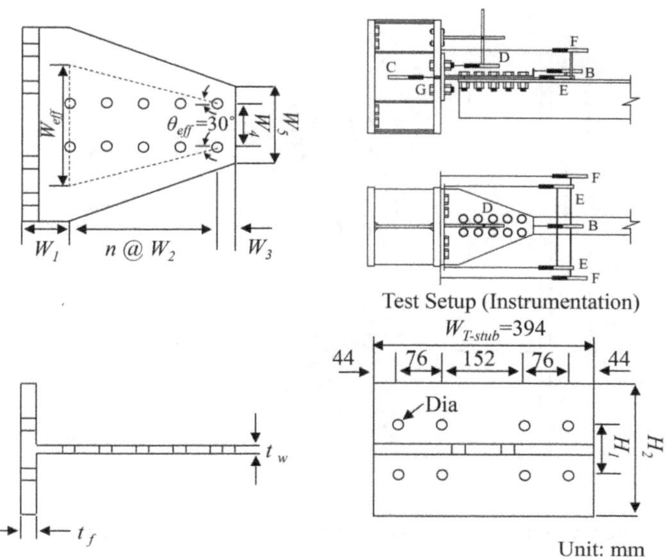


*Figure 5.2* Details of T-stub components and instrumentation for experiments.

*Table 5.2* Geometric of T-stub components.

| *Model* | $W_1$ | $W_2$ (*n*) | $W_3$ | $W_4$ | $W_5$ | $t_f$ | $t_w$ | $H_1$ | $H_2$ | *Dia.* | *Bolt Size* |
|---|---|---|---|---|---|---|---|---|---|---|---|
| TA01 | 84 | 67 (4) | 33 | 89 | 152 | 25 | 14 | 102 | 264 | 24 | 22 × 114 mm (T)*<br>22 × 76 mm (S)** |
| TA09 | 89 | 76 (3) | 38 | 89 | 152 | 25 | 14 | 102 | 264 | 27 | 25 × 114 mm (T)<br>25 × 89 mm (S) |
| TB01 | 84 | 67 (4) | 33 | 102 | 178 | 24 | 14 | 102 | 213 | 24 | 22 × 114 mm (T)<br>22 × 89 mm (S) |
| TB05 | 89 | 76 (4) | 38 | 102 | 178 | 24 | 14 | 102 | 213 | 27 | 25 × 114 mm (T)<br>25 × 89 mm (S) |
| TC01 | 90 | 67 (5) | 33 | 127 | 241 | 32 | 17 | 127 | 292 | 24 | 22 × 127 mm (T)<br>22 × 89 mm (S) |
| TC09 | 95 | 76 (5) | 38 | 127 | 241 | 32 | 17 | 127 | 292 | 27 | 25 × 127 mm (T)<br>25 × 102 mm (S) |
| TD04 | 84 | 67 (3) | 33 | 89 | 152 | 14 | 10 | 102 | 178 | 24 | 22 × 95 mm (T)<br>22 × 76 mm (S) |
| TD08 | 89 | 76 (3) | 38 | 89 | 152 | 14 | 10 | 102 | 178 | 27 | 25 × 95 mm (T)<br>25 × 83 mm (S) |

*(T) = Tension bolts, **(S) = Shear bolts, FE 350(Grade 50) steel, Grade for all used bolts = M10.9 (ASTM A490)

where $V$ is the shear force carried by the beam; $M_r$ is the internal resistant moment; $M_x$ is the moment of the framing beam; $x$ is the distance from the column surface to the position of the moment; and $L_b$ is the length of the beam. The strength for connection design is determined by the full plastic moment of the beam at the plastic hinge defined as:

$$M_p = Z_b F_y \tag{5.2}$$

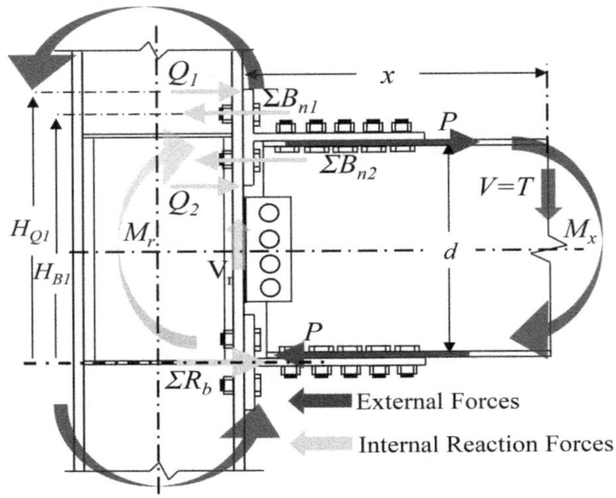


*Figure 5.3* Internal and external force mechanism acting on full scale T-stub connections.

where $Z_b$ is the plastic section modulus; and $F_y$ is the plastic yield stress of the base material. The internal reactions in the component members correspond to the external forces on account of the static equilibrium. They have the following relationship as shown in Fig. 5.3:

$$M_r = \sum B_{n1} H_{B1} + \sum B_{n2} H_{B2} - Q_1 H_{Q1} - Q_2 H_{Q2} \tag{5.3}$$

where $\Sigma B_{n1}$ and $\Sigma B_{n2}$ indicate the summation of bolt reaction forces in tension; $Q_1$ and $Q_2$ stand for the prying forces acting on the tip of the T-stub flange; $H_{B1}, H_{B2}, H_{Q1}, H_{Q2}$, and $H_{Q2}$ are the equivalent heights at each position. The prying forces arise from the local bending action associated with the initial bolt pretension $(B_{pre})$ in the T-stub flange. The rotations are derived from linear displacements that are measured at the tip of the beam. The total rotation of the connection is calculated by:

$$\phi_{total} = \arctan(\Delta_{beam}/L_b) \tag{5.4}$$

where $\Delta_{beam}$ denotes the beam tip displacement.

### 5.3.2 Component strength models

The bending moment $(M)$ transformed from the total applied force at the tip of the beam is transmitted to the connection as the converted axial forces $(P)$. They are computed by dividing the external moment force into the depth of the beam $(d)$ as follows:

$$P = \frac{M}{d} \tag{5.5}$$

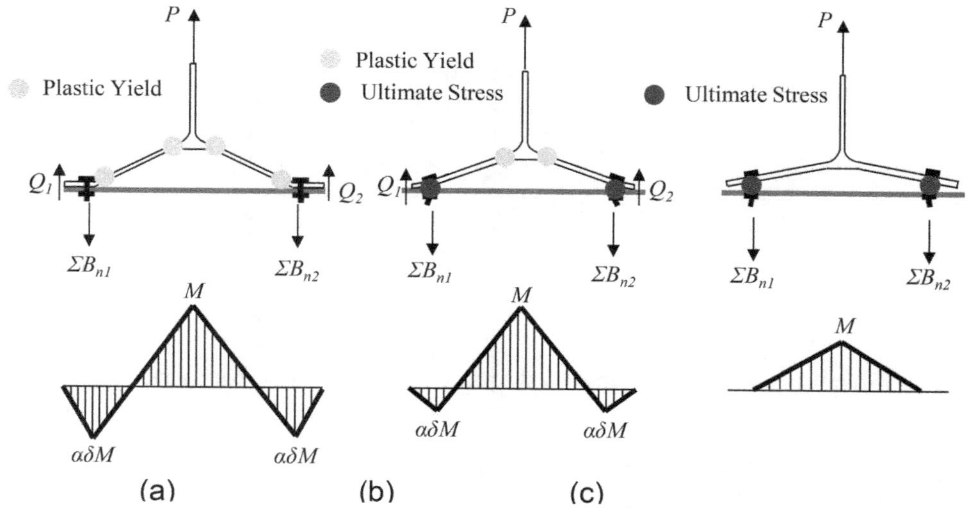


*Figure 5.4* Three possible failure modes and their moment distribution due to bolt/flange prying action.

The internal reactions in the T-stub component act against these axial forces to satisfy the fundamental static equilibrium. They show the following relationship as illustrated in Fig. 5.3:

$$P = \sum B_{n1} + \sum B_{n2} - Q_1 - Q_2 \tag{5.6}$$

Thus, the T-stub components mostly deform by the converted axial forces transferred from the beam flange. The major response mechanisms that create the overall deformation of T-stub components are made up of the flange deformation, tension bolt elongation, T-stem deformation, and relative slip (Swanson & Leon, 2000; Swanson & Leon, 2001; Hu, 2008). These response mechanisms are derived independently from the component test results and correlate well with simplified strength models.

The mechanisms from tension bolts under stretching and a T-stub flange under bending should be treated together to estimate the strength capacity of the T-stub flange because their behaviors are inherently coupled during the force transfer (Swanson & Leon, 2000; Hu, 2008). Especially, the prying action mechanism is accompanied by the ultimate failure of the tension bolts or the yield failure of the T-stub flange. The established prying model accepted in the design guideline (AUSC, 2001) has been used to assess the ultimate strength capacity of the T-stub flange ($P_{n,flange}$). This prying model was based on one of most widely used models proposed by Kulak *et al.*, (1987)

Three possible failure modes occurring at the T-stub flange and tension bolts are equationally expressed as Eqs. (5.7) to (5.9) and illustrated in Fig. 5.4. These failure modes are relevant to the formation of a plastic mechanism on the T-stub flange (Eq. (5.7) and Fig. 5.4(a)), the plastic yielding of the T-stub flange combined with bolt

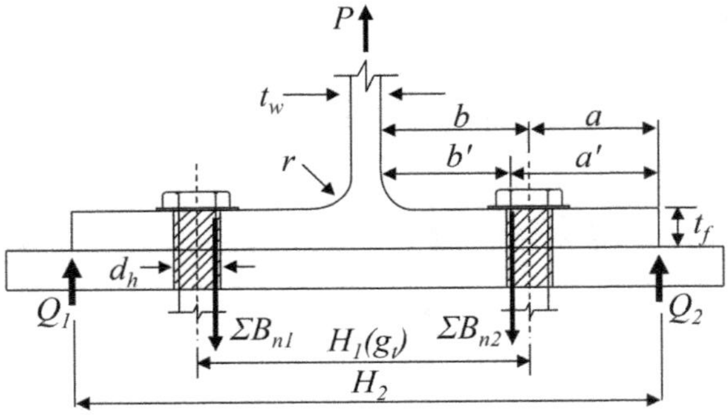


*Figure 5.5* Geometric notations for the prying action model of T-stub components.

prying (Eq. (5.8) and Fig. 5.4(b)), and the fracture of tension bolts without prying forces (Eq. (5.9) and Fig. 5.4(c)) as follows:

$$P_{n,flange} = \frac{(1+\delta)W_{T\text{-}stub}F_y t_f^2}{2b'} \tag{5.7}$$

$$P_{n,flange} = \frac{\sum B_{n,tension}a'}{a'+b'} + \frac{W_{T\text{-}stub}F_y t_f^2}{2b'} \tag{5.8}$$

$$P_{n,flange} = \sum B_{n,tension} \tag{5.9}$$

where $W_{T\text{-}stub}$ is the width of the T-stub flange (see Fig. 5.2); $a'$ and $b'$ substituted for $a$ and $b$ (see Fig. 5.5) indicate the geometric parameters resulting from the precondition that most of bolt reaction forces are assumed to be transferred into the inside edge of the bolt shank rather than the centerline of the bolt shank (Kulak *et al.*, 1987; Thornton, 1985); $\Sigma B_{n,tension}$ is the tensile capacity of tension bolts used for design (refer to Table 5.1); $\delta$ denotes the ratio of the net section area to the gross section area and is written as:

$$\delta = 1 - \frac{n_{tb}d_h}{W_{T\text{-}stub}} \tag{5.10}$$

where $d_h$ is the diameter of the bolt hole including the clearance to the diameter of the bolt ($\Delta_c$); and $n_{tb}$ is the number of tension bolts per each row. The geometric notations for the prying model are shown in Fig. 5.5.

The parameter $\alpha$ indicates the index of the level of the prying action present (Kulak *et al.*, 1987). It is defined as the ratio of the moment at the inside bolt edge to the moment at the face of the T-stem (see Fig. 5.4) and is expressed as follows:

$$\alpha = \left(\frac{1}{\delta}\right)\left(\frac{2n_{tb}Pb'}{W_{T\text{-}stub}t_f^2F_y} - 1\right). \tag{5.11}$$

When the value of $\alpha$ is equal to 1.0, Eq. (5.7) is derived from Eq. (5.11). When $\alpha \geq 1.0$, T-stub flanges are more susceptible to the plastic hinge and prying forces are maximized. Accordingly, the T-stub flange is considered to be a fixed-fixed beam in this case (Failure Mode 1 shown in Fig. 5.4(a)). When $0 < \alpha < 1.0$, the combination of flange yielding and bolt fractures due to bolt prying may occur at the T-stub flange (Failure Mode 2 shown in Fig. 5.4(b)). Finally, if $\alpha \leq 0$, then the T-stub flange completely separates from the column surface, prying forces become zero, and then tension bolts are theoretically subjected to pure stretching (Failure Mode 3 shown in Fig. 5.4(c)). However, stress concentration developed as a consequence of the local bending moment occurs underneath the one-sided bolt head in practice. The prying forces are minimized by augmenting the thickness of the T-stub flange or by decreasing the geometric ratio of $H_1$ to $H_2$ (Kulak *et al.*, 1987; Thornton, 1985). As illustrate in Eq. (5.6) referred to as a static equilibrium, these prying forces make a contribution to increasing the bolt reaction forces ($B_n$) and effectively reducing the applied axial loads ($P$) (Kulak *et al.*, 1987).

The strength model for the T-stem is evaluated with a bilinear representation based on the effective net section strength. As shown in Fig. 5.2, an angle of 30 degree measured from the first row of shear bolts was used to define the effective width of the T-stub ($W_{eff}$). This model was proposed by Whitemore (1952). The effective width of T-stem is written as follows:

$$W_{eff} = s_{sb}(n_{sb} - 2)\tan(\theta_{eff}) + g_s \leq W_{T\text{-}stub} \tag{5.12}$$

where $s_{sb}$ is the shear bolt spacing; $n_{sb}$ is the number of shear bolts; $\theta_{eff}$ is the effective angle of the tensile participation generally taken as 30 degree (see Fig. 5.2); and $g_s$ denotes the gage between two rows of the shear bolts. For the calculation of the section strength, the smaller one between the Whitemore width and the actual width was used to compute the effective net section area of the T-stem ($A_{stem}$). The shear bolt holes were excluded as a part of the net section area.

$$A_{stem} = (W_{eff} - 2d_b)t_w \tag{5.13}$$

where $t_w$ is the thickness of the T-stem. When the applied loads are distributed uniformly along the net section area, the yield and ultimate strength capacity of the T-stem are expressed, respectively, as follows:

$$P_{y,stem} = F_y A_{stem} \tag{5.14}$$

$$P_{u,stem} = F_u A_{stem} \tag{5.15}$$

where $F_u$ is the ultimate stress of the base material.

The slip mechanism is characterized by the simple friction model given in the AISC-LRFD (AISC, 2001). The slip-critical connections are designed with the slip resistance ($P_{slip}$) which shall be equal or over the nominal slip resistance load ($P_{n,slip}$) as follows:

$$P_{n,slip} = 1.13uh_{sc}T_b n_s n_{sb}\left(1 - \frac{P_u}{1.13T_b n_{sb}}\right) \leq P_{slip} \tag{5.16}$$

where $u$ is the mean slip coefficient for Class A, B, or C surfaces (for Class A surface, $u = 0.33$); $h_{sc}$ is the coefficient for the standard bolt holes (i.e., $h_{sc} = 1.0$); $T_b$ is the specified minimum fastener tension used in the AISC-LRFD (e.g. M10.8 bolt with 22 mm has 221 kN); and $n_s$ is the number of slip planes. Unpainted clean mill scale steel surfaces belonging to Class A were used for the mean slip coefficient. The applied ultimate axial force ($P_u$) is computed by the product of the design reduction factor and the ultimate T-stub capacity (e.g., $P_u = \phi P_n$: $\phi = 0.9$ for the plastic flange and $\phi = 0.75$ for the bolt fracture).

Once the shear force transmitted from the applied axial force goes beyond the slip resistance, slippage begins to occur between two contact surfaces. The slip resistance results from the clamping force corresponding to the product of the bolt pretension and the friction coefficient (i.e., $P_{slip} = B_{pre} \cdot u$). The amount of slippage is equal to the clearance of the bolt hole to the diameter of the bolt shank ($\Delta_c$), typically taken as 1.6 mm for construction. After slippage reaches the amount of the bolt clearance, the bolt shank begins to bear around the bolt hole, and then both stiffness and strength increase again.

### 5.3.3 Failure modes

All T-stub connection models were designed to reach yielding on the connection components such as tension/shear/web bolts, shear tab plates, and T-subs when the beam produced its full plastic moment at the plastic hinge. This design criterion fulfills the requirements for connection design given in both the AISC-LRFD manual (AISC, 2001) and the AISC Seismic Provisions (AISC, 2005). In addition, connection components were designed with the intent to avoid the catastrophic losses of stiffness and strength due to brittle failures (e.g., bolt fracture). The design forces ($P_p$) were obtained from the plastic moment of the beam as specified in the Seismic Provisions (AISC, 2005). Accordingly, the ultimate strength capacity of T-stub connections ($P_u$) shall be equal or over the design force as follows:

$$P_p = \frac{M_p}{d} \leq P_u = \frac{M_u}{d}. \tag{5.17}$$

Table 5.3 shows the strength capacities of T-stub components. Both peak loads ($P_{u,Exp}$) and dominant failure modes, which were observed at experimental tests (Smallridge, 1999; Schrauben, 1999), are also summarized in this table. The design strength ($P_p$) is governed by the plastic hinge mechanism of the beam, followed by T-stem yielding ($P_{y,stem}$), and finally followed by the T-stem fracture at considerably large deformations. The experimental connection models can be verified by the fact that their ultimate capacities exceed the design strength as presented in the table (e.g., $P_{u,Exp} > P_p$). The relatively thin member thickness of the T-stem can lead to the major energy dissipation. This situation will provide a balanced failure which maximizes the deformation capacity of the T-stem. In an effort to achieve the balanced failure mode, most of the connection models were designed with the ultimate strength capacities for the T-stubs that finally failed with net section failures (see Eq. 5.18(a) and Table 5.3). The fractures have conventionally classified as the brittle failure but the

*Table 5.3* Strength capacities of T-stub components.

| Model ID | Beam Plastic Hinge | Bolt/Flange Capacity | | T-Stem Capacity | | Experimental Results | |
|---|---|---|---|---|---|---|---|
| | $P_p^*$ | $P_{n,flange}^*$ ($\alpha$) | $\Sigma B_{n,tension}^*$ | $P_{y,stem}^*$ | $P_{u,stem}^*$ | Peak Load ($P_{u,Exp.}$)* | Failure Mode** |
| FS03-TD04 | 1189 | 1745 (3.74) | 2667 | 1106 | 1476 | 1270 | Net Section |
| FS04-TD08 | 1189 | 2300 (5.27) | 3510 | 1214 | 1620 | 1316 | Net Section |
| FS05-TA01 | 1551 | 2173 (1.10) | 2667 | 1742 | 2317 | 1790 | Net Section |
| FS06-TA09 | 1551 | 2788 (1.71) | 3510 | 1504 | 2000 | 1738 | Net Section |
| FS07-TB01 | 1551 | 2185 (1.04) | 2667 | 1870 | 2473 | 1768 | S-Bolt |
| FS08-TB05 | 1551 | 2800 (1.63) | 3510 | 1836 | 2418 | 1643 | Net Section |
| FS09-TC01 | 2428 | 2344 (0.44) | 2667 | 2507 | 3202 | 2787 | T-Bolt |
| FS10-TC09 | 2428 | 2930 (0.83) | 3510 | 2461 | 3144 | 2934 | Net Section |

*: Unit is kN
**: Final Ultimate Failure

level of the deformation associated with the net section fracture of the T-stem provides considerable ductility or energy dissipation compared to other brittle failures. It is due to membrane effects and the isotropic strain hardening of the base material (Swanson & Leon, 2001). On the other hand, the FS09-TC01 model was designed with the relatively thick thickness of the T-stem ($t_w = 17\,\text{mm}$). Accordingly, the capacity of the T-stub flange is less than that of T-stem yielding (see Eq. 5.18(b) and Table 5.3). As a result, the tension bolt fracture combined with the plastic yielding of the T-stub flange was observed at this connection model. Bolt fractures are one of the brittle failure modes that should be avoided because they may cause sudden strength degradation in the whole structure. The sequence of failure modes accompanied with the strength capacities of individual T-stub components can be written as follows:

$$P_{y,stem} \le P_{u,Exp.} < P_{u,stem} \le P_{n,flange} \tag{5.18a}$$

$$P_{n,flange} \le P_{y,stem} < P_{u,Exp.} \le P_{u,stem} \tag{5.18b}$$

It will be also verified through analytical investigation guided by FE analysis results.

## 5.4 3D FE T-STUB CONNECTION MODELS

The ABAQUS nonlinear FE code program (ABAQUS, 2008) was used to predict the response of bolted T-stub connections. FE connection models consisted of several independent parts as shown in Fig. 5.6. The FE connection models were made up of 3D solid elements incorporating fully nonlinear material properties, geometric nonlinearity, symmetric boundary conditions, prescribed displacement, and initial bolt pre-tension. In addition, surface interactions combined with friction and rough condition were assigned between adjacent faying surfaces. They lead to an increase in the slip resistance. Especially, friction occurring at the faying surface between the beam flange and the T-stub became the main source of force transfer before the transmitted axial force arrived at the slip load. The FE connection models incorporated constitutive laws for materials using multi-linear using true stress vs. total true strain curves.

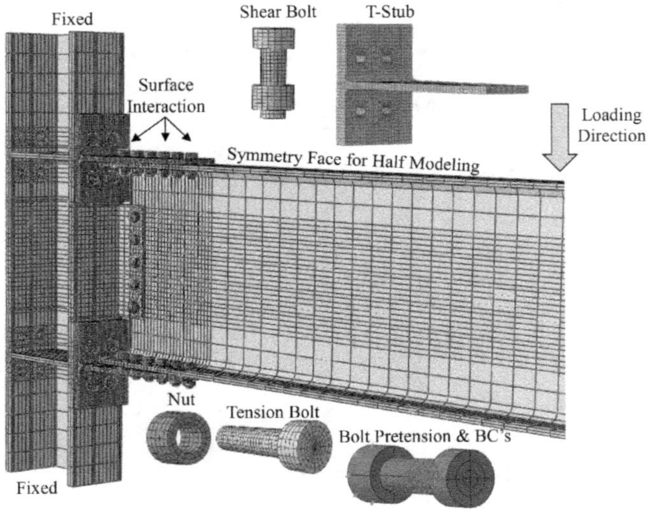


*Figure 5.6* 3D FE T-stub connection model and its independent parts.

The connections were modeled as half models using symmetric boundary conditions for the purpose of shortening the analysis time.

The FE connection models were loaded in two steps. The bolt pretension was generated by applying an adjusted length to the bolt shank during the first step and then propagated to the second step. Prescribed displacements were uniformly adjusted with the value of 0.20 mm at the middle of tension bolt shanks. On the other hand, according to the FE connection models, different adjusted length was imposed at the shear bolts in order that the slip resistance obtained from the FE analysis is accurately calibrated to that from the experimental test. The tip of the bolt was fully fixed during the first step and then released during the second step. To generate a bending moment at the connection, the main displacement loads were imposed on the tip of the beam corresponding to the position of a loading actuator (see Fig. 5.1 (b)) only during the second step. The reaction forces corresponding to these imposed displacement loads were measured by using the history output instrument in the ABAQUS program.

## 5.5 FE ANALYSIS RESULTS

Comparisons between experimental test results and FE analysis results, in terms of applied moment vs. total rotation curves, are presented in Fig. 5.7. Both results comprehensively show good agreements with respect to the initial stiffness, ultimate strength, shape of the envelop, and even level of the slip plateau. All connections models satisfy the design criterion in that they exceed the full plastic strength of the beam ($M_p$). When each pair of connection models with the same size of T-stub components (e.g., FS03-TD04 vs. FS04-TD08) are compared, it is noted that the diameter of tension bolts has much less influence on the increase of the ultimate strength than

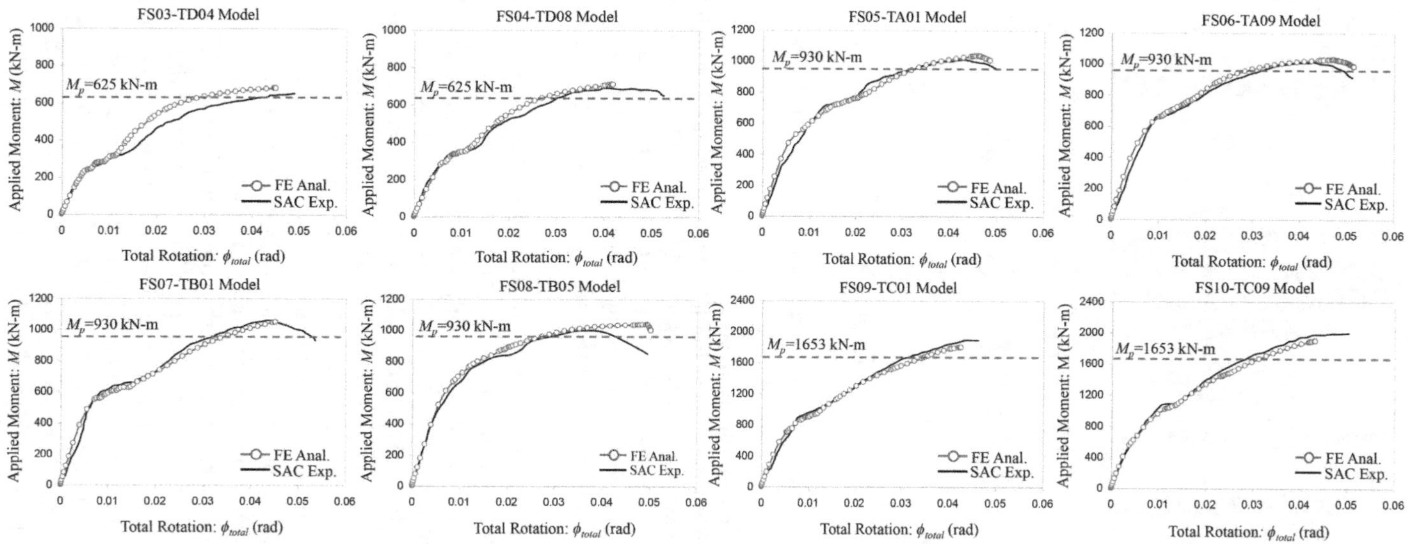


*Figure 5.7* Comparison of overall moment-rotation response between FE analysis and experimental test.

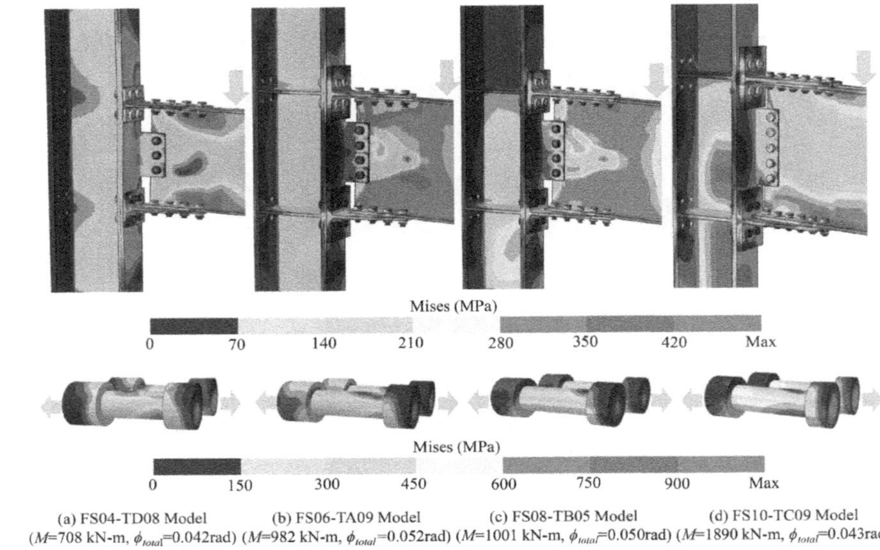


*Figure 5.8* Observation of Mises stress distribution at the final displacement load.

other geometric parameters. The deformable contribution of tension bolts may be negligible as compared to that of desirable component yielding because connection models were designed with the relatively higher strength capacity of tension bolts. Accordingly, the bolt fracture does not occur before the connection components yield. This hypothesis can be verified by observing the stress contour distributed over the connection.

Fig. 5.8 shows the distribution of the Mises stress including the deformed configuration at the final displacement load. The maximum total rotation angles achieved by 3D FE analyses reached more than 0.04 radians. The upper T-stub under tension force pulled out the tension bolts while the lower T-stub under bearing compression pulled back the shear bolts into the steel column. For the beam, the plastic yield stress, which was plotted as the red color at the stress contour, concentrated on the flange, and then spread into the web fast. Moreover, the amount of the plastic yield stress was accumulated at the T-stem near the K-zone.

The stress contours at the tension bolts highlight the prying force as well as the local bending moment. Due to the bending effect in the tension bolts, the concentration of the plastic yield stress occurs underneath a one-sided bolt head. This bending effect can act to reduce their axial strength capacity. The amount of the prying force can be approximately predicted by examining the intensity of the stress contour on the tip of the upper T-stub flange. The more intense level of the stress contour distributed over the upper T-stub flange was observed at the FS04-TD08 model in comparison with the FS10-TC09 model (see Figs. 5.8(a) and (d)). It can be concluded that the FS04-TD08 model produces more prying forces than the FS10-TC09 due to the smaller flange capacity as presented in Table 5.3. T-stems and a beam yielded more than tension bolts as also shown in the figures. As a result, the yielding of such components seems to precede the bolt fracture.

## 5.6 INVESTIGATIONS OF FE ANALYSIS RESULTS

Several general observations from FE connection models after analyses are able to provide valuable and qualitative information on the nonlinear behavior of the T-stub components under actual loading conditions. The first aspect to take into consideration is to investigate the overall force-deformation response of T-stub components. Three selected resulting curves, which were measured from LVDTs F (see Fig. 5.2), are shown in Fig. 5.9. The applied axial forces were converted in accordance with the first order approximation defined as Eq. (5.5). The limits for the plastic yielding of the T-stem ($P_{y,stem}$) are also plotted in the figure. Except for the FS08-TB05 model, the ultimate capacities obtained from experimental and analytical results exceed these limits. The slightly larger over-strength was achieved owing to the hardening range found between the yield stress ($F_y$) and the ultimate tensile stress ($F_u$) of the base T-stem material. The connection models are susceptible to the net section fracture. Furthermore, the main plastic deformation of the T-stub component occurs at the state adjacent to the yielding of the T-stem. Thus, the T-stem contributes to increasing energy capacity and ductility in the connection behavior.

In addition to the deformable contribution, the response mechanism for other component members can be also investigated through FE analysis results. Fig. 5.10 shows the capacity of the T-stub flange plotted as the function of the flange thickness. Two T-stubs, TB05 and TC01, are selected to investigate the capacity of the T-stub flange according to the geometric parameters concerning the T-stub flange thickness ($t_f$), the ratio of $H_1$ to $H_2$, and the diameter of the bolt ($d_b$). The failure modes are also involved with these geometric parameters. The line section OC, CD, and DE are calculated by means of Eqs. (5.7), (5.8), and (5.9), respectively. On the other hand, the line section OB is computed in accordance with the equation resulting from Eq. (5.11). The point B represents the ultimate strength of the T-stub flange corresponding to the T-stub flange thickness.

For the FS08-TB05 model with $t_f = 24$ mm, $H_1/H_2 = 0.48$, and $d_b = 25$ mm, the line section OB is drawn with $\alpha = 1.63$ (see Table 5.3). The line section ABC indicates a bolt fracture in tension after the plastic hinge has been developed in the T-stub flange. The ultimate strength capacity of the T-stub flange (i.e., $P_{n,flange} = 2800$ kN) is obtained from the Point B anchored to the extended line section ABC. After establishing the plastic hinge mechanism, the T-stub flange can resist additional forces until tension bolts fail by the fracture. It is because that strain hardening was applied to the base material of the T-stub flange. Thus, owing to the value of $\alpha$ exceeding 1.0, the plastic hinge mechanism occurring at the line section OC results in the theoretical failure mode but does not represent the ultimate strength capacity. For the FS09-TC01 model with $t_f = 32$ mm, $H_1/H_2 = 0.43$, and $d_b = 22$ mm, the line section OB is drawn with $\alpha = 0.44$. As a result, the Point B referred to as the ultimate strength capacity of the T-stub flange (i.e., $P_{n,flange} = 2344$ kN) is anchored to the line section CD.

The ultimate capacity of the T-stub component obtained from the FE analysis ($P_{u,Anal.}$) is also plotted as the solid symbol on the solution space in an effort to verify

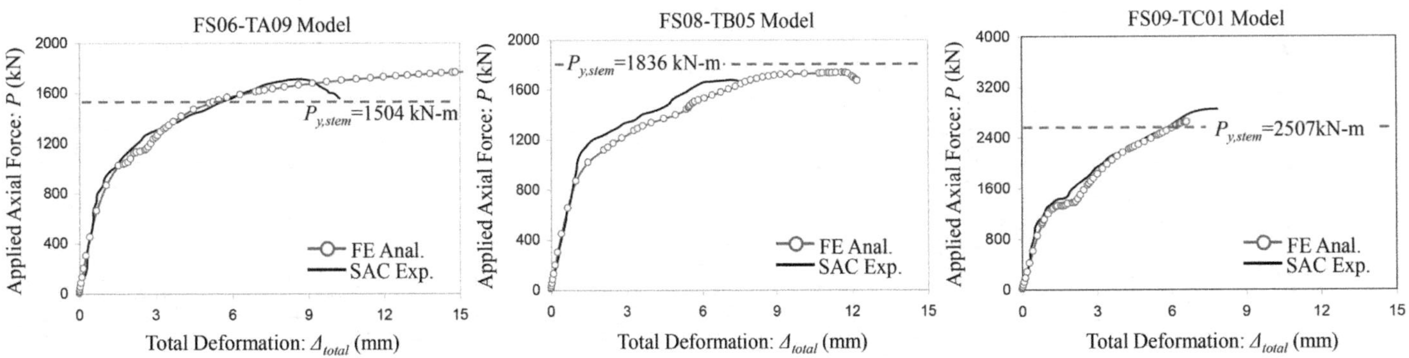


*Figure 5.9* Overall force-deformation response of T-stub components.

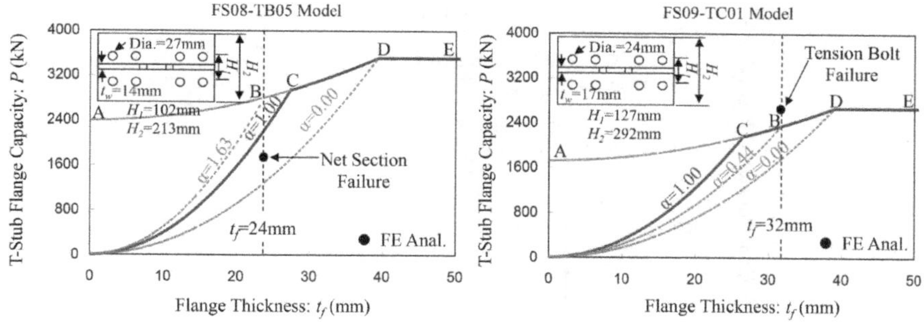


*Figure 5.10* Determination of the strength capacity at the T-stub flange based on the solution space.

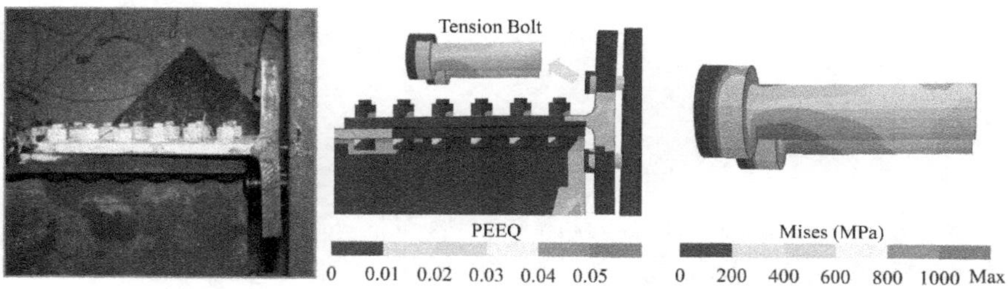


*Figure 5.11* Investigation of the failure mode on connection components with FE analysis results (FS09-TC01 model).

the sequence of failure modes. The ultimate T-stub capacity of the FS08-TB05 model should lie below the capacity curve for the T-stub flange ($P_{n,flange} > P_{u,Anal.} = 1680\,\text{kN}$) on the ground that this connection model failed by the net section fracture preceding the failure of the T-stub flange, as presented in Eq. (5.18a). The bolt fracture also provides insight. For the FS09-TC01 model, the ultimate T-stub capacity lies just above the capacity curve ($P_{n,flange} < P_{u,Anal.} = 2650\,\text{kN}$). This model fails by the tension bolt fracture that follows the failure of the T-stub flange, as presented in Eq. (5.18b). The ultimate T-stub capacity reaches the tensile capacity of bolts very close ($\Sigma B_{n,tension} = 2665\,\text{kN}$).

The observation of field output contours is also needed to investigate not only the dominant failure mode but also the bolt prying mechanism. The Von-Mises stress (Mises) and equivalent plastic strain (PEEQ) field output contours, which were distributed to the tension bolt and the T-stub under the ultimate load, are shown in Fig. 5.11. The FS09-TC01 model is selected for investigation. The intensity of the plastic strain contour demonstrates that bolt shanks underneath the one-sided bolt head undergo plastic deformations on account of the axial force combined with the local bending moment. Particularly, the plastic hinge determined in conformity with the intensity of approximately 4.0 percent plastic strain is found at the section of the

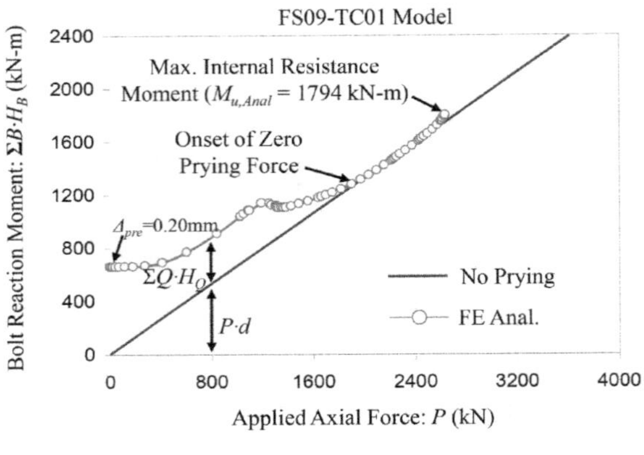


*Figure 5.12* Prying action response of tension bolts.

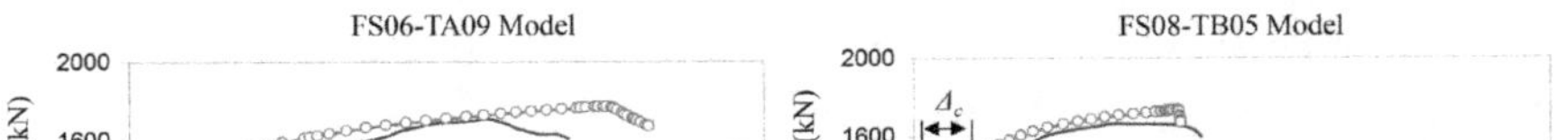


*Figure 5.13* Force-slip and bearing deformation response of T-stub components.

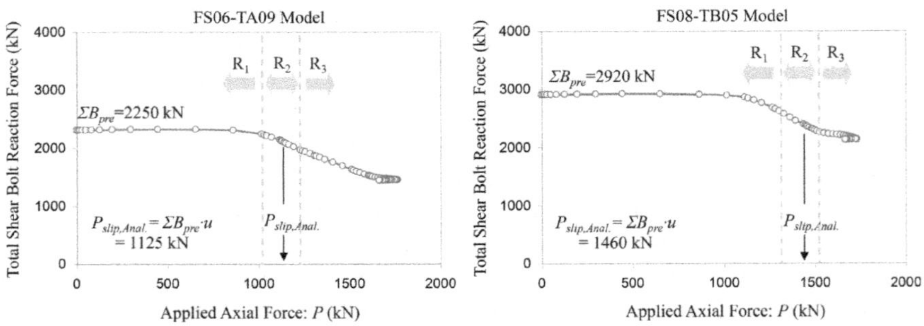


*Figure 5.14* Slip response of shear bolts.

K-zone. This physical phenomenon observed can prove the sequence of failure modes for the FS09-TC01 model as discussed above. Likewise, the final deformed configuration obtained from the experiment fits well into that from the FE analysis in respect that the T-stub flange is completely separated from the column flange. According to the final deformed shape of component members, the ultimate failure for this model can be characterized by the Failure Mode 3 presented in Fig. 5.4 (c). Therefore, the prying force becomes zero at the ultimate loading state.

Bolt reaction forces ($B_n$) measured from the FE models after analyses are also utilized to reliably examine the prying response of the FS09-TC01 model beyond the experimental test results. The detail investigation into the prying response of tension bolts is presented in Fig. 5.12. The blue solid line shown in the figure represents the behavior of the connection where no bolt prying and zero initial bolt pretension are taken into account. However, most of the connection behavior obeys the path of curves reproduced by the FE analyses. The bolt reaction forces are converted into the internal reaction moments defined in Eq. (5.3) (see also Fig. 5.3). The summation of the internal reaction moments starts at a non-zero point because the tension bolts are initially pretensioned. For that reason, the prying force corresponds to the bolt reaction force at the beginning of the loading state. As the converted axial force ($P$) increases, the internal reaction moment in the tension bolts also increases. Note that the solid arrows displayed above $P = 800$ kN in Fig. 12 are employed to emphasize the static equilibrium derived from Eqs. (5.3) and (5.6), such that $M_r = \Sigma B_n H_B = Pd + \Sigma Q H_Q$. When the loading state proceeds to approximately $P = 2000$ kN, the prying force converges to zero, and then separation starts to occur between T-stub flange and column surface. Thenceforth, the zero prying force continues until the ultimate load ($P_{u,Anal.} = 2650$ kN and $M_{u,Anal.} = 1794$ kN-m), and then the converted axial force is directly transmitted into the tension bolt. The bolt fracture is easy to arise for this reason.

Other behavioral characteristics such as slip plateau and bolt bearing deformations can be also elucidated through FE analyses. Fig. 5.13 shows the slip response measured in the experiments with LVDTs B (see Fig. 5.1) and the corresponding FE analysis results. The coefficient of friction and the initial pretension of shear bolts result in the critical parameters to determine the load at the initial slip. The elastic deformation is observed at the curve before slippage occurs. The friction coefficient was uniformly applied to shear faying surfaces between the beam flange and the T-stub with the value of 0.5. Due to the variable displacement used for the initial shear bolt pretension, FE analyses provide the best match into the experimental test data where the level of the slip plateau referred to as the slip resistance was various (see A and B in Fig. 5.13). One would expect very consistent results in this characteristic from the FE models since the uniform clearance between bolt shanks and around bolt holes ($\Delta_c = 1.6$ mm) were made in all FE analyses. Sliding continues to occur until the shanks of shear bolts come into contact with the inner surface of bolt holes. The shear bolts come into bearing thereafter. Bearing deformations start to increase the slope of the loading curves again. The FS06-TA09 model shows a clear slip plateau and bearing deformation, while the FS08-TB05 model has a combination of slip and yielding around 1460 kN. Overall, FE analysis results show satisfactory agreement with the experimental results.

Additional observations for the shear bolt reaction forces that have influence on the slip response are illustrated in Fig. 5.14. The variable adjustment length of the shear bolts causes to generate different initial bolt pretension forces according to the connection model (e.g., $\Sigma B_{pre} = 2250$ kN and $\Sigma B_{pre} = 2920$ kN for the FS06-TA09 model and the FS08-TB05 model, respectively). The path of the curve shown in the figure can be divided into three stages defined as the dashed lines so as to evidently explain the procedure of the slip behavior with more details. The reaction force preserves the initial bolt pretension prior to slippage ($R_1$ in Fig. 5.14). During the section of the slip ($R_2$), the reaction force is on the decrease due to the loss of the bolt pretension. The applied axial force continuously increasing exceeds the slip resistance which results in the product of the initial bolt pretension and the friction coefficient ($P > P_{slip} = \Sigma B_{pre} \cdot u$). The bolt reaction force is continuously decreasing during the bearing deformation ($R_3$). Finally, it starts to maintain the constant level again at the onset of yielding around shear bolt holes.

## 5.7 CONCLUDING REMARKS

The refined 3D FE models successfully predicted the complete behavior of full-scale connections. The local behavior of individual T-stub component members was also captured well by the FE models, which accurately reproduce the development of a prying mechanism, the sequence of sliding, and bearing contact. Moreover, FE analysis results, which were found to be in good agreement with the actual connection behavior, are used to review behavioral changes as the consequence of geometric variations.

The prying action mechanism tends to be more influenced by flange thickness and tension bolt location. The amount of the prying force according to these geometric variations can be clearly identified by observing stress distributions and bolt force tracks. In addition, computations on the strength capacity for the T-stub flange were made, presenting the T-stub failure modes found at FE connection models after analyses, and emphasizing the interaction between the T-stub flange and the tension bolts. Except for the FS09-TC01 model, the tension bolts were not sufficient to develop the plastic deformation in the connection.

The slip model, which determines the slip resistance, is sensitive to bolt pretension as well as a friction coefficient determined in accordance with surface condition. In the same manner of the experimental test, the value of the adjusted length for generating initial bolt pretension in the FE model remains a random variable that has a significant influence on the local behavior of T-stub components. However, the uniform friction coefficient to define the surface interaction between T-stem and beam flange was applied to all FE models. Consequently, the level of the slip plateau evaluated from the FE analysis fits into that obtained from the corresponding experimental test very well.

Based on the failure mode, ideal T-stub connection design is one where the ultimate state is governed by the yielding to the T-stem. The ultimate strength of the T-stem is followed by that of the T-stub flange and finally followed by that of the tension bolts. This failure sequence offers a balanced failure which utilizes the deformation capacity of both stem and flange to the maximum. Especially, the T-stem member was allowed not only to behave in ductile manner but also to achieve substantial deformation.

Finally, it is concluded that FE models proposed herein can be considered as reliable tool to reproduce the complex connection behavior and to estimate the response mechanism according to the parametric effect. They also provide valuable data to assess failure modes and strength capacities.

Chapter 6

# Bolted connections (component models)

## 6.1 GENERAL OVERVIEW

The behavior of full-scale connections typically represented by a nonlinear moment-rotation curve is based on the various response mechanisms of individual connection components (Leon, 1997; Mehrabian *et al.*, 2009; Shi *et al.*, 1996; Faella *et al.*, 2000; Kim *et al.*, 2008). The behavior of each connection component under monotonic loads is fairly simple to simulate with bi-linear or tri-linear stiffness expressions. However, the interaction between connection components is not always easy to predict because they are not behaving independently anymore within the connection. Moreover, this problem is compounded for the case of large cyclic deformations where careful checking on permanent deformations is needed (Rassati *et al.*, 2004). Therefore, the connection models are very complex and require a large number of stiffness components. For the computational convenience, structural connections designed in the past were assumed to be extreme behavioral ones, i.e., the simple pinned connections and the ideally welded connections, and therefore the necessity for the actual moment-rotation response of the connection was very limited (Leon *et al.*, 1998; Green *et al.*, 2004). In reality, most connections including bolted connections exhibit the intermediate behavior between two extreme cases.

The mechanical modeling of steel bolted connections is based on the characterization of individual component members, with a well-defined behavior inside the connection (Pucinotti, 2001; Clemente *et al.*, 2004). The connection components need to schematize the deformability contributions so as to take their interaction into consideration. They can be modeled as the nonlinear component springs with their own stiffness properties. Therefore, mechanical models, which are formed as an assembly of component springs and rigid elements, are suitable for the simulation of complex connection behavior under either static or dynamic loads.

This modeling approach provides the flexibility which is able to accommodate different connection configurations with the same basic component spring theory (Clemente *et al.*, 2004; Eurocode 3, 2003). In this case, the connection can be decomposed into the proper component springs, and then their individual responses are assembled to reproduce the behavior of the full-scale connection. It has the clear advantage of being easily scalable to the modeling of bolted connections. In addition, the mechanical models can relieve high computational complexity and cost in comparison with exiting finite element (FE) models commonly used for calibration.

For these advantages, the mechanical modeling has been accepted for the reliable estimation of their nonlinear response. The mechanical models in the established researches (Clemente *et al.*, 2004; Eurocode 3, 2003; Beg *et al.*, 2003; Coelho & Silva, 2005; Piluso & Rizzano, 2008; Hu *et al.*, 2011; Hu *et al.*, 2012) lay more focus on predicting the initial stiffness and ultimate strength rather than the entire moment-rotation curve, especially, the rotational capacity. Therefore, the mechanical models, which are able to simulate the complete moment-rotation curve of full-scale bolted connections, will be mainly dealt with under cyclic loads.

First, a method of assembling the T-stub connection components, which are converted to nonlinear component springs in the mechanical model, is presented in this study. The contributions of deformability mechanisms between T-stub component members are also investigated by first quantifying their force-deformation responses. These mechanisms are then combined in parallel or in series according to the interaction of the force transfer. The stiffness model predictions of T-stub connection components with respect to initial stiffness, ultimate capacity, and dominant failure mode are built into the multi-linear models for each force-deformation mechanism. The component springs are installed on a mechanical joint model in order to accurately simulate the moment-rotation curve of structural connections apart from experimental testing. The numerical test results obtained from the mechanical models are compared with the corresponding experimental test results to validate that the mechanical modeling proposed herein is adequate to predict the resulting responses for T-stub components and connections.

## 6.2 EXPERIMENTAL STUDY

As a part of the SAC Task 7.03 project (Swanson, 1999; Smallidge, 1999), Swanson tested 48 T-stub components individually under cyclic loads (Swanson, 1999). Additionally, 8 full-scale bolted beam-to-column T-stub connection tests were performed by Smallidge (1999). In this study, 8 T-stub components, which were used in the full-scale connections, were selected to calibrate comprehensive mechanical joint models. The instrumentation scheme and typical dimensions of individual T-stub components tested cyclically are given in Fig. 6.1. The geometric details of selected T-stub components are also summarized in Table 6.1, along with the ultimate load capacity obtained from the experimental component tests.

T-stubs were cut from four standard W-shape sections so as to accommodate more than 8 shear bolts in the stem and were designed to be utilized in full strength (PS) connections for W530 to W690 beam sections. The width of T-stems was tapered. The size, number, gauge, and spacing of the bolts were changed to investigate the bolt prying on the T-stub flange, slippage on the T-stem, and bearing compression according to these parametric effects. T-stub models were also designed with the flange thickness ($t_f$) ranged from 14 mm to 32 mm. The test series, which is TA, TB, TC, and TD listed in Table 6.1, was classified by this flange thickness. All of the T-stub components were fabricated with FE 350 (A572 Gr. 50) steel. 22 and 25 mm diameter M10.8 (ASTM A490) tension bolts with various configurations were used in a pair of test series (e.g. TA01 vs. TA09). The oversize bolt holes (bolt clearance of 1.6 mm) were applied to all specimens.

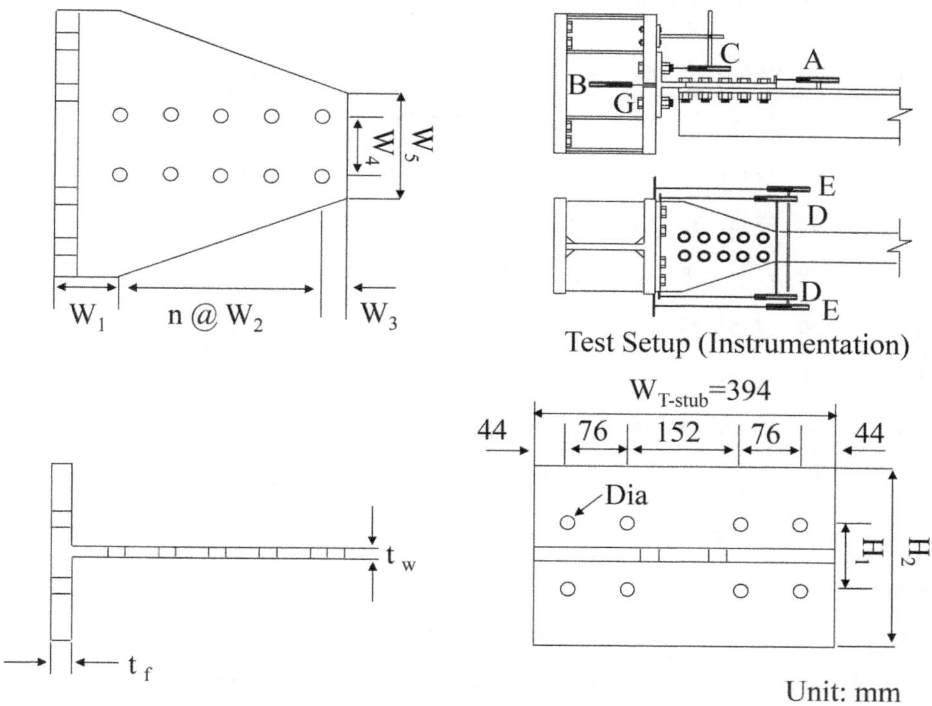


*Figure 6.1* Component details and instrumentation scheme for experimental tests.

The behavior of T-stub components can be predicted rapidly by applying axial loads to the T-stub components. The axial loads correspond to the beam-flange forces in actual connections. The component tests were extensively instrumented with linear variable displacement transducers (LVDTs). Each pair of LVDTs shown in Fig. 6.1 was used to measure individual displacements. LVDTs A measured the relative slip displacement between the T-stub and the beam-flange. LVDTs B monitored the uplift of the T-stub flange from the face of a column, while LVDTs C monitored that of tension bolts. LVDTs D measured the elongation of the T-stem and LVDTs E measured the overall component deformation.

The results of 8 full-scale connection tests were also used to calibrate the mechanical joint models. The connection details are illustrated in Fig. 6.2. Cyclic loads were generated by imposing the support displacements to the tip of the steel beam. The typical test configuration consisted of a $360 \times 216$ column with pinned end supports, a 4.5 m beam which varies from a $W530 \times 66$ to a $W690 \times 125$, and three connecting members which are one shear tab and two T-stubs. Especially for T-stub members, full connection models were designed with the geometric details which were described in the component test. Thus, the model identifications (IDs) are composed of the full-scale test number and a T-stub classification based on the component tests (e.g. FS05-TA01). Besides the T-stub components,

*Table 6.1* Geometric size of component specimens.

| *Model* | $W_1$ | $W_2(n)$ | $W_3$ | $W_4$ | $W_5$ | $t_f$ | $t_w$ | $H_1$ | $H_2$ | *Dia.* | *Bolt (diameter × length)* | *Peak Load (Experiment)* |
|---|---|---|---|---|---|---|---|---|---|---|---|---|
| TA01 | 84 | 67 (4) | 33 | 89 | 152 | 25 | 14 | 102 | 264 | 24 | 22 × 114 mm (T)*<br>22 × 76 mm (S)** | 2085 kN |
| TA09 | 89 | 76 (3) | 38 | 89 | 152 | 25 | 14 | 102 | 264 | 27 | 25 × 114 mm (T)<br>25 × 89 mm (S) | 1924 kN |
| TB01 | 84 | 67 (4) | 33 | 102 | 178 | 24 | 14 | 102 | 213 | 24 | 22 × 114 mm (T)<br>22 × 89 mm (S) | 2252 kN |
| TB05 | 89 | 76 (4) | 38 | 102 | 178 | 24 | 14 | 102 | 213 | 27 | 25 × 114 mm (T)<br>25 × 89 mm (S) | 2236 kN |
| TC01 | 90 | 67 (5) | 33 | 127 | 241 | 32 | 17 | 127 | 292 | 24 | 22 × 127 mm (T)<br>22 × 89 mm (S) | 2601 kN |
| TC09 | 95 | 76 (5) | 38 | 127 | 241 | 32 | 17 | 127 | 292 | 27 | 25 × 127 mm (T)<br>25 × 102 mm (S) | 2949 kN |
| TD04 | 84 | 67 (3) | 33 | 89 | 152 | 14 | 10 | 102 | 178 | 24 | 22 × 95 mm (T)<br>22 × 76 mm (S) | 1094 kN |
| TD08 | 89 | 76 (3) | 38 | 89 | 152 | 14 | 10 | 102 | 178 | 27 | 25 × 95 mm (T)<br>25 × 83 mm (S) | 1132 kN |
| BC's | Stiffened W360X382 Column stub, 152 × 25 plate used for a beam flange | | | | | | | | | | | |
| Steel Grade | | | | | | | | | | | | |
| Beam | FE 350(Grade 50) Steel | | | | | | | | | | | |
| Setback | Each 51 mm for beam setback. | | | | | | | | | | | |
| Diameter | Adding clearance (1.6 mm) to the diameter of bolt | | | | | | | | | | | |

*(T) = Tension bolts; **(S) = Shear bolts Grade for all used bolts = M10.8 (ASTM A490)

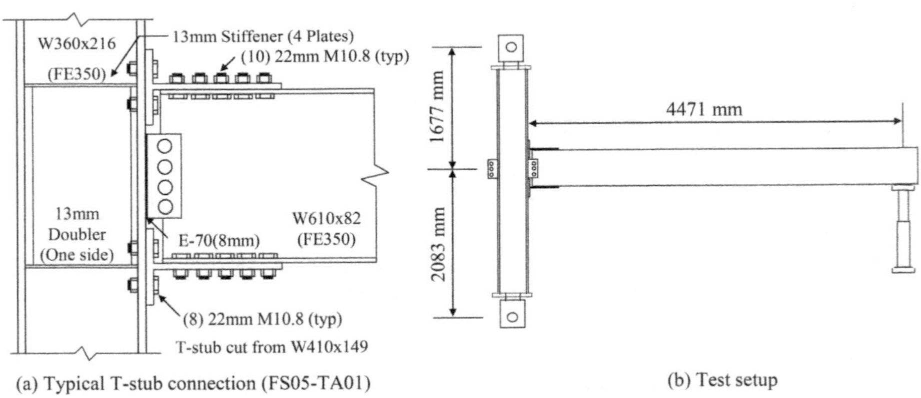


*Figure 6.2* Dimensions of the typical full-scale T-stub connection.

steel components used in the T-stub connections are summarized in Table 6.2. The material properties for the steel components are also provided in this table. The FS05-TA01 model was designed with a W610 × 82 beam connected to a W360 × 216 column by a 10 × 125 × 305 mm shear tab and two T-stubs cut from a W410 × 149 section (see Fig. 6.2). The spacing between two bolts on the shear tab was uniformly taken as

*Table 6.2* Geometric size of full-scale T-stub connections.

| | *T-Stub* | | | | *Beam* | | | *Column* | | | *Bolt* | | |
|---|---|---|---|---|---|---|---|---|---|---|---|---|---|
| *Model ID* | *Model ID* | *Size** | *Yield Stress*** | *Ultimate Stress*** | *Size** | *Yield Stress*** | *Ultimate Stress*** | *Size** | *Yield Stress*** | *Ultimate Stress*** | *Diameter**** | *Tensile Capacity***** | *Shear Tab Size** |
| FS03 | TD04 | W410 × 67 | 396 (F)<br>427 (W) | 557 (F)<br>569 (W) | W530 × 66 | 400 | 489 | W360 × 216 | 386 | 516 | 22 | 334 (T)<br>336 (S) | 10 × 127 × 229 |
| FS04 | TD08 | W410 × 67 | 396 (F)<br>427 (W) | 557 (F)<br>569 (W) | W530 × 66 | 400 | 489 | W360 × 216 | 386 | 516 | 25 | 438 (T)<br>435 (S) | 10 × 127 × 229 |
| FS05 | TA01 | W410 × 149 | 318 (F)<br>352 (W) | 460 (F)<br>469 (W) | W610 × 82 | 420 | 523 | W360 × 216 | 386 | 516 | 22 | 334 (T)<br>336 (S) | 10 × 127 × 305 |
| FS06 | TA09 | W410 × 149 | 318 (F)<br>352 (W) | 460 (F)<br>469 (W) | W610 × 82 | 420 | 523 | W360 × 216 | 386 | 516 | 25 | 438 (T)<br>435 (S) | 10 × 127 × 305 |
| FS07 | TB01 | W530 × 138 | 362 (F)<br>378 (W) | 498 (F)<br>501 (W) | W610 × 82 | 420 | 523 | W360 × 216 | 386 | 516 | 22 | 334 (T)<br>336 (S) | 10 × 127 × 305 |
| FS08 | TB05 | W530 × 138 | 362 (F)<br>378 (W) | 498 (F)<br>501 (W) | W610 × 82 | 372 | 487 | W360 × 216 | 386 | 516 | 25 | 438 (T)<br>435 (S) | 10 × 127 × 305 |
| FS09 | TC01 | W840 × 251 | 390 (F)<br>415 (W) | 523 (F)<br>530 (W) | W690 × 125 | 415 | 510 | W360 × 216 | 386 | 516 | 22 | 334 (T)<br>336 (S) | 10 × 127 × 381 |
| FS10 | TC09 | W840 × 251 | 390 (F)<br>415 (W) | 523 (F)<br>530 (W) | | 415 | 510 | W360 × 216 | 386 | 516 | 25 | 438 (T)<br>435 (S) | 10 × 127 × 381 |

*:Metric Size **: MPa (Coupon Tests) ***:mm ****:kN
F: Flange W: Web T: Tension Bolt S: Shear Bolt

76 mm, so four 22 mm diameter and 58 mm long M 10.8 bolts were used to fasten the beam web to the shear tab. The panel zone was stiffened with four 13 mm thick stiffeners and a 13 mm thick doubler plate on one side.

## 6.3 STIFFNESS MODELING

### 6.3.1 Mechanical joint model

The mechanical spring model initially introduced by the Eurocode 3 (Eurocode 3, 2003) provides an efficient solution for obtaining either continuous or multi-linear moment-rotation curves, as compared with currently existing methods: (1) experimental testing, (2) refined FE modeling, and (3) analytical curve fitting model. In this mechanical model, the individual components of the connection are converted into the component springs. The behavior of these springs can be controlled by the stiffness model which simplifies the actual force-deformation response of connection components. In an effort to properly apply the behavior of bolted T-stub connections to the mechanical modeling, each of component springs should be added to the system in accordance with the force transfer from beam to connection. After considering the modeling philosophy for this spring assemblage, the overall rotational stiffness of the connection is affected by the combined stiffness models.

Figs. 6.3(a) and (b) show the idealization of the force distribution at the T-stub connection and the installation of idealized nonlinear springs in the mechanical joint model, respectively. Generally, the beam develops its flexural strength (i.e. plastic hinge) and carries the bending moment (M) transformed from the total applied force (T) at the tip of the beam. This bending moment is transmitted to the connection as the converted axial forces (P). The internal reactions in the connection component act

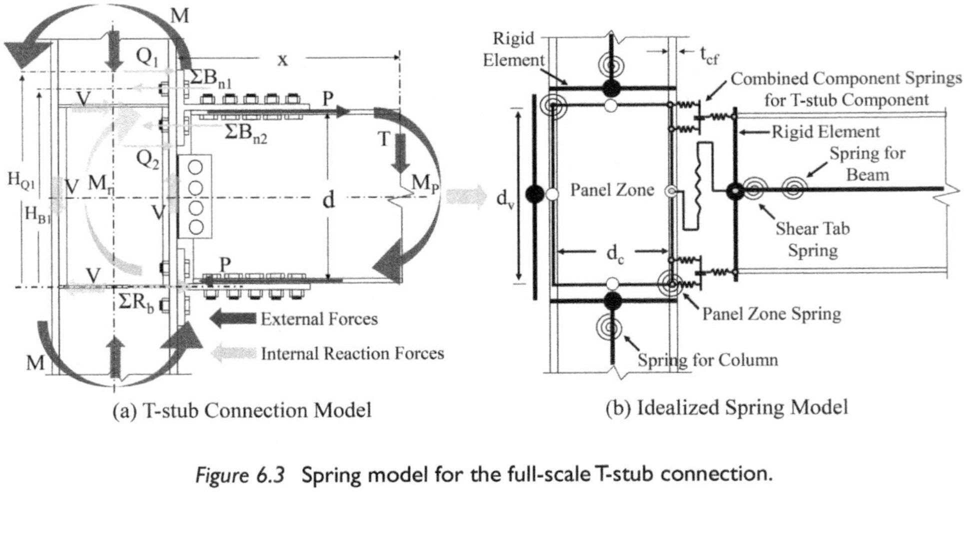


*Figure 6.3* Spring model for the full-scale T-stub connection.

against these external forces in order to satisfy equilibrium. They have the following relationship as shown in Fig. 6.3(a):

$$M_r = M_P + Tx = TL \tag{6.1}$$

$$M_r = \Sigma B_{n1} H_{B1} + \Sigma B_{n2} H_{B2} - Q_1 H_{Q1} - Q_2 H_{Q2} \tag{6.2}$$

$$P = \frac{M_r}{d} \tag{6.3}$$

$$P = \Sigma B_{n1} + \Sigma B_{n2} - Q_1 - Q_2 \tag{6.4a}$$

$$P = \Sigma R_b \tag{6.4b}$$

where $M_r$ is the internal resistant moment; $M_P$ is the plastic moment; $x$ is the distance from the column surface to the position of the plastic hinge; $L$ is the length of the beam; $H_{B1}, H_{B2}, H_{Q1}, H_{Q2}$, and $H_{Q2}$ are the equivalent heights at each position shown in Fig. 6.3(a); $\Sigma B_{n1}$ and $\Sigma B_{n2}$ are the summation of bolt reaction forces in tension; $Q_1$ and $Q_2$ are the prying force acting on the tip of the T-stub flange due to the initial bolt pretension; $d$ is the depth of the beam; and $\Sigma R_b$ represents the bearing force in compression. As shown in Eqs. 6.4(a) and (b), the force equilibriums are established under tension and compression.

Mechanisms that have an effect on the behavior of T-stub connections are classified by five main deformation responses: (1) overall T-stub deformation, (2) panel zone deformation, (3) beam deformation including plastic hinges, (4) shear deformation within the connected region, and (5) shear tab deformation. The mechanical joint model shown in Fig. 6.3(b) can reflect these mechanisms very well on the ground that the internal loads are carried by the component springs corresponding to the component members. The response of the panel zone under the shear deformations resulting from the bending forces occurs at the panel zone spring. It is deformed in a scissors-line manner. In particular, the end face of the beam modeled as the rigid element is assumed to behave as a rigid plate, leading to a linear strain pivoting about the center of bearing.

The combined component springs for the T-stub component are attached between the panel zone and the rigid element (Fig. 6.3(b)). They deform directly by the converted axial forces (P). Their mechanism is very complex and incorporates various types of deformation. However, the behavior of the T-stub component including the shear deformation of the panel zone has a significant effect on the moment-rotation curve. Thus, in spite of difficulty in modeling, both mechanisms are mainly investigated in this section because of this reason.

### 6.3.2 Component stiffness

Fig. 6.4 shows the spring model of the typical T-stub component. The overall T-stub deformation consists of the deformability contribution from the tension bolts, the T-stub flange, the T-stem, the shear bolts, the combined bearing and slip deformations, and the bearing compression. The axial tension-compression loads (P) resulting from the bending moments are transferred into the component springs. As shown in

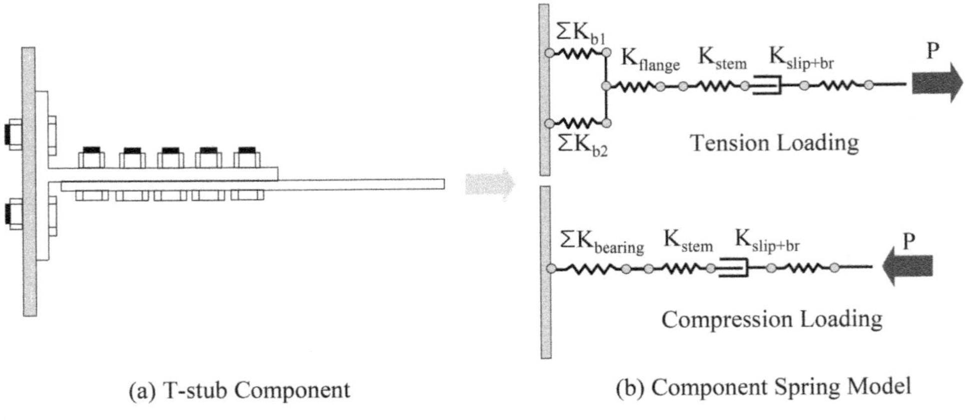


*Figure 6.4* Spring model for individual T-stub components.

Fig. 6.4(b), the different component springs can be assembled in parallel or in series depending on how to interact each other. The corresponding stiffnesses in the T-stub component are summarized as follows:

$$\frac{1}{K_{total,tension}} = \frac{1}{\Sigma K_{b1} + \Sigma K_{b2}} + \frac{1}{K_{flange}} + \frac{1}{K_{stem}} + \frac{1}{K_{slip+br}} \tag{6.5a}$$

$$\frac{1}{K_{total,compression}} = \frac{1}{\Sigma K_{bearing}} + \frac{1}{K_{stem}} + \frac{1}{K_{slip+br}} \tag{6.5b}$$

where $K_{total,tension}$ is the overall stiffness of the T-stub component under tension loads; $\Sigma K_{b1}$ and $\Sigma K_{b2}$ are the summation of the bolt stiffnesses with the parallel system (e.g. $\Sigma K_{b1} = K_{b1,1} + K_{b1,2} + K_{b1,3} + K_{b1,4}$ for four tension bolt installation at the first row); $K_{flange}$ is the stiffness of the T-stub flange subject to bending; $K_{stem}$ is the stiffness of the T-stem subjected to stretching; $K_{slip+br}$ is the stiffness for combining slip and bearing mechanism; $K_{total,compression}$ is the overall stiffness of the T-stub component under compression loads; and $\Sigma K_{bearing}$ is the stiffness resulting from the bearing mechanism between the T-stub flange and the column surface. Except for the individual stiffness of tension bolts, component stiffnesses are assembled into the overall stiffness with the series system.

The force-deformation responses of component springs are summarized in Fig. 6.5. The behavior of the bolts under tension loading is associated with the theoretical equation which is referred to as the force equilibrium (Eq. 6.4(a)). On the other hand, the bolts under compression loading make a negligible contribution to the response mechanism of the component spring model because of the bearing surface (Fig. 6.5(b)). Due to the series system, the overall deformation of the T-stub component can be determined by adding all component deformations as follows:

$$\Delta_{total,tension} = \Delta_{bolt} + \Delta_{flange} + \Delta_{stem} + \Delta_{slip+br} \tag{6.6a}$$

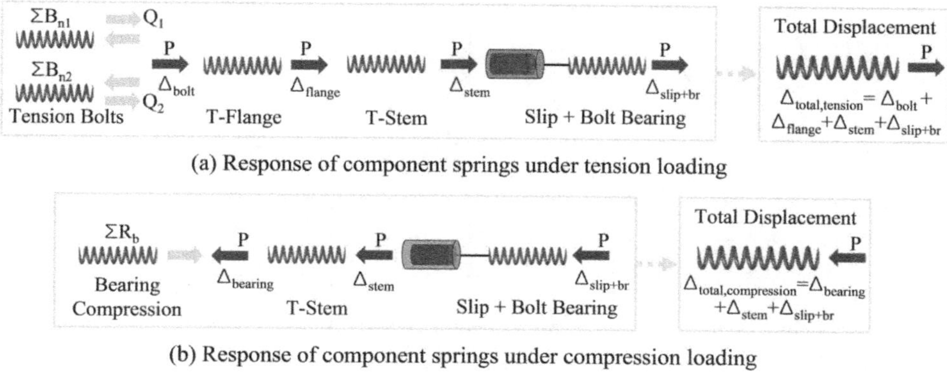


*Figure 6.5* Force-deformation responses of component springs.

$$\Delta_{total,compression} = \Delta_{bearing} + \Delta_{stem} + \Delta_{slip+br} \tag{6.6b}$$

where $\Delta_{total,tension}$ and $\Delta_{total,compression}$ are the overall deformation of the T-stub component; and $\Delta_{bolt}$, $\Delta_{flange}$, $\Delta_{stem}$, $\Delta_{slip+br}$, and $\Delta_{bearing}$ represent individual component deformations. The more detail investigation on the stiffness models will be addressed in the next sub-sections.

#### 6.3.2.1 T-stub flange and tension bolts

The prediction of obtaining the force-deformation relationship for the T-stub flange is a quite complex because several yielding or failure modes interact with one another. Among the possible failure modes, most of studied cases are related to the prying mechanism which arises from the bending action associated with the bolt pretension in the T-stub flange. This prying action is generally accompanied by the ultimate failure of the tension bolts or the yield failure of the T-stub flange. The established prying model accepted in the design guideline (AISC, 2001) has been used to estimate the strength capacity of the T-stub flange. This prying model was based on one of most widely used models proposed by Kulak *et al.* (1987).

The geometric notations used in the prying model are illustrated in Fig. 6.6. The prying forces ($Q$) are additional reaction forces occurring at the tip of the T-stub flange. The fundamental static equilibrium exists in this mechanism as follows:

$$\sum B_n = P + 2Q \tag{6.7}$$

where $\sum B_n$ represents the summation of all reaction forces in the tension bolts. Thus, the prying forces contribute to increasing the bolt reaction forces ($B_n$) and effectively reducing the axial load ($P$). They can be minimized by increasing the flange thickness ($t_f$) or by reducing the tension bolt gage ($g_t$). The geometric parameters $a'$ and $b'$ instead of $a$ and $b$ should be used to calculate the static equilibrium as shown in Fig. 6.6. This precondition is base on the fact that most of bolt reaction forces are assumed to be

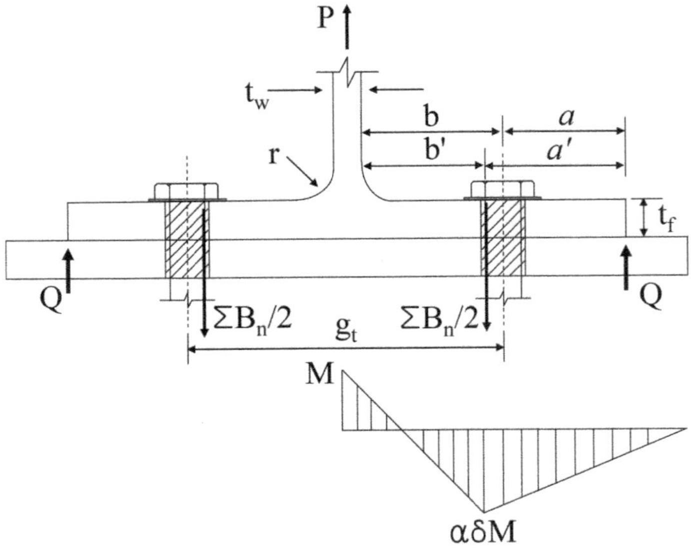


*Figure 6.6* Typical flange prying mechanism.

transferred into the inside edge of the bolt shank rather than the centerline of the bolt shank (Kulak *et al.*, 1987). The magnitude of $a$ is limited to a value no greater than $1.25b$ in this model.

The ultimate strength of the T-stub flange ($P_{n,flange}$) is computed based on three possible failure mode cases which are illustrated in Fig. 6.7. These three failure modes are expressed by Eqs. (6.8) to (6.10). They are relevant to the formation of a plastic mechanism on the T-stub flange (Eq. (6.8); Case 1 shown in Fig 6.7(a)), bolt prying mixed with the flange yielding (Eq. (6.9); Case 2 shown in Fig. 6.7(b)), and the bolt fracture without any prying force (Eq. (6.10); Case 3 shown in Fig. 6.7(c)), respectively.

$$P = \frac{(1+\delta)W_{T-stub}F_y t_f^2}{2b'} \tag{6.8}$$

$$P = \frac{\sum B_{n,tension} a'}{a'+b'} + \frac{W_{T-stub}F_y t_f^2}{2b'} \tag{6.9}$$

$$P = \sum B_{n,tension} \tag{6.10}$$

where $W_{T-stub}$ is the width of the T-stub (see Fig. 6.1); $F_y$ is the yield stress of the base material of the T-stub flange; $B_{n,tension}$ is the tensile capacity of tension bolts; and $\delta$ is the ratio of the net section area to the gross section area and is expressed as follows:

$$\delta = 1 - \frac{n_{tb}d_b}{2W_{T-stub}} \tag{6.11}$$

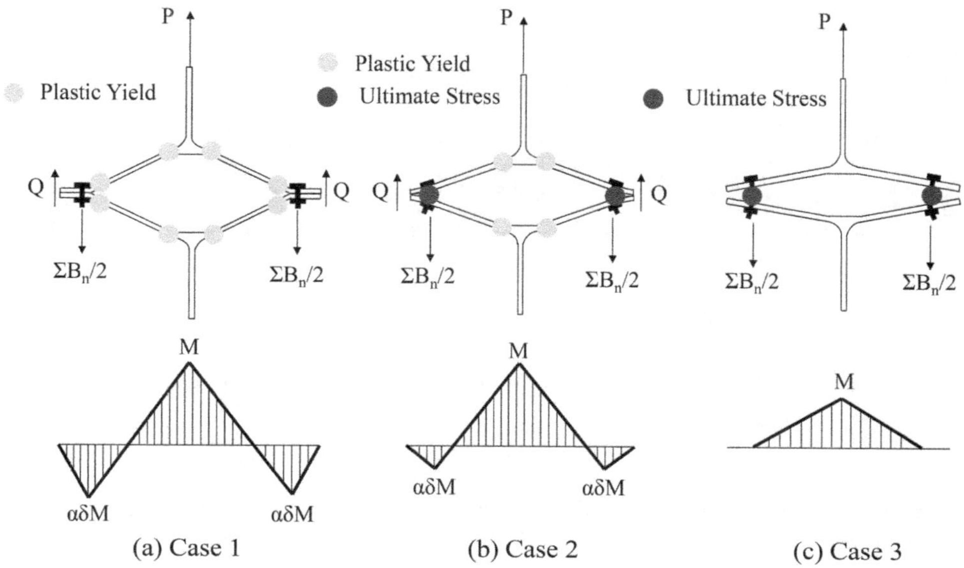


*Figure 6.7* Three possible failure modes under the tension force.

where $d_b$ is the diameter of the bolt hole generally including the clearance to the diameter of the bolt; and $n_{tb}$ is the number of tension bolts.

The parameter $\alpha$ denotes the index of the level of the prying present (Kulak *et al.*, 1987). It is defined as the ratio of the moment at the bolt centerline to the moment at the face of the T-stem (see Fig. 6.7) and is written as:

$$\alpha = \left(\frac{1}{\delta}\right)\left(\frac{2n_{tb}Pb'}{W_{T-stub}t_f^2F_y} - 1\right). \tag{6.12}$$

When the value of $\alpha$ is equal to 1.0, Eq. (6.8) can be derived from Eq. (6.12). If $\alpha \geq 1.0$, then the thickness of the T-stub flange is more sufficient to cause the plastic hinge mechanism and the prying forces are maximized. Accordingly, the flange is assumed to be a fixed-fixed beam (Case 1). If $\alpha \leq 0$, then the T-stub flange separates completely from the column surface, the prying forces are zero, and the tension bolts are subjected to only conventional tension without bending moment (Case 3). Therefore, the bolt fracture results in the dominant failure mode. Finally, when $0 < \alpha < 1.0$ (Case 2), it is likely that the combination of both flange yielding and bolt fracture due to bolt prying occurs at the T-stub flange (Kulak *et al.*, 1987).

The general solution for the determination of the flange capacity can be plotted as a function of the flange thickness ($t_f$). The theoretical solution space for T-stub component models is illustrated in Fig. 6.8. Two T-stub component models, TC01 and TD04, are selected to investigate the prying effect according to the geometric ratio of

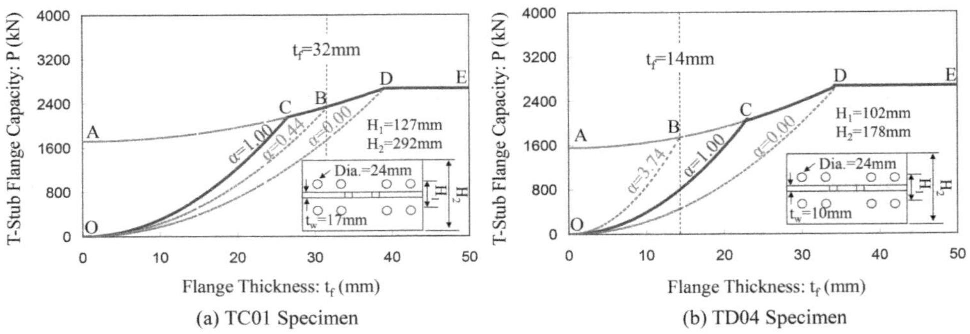


*Figure 6.8* General solution space for the determination of T-stub flange capacity.

$H_1$ to $H_2$ which determines the ultimate capacity, the failure mode, and the extent of the bending mechanism in the T-stub flange.

The line segment OC, CD, and DE are calculated using Eqs. (6.8), (6.9), and (6.10), respectively, whereas the line segment OB is computed using the equation derived from Eq. (6.12). The point B associated with the flange thickness represents the ultimate strength of the T-stub flange ($P_{n,flange}$). For the TC01 model with $H_1/H_2$=0.43 and $t_f = 32$ mm, the line segment OB has the value of $\alpha = 0.44$ (Fig. 6.8(a)). As a result, the point B, which corresponds to $P_{n,flange} = 2344$ kN, is anchored to the line segment CD. On the other hand, the solution space for the TD04 model with $H_1/H_2 = 0.57$ and $t_f = 14$ mm contains the line segment OB computed with $\alpha = 3.74$ (Fig. 6.8(b)). In this case, the expended solution region should be used to determine the ultimate strength owing to the value of $\alpha$ exceeding 1.0 (Kulak *et al.*, 1987; Swanson, 2002; Hu *et al.*, 2009). The dash line segment ABC represents the strength capacity based on the bolt failure after the plastic hinge mechanism is established in the T-stub flange. The stain hardening of the base flange material provides the additional strength until the tension bolts fail by the ultimate fracture. Therefore, the plastic hinge mechanism occurring at the line segment OC results in the preliminary failure mode. The TD04 model has $P_{n,flange} = 1745$ kN on the Point B. Once the ultimate flange strength of the TD04 model is compared with that of the TC01 model, we can check the parametric effect. The larger $H_1/H_2$ leads to the smaller ultimate capacity because the prying force increases quickly. It is also noted that the thicker flange thickness increases the ultimate strength of the T-stub flange.

The stiffness of the T-stub flange in bending ($K_{flange}$) is derived using the analytical model originally used in the Eurocode 3 (2003). The analytical beam model for the initial stiffness calculation under bolt prying is illustrated in Fig. 6.9. The T-stub flange is assumed to be the continuous beam on four supports. Tension bolts, which are modeled as spring elements, interact with the T-stub flange on two elastic supports. The prying forces act on two end supports. In addition, the contributions of the flanges in bending and the bolts in tension can be assembled in series with their associated stiffnesses. Thus, the analytical beam model accurately builds up the prying

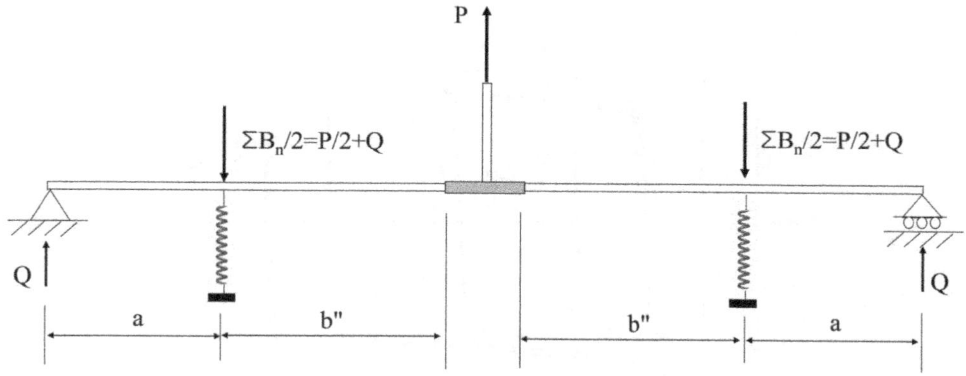


*Figure 6.9* Beam model for the initial stiffness calculation under bolt prying action.

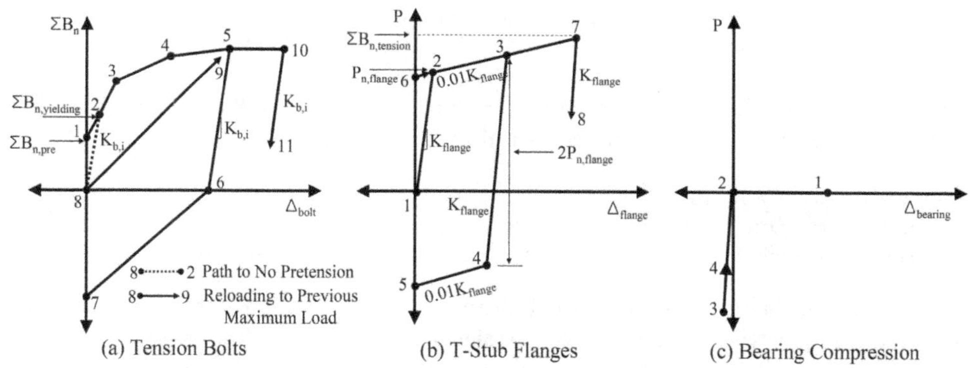


*Figure 6.10* Stiffness models for tension bolts, T-stub flanges, and bearing compression.

mechanism. The stiffness of the T-stub flange obtained from this analytical model is written as follows:

$$K_{flange} = 0.85 \frac{2W_{T-stub} t_f^3}{b''^3} \left( \frac{3b'' + a}{3b'' + 4a} \right) \tag{6.13a}$$

$$b'' = b - 0.8r \tag{6.13b}$$

where $r$ is the radius of the fillet in the K-zone (see Fig. 6.6).

Based on the prying mechanism studied above, the stiffness models for tension bolts, T-stub flanges, and bearing compression are developed as shown in Fig. 6.10. They are available for the cyclic force-deformation curves. The bolt stiffness was obtained by the experimental results taken from the direct tension test of the tension bolt (Fig. 6.10(a)). The initial stiffness is assumed to be infinite until the reaction

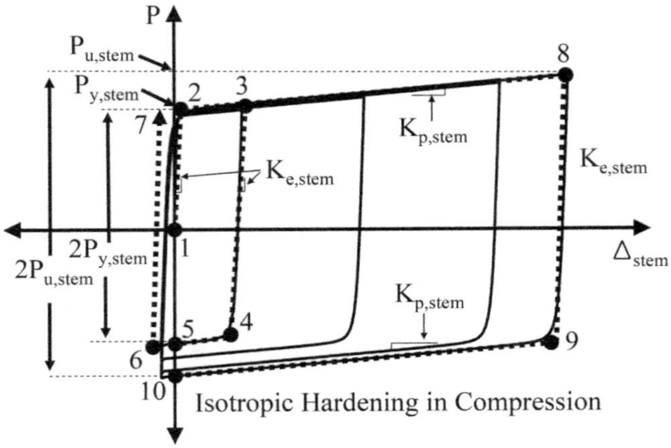


*Figure 6.11* Stiffness model for the behavior of T-stem.

force of the tension bolt exceeds the initial bolt pretension ($B_n \geq B_{n,pre}$). The initial bolt pretension (i.e. Point 1) is calculated as follows:

$$B_{n,pre} = \frac{E_b A_b \Delta_{pre}}{L_b} \tag{6.14}$$

where $E_b$ is the elastic modulus of the bolt material; $A_b$ is the cross section area of the bolt shank;$\Delta_{pre}$ is the prescribed bolt displacement; and $L_b$ is the length of the bolt shank. As shown in Fig. 6.10(a), the path from Point 8 to Point 2 represents the initial bolt stiffness without the initial bolt pretension ($K_{b,i}$). The unloading stiffnesses were taken as equal to the initial stiffness. The reloading paths (i.e. Point 8 to Point 9) headed toward the previous maximum loading points (i.e. Point 5). The stiffness model for the T-stub flange was simulated using the isotropic hardening behavior which is associated with three parameters: (1) the initial stiffness obtained from $K_{flange}$, (2) the yield load obtained from $P_{n,flange}$, and (3) the ultimate load from $\sum B_{n,tension}$. This behavior also includes the value of 0.01 for the strain hardening ratio and the Bauschinger effect. The bearing compression was simulated by the zero tension behavior (Fig. 6.10(c)). Only elastic stiffness with the steep slope exists in this model.

#### 6.3.2.2 T-stem member

The cyclic behavior of the T-stem was similar to the force-deformation curve of the base steel material tested cyclically, so the stiffness model was simulated by the hysteretic behavior with the isotropic hardening in compression. Fig. 6.11 shows the behavior of the T-stem under cyclic loads. This behavior required four parameters: (1) the elastic stiffness ($K_{e,stem}$), (2) the yield load ($P_{y,stem}$), (3) the plastic stiffness ($K_{p,stem}$), and (4) the ultimate load ($P_{u,stem}$). As the number of cycles increases, the yield envelope

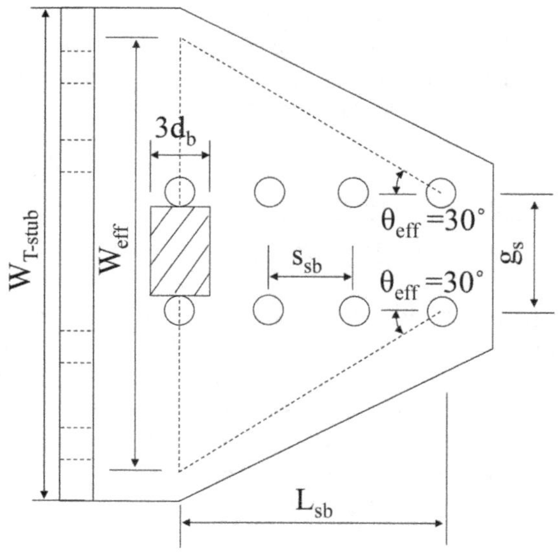


*Figure 6.12* Effective thickness of T-stem based on the Whitmore model.

in compression also increases until the load arrives at the ultimate. The unloading stiffnesses were also taken as equal to the elastic stiffness.

As shown in Fig. 6.12, an angle of 30 degree measured from the first row of shear bolts was used to define the effective width proposed by Whitemore (1952). The effective width of the T-stem ($W_{eff}$) based on the configuration of the shear bolts is written as follows:

$$W_{eff} = s_{sb}(n_{sb} - 2)\tan(\theta_{eff}) + g_s \leq W_{T-stub} \tag{6.15}$$

where $s_{sb}$ is the shear bolt spacing; $n_{sb}$ is the number of shear bolts; $\theta_{eff}$ is the effective angle of the tensile participation generally taken as 30 degree in the Whitemore model; and $g_s$ is the gage between two rows of the shear bolts. The smaller one between the Whitemore width and the actual width was utilized to calculate the net section area of the T-stem ($A_{stem}$) also used in the strength calculation. The shear bolt holes were not taken into consideration as a part of the net section area.

$$A_{stem} = (W_{eff} - 2d_b)t_w \tag{6.16}$$

where $t_w$ is the thickness of the T-stem. When the applied loads are distributed uniformly along the net section area, the yield and ultimate load can be expressed as the product of the material's yield or ultimate stress and the net section area, respectively.

$$P_{y,stem} = F_y(W_{eff} - 2d_b)t_w \tag{6.17a}$$

$$P_{u,stem} = F_u(W_{eff} - 2d_b)t_w \tag{6.17b}$$

The actual T-stem does not have the uniform stress distribution because of the stress concentration that occurs around the bolt hole. However, these approximate equations provide reliable and reasonable predictions. The tapered beam model, which is similar to that presented in Fig. 6.12, was utilized to derive the elastic stiffness of the T-stem. The elastic stiffness for the tapered section of the T-stem is written as follows (Swanson, 1999):

$$K_{e,stem} = \frac{4L_{sb}E_s t_w(\tan\theta_{eff})^2}{2L_{sb}\tan\theta_{eff} + g_s\ln\left(\frac{g_s}{2L_{sb}\tan\theta_{eff}+g_s}\right)} \tag{6.18a}$$

where $L_{sb}$ is the length measured from the first shear bolt to the last one; and $E_s$ is the elastic modulus of the T-stem base material. The angle was limited to a value no greater than $\theta_{eff}$ defined in the gusset plate research performed by Whitemore (1952). The plastic stiffness can be derived on the basis of the observation from component testing (Swanson, 1999). The shaded area between last two bolt holes shown in Fig. 6.12 begins to strain hardening before the rest of the cross section area. The length of the strain hardening area is assumed to be taken as $3d_b$. Accordingly, the plastic stiffness of the T-stem is written as follows:

$$K_{p,stem} = \frac{(g_s - d_b)t_w E_s}{3d_b} \tag{6.18b}$$

where $d_b$ is the diameter of the bolt shank.

The summary of test results is given to Table 6.3. Most of T-stub components failed by the net section fracture (Swanson, 1999). Though the components are designed with the strong bolt and flange capacity, the relatively slender T-stem deteriorates the ultimate capacity of the T-stub component very fast (e.g. TD04 and TD08).

#### 6.3.2.3 *Slip and bearing model*

Slip gives rise to the temporary loss of stiffness that acts as a fuse during reverse cyclic loads (Astaneh-Asl, 1995). Slip limits the force which is transmitted through the

*Table 6.3* Strength of T-stub component in each stiffness model, peak load, and failure mode.

| | *Bolt/Flange Capacity* | | | *T-Stem Capacity* | | *Experimental Results* | |
|---|---|---|---|---|---|---|---|
| *Model ID* | $\Sigma B_{n,yielding}$* | $P_{n,flange}$* | $\Sigma B_{n,tension}$* | $P_{y,stem}$* | $P_{u,stem}$* | *Peak Load** | *Failure Mode*** |
| TA01 | 1822 | 2173 | 2667 | 1742 | 2317 | 2085 | Net Section |
| TA09 | 2383 | 2788 | 3510 | 1504 | 2000 | 1924 | Net Section |
| TB01 | 1822 | 2185 | 2667 | 1870 | 2473 | 2252 | S-Bolt |
| TB05 | 2383 | 2800 | 3510 | 1836 | 2418 | 2236 | Net Section |
| TC01 | 1822 | 2344 | 2667 | 2507 | 3202 | 2601 | T-Bolt |
| TC09 | 2383 | 2930 | 3510 | 2461 | 3144 | 2949 | Net Section |
| TD04 | 1822 | 1745 | 2667 | 1106 | 1476 | 1094 | Net Section |
| TD08 | 2383 | 2300 | 3510 | 1214 | 1620 | 1132 | Net Section |

*: kN
**: Swanson [15]

shear bolts at a given deformation and produces significant energy dissipation. Slip mechanism arises due to the direct shear force converted from the axial force acting on the slip plane. Once the slip force ($P_{slip}$) exceeds the slip resistance provided by the clamping force, slip starts to occur between two contact surfaces. The clamping force is estimated by the product of the friction coefficient and the bolt pretension, such that $u \cdot B_{pre}$. Construction tolerances ($\Delta_c$) require that each bolt hole should be at least 1.6 mm larger than the nominal bolt diameter. Once slippage exceeds this tolerance, the bolt begins to bear on the plate, and then both stiffness and strength increase again. Owing to the sequence of these actions, the slip and bearing mechanism will be treated together in this section.

Expressions for the nominal slip resistance are based on the simple friction models calibrated to the numerous simple shear tests on bolted splices. The slip resistance obtained from the component test shall equal or exceed the nominal slip resistance load ($P_{slip}$) given in the AISC-LRFD (AISC, 2001) as follows:

$$P_{slip} = 1.13 u h_{sc} T_b n_s n_{sb} \tag{6.19a}$$

where $u$ is the mean slip coefficient for Class A, B, or C surfaces, as applicable, or as established by tests (for Class A surface, $u = 0.33$); $h_{sc}$ is the coefficient for the standard bolt holes ($h_{sc} = 1.0$); $T_b$ is the specified minimum fastener tension given in the Table J3.1 of the AISC-LRFD (e.g. M10.8 bolt with 22 mm has 221 kN); and $n_s$ is the number of slip planes. Unpainted clean mill scale steel surfaces belonging to Class A were used for the mean slip coefficient. When a slip-critical connection is subjected to an applied tension force ($P_u$) that mitigates the net clamping force, the slip resistance given to Eq. (6.19a) shall be multiplied by the following reduction factor:

$$1 - \frac{P_u}{1.13 T_b n_{sb}} \tag{6.19b}$$

The applied tension force ($P_u$) is computed by the product of the ultimate T-stub capacity and the design reduction factor ($P_u = \phi P$: $\phi = 0.9$ for the plastic flange and $\phi = 0.75$ for the bolt fracture).

The slip behavior is characterized by the bi-linear model using three parameters: (1) the initial slip stiffness ($K_{i,slip}$); (2) the nominal slip resistant load ($P_{slip}$); and (3) the post-slip stiffness ($K_{p,slip}$). The bi-linear load-slip curve proposed by Rex and Easterling (1996) is shown graphically as a solid line in Fig. 6.13. This curve ascends up to a slip load, and then descends fast with the post-slip stiffness. For that reason, the stiffness $K_{i,slip}$ and $K_{p,slip}$ is determined as the functions of the displacements $\Delta_{slip}$ and $\Delta_{fu}$ as follows:

$$K_{i,slip} = \frac{P_{slip}}{\Delta_{slip}} \tag{6.20a}$$

$$K_{p,slip} = \frac{P_{slip}}{(\Delta_{slip} - \Delta_{fu})} \tag{6.20b}$$

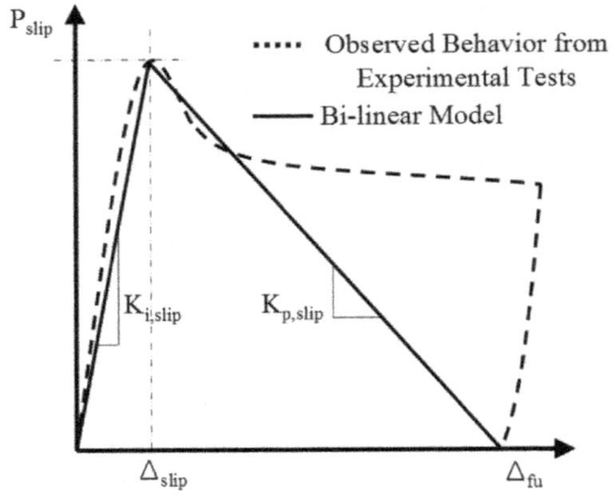


*Figure 6.13* Load-slip relationships.

*Table 6.4* Definition of ultimate slip displacement ($A_{fu}$).

| $(t_w + t_{bf}{}^*)$ | $\Delta_{fu}$ |
|---|---|
| $(t_w + t_{bf}) \leq 12.7$ mm | 10.2 mm |
| 12.7 mm $< (t_w + t_{bf}) \leq 38.1$ mm | 10.2 mm $- 0.3(t_w + t_{bf} - 12.7$ mm) |
| 38.1 mm $< (t_w + t_{bf})$ | 2.5 mm |

$^*t_{bf}$ indicates the thickness of the beam flange.

The slip resistant displacement ($\Delta_{slip}$) is defined as a value of approximately 0.2 mm with a coefficient of variation of 46% recommended by Rex and Easterling (1996). The ultimate slip deformation ($\Delta_{fu}$), which is determined in accordance with the thickness of the jointed plates, is presented in Table 6.4. The load-slip curve obtained from the experimental test is also plotted graphically as the dash line in Fig. 6.13. Though the bi-linear slip model is simple and accurate to predict the pre-slip behavior, there remains a discrepancy between the post-slip behaviors.

The modified model developed by Rex and Eastering was accepted for the bearing mechanism. The elastic bearing stiffness ($K_{e,bearing}$) consisting of three component stiffnesses is by the modified model as follows:

$$\frac{1}{K_{e,bearing}} = \frac{1}{K_{br}} + \frac{1}{K_{be}} + \frac{1}{K_{ve}} \tag{6.21a}$$

where $K_{br}$ is the bolt bearing stiffness; $K_{be}$ and $K_{ve}$ refer to the bending and shearing stiffness, respectively, associated with the end distance of the lap plate. The end of the plate was modeled as a short and deep beam. Three component stiffnesses were combined in series. They are defined as:

$$K_{br} = 120 F_y t_w d_b^{0.8} \tag{6.21b}$$

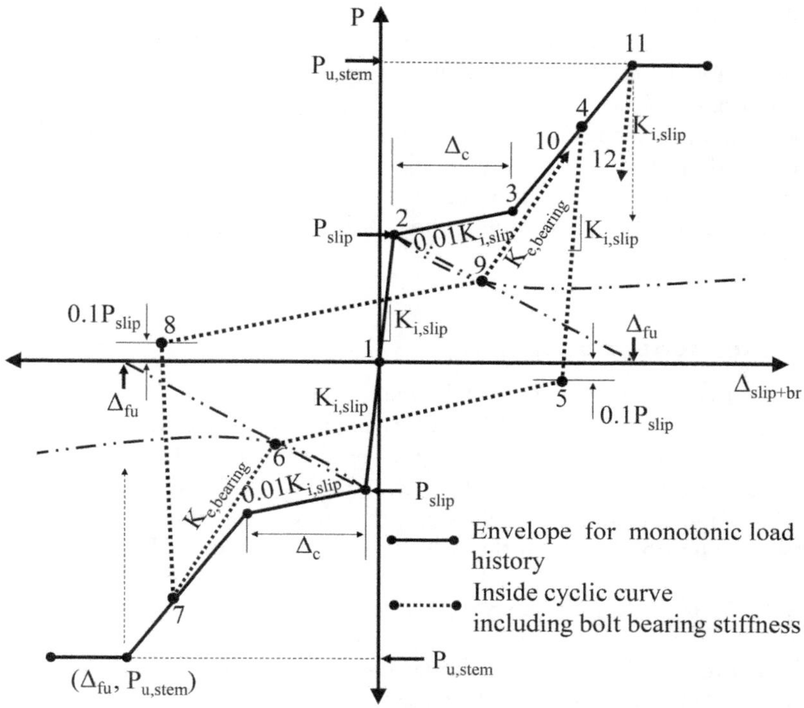


*Figure 6.14* Stiffness model for the behavior of slip and bolt bearing.

$$K_{be} = 32E_s t_w \left(\frac{h}{L}\right)^3 \quad (6.21c)$$

$$K_{ve} = 6.67G_s t_w \left(\frac{h}{L}\right) \quad (6.21d)$$

$$\frac{h}{L} = \frac{L_e}{d_b} - 0.5 \quad (6.21e)$$

where $G_s$ denotes the elastic shear modulus of the T-stem; and $L_e$ denotes the end distance of the last pair of shear bolts. The deformable contribution due to $K_{br}$ dominates, so the influence of both $K_{be}$ and $K_{ve}$ on two end bolts may be neglected to simplify the modeling procedure.

To simulate the combining slip and bearing mechanisms, the bearing stiffness model should be incorporated with the bi-linear slip model as shown in Fig. 6.14. The plateau indicates the slip along the tolerance distance ($\Delta_c$) until the shanks of the shear bolts contact the edge of the bolt holes. The friction between two contact surfaces causes the slowly ascending slope representing $0.01K_{i,slip}$. At the end of the plateau, the slope increases again due to the effect of bolt bearing. If the component is subjected to a monotonic load history, the solid line as shown in Fig. 6.14 results in the actual response history depending on the envelope chosen for the stiffness model. The form of the load-unload-reload response, denoted as the dash line, under the cyclic displacement-load history is defined by inside 6 pinch points. Once the load-displacement point values are located on the degrading post-slip line (e.g. Points 6 and 9), the bearing mechanism starts to occur during the reloading procedure. The bearing mechanism ends in the ultimate strength based on the net section fracture ($P_{u,stem}$). This strength is assumed to be attained at a value of$\Delta_{fu}$. The shear bolts are assumed to be perfectly aligned in the center of the bolt holes, so the stiffness model shows the symmetric behavior.

### 6.3.3 Panel zone model

In this study, the behavior of the panel zone is simulated using the precise manner by means of a panel zone with rigid boundary elements. These rigid elements create a panel zone that deforms into a parallelogram by shear force. More details of modeling panel zone are described well in the FEMA 355c (FEMA-355C, 2000).

For the modern frame, the yield strength of the panel zone ($V_y$) is calculated by the following equation based on the AISC seismic provisions (AISC, 2005).

$$V_y = 0.6F_y d_c t_{cw} \tag{6.22}$$

where $d_c$ denotes the depth of the column (see Fig. 6.3); and $t_{cw}$ denotes the thickness of the column web. The corresponding yield distortion ($r_y$) is given as:

$$r_y = \frac{F_y}{\sqrt{3}G_s}. \tag{6.23}$$

For the modern frame, the plastic shear resistance of the panel zone ($V_p$) is calculated as following equation (FEMA-355C, 2000; AISC, 2005):

$$V_p = 0.6F_y d_c t_{cw}\left(1 + \frac{3b_f t_{cf}^2}{d_v d_c t_{cw}}\right) \tag{6.24}$$

where $d_v$ is the width of the column flange (see Fig. 6.3); and $t_{cf}$ is the thickness of the column flange. This plastic shear strength is assumed to be attained at the value of $10\gamma_y$. The stiffnesses is determined as the functions of the shear distortions as follows:

$$K_{i,PZ} = \frac{V_y}{\gamma_y} \tag{6.25}$$

$$K_{p,PZ} = \frac{V_u - V_y}{9\gamma_y} \tag{6.26}$$

where $K_{i,PZ}$ is the initial stiffness of the panel zone; and $K_{p,PZ}$ is the plastic stiffness of the panel zone. The behavior of the panel zone is simulated using the tri-linear model.

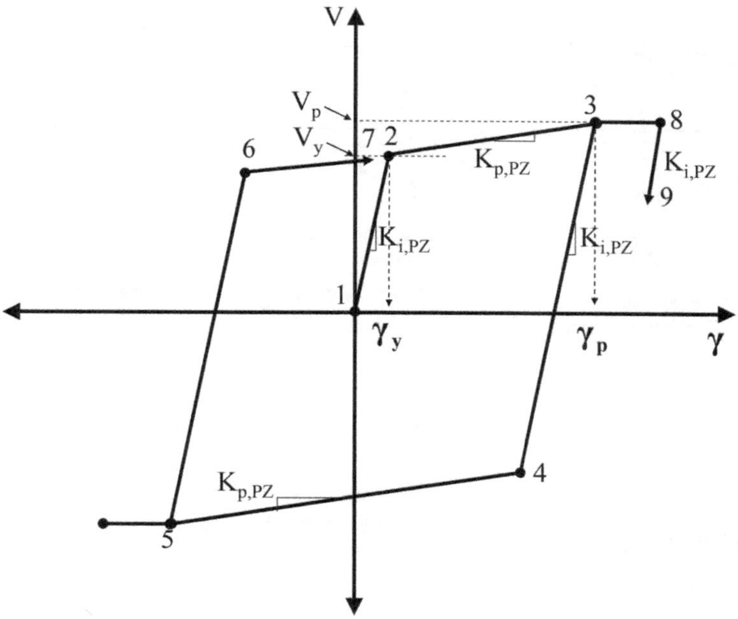


*Figure 6.15* Stiffness model for the behavior of steel panel zone.

Fig. 6.15 shows an example of the tri-linear hardening behavior for the panel zone spring. The parameters are computed by these equations.

## 6.4 NUMERICAL MODELING ATTRIBUTES

In order to implement numerical analyses, the simple mechanical models for T-stub components and full-scale connections are constructed with the OpenSEES program (Mazzoni *et al.*, 2006), a nonlinear finite element (FE) program widely used in the USA for this type of studies. The OpenSEES program automatically produces a finite element model with the appropriate spring and rigid element. Moreover, the stiffness models are simply generated using default 1 dimensional (1D) material models available at this FE program. Details of numerical modeling attributes and nonlinear analyses will be addressed in this section.

### 6.4.1 T-stub components

After each of the different deformation mechanisms is formulated, they can be combined into the total deformation response of the T-stub component. The assembly

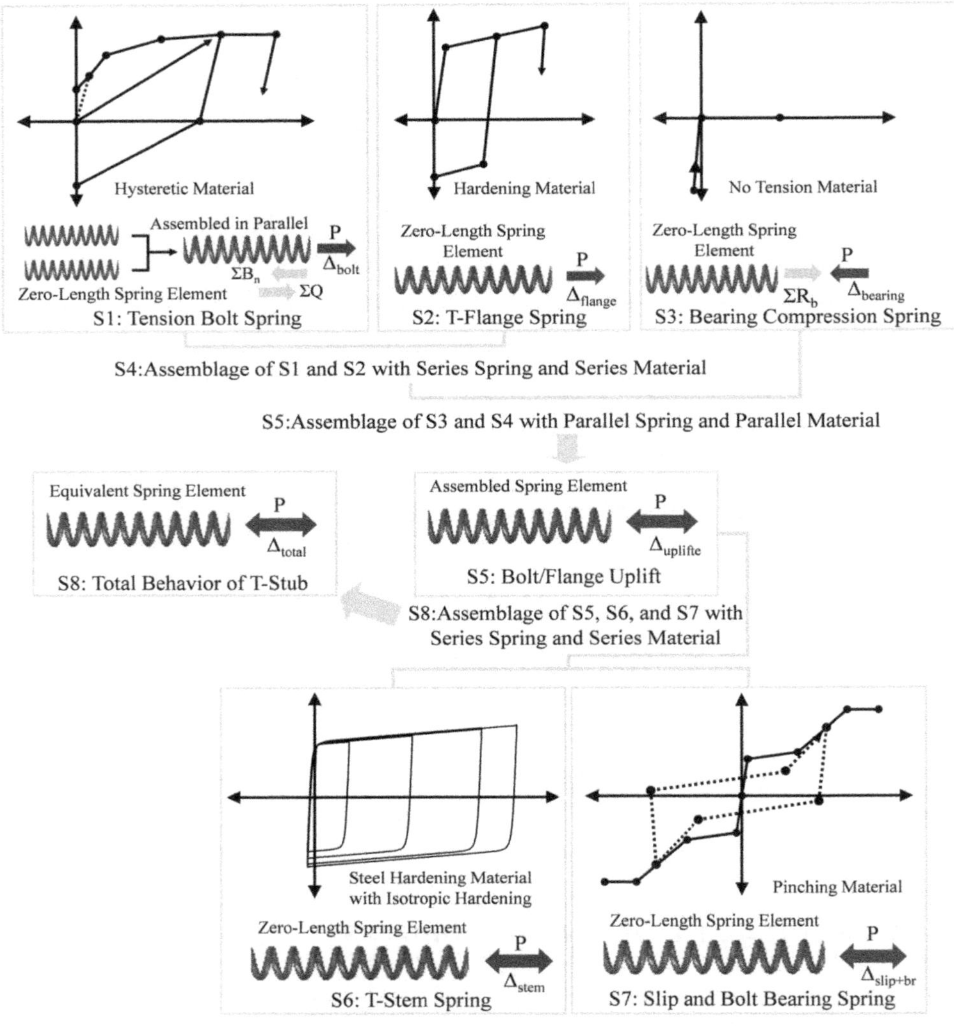


*Figure 6.16* Assembly procedures for component spring elements with their own stiffness models simulated by default material properties in the OpenSEES program.

procedures for connection components are graphically illustrated in Fig. 6.16. This work was conducted by the OpenSEES program.

The real component members were modeled as flexible component springs as shown in the figure. The zero-length spring elements, which connect two nodes at the same position, were used for these component springs. The multi-linear stiffness models which idealize the behavior of individual connection components, were simply simulated by 1D nonlinear material models. The adequate material models (e.g. hysteretic and hardening material) were selected in accordance with the parameters to be required for determining the stiffness models. They were assigned into corresponding component springs.

The overall stiffness model for the T-stub component was constructed by adding the individual stiffness models from the various response mechanisms together under the common axial load (*P*). As shown in the Fig. 6.16, those basic response mechanisms contain tension bolt elongation (S1), bending of the T-stub flange due to the prying action (S2), bearing deformation of the T-stub flange (S3), T-stem elongation (S6), and combining slip and bolt bearing deformation (S7). Each of response mechanisms was assembled with the series and parallel material model also available at the OpenSEES program (e.g. S5 and S8). Therefore, the assembly of the total model proceeded as the flow chart shown in the figure. The total behavior of the T-stub component under cyclic displacement-load history was numerically reproduced using the equivalent spring element, referred to as S8 identification (ID).

### 6.4.2 Full-scale T-stub connections

The mechanical model for the T-stub connection was introduced as 2D joint elements into the OpenSEES program. Fig. 6.17 shows the composition of the user joint element for a general beam-to-column connection. This joint element contains (1) two equivalent spring elements (S8) to reproduce the overall deformations of the T-stub component; (2) 6 internal axial spring elements (S9) to reproduce the elastic axial deformations of the column member; (3) four internal shear spring elements (S10) to reproduce the shear deformations of the beam and column member; and two spring elements (D) to generate the behavior of the shear panel zone. The tri-linear hardening material was assigned into two shear panel spring elements in order to reproduce the behavior of the panel zone. The rotational spring element (F) is used if a shear tab is installed. All spring elements in the joint element were implemented with the interior and exterior rigid planes coincident, using the zero-length spring elements.

The steel column and beam were made up of 1D nonlinear beam-column elements. As shown in Fig. 6.17, the cross sections were modeled as 2D discrete fiber sections which are placed in the integration points of nonlinear beam-column elements (G). The fiber sections include the hysteretic moment-curvature responses as illustrated in Fig. 6.18. The plastic hinge may occur at the section of the beam member under severe loading. To obtain the behavior of the connection numerically, the cyclic displacement-load history was applied to the tip of the beam element corresponding to the position of a loading actuator.

## 6.5 MODEL VALIDATION

The results of cyclic analyses are shown in this section and compared with those of selected experimental tests (Swanson, 1999; Smallidge, 1999) in order to verify the validity of the mechanical spring models. The details of the experiments were already illustrated in Section 6.2.

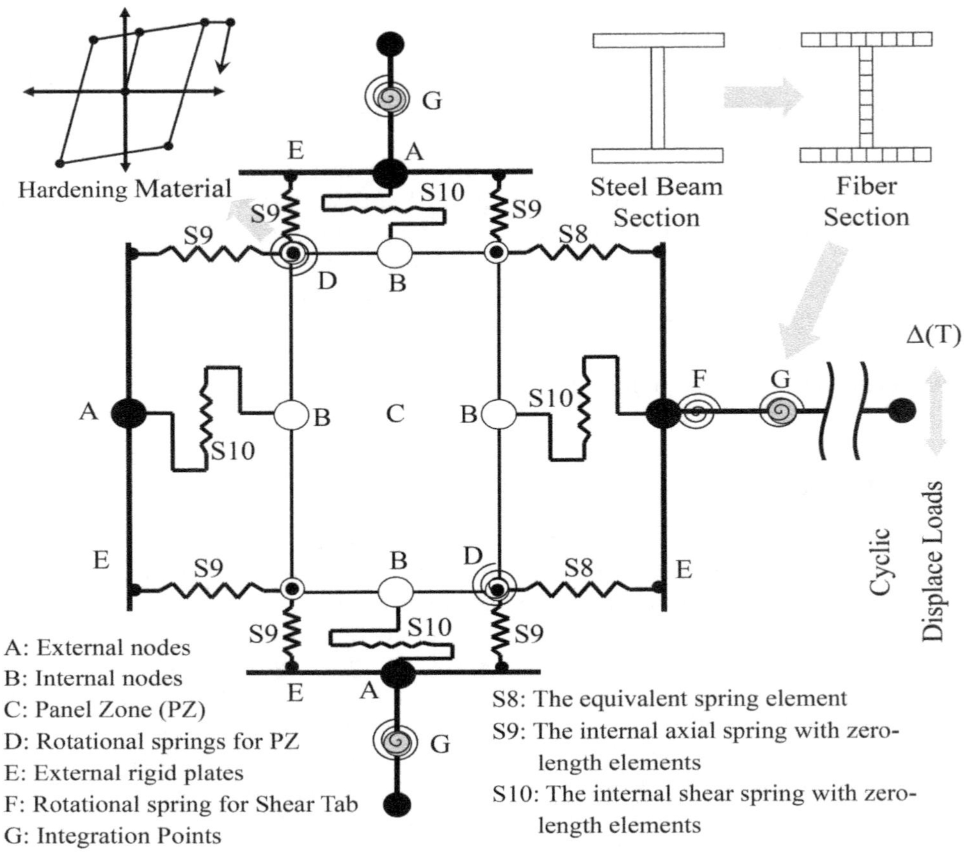


*Figure 6.17* Numerical joint model for the full-scale T-stub connection.

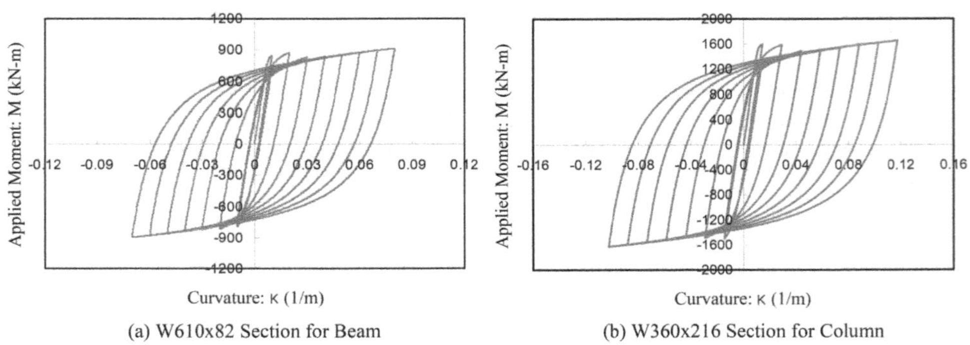


*Figure 6.18* Moment-curvature capacity for beam and column sections under cyclic loads (FS05-TA01 Specimen).

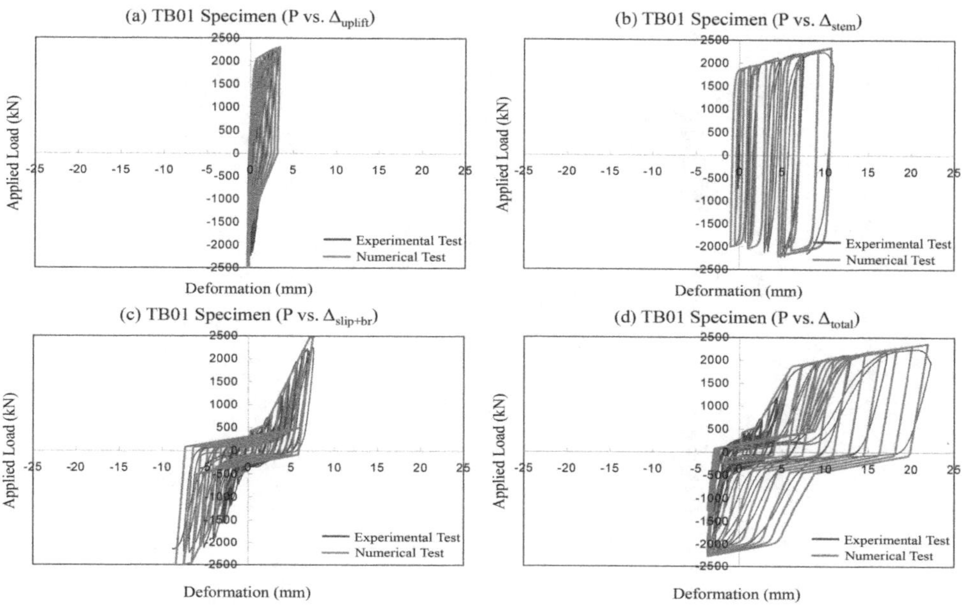


*Figure 6.19* Comparisons of the cyclic behavior of the T-stub component (Experimental test results vs. Numerical test results).

### 6.5.1 Component tests

The cyclic curves obtained from the experimental tests (Smallidge, 1999) were compared with those from the numerical tests. Comparisons between both test results with respect to the force-deformation responses for the T-stub component are shown in Fig. 6.19. The numerical tests were conducted by applying the cyclic displacement loads to each of component spring elements presented in Fig. 6.16.

The overall force-deformation curve for a typical T-stub tested cyclically is shown in Fig. 6.19(d). The bearing compression of the T-stub flange is the cause that the total deformation ($\Delta_{total}$) is larger in tension than in compression. This total deformation is primarily due to three basic mechanisms: (1) the bolt/flange uplift (Fig. 6.19(a)), (2) T-stem deformation (Fig. 6.19(b)), and (3) bearing deformation combined with the slip deformation (Fig. 6.19(c)). The behavior of the bolt uplift considers the prying action accompanied by the tensile yielding of the bolts, so it results in a bi-linear backbone curve. As a result, the tension bolts yield at first, and then the yield line occurs at the T-stub flange. The T-stem component has an isotropic hardening behavior for the base steel material. The relatively thin member thickness of the T-stem ($t_w$) can lead to the major energy dissipation. The combined bearing and slip curve obtained from the experiment lies a little to the left of center. This shift depends on the initial alignment of the bolt holes of the beam flange and T-stem. Because of this, the small discrepancy (approximately 5%) occurs between two slip/bearing curves compared to each other. Generally, the performance of component springs allows

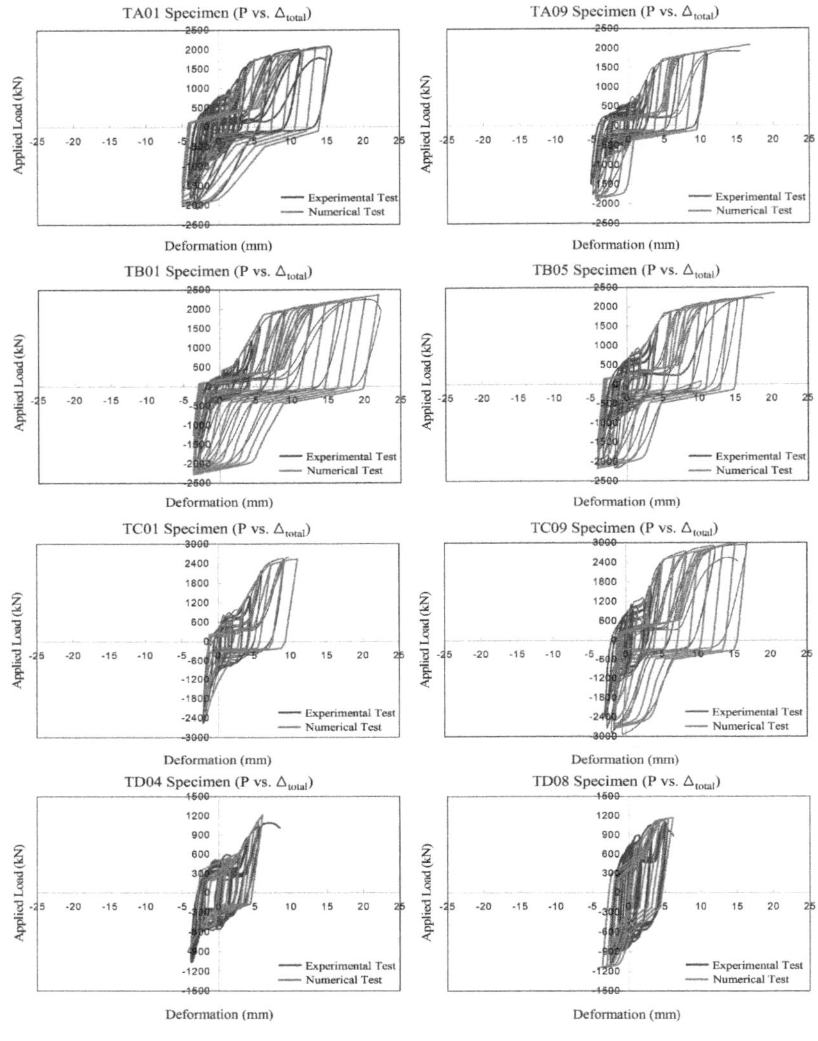


*Figure 6.20* Test results and comparisons for the applied load and total displacement at the T-stub (Experimental test results vs. Numerical test results).

the satisfactory prediction for the actual force-deformation curves as shown in these figures. Again, the results show good agreements in terms of initial stiffness, shape of the envelope, ultimate capacity, and even location of pinching points. The series spring system numerically used for an assemblage of three basic mechanisms is verified by checking the fact that the total deformation of the T-stub is equivalent to the summation of their component deformations under the common axial load level (e.g. $\Delta_{total} = \Delta_{uplift} + \Delta_{stem} + \Delta_{slip+br}$).

As shown in Fig. 6.20, the more overall force-deformation curves for the T-stub components cyclically tested with corresponding equivalent spring elements are

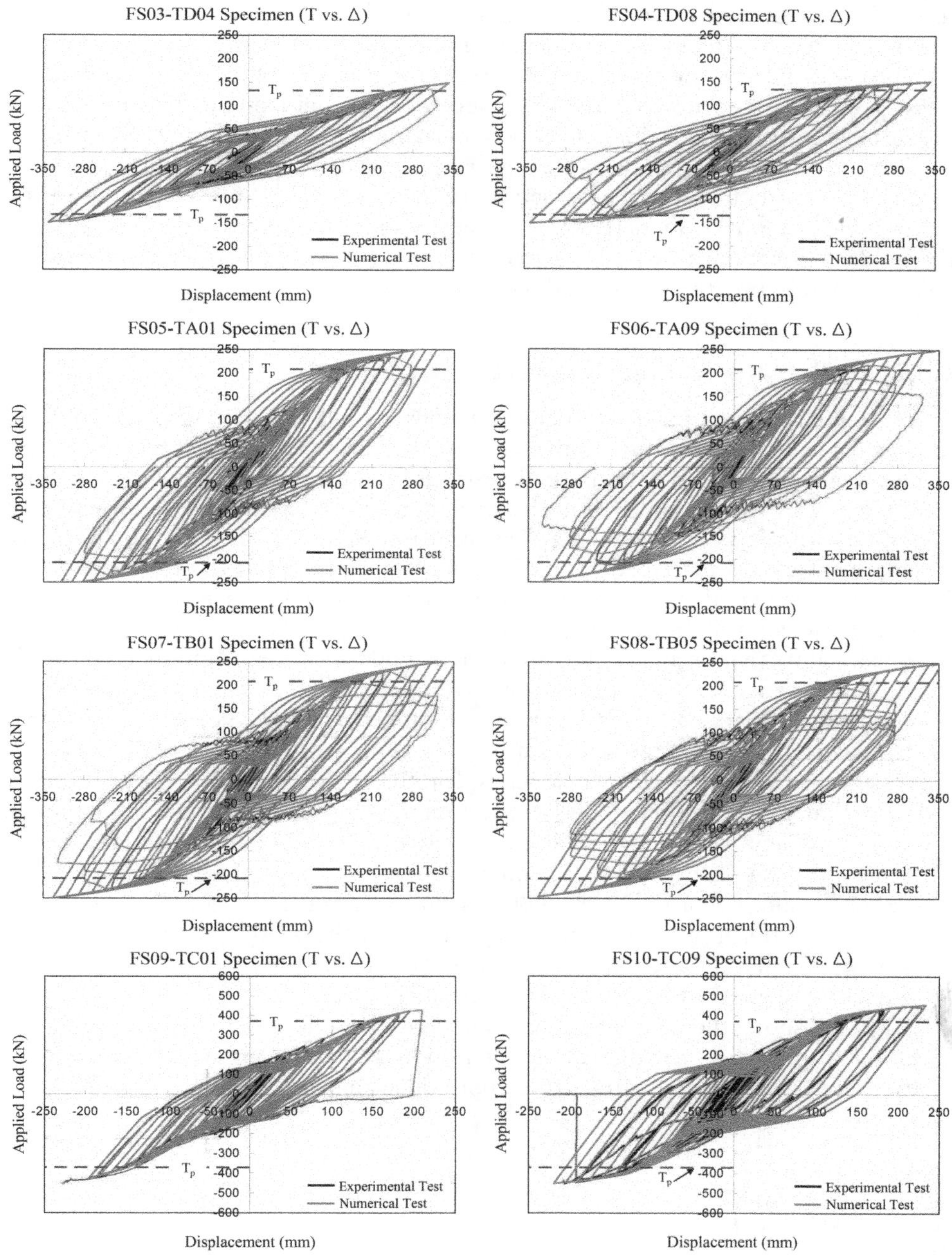


*Figure 6.21* Test results and comparisons for the full connection behavior (Experimental test results vs. Numerical test results).

superimposed on the actual force-deformation curves from experiments in an effort to verify the mechanical modeling, all things considered herein. Comparisons of the simulated curves to the measured deformation curves indicate that the numerical test data provides an accurate prediction for the overall behavior of the T-stub component. Most of equivalent spring elements work very well. Strength and deformable capacity are successfully predicted within narrow error margin. However, these spring elements are not capable of tracking the softening behavior shown in the last cycle of the test as this was mostly due to the propagation of the fracture in the T-stem. This softening behavior cannot be modeled by the simple springs used here. Attempts to model the sudden strength degradation posterior to the ultimate failure were not taken into consideration during the mechanical modeling procedure.

### 6.5.2 Full-scale connection tests

To verify the adequacy of the mechanical models for full-scale connections, the cyclic test results obtained from the numerical tests, which were performed on the joint elements, were compared with those obtained from the experimental tests (Smallidge, 1999). Fig. 6.21 shows comparisons between numerical test results and experimental test results with respect to the force-displacement responses. The total applied force-displacement curves presented in the figures were measured at the tip of the beam. Both curves compared to each other match very well until the fracture began to occur at the experimental specimen. As the strength degraded in the experiment, two curves diverged. Before that occurred, the results are in good agreement in terms of the initial slope, loading envelope, slip plateau, unloading/reloading slope, and even pinching points. The plastic applied loads ($T_P$) were determined in accordance with the reference to the plastic strength of the structural steel beam (e.g. $T_P = M_P/L = Z \cdot F_y/L$). All curves exceed the limit of the plastic loads denoted as the dash lines on the figures. It indicates that all connection models can sustain the full plastic moment of the framing beam ($M_P$). They were designed as the full strength (FS) connection. Further, the T-stub component with the larger geometric size was designed with the bigger size of the steel beam. Accordingly, the TB01specimen, for instance, could accommodate more energy absorption and strength capacity than the TD04 specimen. The good fit between simulation and experiment suggests that the joint element is sufficiently adequate to predict the trend of the connection behavior.

Fig. 6.22 shows the moment-rotation curves for the selected connection specimens. The force-displacement response of the connection is converted to the moment-rotation relationship ($M - \theta$) using a first order approximation:

$$M = TL \tag{6.27a}$$

$$\theta = \arctan\left(\frac{\Delta}{L}\right) \tag{6.27b}$$

In addition to the strength, the connections are also classified by the deformation capacity based on the moment-rotation curve. Connections are classified as brittle or ductile based on their ability to achieve the total rotational angle demand before the occurrence of the strength degradation (Leon, 1997; Hu, 2008). For example, in

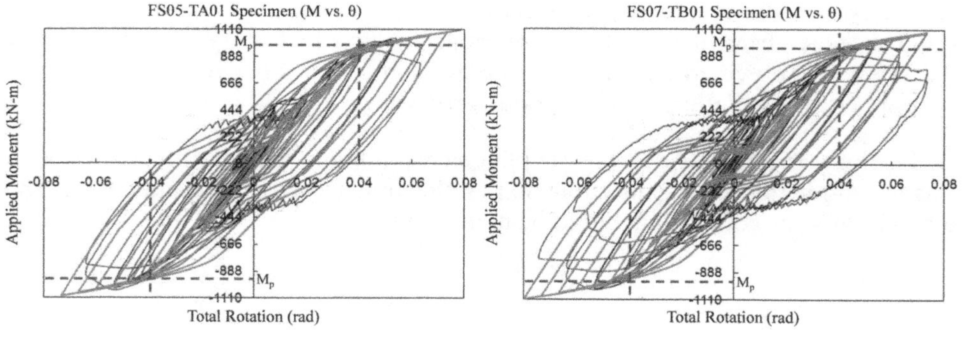


*Figure 6.22* Moment-rotation curves for the selected T-stub connections.

the aftermath of Northridge earthquake, the total rotation of 0.04 radians has been accepted as the rotational angle limit to distinguish between brittle and ductile connections. The connection models presented in Fig. 6.22 exceed both the full plastic strength of the beam and the total rotational angle limit of 0.04 radians required for ductility. The displacement of approximately 180 mm imposed on the tip of the beam is converted into the total rotational angle of 0.04 radians using Eq. 6.27(b). All connection specimens went over this imposed displacement limit without the strength degradation (see Fig. 6.21). Once again, the performance of the mechanical models as an acceptable alternative of the experimental program can also lead to an overall satisfactory estimation of the moment-rotation curve. Moreover, the resulting moment-rotation curve, which was simulated using the user joint element, can be reliably characterized by the T-stub connection with respect to the initial stiffness, strength, and rotational capacity.

## 6.6 CONCLUDING REMARKS

A methodology for the mechanical model to predict the complete moment-rotation curve of bolted beam-to-column connections was mainly discussed in this study. The 2D joint elements were used to actualize the mechanical model for the full-scale connections. The following conclusions are based on the results and observations presented herein.

1. The stiffness model was assigned to a spring element whose property is defined by either its material behavior or the physical action based on the failure mode. Groups of these springs were assembled in series or parallel system to reproduce the overall behavior of the T-stub component.
2. The component spring elements were installed in the 2D joint element. This joint element contains the behavioral properties of the connection components in the form of a simplified stiffness model that aims to reflect the global stiffness, strength, and deformation capacity of the actual connection. This type of the

joint model devotes to significant savings in the running time for nonlinear frame analyses with accuracy and efficiency.

3. The exact behavior of the connections including slippage and Bauschinger effect was observed on the cyclic curves. Comparisons between test results and numerical simulations showed good agreement.

Finally, it is conclude that the 2D joint elements could model sufficiently well the behavior of the full-scale connections so that the number of experimental tests needed to pre-qualify the connections could be minimized.

Chapter 7

# Recentering bolted connections (component models)

## 7.1 GENERAL OVERVIEW

The building structures located on the high seismic regions are susceptible to considerable damage during severe earthquake loads due to large lateral-permanent deformations. These unrecoverable deformations not only cause technical difficulties but also result in high costs for the post-disaster repair work (McCormick *et al.*, 2007; Zhang & Zhu, 2008). Smart systems have been consequently integrated into seismi-cresistant designs in an effort to overcome these problems. The smart systems can automatically adapt their structural characteristics to ambient external loads with the goals of procuring structural safety, extending structural lifetimes, and enhancing system serviceability (Saiid-Saiidi *et al.*, 2007; Zhu & Zhang, 2008). One method of achieving these goals is the utilization of shape memory alloys (SMAs) with self-centering (or recentering) capabilities, which can be integrated into the structural system. For instance, when the superelastic SMAs typically considered to be smart materials are utilized as bolt fasteners in beam-to-column connections, they can alleviate repair costs and allow building structures to sustain usable states even after strong seismic events. In addition, incorporating momentresisting frames with these SMA connections establish additional damping, reduce the effects of the seismic force, and mitigate residual story drift in the behavior of the whole building (Hu, 2008; Hu & Leon, 2010; DesRoches *et al.*, 2004). Since superelastic SMAs can be returned to their original shapes by only load removal without heat treatment as shown in Fig. 7.1, many researchers have been exploring their applications in civil structures. Research efforts have been recently extended to using SMAs for connection design as well (Ocel *et al.*, 2004; Abolmaali *et al.*, 2006; Sepúlveda *et al.*, 2008; Park *et al.*, 2011; Hu *et al.*, 2013).

A new class of connections composed of a shear tab, clip-angles, steel shear bolts, and SMA tension bolts is investigated in this study. Especially, the performance of SMA clip-angle connections proposed herein was rigorously compared to that of traditional clip-angle connections. Nonlinear springs which incorporate the force-deformation response of connection components were calibrated for consistency to the existing component test results. Mechanical joint models comprising these component springs were then created to reproduce the behavior of traditional clip-angle connections as well as the new SMA clip-angle connections. Finally, the behavioral characteristics of both connection types were compared to each other. These results may serve to highlight the benefits of SMAs in bolted connections with respect to recentering capabilities.

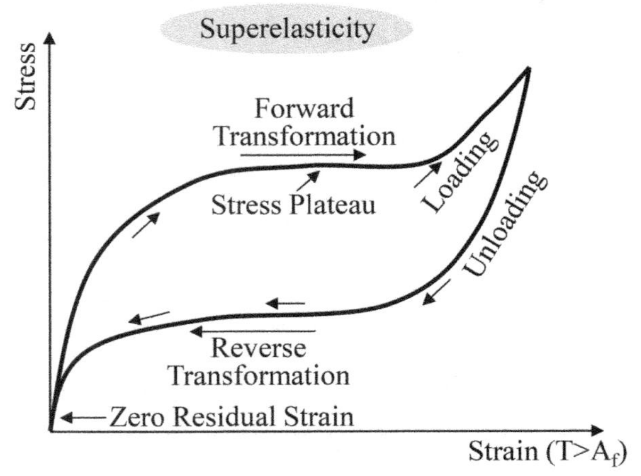


*Figure 7.1* The material behavior of superelastic SMA materials.

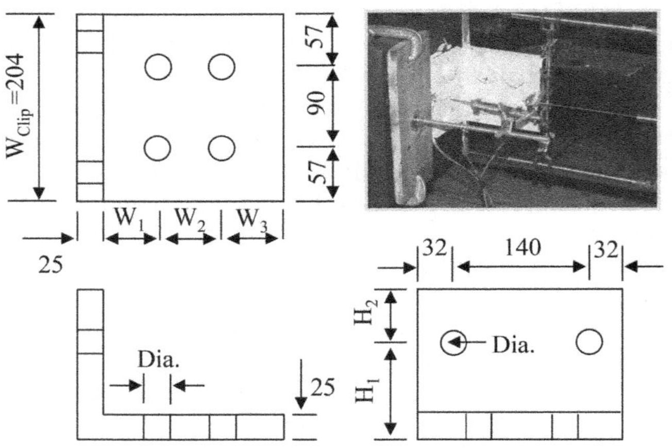


*Figure 7.2* Details of the clip-angle component (Unit: mm) (Swanson, 1999).

## 7.2 EXPERIMENTAL PROGRAM

The former test data from traditional clip-angle connections were used to accurately calibrate the mechanical joint models proposed in this study. As a part of the SAC task project, Swanson (1999) tested 10 heavy clip-angle components and Schrauben (1999) tested two traditional full-scale connections with heavy clip-angle components (e.g., FS01-CA02 and FS02-CA04). The typical dimensions of the specimens utilized in the experimental tests conducted by Swanson and Schrauben are given in Figs. 7.2 and 7.3, respectively. In the present study, two clip-angle components incorporating two full-scale connection models (e.g., CA02 and CA04) were selected for calibration. The details of these clip-angle components are summarized in Table 7.1. Note that the ratio

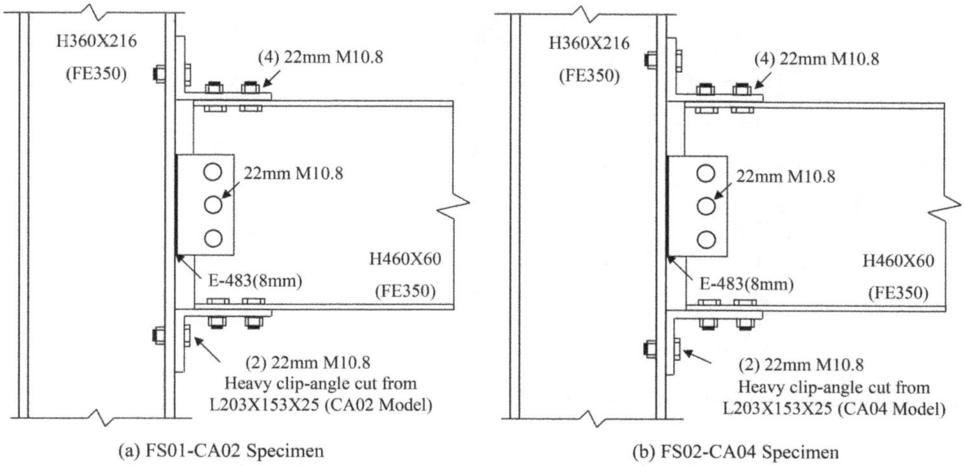


*Figure 7.3* Details of the full-scale clip-angle connection model (Schrauben, 1999).

*Table 7.1* Geometric sizes of the clip-angle component specimens.

| *Model* | $W_1$ | $W_2$ | $W_3$ | $H_1$ | $H_2$ | *Bolt (diameter × length)* | *Bolt Grade**** |
|---|---|---|---|---|---|---|---|
| CA02 | 56 | 67 | 56 | 64 | 89 | 22 × 83 (T)*, 22 × 75 (S)** | M 10.8 |
| CA04 | 56 | 67 | 56 | 102 | 51 | 22 × 83 (T), 22 × 75 (S) | M 10.8 |
| BC's | Stiffened H360 × 382 Column stub, 152 × 25 plate used for a beam flange | | | | | | |
| Steel Grade | FE 350(Grade 50) Steel | | | | | | |
| Beam Setback | Each 48 mm (22 mm Bolt Dia.) and 38 mm (25 mm Bolt Dia.) Beam Setback. | | | | | | |
| Diameter | Adding clearance (2 mm) to the diameter of bolt | | | | | | |

*(T) = Tension bolts
**(S) = Shear bolts
***Grade for used bolts SI M10.8 = ASTM A490

of $H_1$ to $H_2$ (Fig. 7.2) was often the main geometric parameter necessary to model the local bending of the clip-angle because of the prying actions of tension bolts.

As illustrated in Fig. 7.4(a), the component testing device was instrumented extensively with displacement transducers (LVDTs). Individual displacements, used primarily to isolate the different components of overall deformation, were monitored by each pair of LVDTs. For example, LVDT pair A monitored the uplift of the clip angle from the face of a column, LVDT pair B measured the slip between the clip-angle and the beam-flange, LVDT pair C measured the overall deformation of the clip-angle and finally, LVDT pair D measured the elongation of the clip-angle. Axial loads (P) were used to represent the beam-flange forces transferred from beam to column in the actual connection. They were also applied only to the tip of the clip-angle component for the purpose of conducting the component tests. For the full-scale connection tests, the total applied force was measured as a function of relative displacement (T-Δ) at the tip of the framing beam (see Point E; Fig. 7.4(b)).

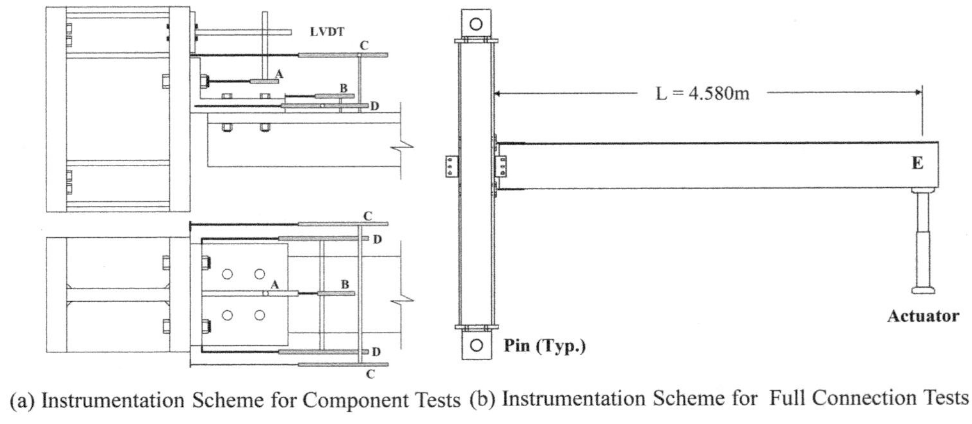


*Figure 7.4* Instrumentation scheme for tests.

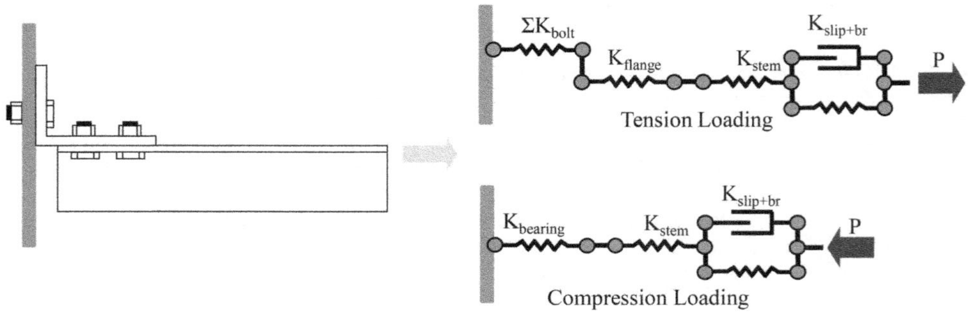


*Figure 7.5* Mechanical spring model for clip-angle components.

## 7.3 MECHANICAL MODELING

The mechanical joint model proposed here, which is entirely composed of component springs and rigid elements was numerically analyzed using the OpenSEES program (Mazzoni *et al.*, 2006), a nonlinear finite element (FE) program that is widely utilized for this type of study.

Fig. 7.5 represents a schematic representation of the clip-angle component as extension springs. The total clip-angle deformation can be decomposed into the deformations of individual connection components: (1) tension bolts ($\Sigma K_{bolt}$); (2) upstanding clip-angle flange ($K_{flange}$); (3) clip-angle stem ($K_{stem}$); (4) combined bolt bearing and connection slip ($K_{slip+br}$); and (5) bearing compression ($K_{bearing}$). The axial loads (P) transmitted from the beam-flange are transferred to the component springs. According to the load transfer, the component springs are assembled in parallel or in series depending on how they physically interact with each other. Note that two systems

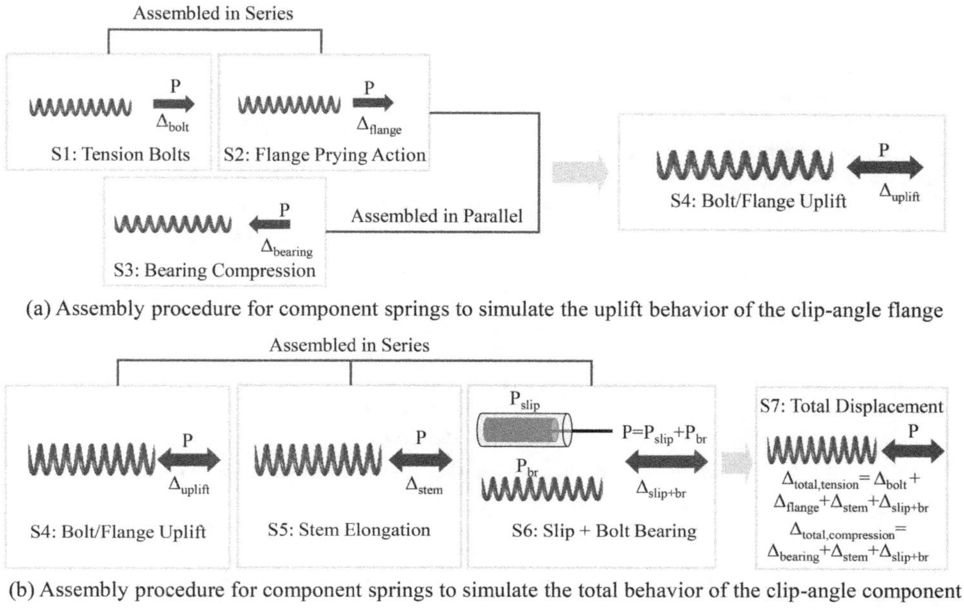


*Figure 7.6* Modeling procedure for assembling component springs and their force-deformation responses.

were used for the mechanical spring models under cyclic loading since the component springs show different responses for tension and compression.

Because the springs were arranged in series, the overall deformation of the T-stub component was calculated by adding all of the component deformations as follows:

$$\Delta_{total,tension} = \Delta_{bolt} + \Delta_{flange} + \Delta_{stem} + \Delta_{slip+br} \tag{7.1}$$

$$\Delta_{total,compression} = \Delta_{bearing} + \Delta_{stem} + \Delta_{slip+br} \tag{7.2}$$

where $\Delta_{total,tension}$ and $\Delta_{total,compression}$ indicate the overall deformation of the clip-angle component in tension and compression, respectively; and $\Delta_{bolt}$, $\Delta_{flange}$, $\Delta_{stem}$, $\Delta_{slip+br}$, and $\Delta_{bearing}$ represent the individual component deformations. The concept for this assembly may be applied to mechanical models as long as the interactions between individual components are considered.

The modeling procedure for assembling the component springs is illustrated as a flow chart in Fig. 7.6. The force-deformation responses are presented in conjunction with the interactions between idealized component springs. The springs were assembled in accordance with the axial load transfer. For numerical analysis, each component spring was modeled as a zero-length spring element along with its corresponding force-deformation response. The total behavior of the clip-angle component, which was simulated using the equivalent spring element referred to as S7 (Fig. 7.6(b)),

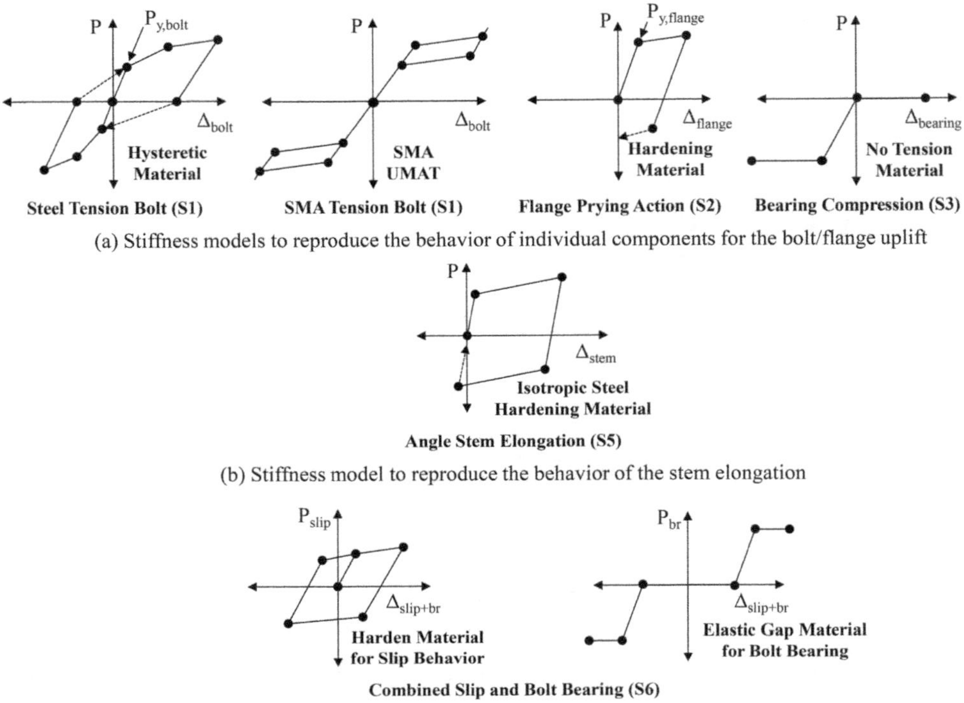


(a) Stiffness models to reproduce the behavior of individual components for the bolt/flange uplift



(b) Stiffness model to reproduce the behavior of the stem elongation



(c) Stiffness models to reproduce the behavior of the combined slip and bolt bearing

*Figure 7.7* Multi-linear stiffness models for simulating the behavior of individual clip-angle components.

was determined by integrating the force-deformation characteristics of three primary mechanisms such as the bolt/flange uplift behavior (S4), the clip-angle stem elongation (S5), and the slip combined with the bolt bearing (S6). In addition, two systems were installed to simulate the uplift behavior of the bolt and clip-angle flange on the basis that this behavior becomes invalid in compression (Fig. 7.6(a)).

The force-deformation responses of the individual components were idealized by multi-linear stiffness models as shown in Fig. 7.7. These stiffness models were obtained via three methods: (1) cyclic material pull-tests; (2) established stiffness models; (3) and curve fitting to existing test data. For example, the stiffness models for tension bolts that were subjected to only tensile loads (i.e., S1) were based on material properties obtained from the cyclic material pull-tests while those for the elastic bearing compression were obtained from curve fitting methods (i.e., S3). The force-deformation curves for the other three components (i.e., S2, S5, and S6) were generated by existing stiffness models (Buehler *et al.*, 1963; Pucinotti, 2001; Hu *et al.*, 2010; Kulak *et al.*, 1987; Astanehasl, 1995; Rex & Easterling, 1996). The stiffness models employed in this study were simulated by using the uni-axial material command of the OpenSEES program. In particular, the stiffness model for A490 steel bolt materials was simulated

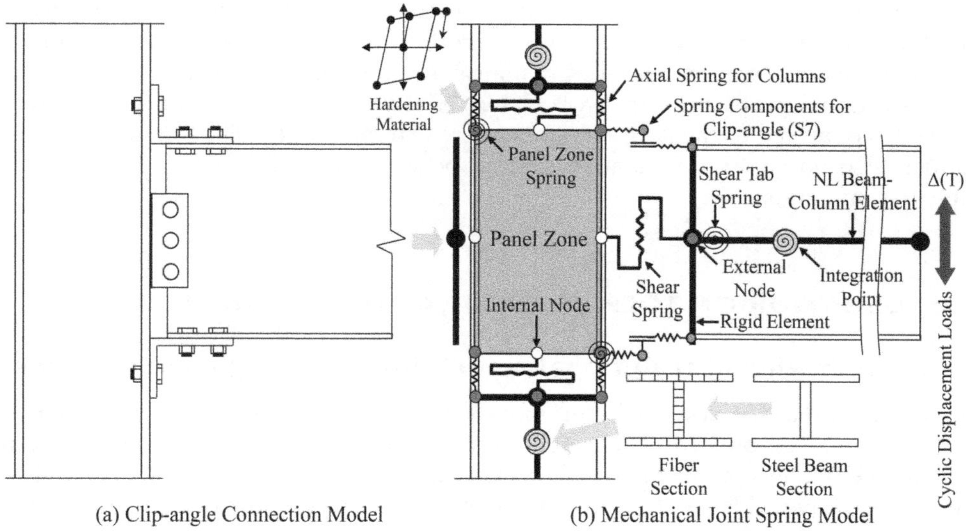


*Figure 7.8* Mechanical joint model for a full-scale clip-angle connection.

by the default hysteretic material commands. The stiffness model for the superelastic SMA materials, on the other hand, was generated by user-defined material code (Auricchio & Sacco, 1997) because an appropriate material command was not available in the OpenSEES program. Finally, all of the uni-axial materials were designated as zero-length spring elements so that they could be used to reproduce the cyclic behavior of clip-angle components.

The mechanical joint models, which were constructed to simulate the behaviors of complete beam-to-column connections, were implemented using 2D joint elements for numerical analyses. Fig. 7.8 shows the modeling attributes of the user-joint element proposed in this study (Hu, 2008). This joint element consists of various zero-length spring elements, rigid elements, and nodes: (1) two equivalent spring elements (S7), which are able to reproduce the overall deformations of clip-angle components; (2) four internal springs, which are able to reproduce the elastic deformations of the column member subjected to axial loads; (3) three internal shear spring elements, which are able to reproduce the elastic shear deformations of both the beam and the column; (4) two rotational spring elements, which cause a panel zone to deform into a parallelogram due to shear force; (5) four external rigid elements, which are able to reproduce the framing behavior of the beam-to-column connection; (6) one rotational spring element, which is able to reproduce the behavior of the shear tab; (7) four internal nodes on the middle edge of the panel zone; (7) and finally, four external nodes on the rigid elements.

The behavior of the panel zone was characterized by the tri-linear stiffness model described in the FEMA 355C report (FEMA, 2000). This stiffness model was simulated using the hardening material command in the OpenSEES program, and then the results were assigned to the two shear panel spring elements. Component springs were employed in the 2D joint element in order to connect the internal and external nodes coincident with the position using zero-length spring elements. Steel beam and column members were modeled as nonlinear beam-column elements with their own cross-sectional properties that were characterized by 2D fiber sections. They were directly connected to the joint elements at the external nodes. The 2D fiber sections, which incorporated the nonlinear material behavior, were located at the numerical integration points.

## 7.4 BEHAVIOR OF BOTH TYPES OF CONNECTION MODELS

### 7.4.1 Behavior of traditional steel bolted clip-angle connections

It is verified that the proposed joint models with component springs are adequate to predict complete connection behaviors as well as local component behaviors by comparing the findings of our model to experimental results (Swanson, 1999; Schrauben, 1999). As shown in Fig. 7.9, numerical analysis results were compared to experimental data so as to assess the behaviors of individual components. The numerical analyses were performed by applying cyclic displacement loads to each of the component spring elements (S4 to S7).

The strength capacities of the steel tension bolts and the clip-angle flanges that were used are summarized in Table 7.2. These values are associated with possible failure modes that are relevant to the formation of plastic mechanisms on upstanding clip-angle flanges. In addition, failures that occur between the clip-angle flange and tension bolts influence the performance of the bolt/flange uplift (S4). For the CA02 model, the plastic yielding of the clip-angle flange ($P_{y,flange}$) indicates the strength capacity between initial bolt yielding and ultimate bolt failure as shown in Table 7.2 (e.g., $P_{y,bolt} < P_{y,flange} < P_{u,bolt}$), and is computed in accordance with the widely used prying mechanism model (Kulak *et al.*, 1987). Therefore, the ultimate strength of the clip-angle component ($P_u = 560$ kN) was balanced between the flange yielding value and bolt fracture limits.

For the CA04 model, in contrast, the plastic mechanism on the clip-angle was followed by initial yielding of the tension bolts owing to bolt strength ($P_{y,flange} < P_{y,bolt}$). The CA04 model was more flexible than the CA02 model, with $H_1/H_2 = 2.0$ (c.f., $H_1/H_2 = 0.72$ for the CA02 model), and failed when the intricate prying mechanism was mixed with flange yielding limitations (Hu *et al.*, 2010; Kulak *et al.*, 1987). In other words, the larger $H_1/H_2$ allowed prying force to build up quickly at the tip of the upstanding clip-angle flange. Then, vigorous prying action increased the strain on the plastic hinge of the clip-angle flange, thereby reducing the ultimate angle capacity ($P_u = 383$ kN).

The clip-angle stem (S5) possesses the stiffest behavior among the three components, since it was designed with a thick angle stem (25 mm). The deformation of the clip-angle stem is concentrated in tension and exhibits a limited amount of hysteresis as shown in Fig. 7.9. The combined slip and bolt bearing deformation of the clip-angle component (S6) are also presented in Fig. 7.9. Prior to slipping, the clip-angle component undergoes elastic deformation. Geometric parameters related to the slip response

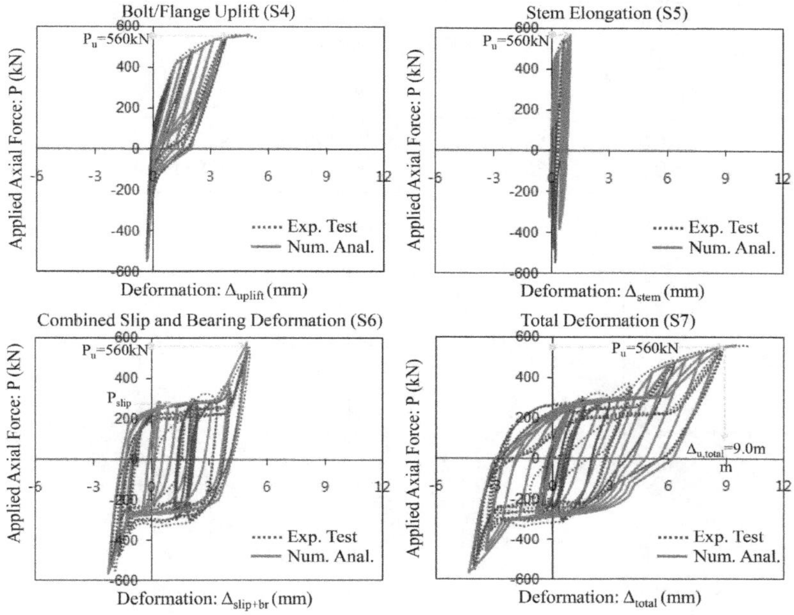


(a) CA02 Model

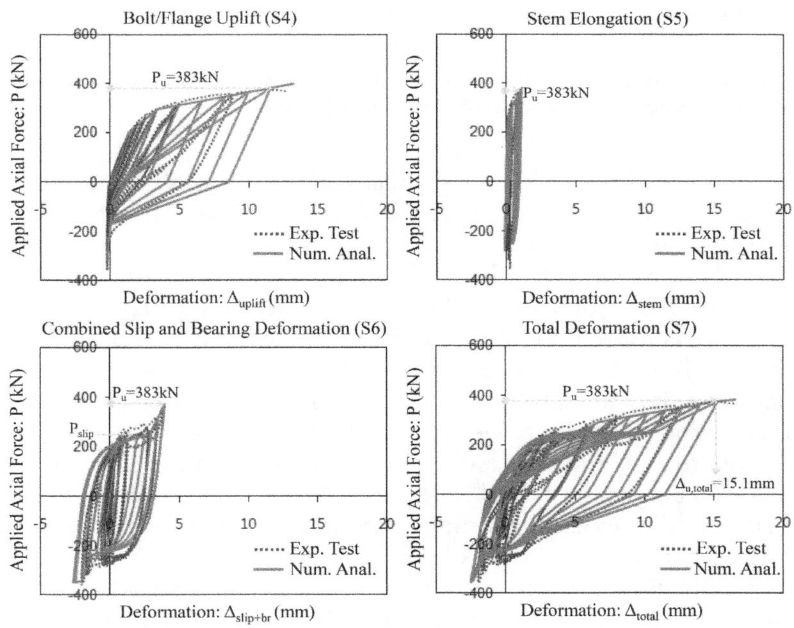


(b) CA04 Model

*Figure 7.9* Comparisons between experimental test (Exp. Test) results and numerical analysis (Num. Anal.) results with respect to individual component behavior.

*Table 7.2* Strength capacities of steel bolts and clip-angle flanges.

| Model ID | *Steel bolt (M10.8) capacity* $P_{y,bolt}$ | $P_{u,bolt}$ | *Flange capacity* $P_{y,flange}$ |
|---|---|---|---|
| CA02 | 456 kN | 667 kN | 540 kN |
| CA04 | 456 kN | 667 kN | 378 kN |

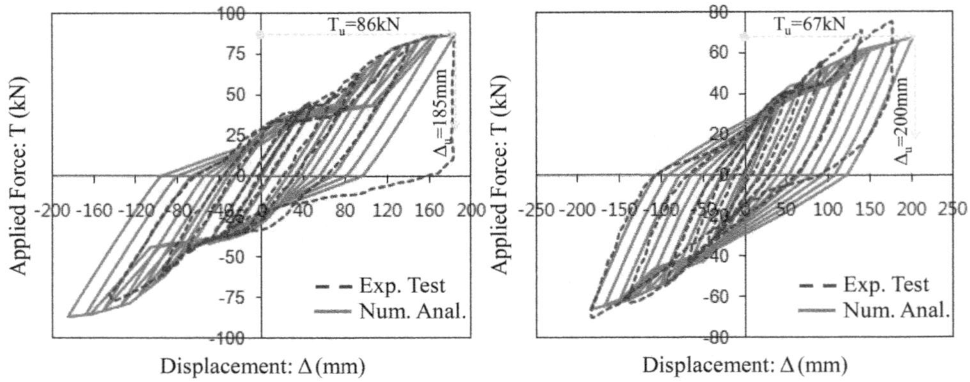


(a) FS01-CA02 Model

(b) FS02-CA04 Model

*Figure 7.10* Comparisons between experimental test (Exp. Test) results and numerical analysis (Num. Anal.) results with respect to the full connection behavior.

are identical between both component models, so they share the same slip resistance ($P_{slip} = 250$ kN). As the number of cycles increases incrementally, the slip resistance gradually deteriorates.

The total deformation of the clip-angle component (S7) includes each of the different deformations under common axial loads (P), thereby ensuring that it includes the deformation mechanisms assembled in series as defined in Eqs. (7.1) and (7.2). Note that the load levels of the slip plateau shown in the total deformation (S7) are identical to those shown in the combined slip and bolt bearing deformation (S6) indicating that the instrumentation worked very well in most cases. The results of numerical analyses were compared to experimental results to verify that the spring elements considered herein accurately predict the overall behavior of the clip-angle component. The results of both analyses show good agreement in terms of initial slope, shape of the envelope, ultimate capacity, and even location of pinching points.

The cyclic curves obtained from the numerical analyses, which were performed on the 2D joint elements, were also compared to cyclic curves from the experimental tests to validate the adequacy of the proposed mechanical models for full-scale clip-angle connections. Fig. 7.10 shows the comparisons between test results for applied total force as a function of relative displacement (T-Δ). The cyclic curves match very well before the ultimate fracture occurs in the experiments.

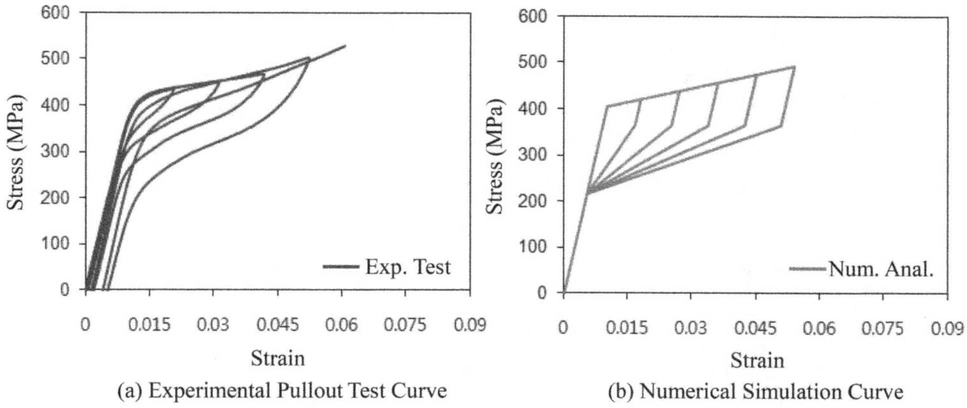


*Figure 7.11* Stress-strain curves for the superelastic SMA material (DesRoches *et al.*, 2004).

### 7.4.2 Behavior of SMA bolted clip-angle connections

In this section, the behaviors of SMA bolted clip-angle components and connections under cyclic loading conditions are discussed to examine not only energy dissipation capacity but also recentering capability provided by superelastic SMA fastener systems. The SMA bolted clip-angle components (CA02-SMA and CA04-SMA) and connections (FS01-CA02-SMA and FS02-CA04-SMA) used for numerical case studies were similar to the corresponding experimental connection models except that the tension bolts were fabricated using superelastic Nitinol SMAs, not M10.8 steel. The numerical studies were conducted because relevant physical testing data were lacking.

First, the calibration of the superelastic material behavior was obtained by comparing the numerical simulation to experimental pullout test data for SMA bars. This information was necessary to reliably model the SMA bolted connections. Stressstrain curves for the superelastic SMA materials are presented in Fig. 7.11. The SMA tension bolts were modeled after a uni-axial test carried out by DesRoches *et al.*, (2004), who researched the cyclic properties of 25-mm diameter superelastic SMA bars (see Fig. 7.11(a)). Although only tensile loadings are taken into consideration in this material test, past numerical studies typically modeled SMA behaviors as symmetric. As shown in Fig. 7.11(b), the cyclic behavior of the superelastic SMA material was simulated using a user defined material (UMAT) code in the OpenSEES program. Its formulation relied on one of the most widely used 1D constitutive material models (Auricchio & Sacco, 1997). The good fit of the material model to the experimental data can be found in Fig. 7.11.

In order to investigate the effects of SMA fastener systems, numerical studies were conducted on component springs and joint elements. Fig. 7.12 shows the behaviors of individual SMA clip-angle components simulated by each of the component spring elements (S4 to S7). Contrary to traditional steel bolted connections (see Fig. 7.9(a)), the uplift behavior of the CA02-SMA model (S4) shown in Fig. 7.12(a) is fully recovered upon unloading, indicating that no residual deformation remains. On the other hand,

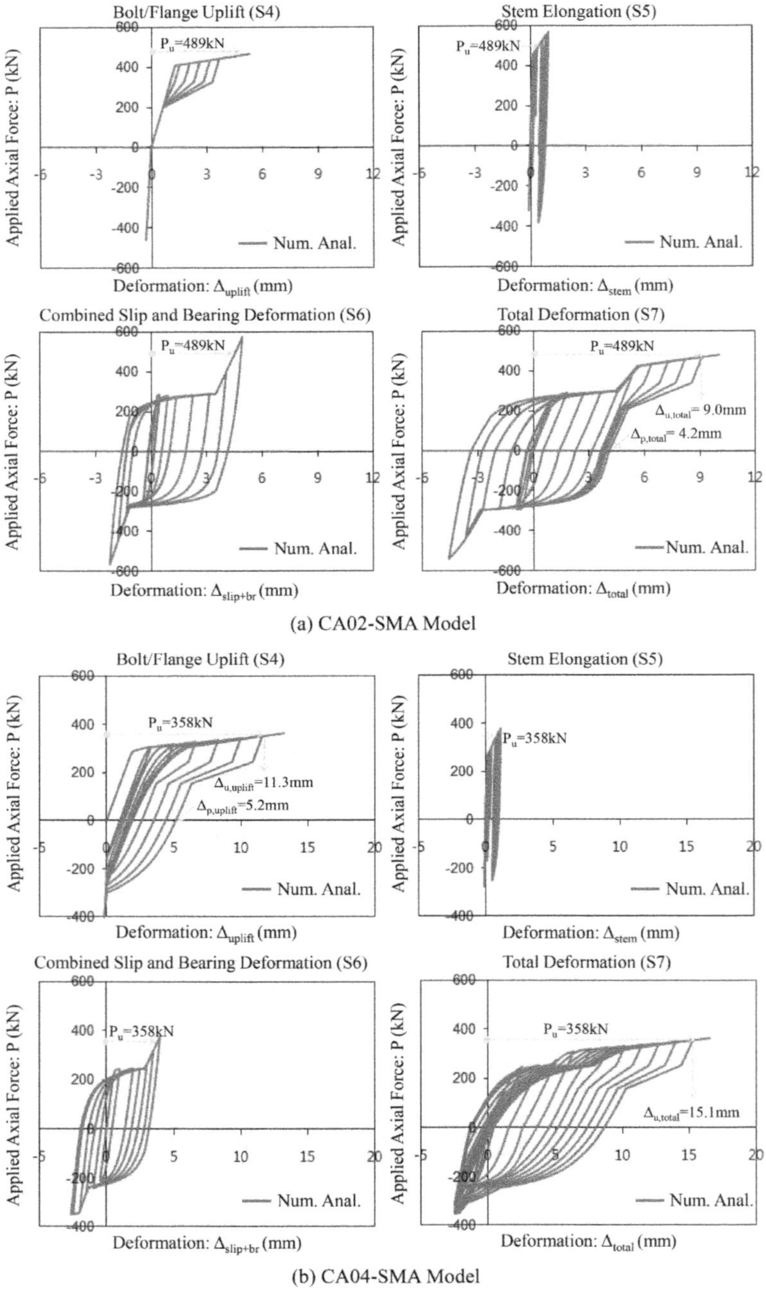


*Figure 7.12* Simulated behaviors of the clip-angle components with superelastic SMA bolts.

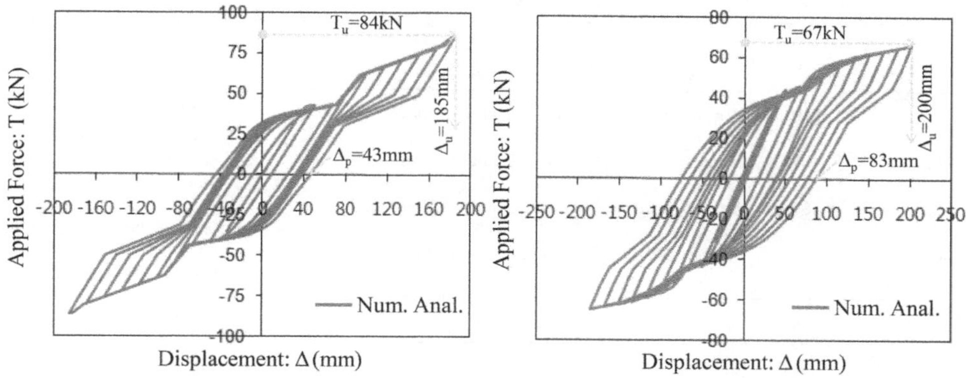


(a) FS01-CA02-SMA Model (b) FS02-CA04-SMA Model

*Figure 7.13* Simulated behaviors of the clip-angle connections with superelastic SMA bolts.

*Table 7.3* Strength capacities of SMA bolts and clip-angle flanges.

| *Model ID* | *SMA bolt capacity* | | *Flange capacity* |
|---|---|---|---|
| | $P_{y,bolt}$ | $P_{u,bolt}$ | $P_{y,flange}$ |
| CA02-SMA* | 420 kN | 544 kN | 520 kN |
| CA04-SMA** | 321 kN | 420 kN | 300 kN |

* = 25-mm tension bolt diameter.
** = 22-mm tension bolt diameter.

the uplift behavior of the CA04-SMA model (S4) shown in Fig. 7.12(b) is only partially recovered upon unloading. For the 11.3 mm displacement loading cycle of the uplift behavior of the CA04-SMA model ($\Delta_{uplift}$), recoverable deformation reached 6.1 mm ($\Delta_{u,uplift} - \Delta_{p,uplift} = 6.1$ mm), accounting for about 54% of the applied deformation. The residual deformation of the uplift behavior was attributable to the formation of the plastic mechanism on the clip-angle flange.

The strength capacities of the SMA tension bolts and clip-angle flanges are summarized in Table 7.3 and indicate whether or not the plastic hinge formed on the clip-angle flange. The CA02-SMA model was designed with the plastic strength capacity of the clip-angle flange, which is greater than the initial yield strength and is immediately followed by ultimate bolt failure as seen in Table 7.3 (e.g., $P_{y,bolt} < P_{y,flange} = 520kN < P_{u,bolt}$). In addition, since $P_u < P_{y,flange}$, the clip-angle flange is still elastic in its final state ($P_u = 489$ kN; see Fig. 7.12(a)).

Accordingly, the uplift deformation of the clip-angle flange even at peak load is completely recoverable upon unloading. In fact, most of its residual deformation is caused by the slip response (S6) which also serves to increase the energy dissipation of the total behavior of the clip-angle component (S7). For the 9.0-mm displacement

*Table 7.4* Summary of numerical analysis results.

| *Model ID* | $P_u$ | $\Delta_{u,total}$ | $\Delta_{p,uplift}$ | $\Delta_{p,stem}$ | $\Delta_{slip+br}$ | $\Delta_{p,total}$ | $(\Delta_{u,total} - \Delta_{p,total})/\Delta_{u,total}$* | *Total EDA*** *(Uplift + Stem + Slip)* |
|---|---|---|---|---|---|---|---|---|
| CA02 | 560 kN | 9.0 mm | 1.9 mm | 0.5 mm | 4.2 mm | 6.4 mm | 0.29 | 5517 kNmm (1503 + 771 + 3243 kNmm) |
| CA02-SMA | 489 kN | 9.0 mm | 0.0 mm | 0.3 mm | 3.9 mm | 4.2 mm | 0.54 | 4367 kNmm (366 + 766 + 3225 kNmm) |
| CA04 | 383 kN | 15.1 mm | 8.8 mm | 0.3 mm | 3.2 mm | 11.3 mm | 0.25 | 5519 kNmm (3058 + 455 + 2006 kNmm) |
| CA04-SMA | 358 kN | 15.1 mm | 5.2 mm | 0.2 mm | 3.0 mm | 8.4 mm | 0.44 | 5344 kNmm (2804 + 451 + 1990 kNmm) |

| *Model ID* | $T_u$ | $\Delta_u$ | $\Delta_p$ | $(\Delta_u - \Delta_p)/\Delta_u$* | *Total EDA*** |
|---|---|---|---|---|---|
| FS01-CA02 | 86 kN | 185 mm | 98 mm | 0.47 | 17088 kNmm |
| FS01-CA02-SMA | 84 kN | 185 mm | 43 mm | 0.77 | 11650 kNmm |
| FS02-CA04 | 67 kN | 200 mm | 127 mm | 0.37 | 17813 kNmm |
| FS02-CA04-SMA | 67 kN | 200 mm | 83 mm | 0.59 | 15525 kNmm |

*Recoverable Deformation Ratio.
**Energy Dissipation Area (EDA).

loading cycle of the total behavior of the CA02-SMA model ($\Delta_{total}$), recoverable deformation reached 4.8 mm ($\Delta_{u,total} - \Delta_{p,total} = 4.8$ mm; see Fig. 7.12(a)), accounting for about 54% of the applied deformation. In the CA04-SMA model, the plastic yielding of the clip-angle flange is followed by the initial yielding of the SMA tension bolts since $P_{y,flange} < P_{y,bolt}$ (see Table 7.3). Accordingly, its residual deformation occurs by means of both plastic yielding on the clip-angle flange and slippage on the shear faying surface.

The applied total force vs. relative displacement behaviors of SMA bolted clip-angle connections simulated by using the 2D joint elements are presented in Fig. 7.13. The FS01-CA02-SMA model shows approximately 25% higher strength capacity under ultimate displacement loading than the FS02-CA04-SMA model, likely owing to the larger clip-angle capacity (Fig. 7.12). Furthermore, the former model exhibits better recentering behavior than the latter (Fig. 7.13). In the SMA bolted connections, the residual deformation of the clip-angle component ($\Delta_{p,total}$) directly influenced on permanent displacement of the connection behavior ($\Delta_p$). Hence, without plastic yielding of the beam, permanent displacement can be predicted by the geometric triangle ratio as follows:

$$\Delta_P = \frac{\Delta_{p,total} L}{d_{beam}} \tag{7.3}$$

where $L/d_{beam}$ is assumed to be 10 in this study. For the FS01-CA02-SMA model, this prediction is a good approximation of the permanent displacement obtained through numerical analyses on the ground of a small gap between them (e.g., $\Delta_p = 43\ \text{mm} \approx 10\Delta_{p,total} = 42$ mm; refer to Table 7.4). Therefore, the permanent displacement is caused primarily by slip response in this model. The prediction of permanent displacement is suitable for other connection models (e.g., $\Delta_p = 84\ \text{mm} \approx$

$10\Delta_{p,total} = 83$ mm for the FS02-CA04-SMA model). The same load level of the slip plateau may be found in traditional steel bolted connection models and SMA bolted connection models because the geometric parameters used for the slip response as well as the surface conditions are identical.

### 7.4.3 Numerical analysis results and observations

The summaries of the resulting cyclic curves from this study are listed in Table 7.4. The values in this table were obtained at the ultimate displacement loading cycles, which start from the maximum displacement loads ($\Delta_{u,total}$ and $\Delta_u$), as depicted in the above figures. For the given maximum total displacement load ($\Delta_{u,total}$), traditional clip-angle components benefit by being able to exploit steel bolts with relatively higher (an average of 9% higher) ultimate strength capacities ($P_u$) compared to SMA bolted clip-angle components. However, as listed in Table 7.4, SMA clip-angle components exhibit much better recoverable deformation ratios (i.e., $(\Delta_{u,total} - \Delta_{p,total})/\Delta_{u,total}$) than traditional clip-angle components during unloads. The degree of recentering capability can be estimated by this recoverable deformation ratio. Aside from residual deformation caused by the slip ($\Delta_{p,slip+br} = 3.0$ mm), the CA04-SMA model undergoes unrecoverable deformation caused by yielding of the clip-angle flange during uplift behavior ($\Delta_{p,uplift} = 5.2$ mm) which indicates that its recoverable deformation ratio deteriorates in comparison with the CA02-SMA model. Additionally, as expected, the residual deformation of the total behavior coincides with the summation of the component's residual deformations at each of the ultimate displacement loading cycles such that $\Delta_{p,total} = \Delta_{p,uplift} + \Delta_{p,stem} + \Delta_{p,slip+br}$.

The values for energy dissipation area (EDA) listed in Table 7.4 were calculated as the enclosed areas of resulting cyclic curves. In general, energy dissipated either by member yield or by slippage causes the cyclic behavior curves to become plump, thereby increasing the EDA. Although SMA bolts contribute to decreasing residual deformation, the full recentering capability obtained from their superelastic behavior without member yield or slippage tends to reduce energy dissipation capacity quickly by pinching the behavior curve. For example, the CA04-SMA model decreased its energy dissipation capacity by approximately 4% compared to the CA04 model. Nevertheless, SMA tension bolts installed can considerably increase the recoverable deformation by up to 46% for the total component behavior. Accordingly, the SMA bolted clip-angle components are anticipated to possess excellent recentering capabilities in addition to adequate energy dissipating characteristics.

Similar to the SMA clip-angle components, full-scale SMA bolted clip-angle connections show moderate energy dissipation and excellent recentering behavior. The deformations of the connections at the given peak displacement loads ($\Delta_u$) are also recoverable upon unloading. For the 185 mm displacement loading cycle of the FS01-CA02-SMA model, recoverable deformation reached 142 mm which was as much as 77% of the applied deformation. This SMA connection model showed moderate energy dissipation capacity with the energy (11650 kN-mm) mostly dissipated by the slip response. Since the ultimate strength capacities of the SMA bolted connection models ($T_u$) were equal to or below the elastic capacity of the H460 × 60 section beam (taken as 84 kN), any plastic hinges are not developed at the connecting beams. In general, more vigorous recentering behavior of the SMA bolted connections is expected by keeping the beam in the elastic state, while the ductility of the SMA bolted clip-angle connection will likely be influenced by the deformation of the installed clip-angle component.

## 7.5 CONCLUSIONS

The following conclusions are based on the results and observations of this study:

(1) The component spring and joint models developed for numerical investigations were carefully back-calibrated to component and connection behaviors, respectively, which were separated in the SAC experiments. These models correctly estimated the sequence of the slip, bolt bearing, and development of the prying mechanism when compared to corresponding experimental test results.
(2) The clip-angle components having more active prying action (e.g. CA04 model series) are susceptible to strength degradation due to larger inelastic deformation resulting from the yield line at the clip-angle flange. The recentering capacity could be increased by reducing the prying forces on the tip of the clip-angle, meaning that clip-angles should be designed with a smaller value of $H_1/H_2$ in order to remain elastic. For full-scale SMA bolted connections, the desired recentering capacity depends on whether their ultimate strengths are below the elastic capacity of the connecting beam.
(3) Aside from the slip response, the inelastic deformation of the clip-angle flange caused the enclosed area of cyclic curves to become fatter. Consequently, the test model series with CA04 clip-angle components showed larger energy dissipation capacity in its cyclic behavior than the test models with CA02 components.
(4) Our numerical analysis results are promising for the practical application of new SMA bolted clip-angle connections, which feature excellent recentering capacity and moderate energy dissipation when compared to traditional steel bolted clip-angle connections. Furthermore, they serve as a good platform for exploring the optimization of recoverable deformation in SMA bolted connections as a function of energy dissipation.

Part 3

# Design and analyses for other smart connections

Chapter 8

# Gusset plate connections

## 8.1 GENERAL OVERVIEW

The interstate highway 35W (I-35W) bridge over the Mississippi river in Minneapolis, Minnesota, collapsed within a matter of seconds at 6:05 p.m. on August 1, 2007. The 508m long bridge with the 8 traffic lane fell into the water, resulting in 13 fatalities, 145 injuries, and 111 vehicles involved in the collapse (see Fig. 8.1). In the worldwide, this type of steel truss bridge has been commonly used for middle-to-long spans that carry heavy loads (Yamamoto *et al.*, 1988). Until the accident occurred, steel truss bridges had made the solid reputation of being reliable and economical (Astaneh-Asl, 2008; Li *et al.*, 2008). However, many scientists were often induced to focus significant attention on reliability and safety for bridge design after I-35W bridge failure. In particular,

(a) I-35W MN Bridge before Collapse (b) Gusset Plate Panel Point (U10)

(c) I-35W MN Bridge after Collapse (d) Gusset Plate Fracture (U10)

*Figure 8.1* Tragic 2007 collapse of the I-35W MN steel truss bridge in USA.

apparent evidence relevant to this bridge failure was pointing toward gusset plates (see U-10 panel point in Fig. 8.1(b)), which were vulnerable to buckling and fracture, as the initial event in the collapse sequence (Holt & Hartmann, 2008; NTSB, 2008; Liao *et al.*, 2011) Therefore, adequate gusset plate design has been requested for the prevention of relapse, including safety assessment. Ordinarily, the gusset plates that connect chord members to compression diagonals and tension diagonals are designed to be stronger than the truss members, and thus shall not control the capacity of the structure.

There are several available approaches to gusset plate design. Engineering principles based on the allowable stress design (ASD) method were previously applied to gusset plate design (Whitmore, 1952; Hardash & Bjorhovde, 1984; Bjorhovde, 1988). In the aftermath of the I-35W bridge failure event, the Federal Highway Administration (FHWA) issued a technical design advisory emphasizing the necessity to check strength capacity along with the connected member into the gusset plate through the reliability-based load rating process (FHWA, 2009; FHWA, 2010). This design advisory provided additional guidance based on the best available modern information regarding gusset plate design for bridge engineers, including load rating design examples (FHWA, 2009). In other words, the FHWA put in place an implementation plan to address recommendations concerning the factored load-carrying capacity of the gusset plates in the non-load path redundant steel truss bridges. Accordingly, in this paper, the load and resistance factor design (LRFD) method adopted in the current American Association of State Highway and Transportation Officials (AASHTO, 2007). Specifications is used for gusset plate design in an effort to keep pace with these technical changes that are recently occurring. The basic strength of the presented gusset plates will be evaluated in accordance with this LRFD method.

To improve gusset plate design, nonlinear analyses should be conducted using finite element (FE) models. For untested specimens, the verified FE models have been used to accurately predict inelastic behavior and stress-strain distribution. They are considered very reliable for the simulation of the complex. Some of these FE analyses have been utilized to perform parametric studies that provide insight and guidance for structural design, due to the advantage of being easily scalable to model (Hu *et al.*, 2010; Hu *et al.*, 2011; Hu *et al.*, 2011, Hu *et al.*, 2012; Hu, 2013). Despite these merits, the advanced FE models that are made up of 3 dimensional (3D) solid elements incorporating fully nonlinear material properties, geometric nonlinearity, contact interactions, and initial bolt pretension are limited to achieving practical design purposes in massive and complicated structures. It is because they require not only high computational cost but also sophisticated preparation process. In the related researches previously established (Liao *et al.*, 2011; Uriz *et al.*, 2008; Yoo *et al.*, 2008). FE analyses were performed only by modeling a local part of gusset plate connections with an intention to avoid these problems. The information gained from these old-fashioned analyses was too restrictive to evaluate the entire gusset plates subjected to multi-axial loads in terms of response mechanisms, failure modes (i.e., strength limit states), and resistance capabilities. The global FE models rendered precise enough to investigate a whole pattern of gusset plate behavior with reasonable memory size are treated for this reason.

In view of the above, this paper is intended to present a state-of-the art study with regard to design approaches for the full-scale gusset plates. For the purpose of determining the capacity of the gusset plate connection, the factored resistance against

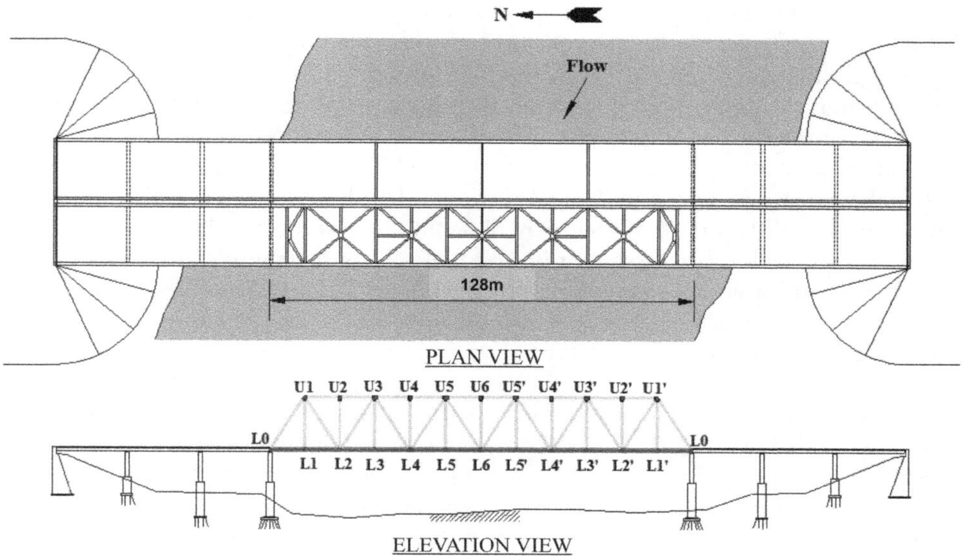


*Figure 8.2* Plan and elevation view of the I-94 IL steel truss bridge.

the patterns of failure observed after the bridge collapse needs to be estimated according to the currently used LRFD method. In addition, the detailed FE analyses are conducted on the critical gusset plate connection in the existing steel truss bridge. The outcomes from these FE analyses have been used to check life safety at the strength limit state as well as to gain visible insight into the causes of possible failure mechanisms.

## 8.2 SELECTION OF REPRESENTATIVE EXAMPLE

The gusset plate connection in the existing steel truss bridge was selected for further study. First of all, scrutinizing original bridge drawings should be required to better understand a design example for the selected gusset plate connection. The plan and elevation views provided from the I-94 steel truss bridge over the Little Calumet River in the US state of Illinois (IL) are presented in Fig. 8.2. In this study, design examples for the I-94 bridges rather than the I-35W one were taken into consideration to avoid overlapping a little with the established researches. This I-94 bridge was designed according to the AASHTO 1989 Standard Specifications based on the ASD method in 1990 and opened to traffic in 1993. The major part of the bridge is 128m (420′) long, 33 m (108′) wide, and single span steel deck superstructure. The steel deck was separated for southbound and northbound. Each deck accommodated three 3.7-m (12′) traffic lanes and two 0.6-m (2′) shoulders. The main truss that rested on the roller supports consisted of 12 panels, each of which was 10.7 m long. The truss members were braced by lateral bracings between upper chords and lower chords, and thus designed in the Warren style. More information on this truss bridge is given by references (Wright, 2009; ILDOT, 2012; Historic Bridge, 2013).

The typical gusset plate connection, the focus of this study, is shown in Fig. 8.3. As can be seen in the figure, the gusset plate connection referred to as the L2 joint (see also Fig. 8.2) was designed on the basis of the original bridge drawings without considering any changes that occurred during the life time (Wright, 2009; ILDOT, 2012). The five load carrying members were connected to the gusset plates and designed to ensure that they remain elastic up to the ultimate load. The chord members and compression diagonal were made up of welded box sections, while the vertical member and tension diagonal were made up of built-up I-sections. The box section members have the same size, i.e., 470 mm (18.5″) wide by 686 mm (27″) deep, but their thicknesses are different from each other as shown in Fig. 8.3(a). The tension diagonal has 533 mm (21″) flange width with 48 mm (1.9″) thickness and 375 mm (14.8″) web height and with 19 mm (0.8″) thickness. The vertical member was built up by the welding of two C type channels. These members lay between double gusset plates each with 29 mm thickness. The filler plates were used between gusset plates and chord members.

The gusset plates were fabricated by SS490 (ASTM A572-Gr. 50) mild steel with a Young's modulus of 200 GPa (29,000 ksi), a Poisson's ratio of 0.3, and a yield strength of 345 MPa (50 ksi). The dimensions of the representative gusset plates (L2) are presented in Fig. 8.3(b). This gusset plate connection belongs to relatively modern design using highstrength bolts (i.e., ASTM A490 bolt). The 25 mm diameter bolts were used in the top and bottom chord splice plates, while the 22 mm diameter ones were used to fasten the chord members and diagonals to the gusset plates. The bolt holes were drilled with 25 mm diameter on the gusset plates and aligned with each of different gage and pitch lengths. They were equal to the nominal diameter of the bolt shank plus 3 mm. The connection between the gusset plates and compression diagonal had only ten rows of fasteners, and so the compression diagonal side was designed with relatively shorter free-edge lengths as compared to other sides. Besides, this compression diagonal had a relatively large distance from the point of five members' convergence. For the gusset plates considered herein, their resistance strength against each failure pattern will be calculated in the next section.

## 8.3 DESIGN RESISTANCE STRENGTH

The capacity of the gusset plate connection shall be determined as the least resistance strength of the gusset plates in compression, shear, and tension. The failure modes feasibly occurring at the gusset plate subjected to such forces are thus investigated, and then design strength to resist each type of failures, such as yielding, buckling, and fracture will be discussed in this section. The LRFD method specified in the AASHTO Specifications is intended to be applied to gusset plate design as mentioned above. Since almost modern truss bridges in the world have either riveted or bolted gusset plates, the focus of this study is on the connection with bolted gusset plates.

The forces that act on the gusset plate shall be preferentially established for design. They are considered as the factored demand in the LRFD method and denoted as follows:

$$\Sigma \eta_i \gamma_i P_i = P_r \tag{8.1}$$

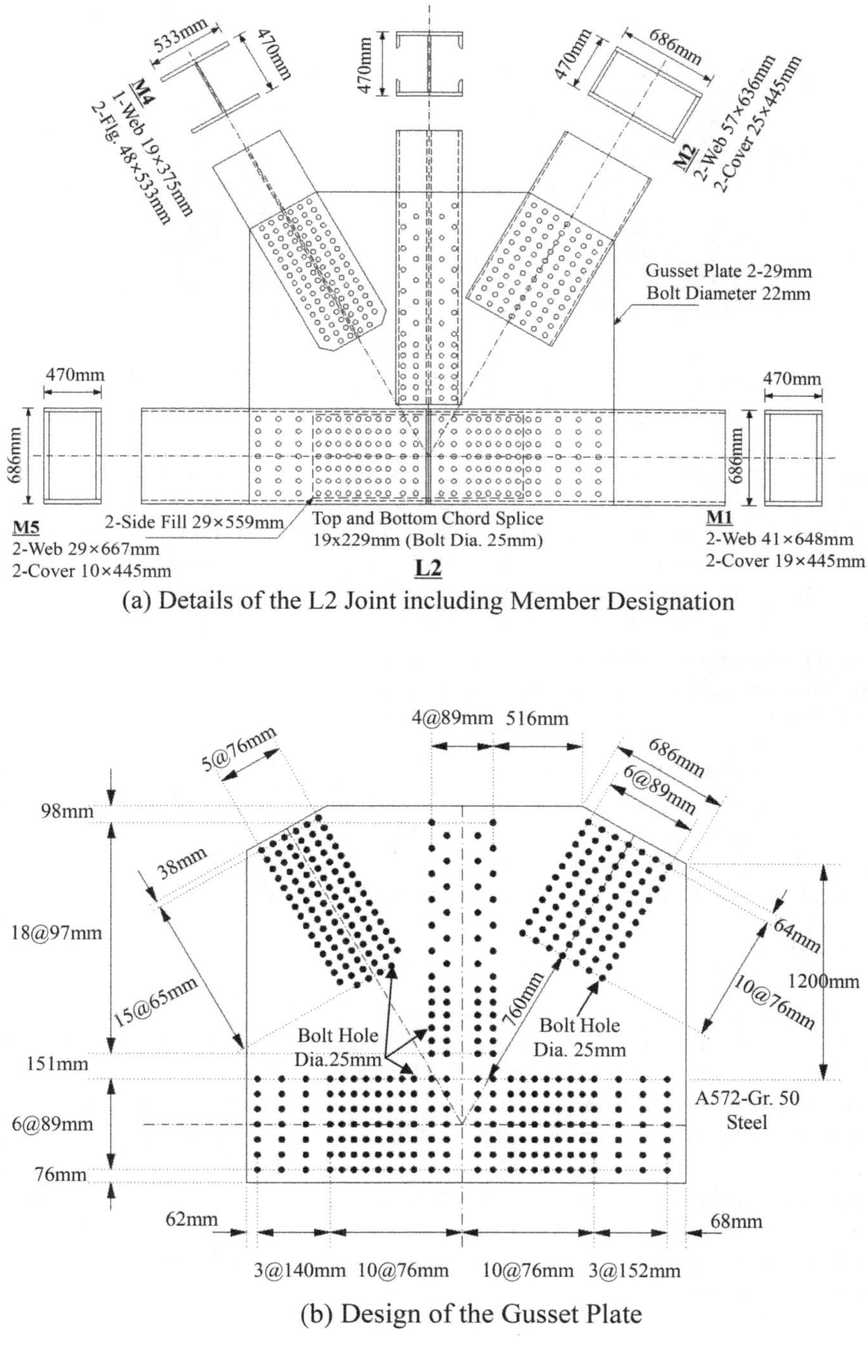


(a) Details of the L2 Joint including Member Designation

(b) Design of the Gusset Plate

*Figure 8.3* Details of the gusset plate L2 joint.

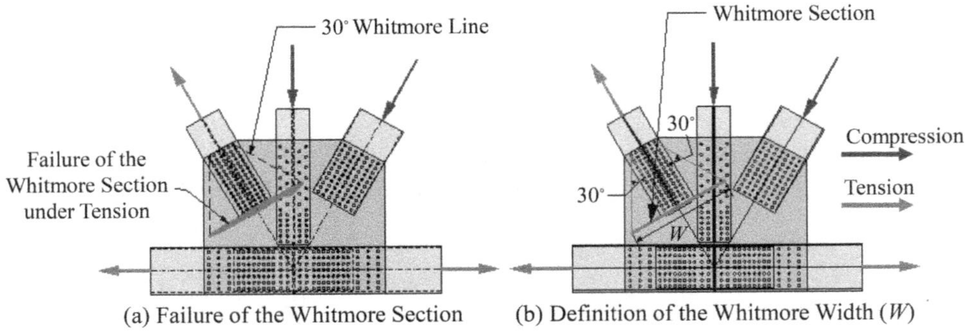


*Figure 8.4* Tension failure of the Whitmore section of the gusset plate.

where, $\eta_i$ is a load modifier factor relating to the ductility, redundancy, and importance of the component being designed and the structure itself, $\gamma_i$ is a statistically based load factor, $P_i$ is the applied force, and $P_r$ represents the factored force. The specific values of these factors are provided by the AASHTO Specifications considering various failure modes. They are typically taken as more than 1.0. On the other hand, the resistance strength of the gusset plate is considered to be the factored capacity for each failure mode and specified as follows:

$$\phi R_n = R_r \tag{8.2}$$

where, $\phi$ is a statistically design resistance factor applied to the nominal resistance of the component being designed ($R_n$) and $R_r$ indicates the factored design resistance. For each failure mode, the factored capacity shall be larger than the factored demand so as to satisfy safe design requirements (e.g., $P_r < R_r$). The rating factor for life safety (RF) is defined as the factored capacity divided by the factored demand, such that

$$\mathrm{RF} = R_r/P_r. \tag{8.3}$$

The gusset plates subjected to axial tension shall be investigated for three conditions: (1) yield on the gross section area, (2) fracture on the net section area, and (3) block shear rupture. The Whitmore method may be accepted to determine gross and net section areas on the gusset plate. Fig. 8.4 provides examples for estimating the effective width in tension according to the Whitmore method (Whitmore, 1952). The effective width can be established by drawing two 30-degree lines from the external fasteners within the first bolt row and obtained by measuring a line perpendicular to the axis of the member, which intersects these two inclined lines at the center of the last bolt row. The section areas determined in accordance with the Whitmore method are given as follows:

$$A_{gw} = Wt_g \tag{8.4}$$

$$A_{nw} = (W - n_b d_h)t_g \tag{8.5}$$

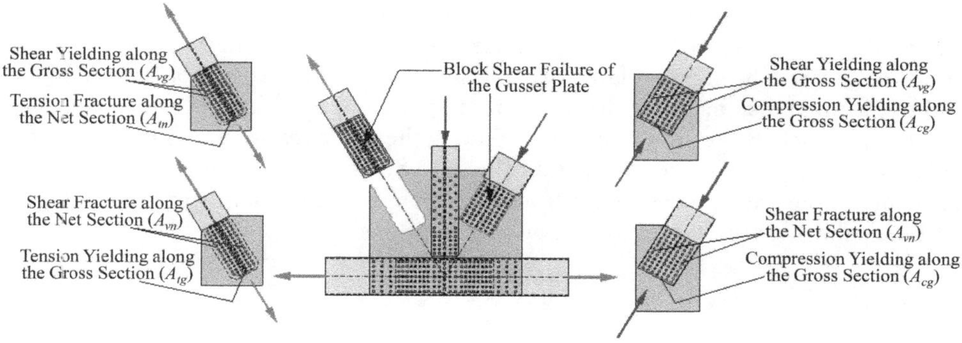


*Figure 8.5* Block shear failure of the gusset plate.

where, $A_{gw}$ is the gross section area, $A_{nw}$ is the net section area, $W$ is the width of the Whitmore section, $n_b$ is the number of the bolts at the last bolt row, $d_h$ is the diameter of the bolt hole, and finally $t_g$ represents the thickness of the gusset plate. Both yielding on the gross section and fracture on the net section shall be simultaneously checked to determine dominating resistance strength, and factored design resistance strength for the gusset plate in tension ($R_{rw}$) shall be taken as the least of the values given by yielding or fracture.

$$R_{rw} = \phi R_{nw} = \min\left[(\phi_y F_y A_{gw}), (\phi_u F_u A_{nw})\right] \tag{8.6}$$

where, $F_y$ and $F_u$ indicate specified yield and ultimate stress, respectively. The design resistance factor for yielding ($\phi_y$) is taken as a value of 0.95, while that for ultimate fracture ($\phi_u$) is taken as a value of 0.80 (FHWA, 2009; AASHTO, 2007). These factors are also applied to the nominal tensile resistance of the gusset plates ($R_{nw}$). After fracture of the net section area, the progressive collapse of the gusset plate connection takes place quite rapidly prior to yielding. To avoid this problem, design resistance strength that becomes governing shall be based on the yielding of the Whitmore section being the desirable ductile failure mode.

For other failure patterns, block shear failure can occur at the gusset plates where tension members are attached as shown in Fig. 8.5. The block shear failure of the gusset plates has been studied by a number of researchers (Hardash & Bjorhovde, 1984; Kulak & Grondin, 2000; Kulak & Grondin, 2001; Huns *et al.*, 2006). The AASHTO Specifications particularly requires that all tension connections including the gusset plates should be investigated for block shear rupture. Similar to tension fracture, block shear rupture is considered to be a brittle-mannered failure mode. The resistance to block shear failure is that resulting from the combined strength of parallel and perpendicular planes; one in axial force and the others in shear force. The following equations are specified in the AASHTO Specifications and used to check this strength limit state.

$$\text{If } A_{tn} \geq 0.58A_{vn}, \quad R_{rbs} = \phi R_{nbs} = \phi_u(0.58F_y A_{vg} + F_u A_{tn}) \tag{8.7}$$

$$\text{Otherwise,} \quad R_{rbs} = \phi R_{nbs} = \phi_u(0.58F_uA_{vn} + F_yA_{tg}) \tag{8.8}$$

where, $R_{rbs}$ is the factored design resistance strength for block shear failure, $A_{vg}$ is the gross section area along the plane resisting shear, $A_{tn}$ is the net section area along the plane resisting tension, $A_{vn}$ is the net section area along the plane resisting shear, and finally $A_{tg}$ indicates the gross section area along the plane resisting tension. The design resistance factor ($\phi_u$) is also taken as a value of 0.80 because of ultimate brittle failure. Then again, the gusset plate connection under compression can be optionally checked for block shear failure to use the following equations.

$$R_{rbsg} = \phi R_{nbsg} = \phi_u 0.58F_yA_{vg} + \phi_c F_yA_{cg} \tag{8.9}$$

$$R_{rbsn} = \phi R_{nbsn} = \phi_u 0.58F_uA_{vn} + \phi_c F_yA_{cg} \tag{8.10}$$

$$R_{rbs} = \phi R_{nbs} = \min\,[\phi R_{nbsg}, \phi R_{nbsn}] \tag{8.11}$$

The definition of notations is illustrated in Fig. 8.5. The design resistance factor for compression yielding ($\phi_c$) is taken as a value of 0.90. In this case, the only gross section area is used along the perpendicular plane resisting compression (i.e., $A_{cg}$).

Buckling occurs within the inner areas of the gusset plates, depending on how many connected members such as diagonals, verticals, and chords are in compression (see Fig. 8.6(a)). In addition to the proximity of the connected members, stress states and boundary conditions have an influence on the resistance of the gusset plates under compression. This buckling failure can be prevented or delayed by adding plate stiffener to the inner compressive areas. The effective width defined by the Whitmore method has been also used to evaluate the buckling capacity of the gusset plates subjected to direct compression. The inner area of the gusset plates in compression is considered as idealized column members in compression due to absence of more rigorous analyses. So, the un-braced length ($L_c$) may be determined as the average of three distances as follows:

$$L_C = \frac{L_1 + L_2 + L_3}{3} \tag{8.12}$$

Including the effective width for a gusset plate in compression, a good example for $L_1, L_2$, and $L_3$ distance is given to Fig. 8.6(b). The following equations also specified in the AASHTO Specifications are used to evaluate the factored design resistance strength ($R_{rcrw}$) according to the slenderness ratio as follows:

$$\lambda = \frac{KL_C}{\pi t_g}\sqrt{\frac{12F_y}{E}} \tag{8.13}$$

$$\text{If } \lambda \leq 1.5, \quad R_{rcrw} = \phi R_{ncrw} = \phi_c 0.658^{\lambda^2} F_yA_{gw} \tag{8.14}$$

$$\text{Otherwise,} \quad R_{rcrw} = \phi R_{ncrw} = \phi_c \frac{0.877}{\lambda^2} F_yA_{gw} \tag{8.15}$$

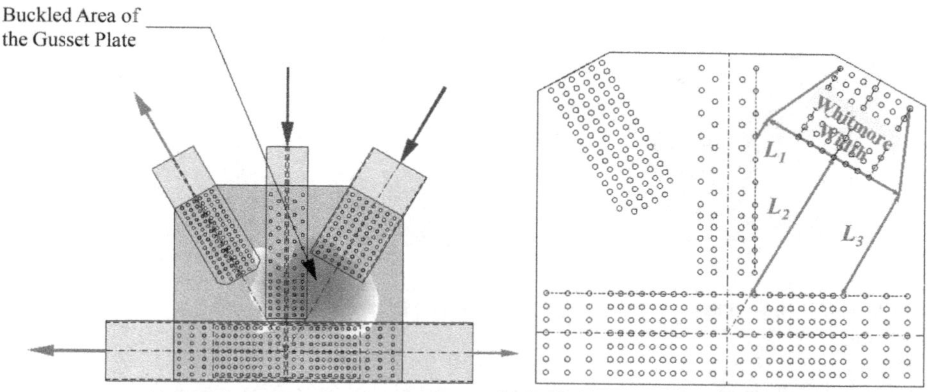


(a) Buckling of the Gusset Plate (b) Definition of the Effective Lengths for Buckling

*Figure 8.6* Buckling of the gusset plate in the compression area below Whitmore width.

where, $K$ indicates the effective length factor stipulated in the traditional column buckling equation and $E$ indicates the elastic modulus of the base steel material. Depending on sway condition and buckled shape, the effective length factor for gusset plate design may be taken as a value of 1.2 (FHWA, 2009; AASHTO, 2007). The value of 0.90 is applied to the design resistance factor for buckling. The gross section area $(A_{gw})$ coincides with that obtained by the Whitmore method (see Eq. (8.4)).

The resistance strength for the gusset plates subjected to flexural shear shall be taken as the lesser one to resist shear yielding on the gross section $(R_{rg})$ and shear fracture on the net section $(R_{rn})$ as follows:

$$R_{rg} = \phi R_{ng} = \phi_{vy} 0.58 F_y A_g U \tag{8.16}$$

$$R_{rn} = \phi R_{nn} = \phi_{vu} 0.58 F_u A_n \tag{8.17}$$

where, $A_g$ and $A_n$ represent the gross area and the net area on the critical section resisting shear, respectively. The design resistant factor for shear yielding is recommended as $\phi_{vy} = 0.95$ to be consistent with that for tension yielding. The reduction factor $(U)$ shall be taken as a value of 0.74 used for flexural shear as specified in the current provisions (FHWA, 2009; AASHTO, 2007). The design resistance factor for shear fracture $(\phi_{vu})$ shall be taken as 0.80 to be consistent with that for block shear fracture. Seeing that this factor already takes into consideration enough safety against fracture, adding an additional reduction factor may be overly conservative to estimate the design resistance strength. Fig. 8.7 shows the failure of the gusset plate along the critical yielding and net section resisting flexural shear force.

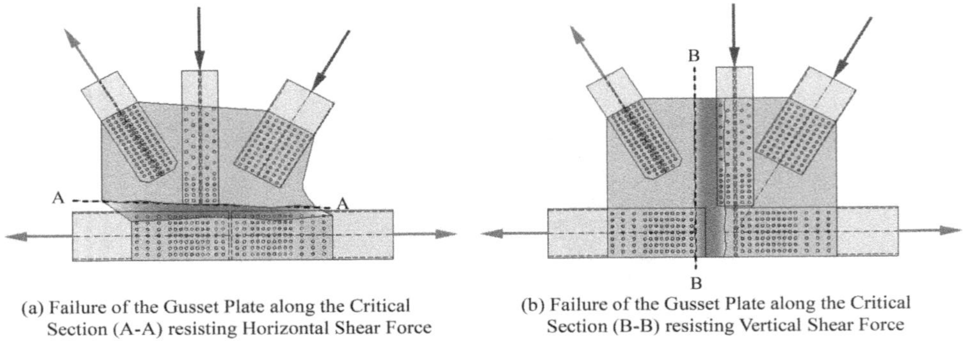


*Figure 8.7* Failure of the gusset plate along the critical section resisting flexural shear force.

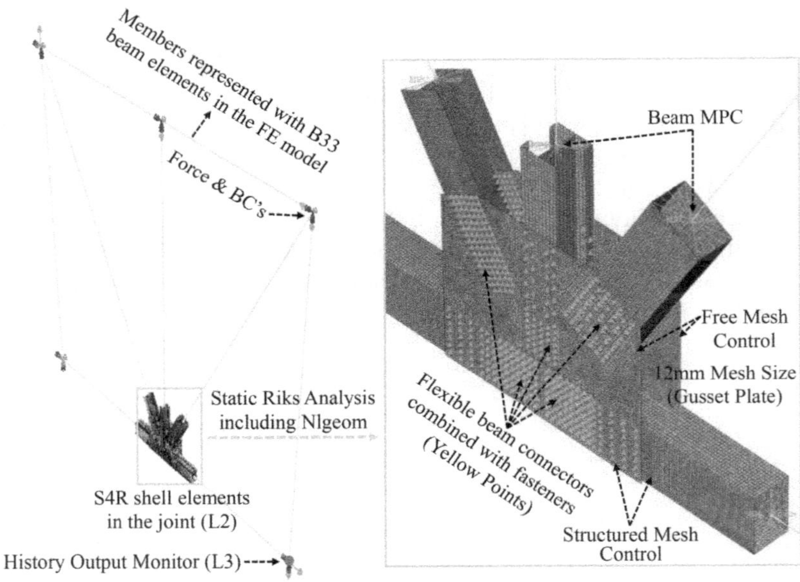


*Figure 8.8* 3D FE gusset plate joint model and its advanced modeling attribution.

## 8.4 3D FE GUSSET PLATE CONNECTION MODELS

The Abaqus nonlinear FE code program (ABAQUS, 2010) was used to investigate the inelastic behavior of the representative gusset plate connection selected from the I-94 bridge (i.e., L2 joint presented in Fig. 8.3). The L2 joint model consisted of several independent parts as shown in Fig. 8.8. The FE model applied typical steel material properties shown in Fig. 8.9 to analysis. The fully nonlinear material property obtained

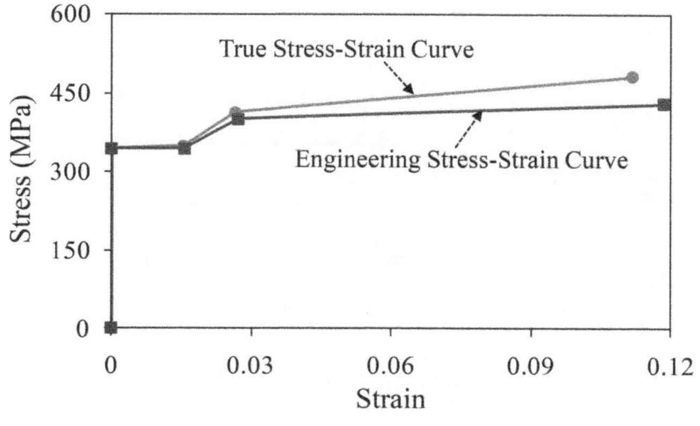


*Figure 8.9* Stress and strain curves for SS490 (ASTM A572-Gr. 50) carbon steel $F_y = 345$ and $F_u = 483$ MPa).

by the true stress-strain curve for SS490 steel was assigned to the plate parts. On the other hand, elastic material behavior was applied to the connected truss members initially designed to remain elastic throughout the test. Hence, in the FE model, we can focus on inelastic behavior and failure modes accepted at the plate parts only.

Both gusset plates and splice plates were made up of S4R (four node shell with linear order and reduced integration) elements incorporating nonlinear material properties and geometric nonlinearity (Nlgeom). These plates were modeled as located at their mid-thickness. The uniform shell element size taken as 12 mm was distributed over the gusset plate with a view to providing a numerical solution of the response contours appropriately converging. The truss members connected to the gusset plates were modeled using shell elements up to a length of approximately two times member depth ($2d$) from the edge of the gusset plate. Other parts of the truss member were modeled with eight B33 (three node beam with nonlinear order) elements. As can be seen in the figure, these beam elements were connected to the edge of shell element sections by means of beam-type multipoint constraints (MPCs) provided in the program.

In the FE model, inelastic bolt behavior can be simulated by multi-linear stiffness models that are derived by fitting into nonlinear curves as shown in Fig. 8.10. These nonlinear curves indicating force and displacement relationships for high-strength bolts subjected to direct shear ($R_{sb}$ vs. $\Delta_{sb}$) are formulated using an empirical equation stipulated in the AISC-LRFD manuals as follows:

$$R_{sb} = R_{ult}(1 - e^{-10\Delta_{sb}})^{0.55} \tag{8.18}$$

where, $R_{ult}$ represents the ultimate capacity of the shear bolts (AISC, 2001). These derived multi-linear stiffness properties were applied to the in-plane shear response parallel to the direction of the applied loads. The out-of-plane response of the bolts was assumed to be elastic behavior reproduced by a stiffness of simple elastic bars.

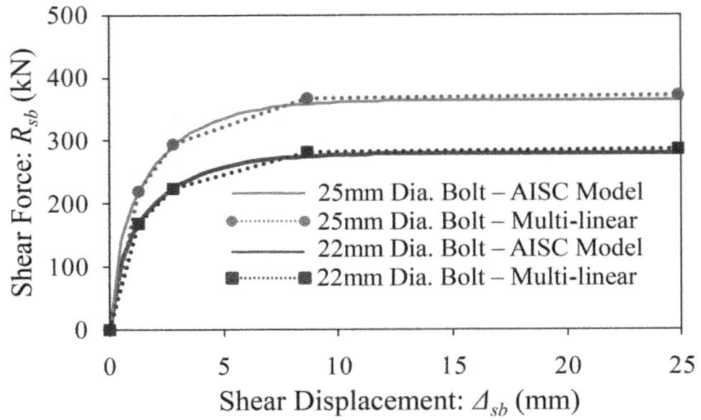


*Figure 8.10* Force and displacement relationships for ASTM A490 steel bolts subjected to shear.

The in-plane shear response behaves more flexibly as compared to the out-of-plane response and so has become dominate in the behavior of the bolt component.

The bolt components were modeled with Cartesian plus Align beam connectors in which three rotational degree of freedoms (DOFs) are restrained and only three translational ones are available. The extra constraints that bolt heads or nuts provide on two connected surfaces are generated using the fastener command available in the program. The fastener property including a radius of influence where displacements are constrained was assigned to the connector element. Regular-sized meshes generated using structured mesh controls help to keep this radius of influence evenly distributed over the connected plate surface. The bolt holes are not necessary to model explicitly owing to the fastener property. As shown in Fig. 8.8, discrete flexible beam connectors that were combined with fasteners connected two nodes on the centerlines of faying plates.

Torsional rotations and out-of-plane displacements were restrained at the ends of the truss members in that in-plane boundary conditions (BCs) were employed in the FE model (see Fig. 8.8). All six DOFs for displacements and rotations were fixed at the end of the left hand chord member in order to make stable conditions. However, translational DOFs along the direction of the applied loads were released at each of the nodal joints. The intent for FE modeling described above (e.g., used element types, beam MPCs, bolt connectors, BCs and so on) is to save memory costs as much as possible, while aiming to increase the possibility of obtaining reliable results for FE analyses and to produce more accurate predictions of gusset plate behavior.

The factored design loads applied to the truss members are shown in Fig. 8.11. Some member forces were changed to satisfy static equilibrium. The FE analysis was implemented in two steps. The factored dead loads (1.2DL) were applied in the first step, whereas the factored live and impact loads (1.75(LL + IL)) were incrementally applied in the second step. The dead loads (DL) included the weight of all concrete and steel components at the time of original construction. The live loads (LL) were evaluated by uniformly distributing total AASHTO HS20-44 vehicle weight, and finally

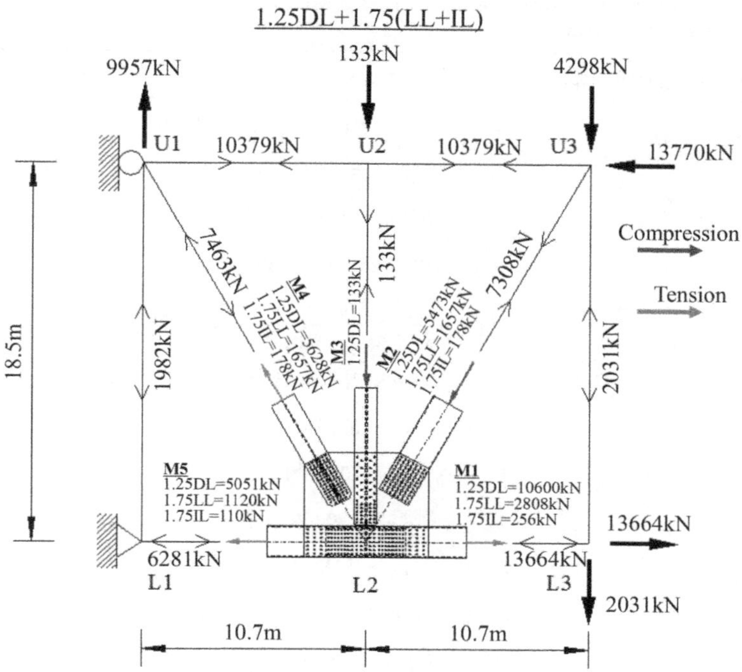


*Figure 8.11* Factored member loads used for bridge design.

the impact loads (IL) were accounted for by adding approximately 10% of the vehicle loads as presented in the original drawings (Wright, 2009; ILDOT, 2012). These loads were imposed on the truss joint nodes using the appropriate factors for LRFD load combination (LC) (see Fig. 8.11, LC = 1.2DL + 1.75(LL + IL)) (FHWA, 2009; FHWA, 2010; AASHTO, 2007). The Riks analysis method was utilized to predict the maximum load at instability by proportionally increasing live and impact loads, while other dead loads were maintained to be constant at the estimated values. The nodal displacements at the L3 joint were measured by using the history output instrument in the program.

## 8.5 STRENGTH EVALUATION

The primary failure mode is difficult to define in the gusset plate connection since multiple failure patterns are often occurring at the same time. Instead, it can be suggested that investigating the RF ratios of the resistance capacity to the applied load for design is one of the most effective ways to determine the dominating failure mode. The gusset plates are very susceptible to the failure mode with the least of the RF values. According to individual failure modes feasibly occurring at the sample gusset plate,

*Table 8.1* Calculation example for the design resistance strength based on the Whitmore section failure of the sample gusset plate.

| *Failure modes of the gusset plate* | *Defining equations* | *Rating factor (RF)* |
|---|---|---|
| 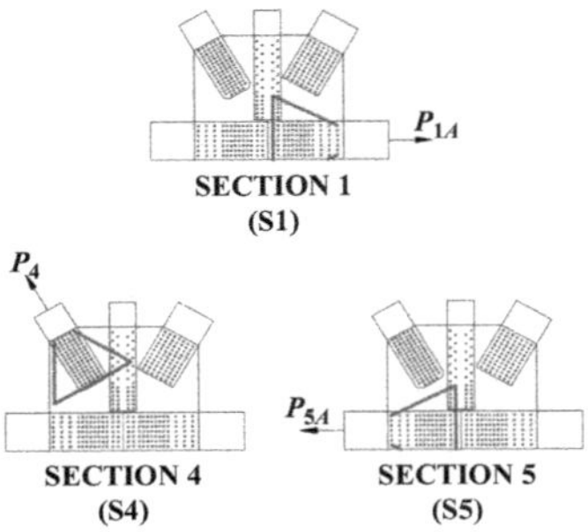 | $R_{rw} = \phi R_{nw} = \min[(\phi_y F_y A_{gw}), (\phi_u F_u A_{nw})]$ | $RF_{w,S1} = \frac{R_{rw,S1}}{P_{1A}} = 2.96$ |
| | $A_{gw} = 2Wt_g, A_{nw} = 2(W - n_b d_h)t_g$ | $RF_{w,S4} = \frac{R_{rw,S4}}{P_4} = 4.41$ |
| | $A_{TCS} = 43.55\ \text{cm}^2, A_{BCS} = 43.55\ \text{cm}^2$ | $RF_{w,S5} = \frac{R_{rw,S5}}{P_{5A}} = 6.45$ |
| | $A_{SCS} = 319.35\ \text{cm}^2, A_{GP} = 391.93\ \text{cm}^2$ | |
| | $RF_A = \frac{A_{GP}}{A_{TCS} + A_{BCS} + A_{SCS} + A_{GP}} = 0.491$ | |
| | $P_1 = 13664\ \text{kN}, P_4 = 7463\ \text{kN}, P_5 = 6281\ \text{kN}$ | |
| | $P_{1A} = P_1 \cdot RF_A = 6709\ \text{kN},$ | |
| | $P_{5A} = P_5 \cdot RF_A = 3084\ \text{kN}$ | |
| | $A_{gw,\,S1} = A_{gw,\,S5} = 703.74\ \text{cm}^2,$ | |
| | $A_{nw,S1} = A_{nw,S5} = 515.03\ \text{cm}^2$ | |
| | $A_{gw,S4} = 1005.82\ \text{cm}^2, A_{nw,S4} = 947.76\ \text{cm}^2$ | |
| | $R_{rw,S1} = R_{rw,S5} = 19879\ \text{kN},$ | |
| | $R_{rw,\,S4} = 32912\ \text{kN}$ | |

*Table 8.2* Calculation example for the design resistance strength based on the block shear failure of the sample gusset plate under tension.

| *Failure modes of the gusset plate* | *Defining equations* | *Rating factor (RF)* |
|---|---|---|
| 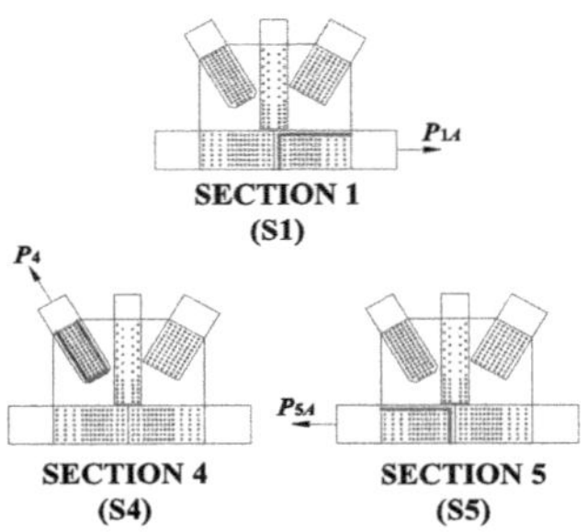 | If $A_{tn} \leq 0.58A_{vn}$, $R_{rbs} = \phi R_{nbs} = \phi_u(0.58F_y A_{vg} + F_u A_{tn})$ | $RF_{bs,S1} = \frac{R_{rbs,S1}}{P_{1A}} = 3.00$ |
| | Otherwise, $R_{rbs} = \phi R_{nbs} = \phi_u(0.58F_u A_{vn} + F_y A_{tg})$ | $RF_{bs,S4} = \frac{R_{rbs,S4}}{P_4} = 2.27$ |
| | $A_{tn,S1} = A_{tn,S5} = 232.26\ \text{cm}^2,$ | $RF_{bs,S5} = \frac{R_{rbs,S5}}{P_{5A}} = 6.53$ |
| | $A_{vn,S1} = A_{vn,S5} = 497.18\ \text{cm}^2$ | |
| | $A_{tg,S1} = A_{tg,S5} = 326.61\ \text{cm}^2,$ | |
| | $A_{vg,S1} = A_{vg,S5} = 678.63\ \text{cm}^2$ | |
| | $A_{tn,S4} = 174.19\ \text{cm}^2, A_{vn,S4} = 453.77\ \text{cm}^2$ | |
| | $A_{tg,S4} = 246.77\ \text{cm}^2, A_{vg,S4} = 678.77\ \text{cm}^2$ | |
| | $R_{rbs,S1} = R_{rbs,S5} = 20135\ \text{kN}, R_{rbs,S4} = 16960\ \text{kN}$ | |

*Table 8.3* Calculation example for the design resistance strength based on the block shear failure of the sample gusset plate under compression.

| *Failure modes of the gusset plate* | *Defining equations* | *Rating factor (RF)* |
|---|---|---|
| 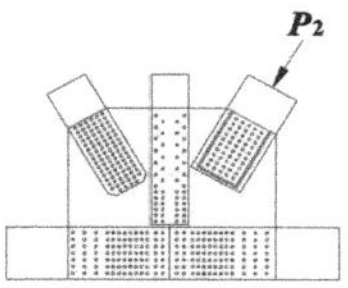  SECTION 2 (S2) | $R_{rbsg} = \phi R_{nbsg} = \phi_u 0.58 F_y A_{vg} + \phi_c F_y A_{cg}$<br>$R_{rbsn} = \phi R_{nbsn} = \phi_u 0.58 F_y A_{vn} + \phi_c F_y A_{cg}$<br>$R_{rbs} = \phi R_{nbs} = \min[\phi R_{nbsg}, \phi R_{nbsn}]$<br>$A_{cg,S2} = 261.29\ \text{cm}^2, A_{vg,S2} = 981.00\ \text{cm}^2,$<br>$A_{vn,S2} = 676.16\ \text{cm}^2$<br>$R_{rbs,S2} = 23232\ \text{kN}$ | $RF_{bs,S2} = \frac{R_{rbs,S2}}{P_2} = 3.18$ |

*Table 8.4* Calculation example for the design capacity of the sample gusset plate susceptible to buckling in the compression area.

| *Failure modes of the gusset plate* | | *Defining equations* | *Rating factor (RF)* |
|---|---|---|---|
| 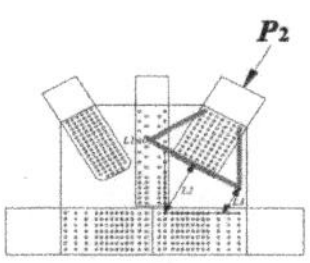  SECTION 2 (S2) | 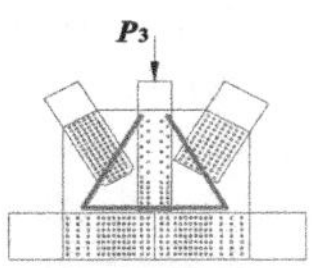  SECTION 3 (S3) | $\lambda = \frac{KL_c}{\pi t_g}\sqrt{\frac{12F_y}{E}}, L_C = \frac{L_1 + L_2 + L_3}{3}$<br>If $\lambda \leq 2.25$, $R_{rcw} = \pi R_{ncrw} = \phi_c 0.658^{\lambda^2} F_y A_{gw}$<br>Otherwise, $R_{rcrw} = \pi R_{ncrw} = \pi_c \frac{0.877}{\lambda^2} F_y A_{gw}$<br>$\lambda_{S2} = 0.179, \lambda_{S3} = 0.031$<br>$P_2 = 7308\ \text{kN}, P_3 = 133\ \text{kN}$<br>$A_{gw,S2} = 763.55\ \text{cm}^2, A_{gw,S3} = 1079.27\ \text{cm}^2$<br>$R_{rcrw,S2} = 21975\ \text{kN}, R_{rcrw,S2} = 31654\ \text{kN}$ | $RF_{crw,S2} = \frac{R_{rcrw,S2}}{P_2} = 3.00$<br>$RF_{crw,S3} = \frac{R_{rcrw,S3}}{P_3} = 238$ |

factored design resistance strengths, factored forces transferred into the plates, and corresponding RF values are summarized in Table 8.1 to Table 8.5.

The forces which truss members transmit are assumed to be transferred into all splice and gusset plates in proportion to their area (see Table 8.1). Accordingly, the forces transferred into the gusset plates ($P_A$) are estimated to multiply the member forces ($P$) by a fraction of the gusset plate area occupied at the plates ($RF_A$).

$$P_A = RF_A \cdot P \tag{8.19}$$

$$RF_A = \frac{A_{GP}}{A_{TCS} + A_{BCS} + A_{SCS} + A_{GP}} \tag{8.20}$$

*Table 8.5* Calculation example for the design resistance strength of the sample gusset plate along the critical section resisting shear force.

<table>
<tr><th>Failure modes of the gusset plate</th><th>Defining equations</th><th>Rating factor (RF)</th></tr>
<tr><td>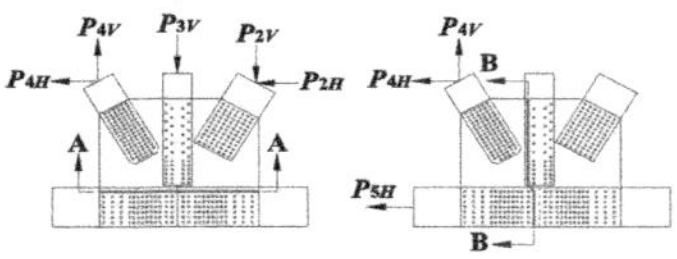

SECTION A-A<br>(A-A)<br>SECTION B-B<br>(B-B)</td>
<td>$R_{rg} = \phi P_{ng} = \phi_y 0.58 F_y A_g U$ (Gross Yield)<br>
$R_{rn} = \phi R_{nn} = \phi_u 0.58 F_u A_n$ (Net Fracture)<br>
$A_{g,A\text{-}A} = 1487.32\ \text{cm}^2, A_{n,A\text{-}A} = 1109.90\ \text{cm}^2$<br>
$A_{g,B\text{-}B} = 1318.93\ \text{, cm}^2,$<br>
$A_{n,B\text{-}B} = 1128.61\ \text{cm}^2, U = 0.74$<br>
$P_{H,A\text{-}A} = P_{4H} + P_{2H} = 7454\ \text{kN}$<br>
$P_{V,B\text{-}B} = P_{4V} = 64543\ \text{kN}$<br>
$R_{rgs,A\text{-}A} = 20916\ \text{kN}, R_{rns,A\text{-}A} = 24867\ \text{kN}$<br>
$R_{rgs,B\text{-}B} = 18539\ \text{kN}, R_{rns,B\text{-}B} = 23044\ \text{kN}$</td>
<td>$RF_{gs,A\text{-}A} = \frac{R_{rgs,A\text{-}A}}{P_{H,A\text{-}A}} = 2.81$<br>
$RF_{ns,A\text{-}A} = \frac{R_{rns,A\text{-}A}}{P_{H,A\text{-}A}} = 3.34$<br>
$RF_{gs,B\text{-}B} = \frac{R_{rgs,B\text{-}B}}{P_{V,B\text{-}B}} = 2.87$<br>
$RF_{ns,B\text{-}B} = \frac{R_{rns,B\text{-}B}}{P_{V,B\text{-}B}} = 3.87$</td></tr>
</table>

where, $A_{GP}$ is the cross-section area of the gusset plates. $A_{TCS}$, $A_{BCS}$, and $A_{SCS}$ indicate that of top, bottom, and side chord splice plates, respectively.

The RF values range from 2.27 to 2.38 as presented in the tables. Block shear failure occurring at the connection between gusset plates and tension diagonal shows the least RF value (RF = 2.27), meaning that the strength limit state to resist this failure mode may be firstly reached at the gusset plates subjected to incrementally increasing loads. The RF value of 2.27 theoretically represents the ability of the gusset plates to plastify along the entire plane resisting tension prior to the ultimate load. This block shear failure is succeeded by shear yielding on the gross section, indicating the RF value of 2.81 (see Table 8.5).

Overall, factored design strength is much larger than factored load demand. When the I-94 bridge was designed on the basis of the ASD method, the capacity of the gusset plates might be overly underestimated. In other words, the steel structures to meet design requirements for the ASD method can just accommodate a limited amount of applied loading (e.g., ASD LC = DL + LL + IL) long before arriving at static yield strength. Instead, economic design strength limits can be established by the new LRFD method, which reliably permits more load demand. In the following section, this statement will be verified by observing FE analysis results.

## 8.6 ANALYSIS RESULTS

From the FE analysis, the factored loads applied over two steps are expressed as follows:

$$P_r = 1.2\text{DL} + \text{ALF}[1.75(\text{LL} + \text{IL})] \tag{8.21}$$

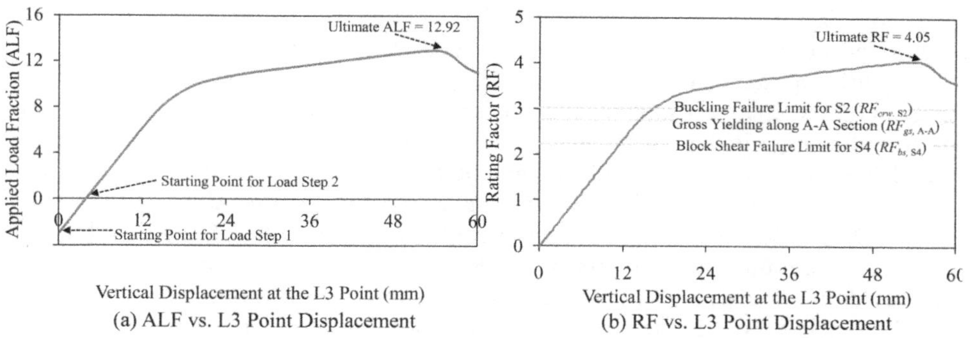


*Figure 8.12* Applied load vs. vertical displacement measured at the L3 point.

After factored dead loads are completely imposed on the truss joints at the first step, factored live and impact loads which start to increase at the second step are applied multiple times with applied load fractions (ALFs) (see Fig. 8.12(a)). Therefore, the bridge's structural capacity to accommodate the number of vehicles exceeding design limits is estimated through this analysis method. The RFs for life safety are converted into the ALF using the equation given to below.

$$\mathrm{ALF} = \frac{1.2\mathrm{DL}(\mathrm{RF}-1)}{1.75(\mathrm{LL}+\mathrm{IL})} + \mathrm{RF} \tag{8.22}$$

The ALFs and the RFs can be changed from total applied loads ($P_r$) depending on what FE analysis results are used for and plotted versus vertical displacements measured at the L3 joint in Figs. 8.12(a) and (b), respectively. The computed RF limits for dominant failure modes occurring at the sample gusset plate are also plotted as the dotted sky-blue lines in Fig. 8.12(b). The ultimate ALF of 12.92 corresponds to the ultimate RF of 4.05 at the vertical displacement of approximately 54 mm. The FE model still remains elastic up to the RF limit for buckling failure under compression (i.e., RF = 3.0 and ALF = 8.8). It is shown that design strength limits derived from the LRFD method are a little conservative for failure and capacity predictions. The FE model shows strength degradation resulting from instability after reaching its ultimate load.

After FE analyses, both von-Mises stress and equivalent plastic strain distributed over the gusset plate are also investigated at each of the strength limits in order to check the failure state predicted by the RF ratio. The distributions of von-Mises stress and equivalent plastic strain (PEEQ) filed contours are shown in Fig. 8.13 and Fig. 8.14, respectively, at the given stages of the FE analysis. The accompanied-deformed configurations that need to be examined to confirm final failure shape are also found in both figures with a deformed scale factor (DSF) of 5.0.

The gusset plate areas shown with the yellow contours have reached the onset of plastic yielding in the base steel material. Though the global behavior of the gusset plate connection is still elastic at the strength limit for block shear failure (RF = 2.27), plastic yielding starts to occur at the part of the gusset plates where member forces converge. This von-Mises stress filed contour shows that yielding is concentrated around the plate areas adjacent to the free edges of truss members. After exceeding the strength limit of gross yielding along the A-A chord plane (RF = 2.81), the plastic stress fields spread to the gusset plate area very quickly as increasing the total applied loads. A considerable amount of plastic yielding has been subsequently observed at the buckling failure limit state (RF = 3.0). The instant of plate buckling at the compression diagonal is shown in the FE model when RF = 3.57. The strength limit state determined in accordance with the LRFD method is somewhat conservative to predict buckling failure, and thereby possible to guarantee design safety. When RF = 4.05 corresponding to the peak of the curve as shown in Fig. 8.12(b), the red colored contours indicating that the base steel material reaches its ultimate stress are observed at the plate areas contiguous to the edges of the members. After the ultimate state (RF = 3.56), the gusset plates undergo extensive plastification and instability. At the last stage, the significant increase in out-of-plane deformation as compared to in-plane deformation confirms the evidence of buckling failure.

As shown in Fig. 8.14, corresponding PEEQ field contours match the von-Mises contours very well in that they are also presented to show yielded regions and plasticity patterns. The materials that are still elastic are shown with blue colored contours, while field contours that vary from light green to orange are presented to display variations in the magnitude of plastic deformations. The FE models composed of compatibility-based elements cannot capture progressive fractures. One standard indicating a point in time when fractures may begin to be concerned results in the maximum PEEQ at the mid-thickness of the plate. The maximum field contours exhibited in the areas very closed to ultimate failure are arbitrarily set up at 4% PEEQ shown with intense red color. For this model, 4% PEEQ is just reached at the peak load (RF = 4.05), and so a considerable amount of plastic deformation is assumed to happen at the last stage of the FE analysis.

## 8.7 INVESTIGATIONS OF FE ANALYSIS RESULTS

The general observations from FE analysis results are able to provide valuable and qualitative information on the nonlinear behavior of the gusset plate connection. Furthermore, the adequacy of the presented design method can be validated by examining the load-carrying capacity of the gusset plates after FE analysis, and then comparing with design resistance strength. The first aspect to take into consideration is to investigate normal and shear stress along the critical path where dominant failure modes take place, according to an increase in the applied load (see Figs. 8.15 and 8.16).

Fig. 8.15 shows normal and shear stress along the path where block shear failure is likely to occur. The normal stresses are measured along the tensile load-carrying path, while the shear stresses are measured along the shear load-carrying path. As far as considering area conditions, the factored design resistance strength is determined in accordance with Eq. (8.8). Therefore, the strength limit to resist this failure mode is set up based on tensile yielding and ultimate shear fracture, which are depicted as the dotted sky-blue lines in Fig. 8.15(a) and Fig. 8.15(b), respectively. The normal stresses are more uniformly distributed over the gusset plate as compared to the shear

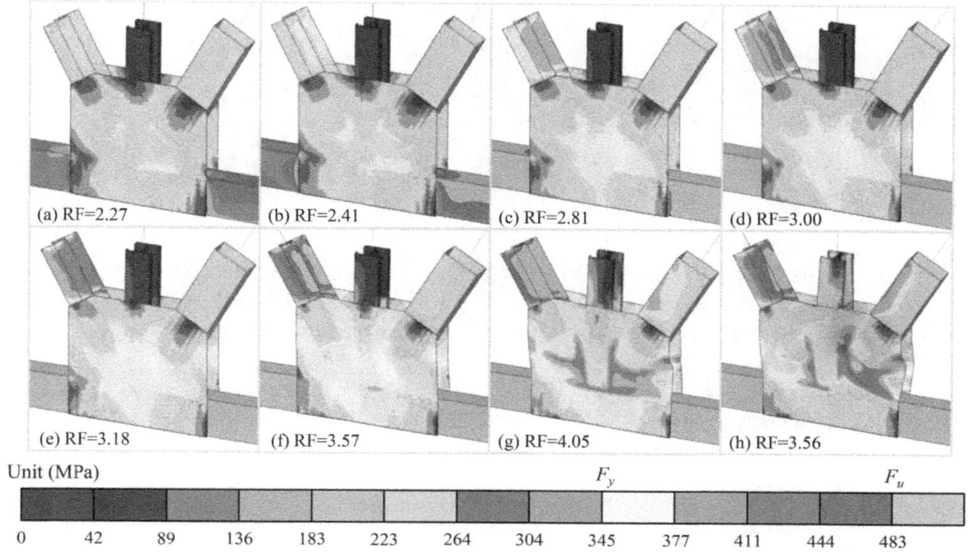


*Figure 8.13* Observation of Mises stress field contours on deformed shape at the specific RF point (DSF = 5).

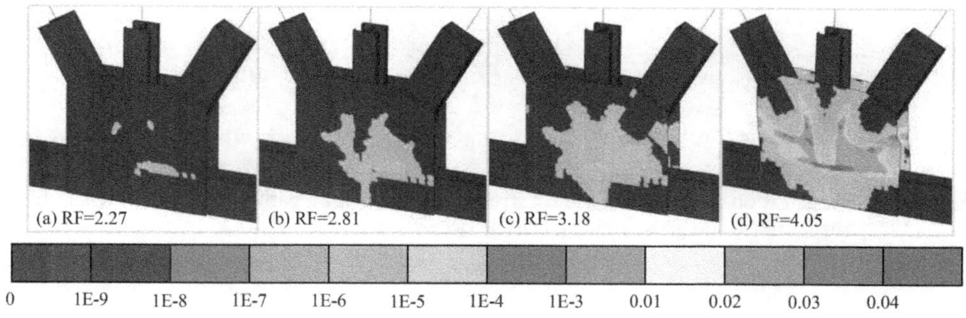


*Figure 8.14* Observation of plastic strain field contours (PEEQ) on deformed shape at the specific RF point (DSF = 5).

stresses. The maximum normal and shear stresses do not exceed the strength limit lines as can be seen in the figures when the gusset plates are subjected to the factored load corresponding to the RF limit for block shear failure (RF = 2.27). They are going over these lines at the next RF stage (RF = 2.81). It is also shown that the design limit state determined by the RF value is slightly conservative to predict the corresponding failure mode. The same trend toward stress distribution is also observed in Fig. 8.16.

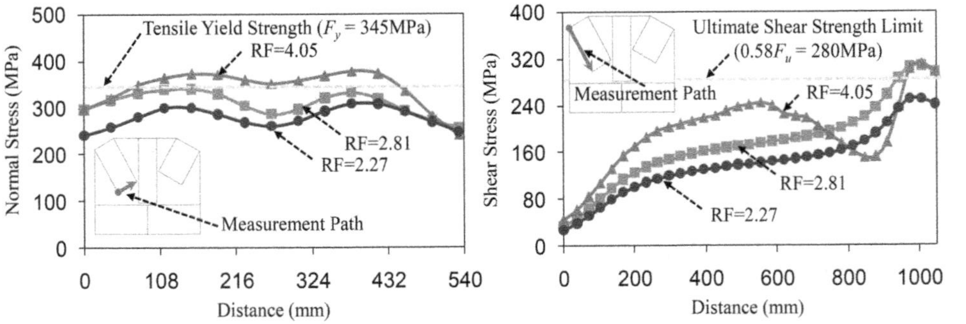


(a) Tensile Stress normal to the Measurement Direction (b) Shear Stress parallel to the Measurement Direction

*Figure 8.15* Investigation of normal and shear stress along the critical path (S4) where block shear failure feasibly occurs.

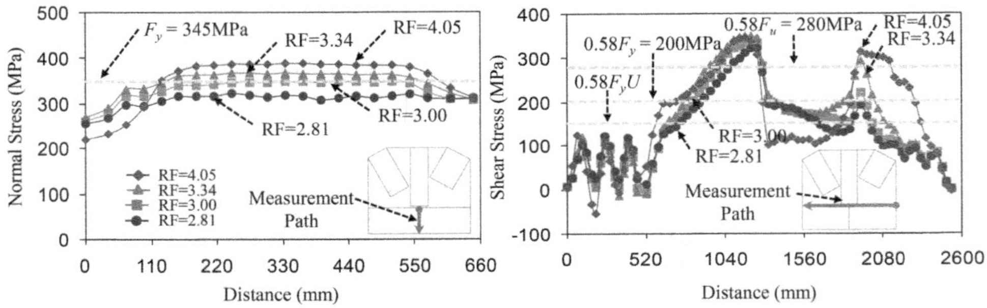


(a) Tensile Stress normal to the Measurement Direction (b) Shear Stress parallel to the Measurement Direction

*Figure 8.16* Investigation of normal and shear stress along the critical paths (S1 and A-A) where block shear failure and gross yielding feasibly occur.

In addition to the investigation of normal and shear stresses along the critical path, out-of-plane deformations along the critical path where compression buckling is likely to occur are also examined to check the instability of the gusset plate after the peak load. Out-of-plane deformations measured along the critical path are shown in Fig. 8.17(a), while those measured at the critical point are shown in Fig. 8.17(b). The RF limit for compression buckling failure (RF = 3.0) is also depicted as the dotted sky-blue line in the Fig. 8.17(b). Out-of-plane deformations are linearly increasing before the RF limit for compression buckling (see Fig. 8.17(b)). The maximum out-of-plane deformation is approximately 4.5 mm at the critical load (RF = 3.73). The critical load to raise instable buckling is larger than the RF limit state for compression buckling. It is concluded that the strength limit state used for design is conservatively estimated to be less than the actual capacity of the gusset plate. Accordingly, reasonably enough safety against

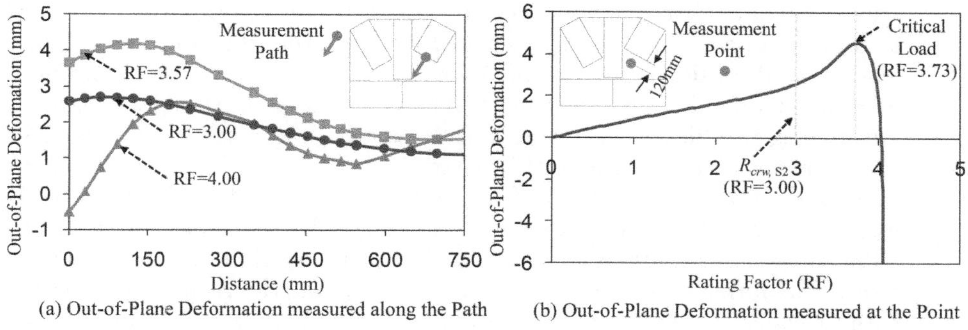


*Figure 8.17* Investigation of out-of-plane deformation along the critical path (S2) where compression buckling feasibly occurs.

buckling failure can be taken into consideration for gusset plate design. Finally, the out-of-plane deformations dramatically increase toward the opposite direction immediately after the critical load of the curve (see also Fig. 8.17(b)).

## 8.8 CONCLUSIONS

In the aftermath of the I-35W tragic event, many engineers have been asked to develop reliable and rational recommendations used for new gusset plate design. For this motivation, the LRFD method available in the current FHWA guidance was investigated in this study. The analytical studies were also conducted so as to inspect this design recommendation. The presented design checks suggest a comprehensive methodology for determining the resistance capacity of the gusset plates when taking their dominant failure mode into account. More conclusions are summarized as follows:

(1) The detailed design procedures presented in the current AASHTO Specifications were mainly treated in order that resistance strength capacities for the gusset plates should be estimated as consistent with the LRFD method. The gusset plates subjected to axial tension or shear were checked for three failure conditions such as yielding on the gross section, fracture on the net section, and block shear rupture, whereas those subjected to compression were investigated for buckling. The factored design resistance strength for the gusset plates was taken as the least of the values given among strength capacities to resist individual failure types.

(2) The FE gusset plate connection model successfully predicted the nonlinear behavior of the gusset plate connection. The failure patterns occurring at the gusset plates were also captured well by this refined FE model, which are able to accurately reproduce the mechanism of yielding or ultimate stress along the critical load-carrying path. In addition, predictions on the resistance capacity for the gusset plates subjected to static factored loads were possibly made as observing the sequence of failure shape after analyses.

(3) The gusset plate connection model presented in this study was designed based on the previous ASD method. The strength capacity determined by the ASD method had been overly underestimated by restricting the amount of applied loading used for design long before base materials reach yield strength. On the other hand, the LRFD method can permit an increase in the strength capacity by introducing reliable design factors. For this reason, this gusset plate connection model shows relatively large LRFD-based RF values ranged from 2.27 to 2.38.
(4) The von-Mises stress and PEEQ filed contours distributed on the gusset plates were investigated for the purpose of checking individual failure states at the given stages of the FE analysis. They clearly presented yielding and plasticity patterns. The normal and shear stresses measured along the critical load-carrying path provided good information on the failure state as well. The out-of-plane deformations increasing at the end of the FE analysis very quickly demonstrated evidence that the presented gusset plates ultimately failed by compression buckling. After observing these resulting values, it is concluded that design strength limits determined by the LRFD method are somewhat conservative for capacity predictions due to the reduced design factors.

Chapter 9

# Recentering slit damper connection

## 9.1 GENERAL OVERVIEW

The steel slit dampers that can be integrated to general seismic resistant systems such as special and ordinary moment-resisting frames have been utilized as easily replaceable energy dissipation devices with an intention to protect main structural members (e.g., beam and column) (Karavasilis *et al.*, 2012; Soong & Spencer, 2002; Chan & Albermani, 2008; Kim *et al.*, 2013). Inelastic deformations in the main structural members make it difficult to repair seismic damage, and hence cause to rebuild the building structure (Hu, 2013; Sabelli *et al.*, 2003). Therefore, these devices dissipating energy based on the yielding of standard base steel sections are designed to concentrate significant inelastic deformations under severe earthquake events. Such a design methodology takes advantage of acceptable seismic performance with respect to economy and safety (Hu *et al.*, 2010; Hu *et al.*, 2013; Hu, 2014). In spite of damage control obtained from energy dissipation devices (i.e. steel slit dampers), their permanent deformations still give rise to residual inter-story drifts in the whole moment-resisting frame. The conventional passive control systems with steel energy dissipation devices cannot adequately supply a demand for harmonization between structural and non-structural damage, and thus adding strength and stiffness to the frame structure shall be required for aseismic design in order to reduce storey drifts. Some scientists emphasize that non-structural damages related to residual inter-story drifts are more dangerous than damages related to structural member failure (McCormick *et al.*, 2008; Hu, 2013). In particular, a recent report study highlights that if the frame systems undergo the residual inter-story drift greater than 0.5%, the owners of buildings in Japan had better rebuild the whole structures from an economic point of view rather than repair them (McCormick *et al.*, 2008). For this motivation, this study mainly focuses on the slit damper device with recentering capability so as to considerably decrease permanent deformation in the steel frame structure.

One of the best ways to improve seismic performance as regards vibration control and self-centering effect can be achieved by the utilization of smart materials to aseismic design. Superelastic shape memory alloys (SMAs) have been currently prevalent as smart materials used for seismic control devices in that they exhibit unique material behavior characterized by a flag-shape hysteresis under cyclic loading. The hysteretic behavior of superelastic SMA materials is illustrated in Fig. 9.1. The general SMA composed of a metallic alloy of nickel and titanium, which is referred to as Nitinol,

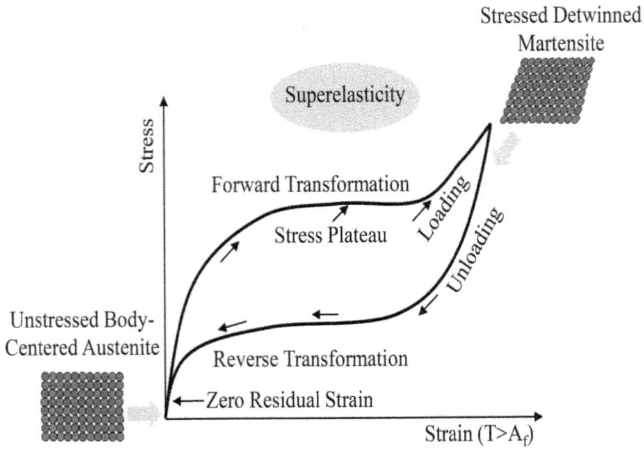


*Figure 9.1* Stress and strain curve for superelastic shape memory alloy (SMA) materials.

shows superelasticity (or pseudo-elasticity) that is able to recover original shape only by the removal of stress upon unloading. As shown in the figure, superelastic Nitinol SMAs that typically occur at a temperature limit above the austenite phase transformation (Af) do not exhibit any residual deformation without additional heating even after applying substantial strain ranged from 6% to 8%. This material behavior makes a significant contribution to providing excellent recentering capability as well as supplemental energy dissipation for the entire frame structure when such superelastic SMA materials are used in the damper device (Tobushi *et al.*, 2009; Lantada *et al.*, 2013). In this study, the slit damper devices fabricated with superelastic SMA materials are consequently introduced to attain both the establishment of additional damping and the mitigation of residual inter-story drifts. The behavior of superelastic SMA slit dampers are compared with that of conventional steel slit dampers after performing finite element (FE) analyses. In addition to the user-material (UMAT) model for reproducing the material behavior of superelastic SMA materials, FE models are additionally calibrated to experimental results with an aim to obtain reliable prediction. Finally, both types of slit damper devices, which are compared to each other, are simultaneously evaluated for ultimate strength and recentering capability in order to verify SMA's superior effect.

## 9.2 SAMPLE SLIT DAMPER SPECIMENS

The typical slit damper devices can be installed on top of an inverted-V brace at the concentrically braced frame structure and connected to the middle of the beam member as shown in Fig. 9.2. The detailed drawings of the slit damper devices are also presented in Fig. 9.3. They are manufactured from the short length of standard I-shape sections with a number of slits cut from the web and leaving strips between two flanges. The strips are fabricated to be circular at their ends for the purpose of mitigating stress concentration at the corners. The flange of the slit damper device, where four bolt holes are drilled, is attached to the frame by using weld-free bolts and nuts, thereby

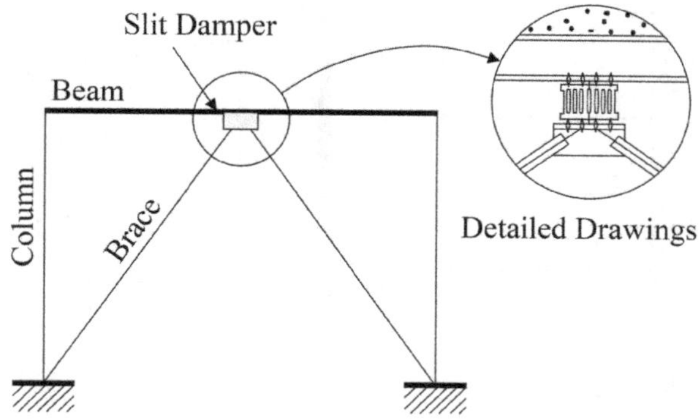


*Figure 9.2* Installation of the slit damper at the concentrically braced frame structure.

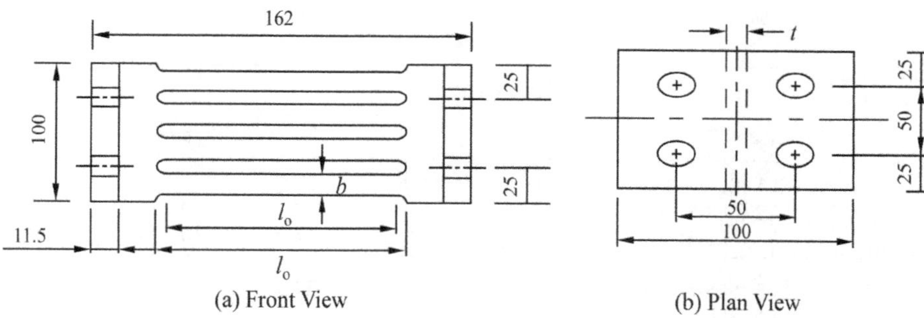


*Figure 9.3* Geometric details for experimental slit damper models.

eliminating failure uncertainties due to welding (Karavasilis *et al.*, 2012; Chan & Albermani, 2008).

This device directly copes with shear forces transferred from the frame members ($P$) and corresponding deformation ($\delta$). The strips behave as fixed-end beams under relatively large displacement between two supported flanges. The bending mechanism of the strips is shown in Fig. 9.4. The plastic hinges are likely to form at both ends of individual strips subjected to sufficient displacement. Thus, a significant amount of energy can be dissipated owing to these plastic hinges under bending mechanism. The required parameters to describe the mechanical response of the slit damper i.e., strip length ($l_o$), strip depth ($b$), and web thickness ($t$) are also presented in Fig. 9.3. The yield load of the slit damper ($P_y$) can be defined under plastic bending mechanism with the assumption of perfectly elasto-plastic material behavior as follows:

$$M_P = \sigma_y \frac{tb^2}{4} \tag{9.1}$$

$$P_y = \frac{2nM_P}{l_0} = \frac{n\sigma_y tb^2}{2l_0} \tag{9.2}$$

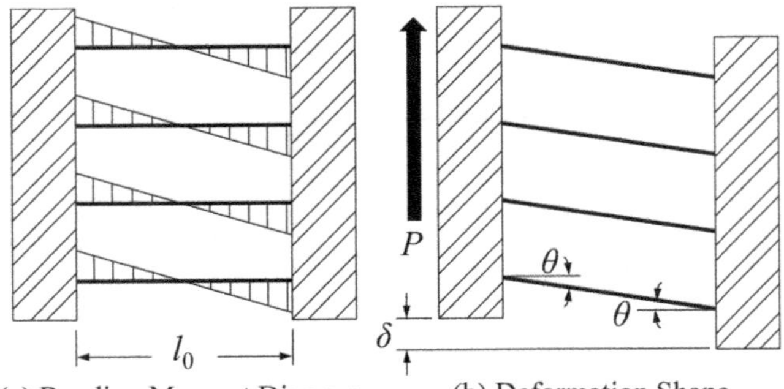


*Figure 9.4* Response mechanism of the slit damper model.

*Table 9.1* Geometric design parameters for individual slip damper models (Unit: mm).

| | *Dimensions* | | | |
|---|---|---|---|---|
| *Model ID* | *t* | *b* | $l_0$ | $b/l_0$ |
| SL1 (SL1-SMA) | 8.0 | 14.9 | 97.0 | 0.155 |
| SL2 (SL2-SMA) | | 15.0 | 87.1 | 0.172 |
| SL3 (SL3-SMA) | | 15.1 | 77.0 | 0.195 |
| SL4 (SL4-SMA) | | 16.9 | 99.2 | 0.172 |
| SL5 (SL5-SMA) | | 16.8 | 88.3 | 0.191 |
| SL6 (SL6-SMA) | | 16.5 | 79.0 | 0.215 |

where $M_P$ indicates the full plastic moment when plastic hinges form at the both ends of each strip with the rotation of $\theta_p$ and $n$ indicates the number of strips in the damper device. The stiffness of the slit damper device can be defined on the basis of an assumption that individual strips are fully constrained at their ends. It is determined as follows:

$$K = n\frac{12EI}{l_0^3} = n\frac{Etb^3}{l_0^3} \tag{9.3}$$

where $I$ is the moment inertia of the prismatic strip.

The experimental tests related to the slit damper devices were conducted by Chan and Albermani (2008) with an intention to examine not only cyclic responses but also structural characteristics, and then the effects of geometric design parameters were also investigated to identify changes in stiffness and strength. The FE models used for simulating the behavior of the slit damper devices are calibrated to these experimental test results so as to verify the adequacy of modeling. A summary of experimental specimens are given to Table 9.1. The design parameters defined as the measured dimensions in the table are similar to ones presented in Fig. 9.3. In this paper, six

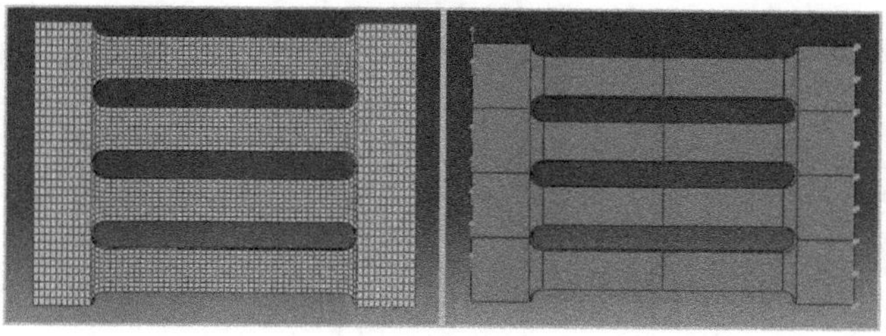

(a) Element Mesh (b) Loading and BC's

*Figure 9.5* 3D finite element (FE) models for nonlinear analyses.

specimens out of total nine specimens are selected for calibration and parametric study. All presented specimens were fabricated cut from a standard steel wide-flange section, i.e., 161.8 mm (depth) × 152.2 mm (flange width) × 8 mm (web thickness) × 11.5 mm (flange thickness) (Chan & Albermani, 2008). The standard coupons used to determine material properties were obtained from the web. After coupon tests, average yield stress and average elastic modulus are taken as 316.5 MPa and 206.1 GPa, respectively. As presented in the table, the specimens are classified according to varied $b/l_o$ ratios ranging from 0.155 to 0.215. The original specimens made by Gr. 50 carbon steel are labeled from SL1 to SL6. On the other hand, the proposed specimens fabricated with superelastic SMA materials are additionally labeled as '-SMA' in the last acronym of the model identification (ID).

## 9.3 FINITE ELEMENT MODELS

The ABAQUS nonlinear FE code program (ABAQUS, 2010) was used to predict the cyclic response of slit damper devices. FE models were made up of 3D solid elements (i.e., C3D8 in the ABAQUS program) incorporating fully nonlinear material properties, geometric nonlinearity, and displacement-controlled loading. Fig. 9.5 shows 3D FE models concerning element mesh, displacement loading, and boundary conditions (BC's). The structural meshes generated by dividing the part were used to make uniform element size in the FE model. The flange of the slip damper was assumed to be rigid, and accordingly detailed modeling for a supported flange was replaced with BC's. Displacement loading was directly imposed on the end of the web as well instead of flange modeling. The history of cyclic displacement loading for quasi-static FE analyses was simulated using the static step and the default amplitude function associated with BC's in the program. For each specimen, FE analyses were carried out with similar loading histories to the experimental tests.

The elasto-plastic material behavior with the combination of isotropic and kinematic strain hardening was assigned into FE models for steel slit damper devices. The nonlinear isotropic/kinematic hardening material model which includes some physical features such as Bauschinger effect, plastic shakedown, ratcheting, and stress relaxation (ABAQUS, 2010) was selected to simulate the behavior of steel materials in the cyclic loading condition. On the other hand, to simulate the cyclic behavior of superelastic SMA materials, the user material (UMAT) subroutine based on Aurrichio's model (Auricchio & Sacco, 1997; Auricchio & Taylor, 1997) was employed in the absence of adequate built-in material models provided by the program. This Aurrichio's material model reflects forward and reverse phase transformation involved in superelasticity under isothermal conditions. It was also based on the concept of generalized plasticity (Lubliner & Auricchio, 1996; Crisfield, 2012; Hu & Park, 2013).

## 9.4 UMAT EQUATIONS AND SIMULATION

In the UMAT subroutine, the degree of phase transformation was represented by an internal variable which may track the fraction of martensite distribution. The internal variables also include transformation strain and equivalent stress-strain relation. Two phase transformation processes which are divided according to the martensite fraction ($v_S$) ranged from 0 to 1 are necessary to define such that (1) transformation from austenite to martensite (A → S) and (2) transformation from martensite to austenite (S → A). The linear kinetic rules with respect to the uniaxial stress ($\sigma$) are applied to forward transformation (A → S) as follows:

$$\sigma_s^{AS} < |\sigma| < \sigma_f^{AS} \quad \text{and} \quad |\dot{\bar{\sigma}}| > 0 \tag{9.4}$$

where $\sigma_s^{AS}$ indicates martensite start stress, $\sigma_f^{AS}$ indicates martensite finish stress, | | represents an absolute value, and a superpose dot denotes a time derivative. The corresponding time continuous evolution equation can be defined by the relation, as follows:

$$\dot{v}_S = -(1 - v_S)\frac{|\dot{\bar{\sigma}}|}{|\sigma| - \sigma_f^{AS}} \tag{9.5}$$

The condition for reverse transformation (S → A) and corresponding evolution equation can be defined as follows:

$$\sigma_f^{SA} < |\sigma| < \sigma_s^{SA} \quad \text{and} \quad |\dot{\bar{\sigma}}| < 0 \tag{9.6}$$

$$\dot{v}_S = v_S\frac{|\dot{\bar{\sigma}}|}{|\sigma| - \sigma_f^{AS}} \tag{9.7}$$

where $\sigma_f^{SA}$ indicates austenite start stress, $\sigma_s^{SA}$ indicates austenite finish stress. Total strain can be decomposed into two components, (a) a purely linear-elastic component ($\varepsilon^e$) and (b) a transformation strain component ($\varepsilon_L$) as follows:

$$\varepsilon = \varepsilon^e + \varepsilon_L v_S \,\mathrm{sgn}(\varepsilon) \tag{9.8}$$

where sgn() is the sign function. As shown in Eq. 9.8, the amount of plastic strain is proportional to the martensite fraction. Total strain is assumed to be a control variable.

The elastic stress is linearly related to elastic strain. The constitutive equation is written with elastic modulus ($E$).

$$\sigma = E\varepsilon^e \tag{9.9}$$

The increment of the martensite fraction within discrete time ($\lambda_S$) is obtained by integrating the ration of the fraction as follows:

$$v_S = v_{S,n} + \lambda_S \text{ or } \lambda_S = \int_{t_n}^{t_{n+1}} \dot{v}_S dt \tag{9.10}$$

where the subscript $n$ denotes a quantity estimated at time ($t$) and $t_{n+1}$ is the time value of interest immediately after $t_n$. Eq. 9.10 can be rewritten based on the linearization of the strain components as follows:

$$\sigma = E[\varepsilon - \varepsilon_L \operatorname{sgn}(\varepsilon) v_S] \tag{9.11}$$

$$d\sigma = E[d\varepsilon - \varepsilon_L \operatorname{sgn}(\varepsilon) d\lambda_S] \tag{9.12}$$

The quantity of $\lambda_S$ is proportional to that of plastic strain after yielding, thereby defining such that:

$$d\lambda_S = H d\varepsilon \tag{9.13}$$

where $H$ indicates the scalar quantity for the tangent modulus after yielding. Using this relation between plastic strain and martensite fraction increment, Eq. 9 can be converted as follows:

$$d\sigma = E^T d\varepsilon \tag{9.14}$$

The tangent modulus ($E^T$) can be rewritten as blow.

$$E^T = E[1 - \varepsilon_L H \operatorname{sgn}(\varepsilon)] \tag{9.15}$$

The scalar quantity ($H$) used to evaluate the tangent modulus can be computed using the linearization of evolution equations consistent with phase transformation (Eqs. 9.8 and 9.10) and defined as follows:

$$H = H^{AS} = \frac{-\operatorname{sgn}(\varepsilon)(1 - v_{S,n})E}{(1 - v_{S,n})[-\operatorname{sgn}(\varepsilon)E\varepsilon_L] + \sigma_n - \operatorname{sgn}(\varepsilon)\sigma_f^{AS}} \tag{9.16}$$

$$H = H^{SA} = \frac{\operatorname{sgn}(\varepsilon) v_{S,n} E}{v_{S,n}[\operatorname{sgn}(\varepsilon)E\varepsilon_L] + \sigma_n - \operatorname{sgn}(\varepsilon)\sigma_f^{SA}} \tag{9.17}$$

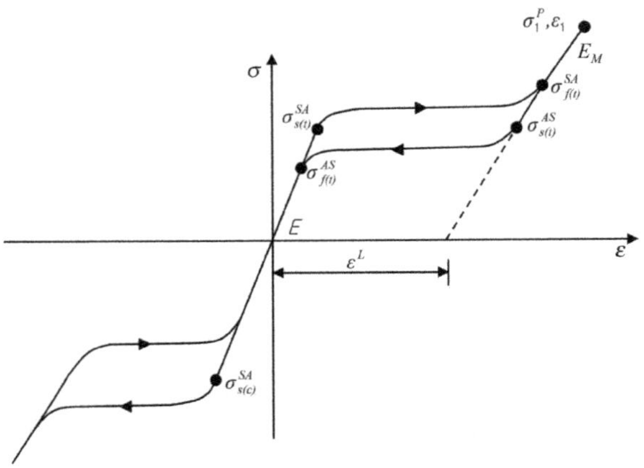

(a) UMAT for simulating the superelastic-plastic behavior of SMA materials under axial loading

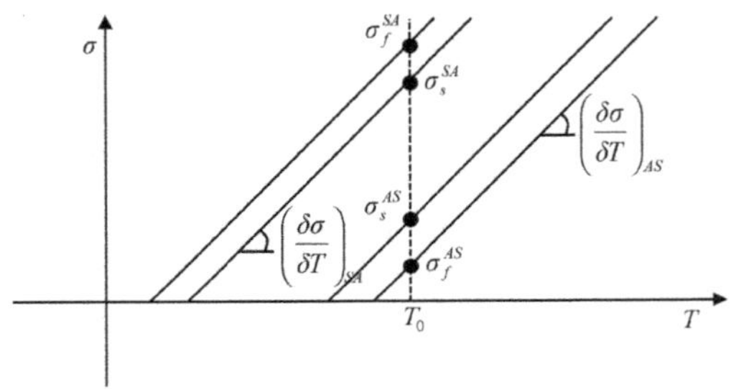

(b) Temperature-dependent phase transformation

*Figure 9.6* Required parameters used to define the behavior of superelastic SMA materials on the UMAT subroutine.

Using time-discrete evolutionary equations (Auricchio & Sacco, 1997), martensite fractions during each phase transformation process are obtained as follows:

$$v = v_S^{AS} = \frac{v_{S,n}E\varepsilon - \text{sgn}(\varepsilon)v_{S,n}\sigma_f^{AS} - E\varepsilon + \sigma_n}{-\text{sgn}(\varepsilon)\sigma_f^{AS} + \text{sgn}(\varepsilon)v_{S,n}E\varepsilon_L - \text{sgn}(\varepsilon)E\varepsilon_L + \sigma_n} \tag{9.18}$$

$$v = v_S^{SA} = \frac{v_{S,n}E\varepsilon - \text{sgn}(\varepsilon)v_{S,n}\sigma_f^{SA}}{-\text{sgn}(\varepsilon)\sigma_f^{AS} + \text{sgn}(\varepsilon)v_{S,n}E\varepsilon_L + \sigma_n} \tag{9.19}$$

Finally, the critical strains at the start of martensite, finish of martensite, start of austenite, and finish of austenite are determined as follows:

$$\varepsilon_s^{AS} = \text{sgn}(\varepsilon)\frac{\sigma_s^{AS}}{E} + \text{sgn}(\varepsilon)v_{S,n}\varepsilon_L \tag{9.20}$$

$$\varepsilon_f^{AS} = \text{sgn}(\varepsilon)\frac{\sigma_s^{AS}}{E} + \text{sgn}(\varepsilon)\varepsilon_L \tag{9.21}$$

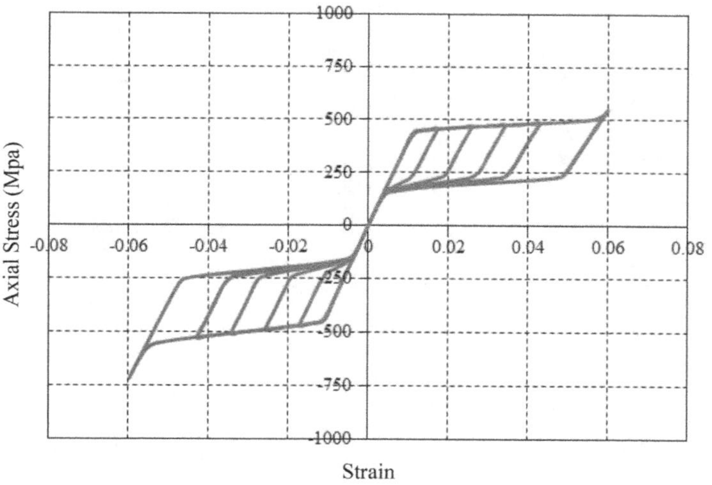


*Figure 9.7* Simulated stress-strain curve for the superelastic SMA material (Engineering measurement).

$$\varepsilon_s^{SA} = \text{sgn}(\varepsilon)\frac{\sigma_s^{SA}}{E} + \text{sgn}(\varepsilon)\upsilon_{S,n}\varepsilon_L \tag{9.22}$$

$$\varepsilon_f^{SA} = \text{sgn}(\varepsilon)\frac{\sigma_f^{SA}}{E} \tag{9.23}$$

The material data required as input values to the UMAT subroutine are obtained from the observation of uniaxial tests with respects to loading, unloading, and reloading under constant temperature. The required parameters used to define the behavior of superelastic SMA materials on the UMAT subroutine are illustrated in Fig. 9.6. The general plasticity was applied to the UMAT algorithm so that material data in the uniaxial curve should be available at the 3D state during FE analyses. The UMAT code was built in the ABAQUS program associated with a FORTRAN computer language with a view to numerically simulating the behavior of superelastic SMAs. The simulated stress and strain curve for the superelastic SMA material is shown in Fig. 9.7. In this study, the required material parameters used for simulation – i.e., elastic modulus (40 GPa), Poisson's ratio (0.33), martensite start stress (440 MPa), martensite finish stress (540 MPa), austenite start stress (250 MPa), austenite finish stress (140 MPa), transformation strain (0.042), temperature (22°C), and so on – straightforwardly obtained from uniaxial pull-out tests carried out by DesRoches *et al.* (2004).

## 9.5 ANALYSIS RESULT AND VERIFICATION

The FE analyses where both refined solid elements and material nonlinearities are taken into consideration are able to accurately predict the behavior of slit damper devices subjected to cyclic loading. Fig. 9.8 shows applied force vs. corresponding displacement hysteresis curves for steel slit damper models. The detail about force ($P$)

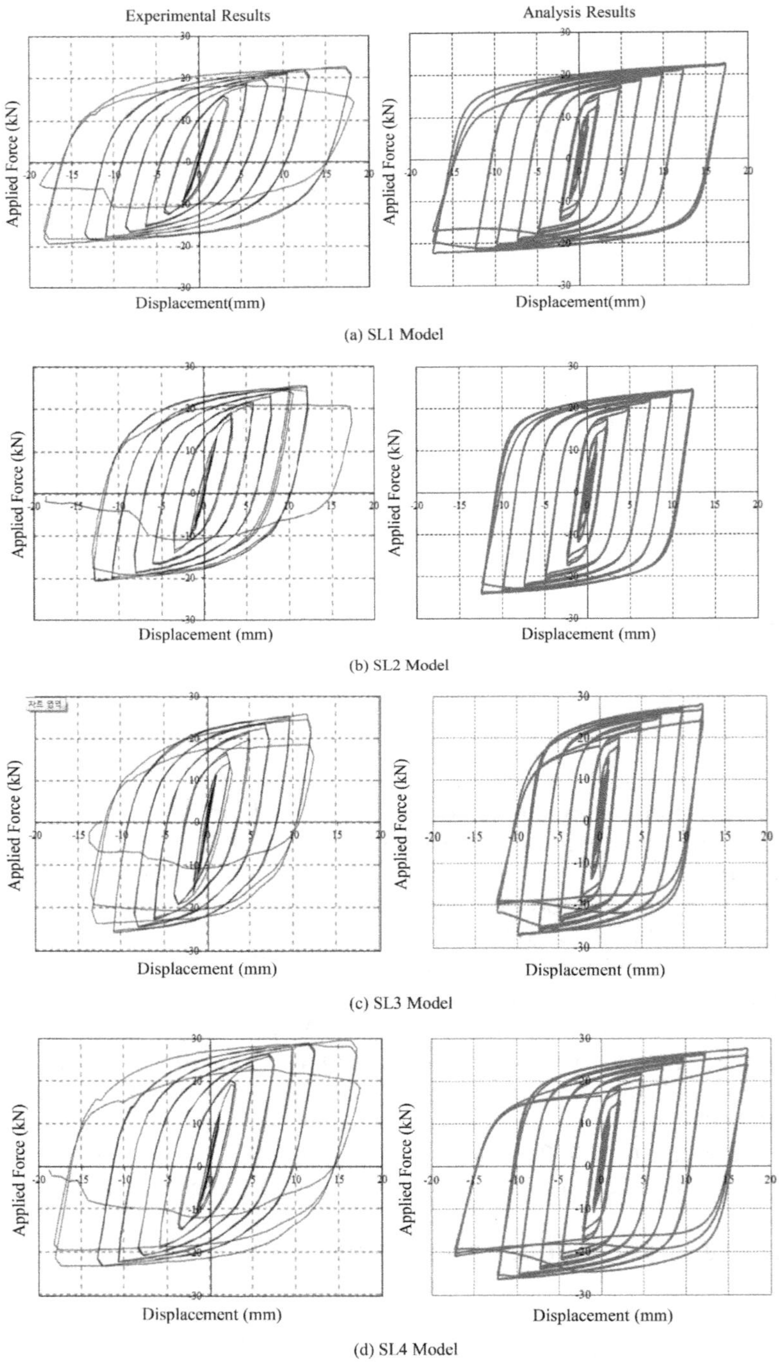


*Figure 9.8* Applied force vs. displacement curves for conventional steel slit damper models.

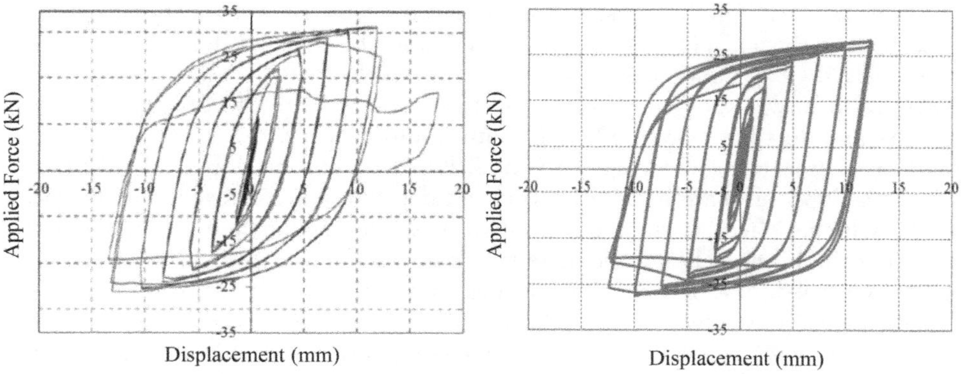


(e) SL5 Model

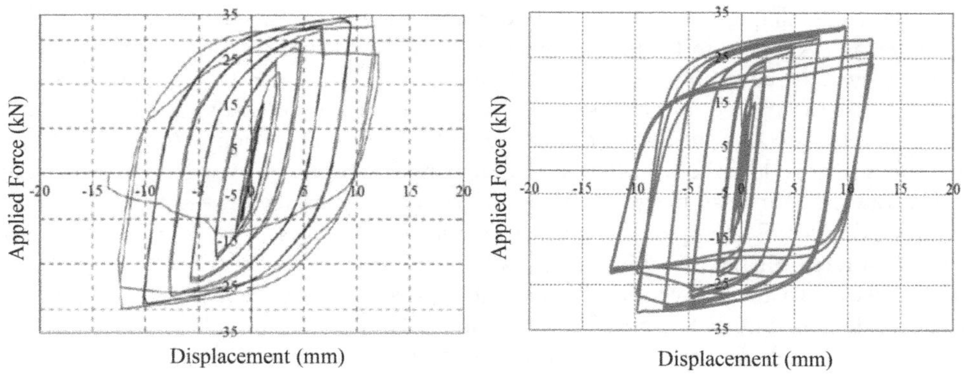


(f) SL6 Model

*Figure 9.8* Continued.

and displacement ($\delta$) measurement is presented in Fig. 9.4(b). According to individual specimens, analysis results are compared with experimental results in an effort to verify the adequacy of FE modeling under the same displacement loading history. Three cycles were conducted at each of amplitude. The experimental tests were carried out until specimens completely failed by fracture.

All specimens for steel slit damper devices have yielded under small displacement loading owing to the inherent characteristics of base steel materials, thereby dissipating a huge amount of energy. In addition, they exhibit stable hysteretic behavior including gradual transition and Bauschinger effect. The SL1 specimen with the smallest $b/l_o$ ratio withstands the lowest shear force while the SL6 specimen with the highest $b/l_o$ ratio sustains the largest shear force. The SL4 specimen with relatively long $l_o$ length exhibits excellent ductility, meaning that it has the ultimate displacement of approximately 17.5 mm prior to strength degradation. For the experimental results, strength degradation begins to appear when facture gradually occurs at the ends of the strips due to stress concentration. The FE models consisting of compatibility-based solid elements with the continuous displacement fields do not include the ability to track the propagation of crack and fracture. For this reason, strength degradation also observed at the FE analysis results forms due to geometric nonlinearity rather than fracture after large displacement is imposed on the FE models. Before that occurs, the FE models show symmetric-shaped loops with stable energy dissipation. The parametric ratio of $b/l_o$ has an influence on the capacity of the FE models as regards strength and ductility as well. Overall, both resulting curves compared to each other are in good agreement with respect to initial slope, loading envelope, unloading slope, reloading slope, ultimate strength, permanent deformation, and even pinching points for Bauschinger effect. Further, this good fit between experiment and simulation suggests that the FE models are adequate to predict the behavior of slit damper devices cyclically loaded. Not only the effect of design parameters but also that of used base materials will be investigated through the observation of the FE analysis results.

Fig. 9.9 shows applied force vs. corresponding displacement hysteresis curves for superelastic SMA slit damper models. It is interesting to note that all of superelastic SMA specimens behave in a similar pattern. They show unique behavior characterized by a flag-shape hysteresis loop under cyclic loading. As we expected, excellent recentering responses indicating nearly zero permanent deformation upon unloading are observed at the simulating curves. Owing to the restoration of superelastic SMAs, strength degradation does not take place regardless of geometric nonlinearity. Besides, the superelastic SMA slit damper devices display higher post-yield strength and more flexible stiffness than the steel slit damper devices. It can be clearly shown that mechanical properties for base materials have a significant influence on the behavior of slit damper devices. The superelastic SMA slit damper devices possess superior performance in terms of flexible initial slope, post-yield strength, and recentering behavior as compared to the conventional steel slit damper devices. Similar to the steel slit damper devices, superelastic SMA slit damper devices with the relatively higher $b/l_o$ ratio (i.e., SL6-SMA specimen) can sustains larger shear forces.

The test results are summarized in Table 9.2. The subscripts of 'exp' and 'anal' in the table denote experiment and analysis, respectively. The yield strength obtained from experimental results and analysis results ($P_{y,exp}$ and $P_{y,anal}$) can be determined by finding the intersection of the force-displacement curve with a secant line parallel to the initial slope of the curve ($K_{exp}$ and $K_{anal}$). The values of yield strength ($P_y$) calculated by Eq. 9.2 are also tabulated for comparison to measured properties. The coefficient $c$ is defined as the ratio of measured stiffness to theoretical stiffness for the fixed-end beam (see Eq. 9.3), as follows:

$$c = \frac{K_{exp}}{K} \text{ or } \frac{K_{anal}}{K} \tag{9.24}$$

Regardless of loading direction, both positive ($P_{max}$ upward) and negative ($P_{min}$ downward) peak strengths are simultaneously tabulated at the analysis results for superelastic SMA slit damper devices because their simulated hysteresis loops exhibit perfectly symmetric shape due to the absence of strength degradation (see also Fig. 9.9). The ductility ratio is defined as maximum displacement divided by yield displacement, such that $u = \delta_{y,exp}/\delta_{max}$.

As surmmaized in the table, the resulting value theoretically calculated is closed to the finding obtained from experimental tests or FE analyses. Therefore, stiffness coefficients (c) as well as normalized yield strength ratios ($P_y/P_{y,exp}$ or $P_y/P_{y,anal}$) have

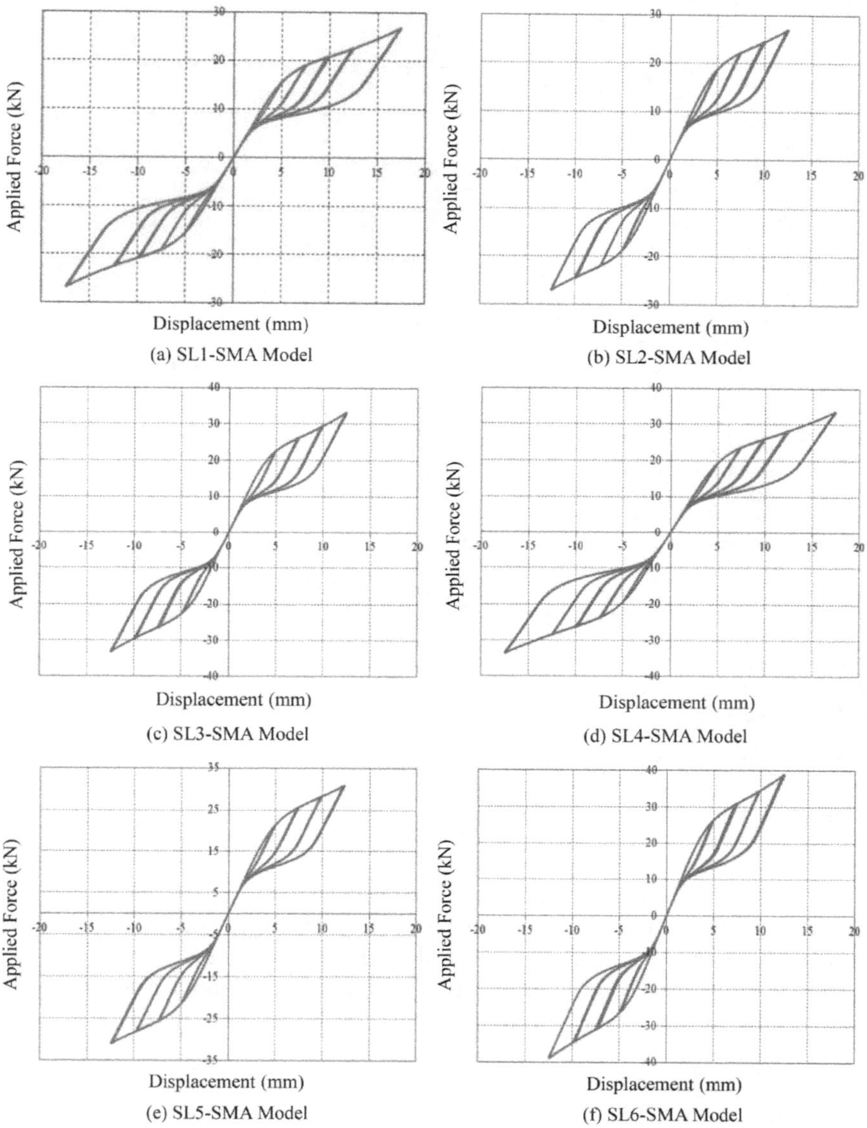


*Figure 9.9* Applied force vs. displacement curves for superelastic SMA slit damper models.

the value of approximately 1.0. The value of maximum peak strength is average 2.0 times larger than that of yield strength because of material's strain hardening. For steel slit damper specimens, the yield displacements obtained from experimental tests ($\delta_{y,exp}$) are identical to those from FE analyses ($\delta_{y,anal}$). Moreover, experimental results for maximum positive strength ($P_{max}$) are in good agreement with analytical results for post-yield strength under approximately identical target displacement

*Table 9.2* Summarization of the test results (Unit: kN and mm).

| *Specimen* | $K_{exp}$ | $c$ | $P_y$ | $P_{y,exp}$ | $P_y/P_{y,exp}$ | $P_{max}$ | $P_{min}$ | $\delta_{y,exp}$ | $\delta_{max}$ | $\mu$ |
|---|---|---|---|---|---|---|---|---|---|---|
| SL1 | 23.49 | 0.98 | 11.59 | 11.51 | 1.01 | 22.61 | −19.37 | 0.49 | 17.32 | 35.42 |
| SL2 | 33.56 | 1.00 | 13.08 | 13.09 | 1.00 | 25.54 | −20.59 | 0.39 | 12.05 | 30.86 |
| SL3 | 50.07 | 1.01 | 15.00 | 15.02 | 1.00 | 25.81 | −25.98 | 0.30 | 11.66 | 38.49 |
| SL4 | 32.49 | 1.00 | 14.58 | 14.62 | 1.00 | 29.61 | −23.28 | 0.45 | 16.47 | 36.69 |
| SL5 | 44.75 | 0.99 | 16.19 | 16.11 | 1.00 | 31.26 | −26.40 | 0.36 | 11.92 | 32.83 |
| SL6 | 60.24 | 1.00 | 17.45 | 17.47 | 1.00 | 35.68 | −29.79 | 0.29 | 11.44 | 39.19 |

(a) Experimental results for the steel slit damper models

| *Specimen* | $K_{anal}$ | $c$ | $P_y$ | $P_{y,anal}$ | $P_y/P_{y,anal}$ | $\delta_{y,anal}$ | $+P_{anal}$ $(+\delta_{anal})$ |
|---|---|---|---|---|---|---|---|
| SL1 | 23.77 | 0.99 | 11.59 | 11.41 | 1.02 | 0.48 | +22.30 (+17.5) |
| SL2 | 34.00 | 1.01 | 13.08 | 12.92 | 1.01 | 0.38 | +24.21 (+12.5) |
| SL3 | 52.48 | 1.06 | 15.00 | 15.22 | 0.99 | 0.29 | +27.92 (+12.5) |
| SL4 | 32.49 | 1.00 | 14.58 | 14.62 | 1.00 | 0.45 | +27.82 (+17.5) |
| SL5 | 44.08 | 0.97 | 16.19 | 16.31 | 0.99 | 0.37 | +28.82 (+12.5) |
| SL6 | 60.52 | 1.01 | 17.45 | 17.55 | 0.99 | 0.29 | +31.89 (+10.0) |

(b) Analysis results for the steel slit damper models

| *Specimen* | $K_{anal}$ | $c$ | $P_y$ | $P_{y,anal}$ | $P_y/P_{y,anal}$ | $\delta_{y,anal}$ | $\pm P_{anal}$ $(\pm\delta_{anal})$ |
|---|---|---|---|---|---|---|---|
| S1L-SMA | 3.55 | 1.02 | 16.11 | 15.80 | 1.02 | 4.45 | ±26.80 (±17.5) |
| S2L-SMA | 5.05 | 1.03 | 18.19 | 18.23 | 1.00 | 3.61 | ±27.02 (±12.5) |
| S3L-SMA | 7.28 | 1.00 | 20.85 | 21.90 | 0.95 | 3.01 | ±33.29 (±12.5) |
| S4L-SMA | 4.34 | 0.91 | 20.27 | 21.80 | 0.93 | 5.12 | ±33.69 (±17.5) |
| S5L-SMA | 6.25 | 0.95 | 22.50 | 22.81 | 0.99 | 3.65 | ±31.05 (±12.5) |
| S6L-SMA | 8.46 | 0.97 | 24.26 | 24.12 | 1.01 | 2.85 | ±38.95 (±12.5) |

(c) Analysis results for the SMA slit damper models

($+P_{anal}$ $(+\delta_{anal})$). The dissipated energy capacity according to individual specimens can be also treated in the same manner. These findings indicate that FE models presented herein are adequate to predict the behavior of slit damper devices. The specimens fabricated with superelastic SMA material show more flexible stiffness and larger post-yield strength than those fabricated with conventional steel material (e.g., $K_{anal} = 3.55$ kN/mm and $P_{anal} = 26.80$ kN for SL1-SMA specimen vs. $K_{anal} = 23.77$ kN/mm and $P_{anal} = 22.30$ kN for SL1-SMA specimen), meaning that important characteristics for the behavior of slit damper devices are deeply affected by used material's properties. More investigation on field contours and history outputs will be conducted in the next section.

## 9.6 COMPARISON AND OBSERVATION

The test setup for collecting analysis data was developed on the basis of monitoring conditions. Individual target displacements for filed contour observation and measurement points for monitoring stress-strain curves are described in Fig. 9.10(a) and Fig. 9.10(b), respectively. Four target displacements (e.g., S1 = 5 mm, S2 = 10 mm, S3 = 17.5 mm,

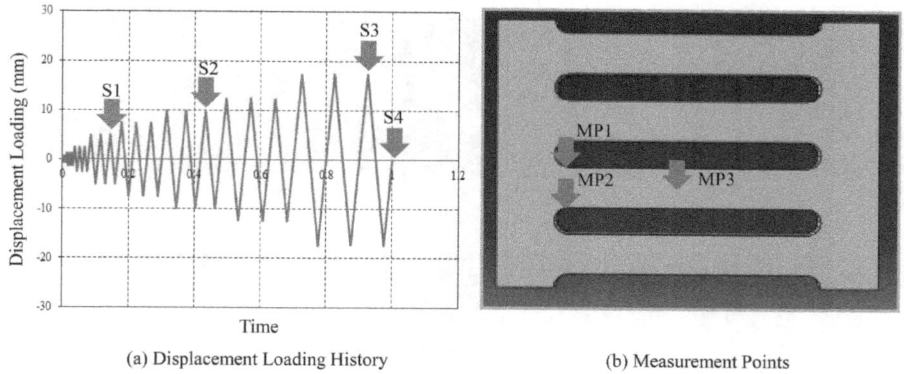


*Figure 9.10* Target displacement for field contour observation (a) and measurement points (b).

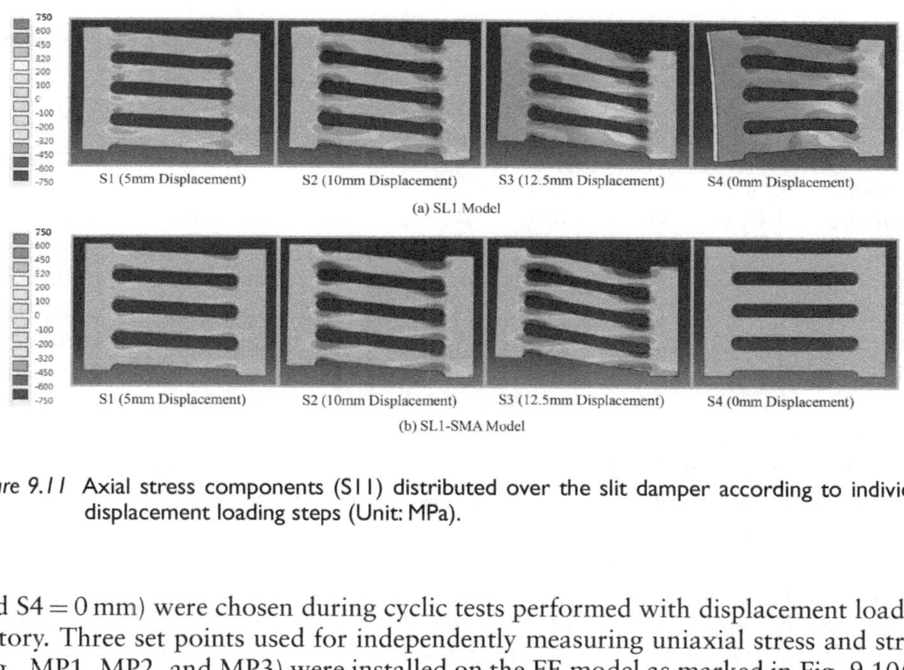


*Figure 9.11* Axial stress components (S11) distributed over the slit damper according to individual displacement loading steps (Unit: MPa).

and S4 = 0 mm) were chosen during cyclic tests performed with displacement loading history. Three set points used for independently measuring uniaxial stress and strain (e.g., MP1, MP2, and MP3) were installed on the FE model as marked in Fig. 9.10(b). The measured data are collected using the history output function provided in the program (ABAQUS, 2010). Contrary to MP1 that installed at the middle of the strip, both MP1 and MP2 can detect plastic hinges that generally form at the end of the strip. In particular, the stress-strain curves measured from these set points confirm the validity of the bending mechanism as elucidated in Fig. 9.4.

The axial stress field contours (S11) distributed over the slit damper according to individual displacement loading steps are shown in Fig. 9.11. The logarithmic axial strain field contours (LE11) are also presented in Fig. 9.12. The SL1 and SL1-SMA

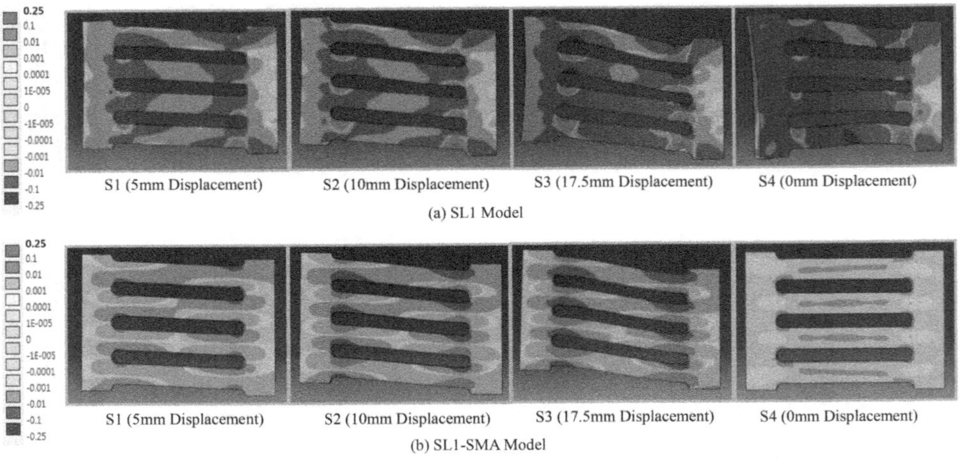


*Figure 9.12* Axial strain components (LE11) distributed over the slit damper according to individual displacement loading steps.

specimens are selected for this investigation. The deformed configurations that are particularly necessary to confirm permanent deformation at the final loading step are also found in the figures with a unit deformed scale factor. The colored graph legends are plotted to easily distinguish the magnitude of axial stress and strain contours. The slit damper areas displayed with orange- (for tension) and light blue-colored (for compression) contours have already reached the onset of plastic yielding. Once the amplitude of the loading history exceeds the limit of the yield displacement ($\delta_y$), plastic yielding starts to occur at the strip. For this reason, stress filed contours greater than the level of plastic yielding are observed under the first loading step (S1). These axial stress field contours demonstrate that tension and compression yielding are concentrated around the both ends of the strips. As increasing the displacement loading step, plastic stress filed areas spread toward the middle of the strips. The red and blue-colored contours indicating that base materials reaches their ultimate tension and compression stress are found at the both ends of the strips when both specimens compared to each other (SL1 and SL1-SMA specimen) are subjected to the third loading step (S3). At the last loading step (S4), the SL1-SMA specimen completely recovers original shape without any residual stress distributed over the strips. On the other hand, the SL1 specimen obviously displays out-of-plane deformation that confirms the evidence of instable failure as well as a considerable amount of residual stress. It may thus be concluded that superelastic SMAs make a good contribution to decrease both permanent deformation and residual stress without additional treatment for repair in case of utilizing in the slit damper device.

As shown in Fig. 9.12, corresponding logarithm axial strain field contours match the axial stress field contours very well in that they can capture similar yielded regions and plasticity patterns under the same loading step. The base materials that are under plastic yielding are shown with red- (for tension) and blue-colored (for compression) contours. The SL1-SMA specimen has nearly-zero residual strain at the final loading

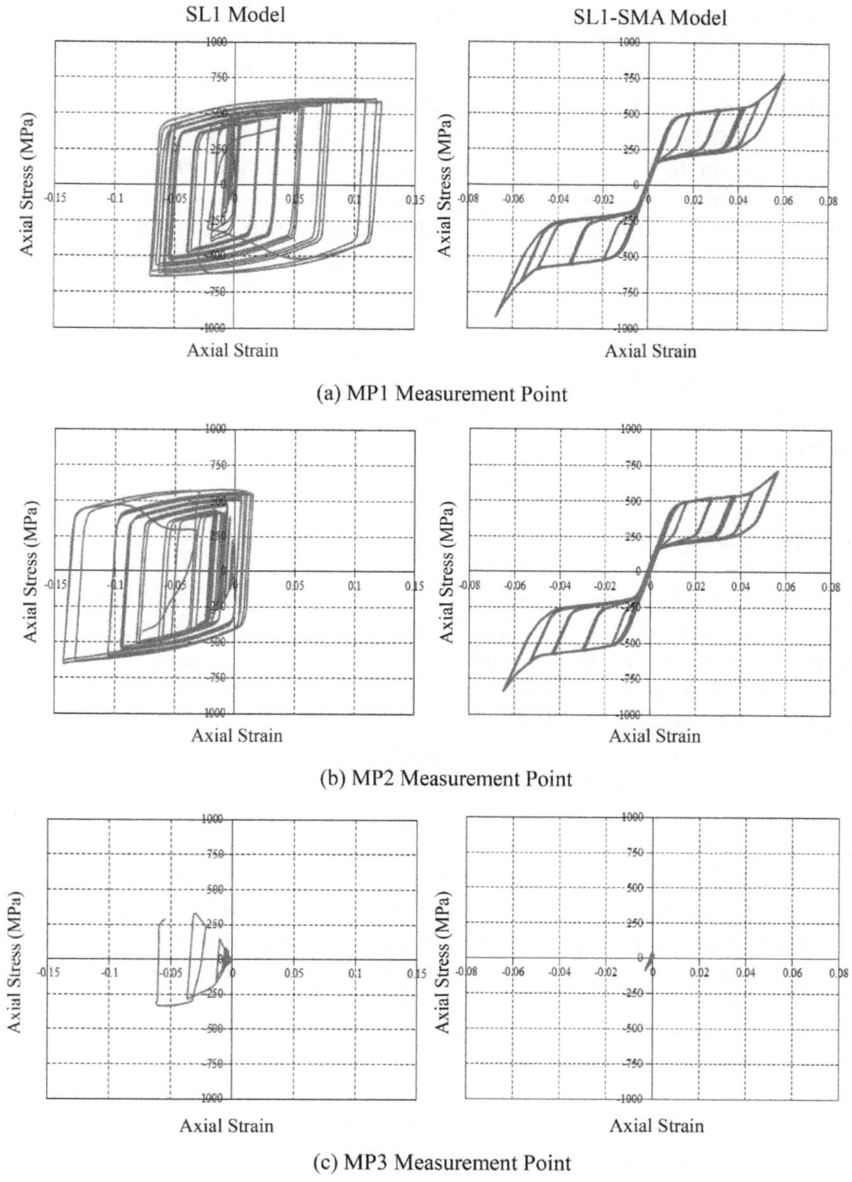


*Figure 9.13* True axial stress and strain curves at the measurement points.

step (S4) while the SL1 specimen shows a lot of residual strain generated due to out-of-plane deformation.

Fig. 9.13 shows true axial stress and strain curves obtained from the measurement points. For the SL1 specimen, the axial stress and strain curve measured from the MP1 point shifts its center to the right hand side of the graph. Thus, the MP1 measurement point undergoes tension stress during most of loading cycles. On the other hand, the MP2 measurement point is under compression stress as shown in Fig. 9.11(a). Both curves exceed the limit of plastic yielding (316.5 MPa). The stress and strain curve measured from the middle of the strip (MP3 point) is under elastic condition during all loading cycles. However, residual stress taken as 260 MPa exists at this measurement point. For the SL1-SMA specimen, both axial stress and strain curves measured from the MP1 and MP2 points are the symmetric flag-shape hysteresis loop similar to inherent material behavior. Although both measurement points undergo a considerable amount of plastic deformation upon loading, this specimen can return to the original state upon unloading owing to recentering capability. Finally, the MP3 point is also under elastic condition during all loading cycles. Contrary to the SL1 specimen, the SL1-SMA specimen shows almost zero residual stress as shown in Fig. 9.13(c). It can be thus shown that the UMAT subroutine implemented herein is able to accurately predict mechanical stress as well as entire behavior for structures made by superelastic SMA materials.

## 9.7 CONCLUSIONS

The superelastic SMAs as innovative smart materials have been widely applied to passive vibration control devices because they possess unique and ideal properties such as self-healing capacity attributed to the superelastic effect, supplemental damping guaranteed by the flag-shape hysteresis, and outstanding metal fatigue. This paper describes new superelastic SMA slit damper devices with recentering capability and energy dissipation. For the purpose of simulating the stress-strain curves of superelastic SMA materials, the UMAT subroutine that can be implemented on the ABAQUS program is also mainly treated in this study. The behaviors of individual slit damper devices are reproduced by FE analyses. The FE models are calibrated to the established test data aiming for reliable prediction. After obtaining FE analysis results, the slit damper devices fabricated with superelastic SMAs are compared to those made by steel materials with respect to initial flexibility, permanent deformation, strength capacity, and residual stress in order to prove that superelastic SMA slit damper devices have superior performance. The FE analysis results demonstrate that the conventional steel slit damper devices are susceptible to permanent displacement, residual stress and instability resulting from out-of-plane deformation. However, the proposed SMA slit damper devices overcome these problems. These FE analysis results are promising for the practical application of superelastic SMA slit damper devices, which feature excellent recentering, moderate energy dissipation, nearly zero residual stress, and relatively larger post-yield strength. Therefore, the outstanding performance as the self-centering and vibration control device can be expected in case of utilizing such smart materials for aseismic design.

Chapter 10

# Smart damping connectors

## 10.1 GENERAL OVERVIEW

The 1994 Northridge and 1995 Kobe earthquakes caused unanticipated damages to steel moment-resisting frames due to quasi-brittle fractures that developed in and around the welded beam-to-column connections (Leon *et al.*, 1998; Green *et al.*, 2004; Rassati *et al.*, 2004; Hu *et al.*, 2010; Park *et al.*, 2011; Hu, 2008; Hu & Leon, 2011; Hu *et al.*, 2011; Hu, 2014). Therefore, these frames require special attention in order to limit the excessive non-structural damage resulting from unacceptably large lateral displacements, as well as to avoid the problems associated with P-delta effects (Rassati *et al.*, 2004; Hu *et al.*, 2010; Park *et al.*, 2011; Hu, 2008). As a result, many engineers have increasingly started using concentrically braced frames (CBFs) as an economical load-resisting system that promises good seismic performance with reduced lateral displacement. Compared to the moment-resisting frames, braced frames are expected to offer higher lateral stiffness for drift control in the benefit of diagonal bracing members installed (Sabol, 2004; Sabelli, 2001; Sabelli *et al.*, 2003). However, individual braces in the CBF often exhibit limited energy dissipation under cyclic loading because they are susceptible to buckling prior to gross section yielding. In other words, the hysteretic behavior of these braces is unsymmetric in tension and compression, and the braces show rapid strength deterioration when they buckle under compression (Inoue *et al.*, 2001; Watanabe *et al.*, 1988). Moreover, brace buckling may cause the sudden collapse of the entire frame structure. The disadvantages of this CBF system can be overcome in case the brace can yield during tension and compression without buckling.

As the replacement of the CBF system, buckling-restrained braced frames (BRBFs) have been used over the past decade to increase energy dissipation as well as ductility capacity because they preclude brace buckling in compression (Inoue *et al.*, 2001; Watanabe *et al.*, 1988; Black *et al.*, 2002; Aiken *et al.*, 2002). The steel core is placed inside a steel casing configured as a hollow structural section (HSS) tube, and then the steel casing is filled with concrete or mortar to prevent the global buckling of the brace member as well as the local buckling of the steel core (Watanabe *et al.*, 1988; Black *et al.*, 2002; Aiken *et al.*, 2002; Tsai *et al.*, 2002). As a new type of the concentrically braced system, the BRBFs are characterized by the utilization of brace members that yield inelastically under both tension and compression, thereby providing stable hysteretic behavior and high energy dissipation to the behavior of the frame structure (Clark *et al.*, 1999; Reina & Normile, 1997; Sabelli, 2004).

In addition, the ductility of the brace can be ameliorated by embedding a steel core of adequate length in the steel casing.

The braces suffer large permanent deformation under strong seismic excitation on the ground that they are usually designed with the concept of a damage tolerant structure (Wada, 1992). Acceptable damage mostly occurs in the braces that are added to the main frame structures such as beams, joints, and columns. These braces function as a replaceable structural fuse to minimize the damages to these main frame structures. The deformed braces can thus be viewed as sacrificial members. Nevertheless, permanent deformation generally occurring in the braces has a tendency to develop a residual inter-story drift in the entire frame structure. Indeed, building residents are able to perceive the residual inter-story drift that exceeds 0.5%. Above all, they feel dizziness and nausea as these drift levels approach 1.0%. Some scientists suggest that if the residual inter-story drift exceeds 0.5% the owners of buildings in Japan should rebuild the structures rather than repair them (McCormick *et al.*, 2008). Although the braced frame buildings experience residual inter-story drifts less than 0.5%, extra repair costs are required to restore the entire laterally deformed structure to its original state. Criteria also need to be established for the detection and changing of damaged bucking-restrained braces (BRBs) because replacement work is carried out on the basis of an engineer's subjective decision. Consequently, in this study, the authors suggest that a recentering device should be supplemented to the braced frames including CBFs and BRBFs in order to considerably decrease the degree of permanent deformation even during high-level seismic events.

One of the ways to clearly improve the seismic performance of the braced frames concerning response reduction or recentering capacity is through the utilization of smart materials, which can be integrated into the bracing system. In particular, superelastic shape memory alloys (SMAs) have currently emerged as one of the most popular smart materials for seismic control devices because of their unique behavior characterized by a flag-shape hysteresis under cyclic loading. This hysteresis behavior provides supplemental energy dissipation and excellent recentering capability for the entire frame structure in case such smart materials are utilized in the braces. Therefore, the passive seismic control achieved exploiting superelastic SMAs contributes to the establishment of additional damping and the mitigation of residual inter-story drifts as regards the behavior of an entire building. The primary purpose of this study is to evaluate the seismic performance of both conventional braced frames with the steel bracing systems and innovative braced frames with the superelastic SMA bracing systems in terms of residual inter-story drifts and recentering ratios through a series of nonlinear time-history analyses. The analytical results for both braced frame types are compared to verify that the SMAs can be used to more effectively control the response of the frame structure under moderate-to-strong seismic loading.

## 10.2 SHAPE MEMORY ALLOYS

SMAs exhibit two distinct crystal phases: (a) a weaker martensite phase, which is stable at low temperatures and at high stress values, and (b) a stronger austenite phase, which is stable at high temperatures and at low stress values and these phases account for the unique physical properties such as shape memory effect (SME) and superelastic

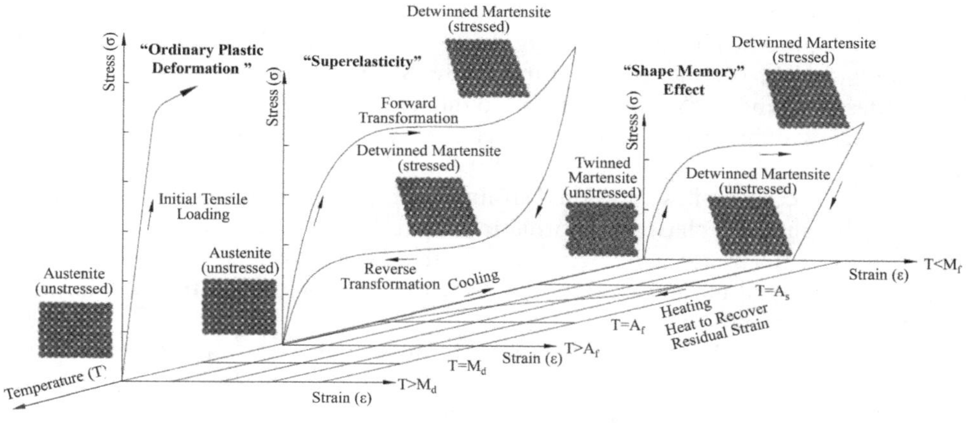


*Figure 10.1* Behavioral characteristics of temperature- and stress-dependent SMA materials.

effect (SE), respectively. SME is related to these alloys' ability to regain their original shape upon heating, while SE refers to their ability to automatically recover from large deformations after the removal of load (Ocel *et al.*, 2004; DesRoches, 2004). The driving force for phase transformations is attributed to the energy difference between the two phases, which can be provided by either a temperature gradient or mechanical loading. Therefore, the forward phase transformation from austenite to martensite (or martensitic phase transformation) is generated by decreasing the temperature or by applying loads. The reverse phase transformation, on the other hand, is obtained by increasing the temperature or by removing the load.

An idealized plot of the stress-strain-temperature relationships in typical SMAs is shown in Fig. 10.1. SMAs exhibit SME for temperatures below the martensite finish temperature ($M_f$). As can be seen in this figure, the behavior of martensitic SMAs is characterized by a relatively low stress plateau and a large value for residual strain. Only elastic strain recovers spontaneously as a result of removal of the load. On the contrary, superelastic (or austenitic) behavior, which is sometimes referred to as a flag-shape hysteresis with zero residual strain upon unloading, occurs at temperatures exceeding the austenite finish temperature ($A_f$). The zero residual strain in austenitic phase SMAs is expected to gain a range of up to 6 percent strain (Hu, 2008; Hu & Leon, 2011; Hu *et al.*, 2011; DesRoches *et al.*, 2004). Additionally, the resulting SMAs display higher stress plateaus when the temperature rises above $A_f$. The martensitic phase transformation that there mostly exists a critical stress for the slip plateaus results in the main contribution to the formation of both SME and SE at the different thermomechanical process (Duerig *et al.*, 1990; Miyazaki *et al.*, 1990; Miyazaki *et al.*, 1986). Once the stress to induce the martensitic transformation lies above the critical slip stress of SMA materials, permanent plastic deformation even occurs during the reverse phase transformation. This phenomenon is labeled "ordinary plastic deformation" in Fig. 10.1. Crystallographic reversibility (or pseudoelasticity) is lost at the room temperature above $M_d$. Therefore, the thermoelatic nature of the material needs to be taken into consideration for the application of SMAs that may experience a broad temperature range.

Crystallographic changes occurring during the cycle of phase transformation are also shown in Fig. 10.1. At temperatures below $M_f$, the heavy twin martensite begins to manifest a change in shape via twin boundary movement, by a process generally known as detwinning. Detwinning, in which the crystal phase has a flaw, occurs mostly at the stress plateau during deformation. When the load is removed, the residual strain remains in the crystal phase. Residual strains begin to recover when SMAs are heated above $A_f$ because the phase transformation from the weak detwinned martensite to strong body-centered austenite occurs during this heat treatment. The materials then return to a twinned martensite upon cooling, thereby completing the cycle of phase transformation due to SME.

The crystallographic changes that occur during a superelastic effect cycle are shown at temperatures above $A_f$. Once sufficiently high loads are applied to the SMA specimen, the phase transformation from austenite to stress-induced martensite occurs mostly in the stress plateau. Though the temperature remains constant throughout this process, the reverse transformation back to austenite occurs upon unloading. This stress-induced transformation causes a superelastic (or pseudoelastic) effect. Superelastic SMAs have shown good prospects in engineering applications because of their unique and ideal properties. These properties include repeatable recentering capability (guaranteed by the superelastic effect), supplemental damping (attributed to the flag-shape hysteresis), stress plateaus (which limit the force transmission at intermediate strain levels), stress stiffening at large strain levels, excellent corrosion resistance, and outstanding fatigue properties, among others (DesRoches *et al.*, 2004; Auricchio & Sacco, 1997). In this paper, we primarily discuss the inelastic behavior of recentering BRB frames in terms of their performance in case studies and numerical analyses with a view to highlighting the effectiveness of the use of superelastic SMA materials in a bracing system.

## 10.3 FRAME DESIGN AND DESCRIPTION

All prototype building structures presented herein were designed in accordance with the ASCE 7-05 (ASCE, 2005). The steel members were designed according to the AISC-LRFD manual (AISC, 2001). An ordinary office building with an important factor of 1.5 was constructed on a stiff soil site (site class D according to the ASCE 7-05) in the LA area. The seismic design category (SDC) was assumed to be a high seismic area (i.e., SDC D class). A 10% probability of exceedance in 50 year seismic hazard corresponding to design-based earthquake (DBE) force level was applied to the design of braced frames. The mapped spectral accelerations for a 0.2-second period and a 1-second period are 2.35g and 1.41g, respectively. The response modification coefficients for building frame systems were taken as 6 consistent with the special steel CBFs and 8 consistent with the moment-resisting BRBFs (ASCE, 2005; ICC, 2006). A set of lateral loads corresponding to design-based earthquake force was generated based on the equivalent lateral load procedure also stipulated in the ASCE 7-05 in order to conduct a trial frame design. Starting with a preliminary design for each building, complete design was achieved by 2D pushover analyses conducted with

*Table 10.1* Basic information for building design.

| *Located Area* | *Loads (Other)* | *Loads (Roof)* | *SDC* | *Site condition* | *Occupancy category* |
|---|---|---|---|---|---|
| LA Area | DL: 4.12 kPa<br>LL: 2.39 kPa | DL: 4.50 kPa<br>LL: 0.96 kPa | D Class | Stiff Soil (Class D) | Ordinary Structures |

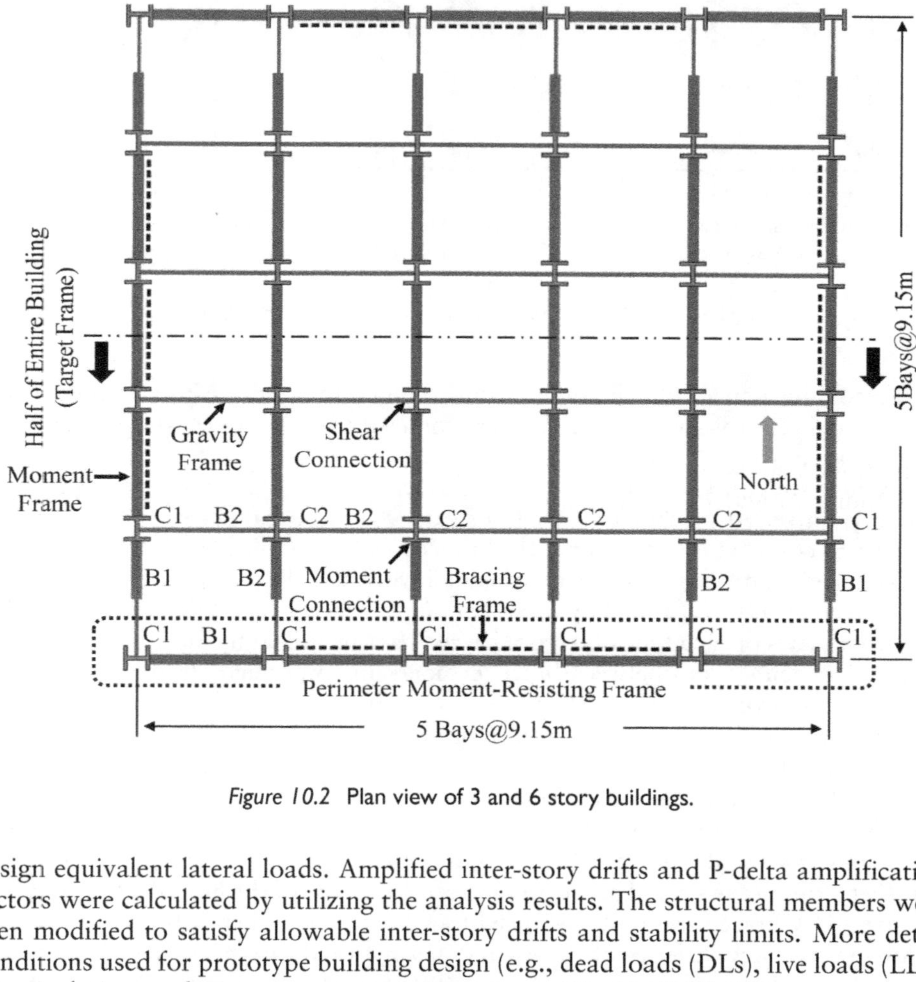


*Figure 10.2* Plan view of 3 and 6 story buildings.

design equivalent lateral loads. Amplified inter-story drifts and P-delta amplification factors were calculated by utilizing the analysis results. The structural members were then modified to satisfy allowable inter-story drifts and stability limits. More detail conditions used for prototype building design (e.g., dead loads (DLs), live loads (LLs), seismic design, and occupancy category) are summarized in Table 10.1.

The prototype frame structures presented herein consisted of 3 story buildings and 6 story buildings. Both types of buildings were designed with the braced bays along the center of the perimeter so as to provide resistance against lateral loads during a seismic event. A plan view of 3 and 6 story prototype square frame buildings with five 9.15 m bays is presented in Fig. 10.2. Six braced bays denoted by dashed lines in the figure were built in each direction. The moment-resisting frames, which are described as thick lines in the building plan, were connected by the welded moment connections

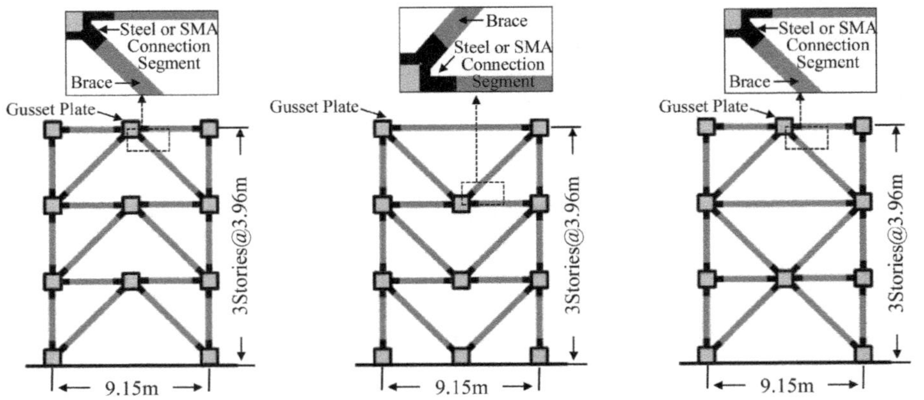


(a) Chevron Braced Frames (3-SB-S or -A; 3-SC-S or -A)

(b) V-Braced Frames (3-VB-S or -A; 3-VC-S or -A)

(c) 2 Story X-Braced Frames (3-XB-S or -A; 3-XC-S or -A)

*Figure 10.3* Details and front views of 3 story braced frames.

to the column flange. On the other hand, the gravity load-resisting frames, which mostly run in the east-west direction, were connected by the pinned shear connections to the column web.

Among the concentrically braced systems feasibly adopted for the design of frame structures, inverted V- (chevron), V-, and 2 story X-braced frame systems were selected in order to compare the traditional steel-braced frame buildings with the superelastic SMA-braced frame buildings. More details involved in braced frame design can be found in the study performed by Sabelli *et al.* (Sabelli *et al.*, 2003; Hu *et al.*, 2013). The front views of each type of 3 story braced bays including the detailed drawing for gusset plate connections between brace members and other main frame members (i.e. beams) are shown in Fig. 10.3. Those of 6 story braced bays are shown in Fig. 10.4. Either superelastic SMA segments or steel segments were installed at the end connector representing the part where large deformations are feasible to occur, and were then connected in series to the brace member.

The model identifications (IDs) are also presented in Figs. 10.3 and 10.4. The first acronym shown in the model ID (i.e., 3- or 6-) indicates 3 or 6 story braced frames. The first letter of the second acronym represents the type of the braced framing system (S: inverted V-braced frame, V: V-braced frame, and X: 2 story X-braced frame) and following, the second letter of the second acronym indicates the type of the braced frame (B: buckling-restrained braced frame (BRBF) and C: concentrically braced frame (CBF)). The braced frames with the conventional steel bracing systems are labeled as "-S" while those with the superelastic SMA bracing systems are additionally labeled as "-A" in the last acronym. For example, the model ID for the 3 story BRBF with the conventionally inverted V-braced steel framing system is denoted as 3-SB-S.

All frame structures have the uniform story height of 3.96-m. The complete description of design results for 3 and 6 story frame structures are given in Tables 10.2 and 10.3, respectively. The steel members were fabricated with standard shapes

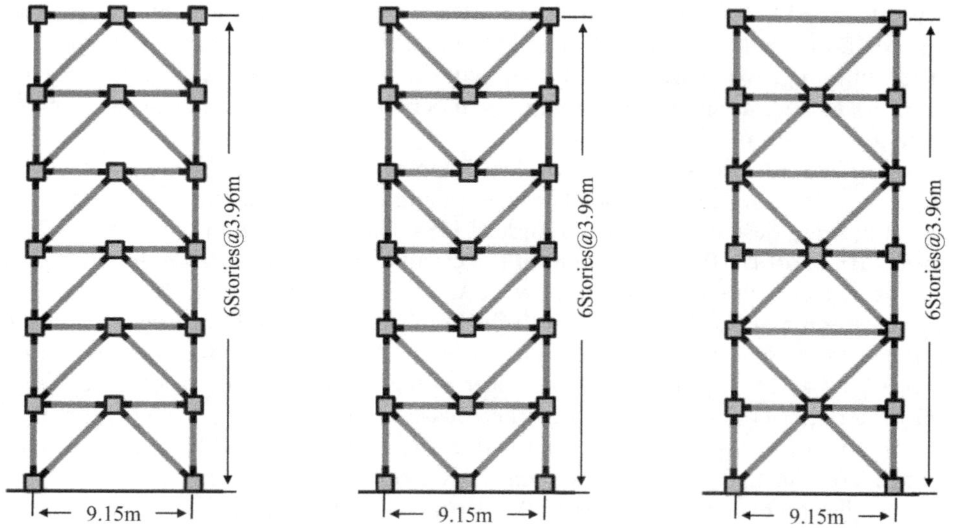


(a) Chevron Braced Frames (6-SB-S or -A; 6-SC-S or -A)

(b) V-Braced Frames (6-VB-S or -A; 6-VC-S or -A)

(c) 2 Story X-Braced Frames (6-XB-S or -A; 6-XC-S or -A)

*Figure 10.4* Front views of 6 story braced frames.

*Table 10.2* Design result of 3 story building.

| | | | *BRBF* | | | | |
|---|---|---|---|---|---|---|---|
| *Story* | *Column (C1)* | *Beam (B1)* | *Core area** *($mm^2$)* | *Casing Tube*** | *CBF**** | *Int. column (C2)* | *Int. beam (B2)* |
| 1 | W12 × 79 | W24 × 68 | 2580 | HSS6 × 1/4 | HSS6 × 6 × 3/8 | W12 × 65 | W18 × 50 |
| 2 | W12 × 79 | W24 × 68 | 2580 | HSS6 × 1/4 | HSS6 × 6 × 3/8 | W12 × 65 | W18 × 50 |
| 3 | W12 × 79 | W18 × 50 | 2027 | HSS6 × 1/4 | HSS6 × 6 × 1/4 | W12 × 65 | W18 × 50 |

*Gr.50 Carbon Steel; **Gr.B Carbon Steel for Round Shape; ***Gr.B Carbon Steel for Rectangular Shape.

*Table 10.3* Design result of 6 story building.

| | | | *BRBF* | | | | |
|---|---|---|---|---|---|---|---|
| *Story* | *Column (C1)* | *Beam (B1)* | *Core area** *($mm^2$)* | *Casing Tube*** | *CBF**** | *Int. column (C2)* | *Int. beam (B2)* |
| 1 | W14 × 109 | W24 × 84 | 2580 | HSS6 × 1/4 | HSS6 × 6 × 3/8 | W12 × 87 | W24 × 68 |
| 2 | W14 × 109 | W24 × 84 | 2580 | HSS6 × 1/4 | HSS6 × 6 × 3/8 | W12 × 87 | W24 × 68 |
| 3 | W14 × 109 | W24 × 68 | 2580 | HSS6 × 1/4 | HSS6 × 6 × 3/8 | W12 × 87 | W24 × 68 |
| 4 | W14 × 109 | W24 × 68 | 2580 | HSS6 × 1/4 | HSS6 × 6 × 3/8 | W12 × 87 | W24 × 68 |
| 5 | W14 × 109 | W18 × 50 | 2027 | HSS6 × 1/4 | HSS6 × 6 × 1/4 | W12 × 87 | W24 × 68 |
| 6 | W14 × 109 | W18 × 50 | 2027 | HSS6 × 1/4 | HSS6 × 6 × 1/4 | W12 × 87 | W24 × 68 |

*Gr.50 CarbonSteel; **Gr.B Carbon Steel for Round Shape; ***Gr.B Carbon Steel for Rectangular Shape.

available in the current design manual (AISC, 2001). The braced frames with the superelastic SMA bracing systems have the same member sizes as conventional steel-braced frames being compared. The column sections were uniform over the height of the building, whereas, except for the interior floor beams (B2), the exterior floor beams (B1) were designed with smaller member sizes for the higher stories in an effort to achieve economical construction.

The BRB members are generally made up of a circular metal core encased in a concrete-filled steel casing tube. The steel casing tubes were circular in cross-section, and thus were designed with the standard HSS shapes fabricated with Gr.B steel (see Tables 10.2 and 10.3). The segment of the circular metal core surrounded by concrete and steel casing tube was fabricated with A572-Gr.50 carbon steel, thereby providing high strength and excellent ductility to the bracing system. This encased steel core acts as the primary load-carrying element in the brace member. Accordingly, as presented in Tables 10.2 and 10.3, BRB members placed at the low-to-middle stories were designed with the larger cross-section area of the encased steel core (e.g., 1st to 2nd story for 3 story building) so that the braced frame structures could withstand more lateral forces. For the CBF structure, standard rectangular HSS tubes were only used for the installation of brace members in the perimeter moment-resisting frames. Similarly, brace members at the low-to-middle stories were designed with the larger HSS tube sections so as to resist more lateral loads in the braced frame system.

## 10.4 ANALYTICAL FRAME MODELS

In this study, the seismic performance of braced frame structures was investigated through nonlinear dynamic time-history analyses conducted on 24 frame models. These dynamic analyses were performed by using the Open System for Earthquake Engineering Simulation (OpenSEES) (Mazzoni *et al.*, 2006), an open-source program widely used in the USA for this type of studies.

All prototype building models presented herein are symmetrical to the center axes with a uniform distribution of mass and stiffness (see Fig. 10.5), and thus can be modeled as 2D frame structures that are of essentially regular condition. As shown in Fig. 10.5, the 2D frame models composed of perimeter moment-resisting frames and gravity load-resisting frames were used for nonlinear frame analyses. In plane-torsion effects may not be taken into consideration during these analyses. The brace members installed on the perimeter moment-resisting frame are necessary to effectively withstand lateral forces (Sabelli, 2001; Sabelli *et al.*, 2003; Sabelli, 2004). On the other hand, the leaning columns in the interior gravity load-resisting frame are required to mainly resist half of entire building weight (Kim & Leon, 2007). These leaning columns were subjected to dead loads plus live loads. Likewise, P-delta effects resulting from second-order behavior due to large deformation were also considered during analyses.

The main frame members were modeled as nonlinear beam-column elements containing 2D fiber sections for the purpose of reproducing the inelastic behavior of beams and columns. In the program, P-delta coordinate transformations were used to generate geometric nonlinearities involving with second-order behavior on the frame model. The structural beams were assumed to be rigidly connected to the

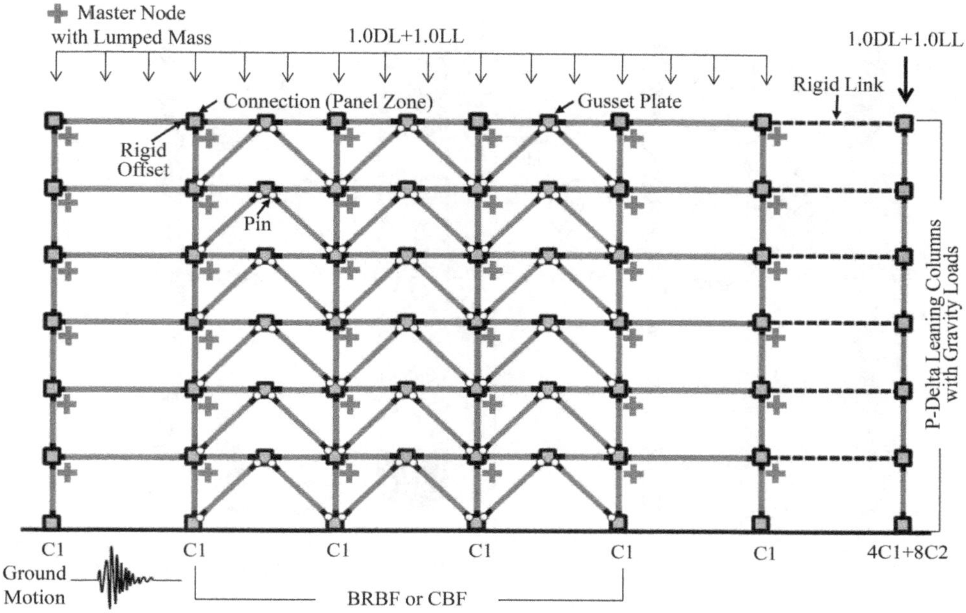


*Figure 10.5* Modeling attributes for 2D frame models.

columns, thereby applying rigid offsets to beam elements as shown in Fig. 10.5. The stiffening effect of the gusset plates was also constructed by using these rigid offsets as well. The brace members also modeled as nonlinear beam-column elements were pinned at both ends because they were considered to be structural elements to sustain only axial forces. Rigid links with hinge connection were installed to enforce rigid diaphragm constraints between perimeter moment-resisting frames and interior gravity load-resisting frames and thus each floor could translate as a rigid body.

The column bases of the frame structure are considered to be fixed. In this study, all steel members possessed the same strain hardening ratio of 1.5%. The transient equilibrium analysis based on the Newmark method (Newmark, 1959) was conducted to solve the time dependent-dynamic problem. The effective damping, as defined by the Rayleigh command in the OpenSEES program, was added to these analyses. According to common practice for code designed steel structures (Sabelli, 2001; Sabelli *et al.*, 2003), an effective viscous damping coefficient of 5% was applied to the 2D frame models. Lumped masses were assigned to nodes on the moment-resisting frames (see Fig. 10.5) so as to generate the story shear forces resulting from ground acceleration. Lumped masses were composed of 1.0 times dead loads plus 0.2 times live loads (Hu, 2008; Hu & Leon, 2011; Hu *et al.*, 2011; Kim & Leon, 2007). The primary response data including nodal displacements were collected by utilizing the recorder command in the program.

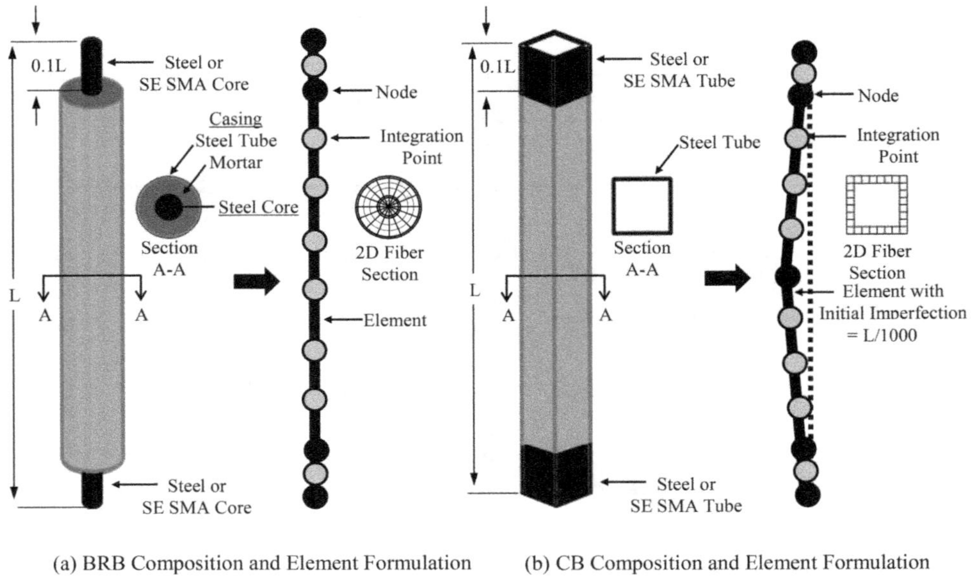


*Figure 10.6* Modeling of brace members.

As shown in Fig. 10.6, the brace members were also modeled as nonlinear beam-column elements with 2D discrete fiber sections that contained nonlinear material behavior. The cross section of the BRB member consisted of composite fiber sections with different material properties for the steel casing tube and the confined core mortar (or concrete), whereas that of the concentric brace (CB) member was composed of only hollow steel tube fiber sections. To reproduce the behavior of the confined concrete material, a uniaxial Kent-Scott-Park concrete model accounting for compressive strength, crushing strength with gentle degradation, and zero tensile strength was implemented into the program. The behavior of the typical steel with strain hardening was simulated using the uniaxial steel material command in the program. The 2D fiber sections were assigned to numerical integration points along the element. Accordingly, the frame element utilized herein accounts for distributed inelasticity through the subsequent integration of material response over the cross section.

The hysteretic load-deformation behaviors of a pin-ended BRB and a conventional pin-ended CB are also presented in Fig. 10.7. The numerical modeling of the BRBs where a point related to global buckling can be disregarded is attained by the composite strut as a nonlinear 1D pin-ended element with zero initial imperfection and appropriate uniaxial material properties. Therefore, the hysteretic behavior of BRB members can be reproduced by using this element where composite fiber sections are employed. As shown in Fig. 10.7(a), the BRB member yields under compression without buckling due to the casing tube. In addition, the behavior of the confined concrete material allows the BRB member to sustain compressive strength that is slightly higher than tensile strength.

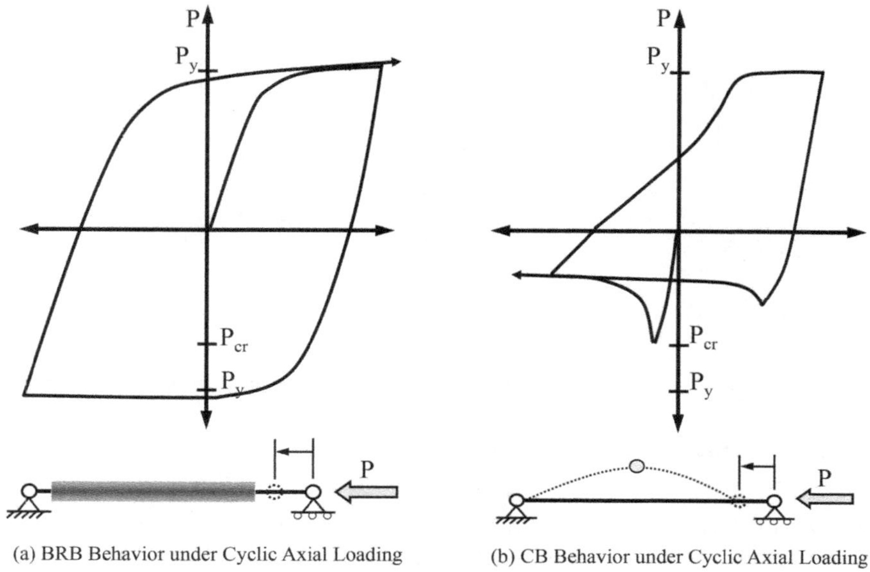


*Figure 10.7* Behavior of frame members.

For the CBs that are generally allowed to buckle in compression, the initial imperfection (camber) representing the ratio of L/1000 was introduced by offsetting the node at the mid-length of the brace member (see Fig. 10.6(b)). The small lateral deflection would help trigger buckling in the CB member. As shown in Fig. 10.7(b), the brace loaded in compression behaves elastically until it arrives at the buckling condition ($P_{cr}$). During a compressive excursion, global buckling occurs at the mid-span of the brace when the peak load is reached. In addition to global bucking, other characteristic branches accounting for negative post-buckling stiffness, unloading, elastic reloading in tension, and uniaxial yielding ($P_y$) are also indicated on the hysteretic behavior curve. The slopes of unloading and reloading lines are less than the initial slope in compression because the deflection of the brace is greater than the corresponding initial imperfection. The branch located posterior to the yielding of the brace develops a small positive slope owing to material strain hardening. The buckling load significantly degrades with respect to the one that corresponds to the first cycle. This physical phenomenon is caused by the Bauschinger effect resulting from the inelastic behavior of the material and by the residual deformation of the brace (Black *et al.*, 1980).

The BRB member needs to be subdivided into at least three nonlinear beam-column elements for modeling both pin-ended bar connectors. One more fine subdivision is necessary for the CB member to model initial imperfection. At both ends of the brace member, the fiber sections contained the material behavior of Gr.50 steel or superelastic SMA according to the model cases (see Fig. 10.6). The analytically simulated curves for corresponding inelastic material behavior are presented in Fig. 10.8. Using the uniaxial steel material command, as mentioned above, the material behavior of Gr.50 steel was simulated with a yield stress ($F_y$) of 345 MPa, an elastic modulus (E) of 200 GPa,

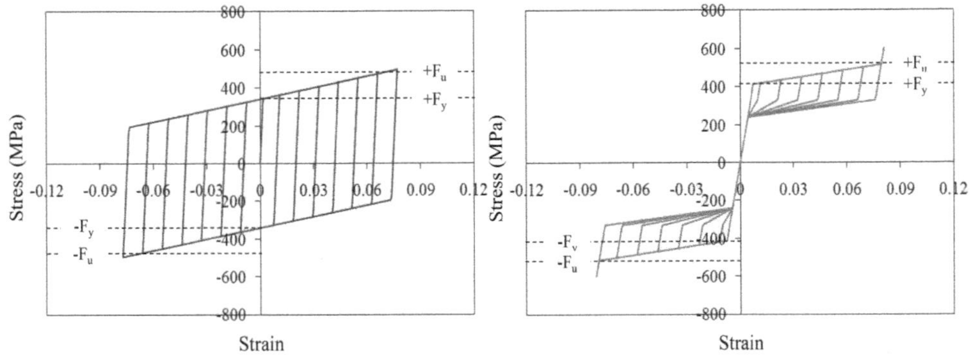


(a) Stress and Strain Curve for Gr.50 Carbon Steel

(b) Stress and Strain Curve for SE SMA

*Figure 10.8* Stress and strain curves for Gr.50 carbon steel and superelastic SMA material.

and a hardening ratio of 1.5% (see Fig. 10.8(a)). However, the material behavior of the superelastic SMA was simulated using the user-defined material (UMAT) code (see Fig. 10.8(b)) (Fugazza, 2003; Auricchio & Sacco, 1997), because an appropriate default material command was absent in the OpenSEES program. The hysteretic behavior of the superelastic SMAs was idealized by the stiffness model, and denoted as a series of straight lines that conform with the process of phase transformations. In particular, the stiffness model implemented in the UMAT code took into account the same elastic properties of martensite and austenite. As a consequence, the loading and unloading branches have the same slope. Details related to the model formulation and the implementation code are also found in Fugazza (2003). In this study, the required material parameters that were used for generating the stiffness model – i.e., elastic modulus (41 GPa), yield stress (413 MPa), recoverable elongation (8%), and ultimate recoverable stress (516 MPa) – were obtained from uniaxial pull-out tests performed by DesRoches *et al.* (2004) utilizing the 25-mm-diameter superelastic SMA bars.

## 10.5 GROUND MOTIONS

The ground motion records used in this study were originally developed as part of the FEMA/SAC steel project (Somerville *et al.*, 1997). Two different seismic hazard levels of seismic ground motions, (a) the design level earthquake (DLE) associated with a 10% probability of exceedence in 50 years (LA01 to LA20), and (b) the maximum credible earthquake (MCE) associated with a 2% probability of exceedence in 50 years (LA21 to LA40), were considered as we conducted a total of 960 nonlinear dynamic time-history analyses. The design level spectrum corresponds to 2/3 of the maximum credible spectrum (Somerville *et al.*, 1997). For each seismic hazard level, 20 earthquake records in Los Angeles (L.A.) were used to obtain more insight into the seismic behavior of individual frame models.

The average acceleration response spectrum curves for each seismic hazard level of ground motion records are compared to the corresponding design response spectrum

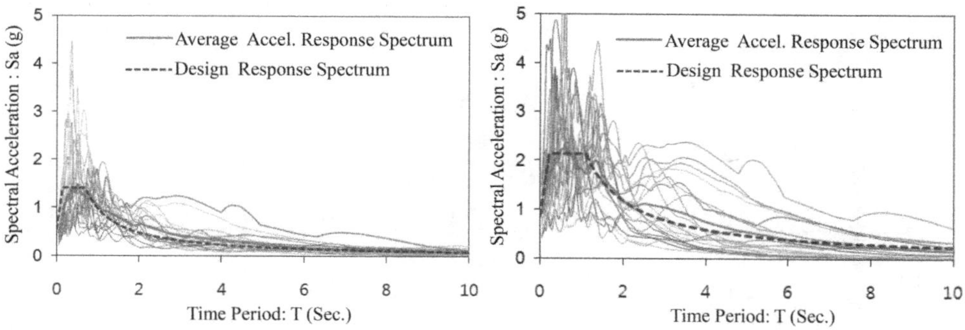


(a) 10% in 50 Years

(b) 2% in 50 Years

*Figure 10.9* Spectral accelerations of LA ground motions.

*Table 10.4* Details of design and average spectral acceleration at the fundamental time period of individual frame models.

| | | *10% in 50 Years* | | *2% in 50 Years* | | | | *10% in 50 Years* | | *2% in 50 Years* | |
|---|---|---|---|---|---|---|---|---|---|---|---|
| *Model ID* | *T* (sec.) | *Sa* (g)* | *Sa* (g)** | *Sa* (g)* | *Sa* (g)** | *Model ID* | *T* (sec.) | *Sa* (g)* | *Sa* (g)** | *Sa* (g)* | *Sa* (g)** |
| 3-SB-S | 0.44 | 1.56 | 1.53 | 2.14 | 2.03 | 3-SC-S | 0.45 | 1.56 | 1.51 | 2.14 | 2.03 |
| 3-SB-A | 0.58 | 1.56 | 1.49 | 2.14 | 2.30 | 3-SC-A | 0.53 | 1.56 | 1.64 | 2.14 | 2.22 |
| 3-VB-S | 0.44 | 1.56 | 1.53 | 2.14 | 2.03 | 3-VC-S | 0.45 | 1.56 | 1.51 | 2.14 | 2.03 |
| 3-VB-A | 0.58 | 1.56 | 1.49 | 2.14 | 2.30 | 3-VC-A | 0.53 | 1.56 | 1.64 | 2.14 | 2.22 |
| 3-XB-S | 0.42 | 1.56 | 1.57 | 2.14 | 2.03 | 3-XC-S | 0.43 | 1.56 | 1.56 | 2.14 | 2.04 |
| 3-XB-A | 0.57 | 1.56 | 1.56 | 2.14 | 2.29 | 3-XC-A | 0.52 | 1.56 | 1.68 | 2.14 | 2.21 |
| 6-SB-S | 0.82 | 1.56 | 1.39 | 2.14 | 1.99 | 6-SC-S | 0.83 | 1.56 | 1.40 | 2.14 | 1.98 |
| 6-SB-A | 1.06 | 1.33 | 1.15 | 2.14 | 1.75 | 6-SC-A | 0.97 | 1.45 | 1.25 | 2.14 | 1.81 |
| 6-VB-S | 0.84 | 1.56 | 1.41 | 2.14 | 1.98 | 6-VC-S | 0.85 | 1.56 | 1.43 | 2.14 | 1.98 |
| 6-VB-A | 1.08 | 1.31 | 1.11 | 2.14 | 1.72 | 6-VC-A | 0.98 | 1.44 | 1.24 | 2.14 | 1.80 |
| 6-XB-S | 0.79 | 1.56 | 1.37 | 2.14 | 2.02 | 6-XC-S | 0.80 | 1.56 | 1.37 | 2.14 | 2.00 |
| 6-XB-A | 1.04 | 1.36 | 1.17 | 2.14 | 1.76 | 6-XC-A | 0.94 | 1.50 | 1.27 | 2.14 | 1.83 |

*Design Response Spectrum; **Average Acceleration Response Spectrum.

curves in Fig. 10.9. Details of design and average spectral acceleration at the fundamental time period of individual frame models are also presented in Table 10.4. These ground motion records were further scaled to match each design response spectrum for the L.A. area where the frame models were located, which corresponds to the site class D in ASCE 7-05. On that account, the average acceleration response spectrums for DLE records were slightly lower than the design response spectrums associated with the first deformation modes for individual frame models. Overall, the superelastic SMA-braced frame models showed relatively longer fundamental time periods than the conventional steel-braced frame model being compared. It indicates that the superelastic SMA bracing systems were initially more flexible than conventional steel bracing

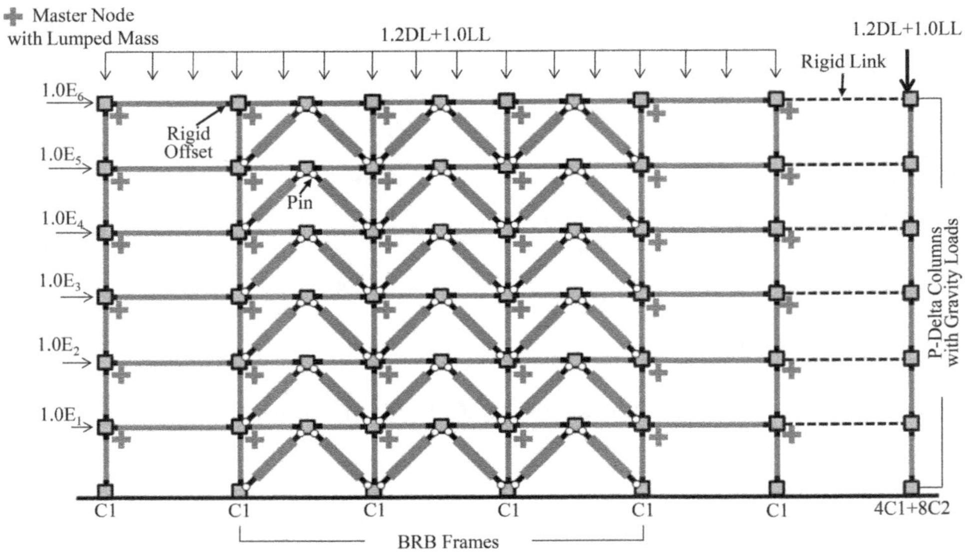


*Figure 10.10* Modeling attributes for 2D frame models.

systems. Hence, the 6 story SMA-braced frame models were designed with smaller seismic design base shear coefficients than those used for the 6 story steel-braced frame models.

## 10.6 NONLINEAR PUSHOVER ANALYSES

### 10.6.1 Analysis procedure and seismic design checks

In the L.A. area, earthquake loads dominate over wind loads. The dominant load combination (LC), which consisted of dead (DL), live (LL), and earthquake (E) loads, i.e., LC5 = 1.2DL + 1.0LL + 1.0E as stipulated in ASCE 7-05, was used to perform a total of 12 nonlinear pushover analyses (see Fig. 10.10). The uniform dead and live loads that were converted into equivalent point loads were applied to the beam elements in the direction of gravity using the constant time series function associated with a load pattern in the OpenSEES program. The equivalent lateral loads on the joints were simulated using the linear time series function, so these loads could be applied in the linearly static or cyclic incremental manner associated with a predefined time step. For each time step, a static or cyclic pushover analysis was conducted using a displacement control algorithm (Mazzoni *et al.*, 2006).

The total effective seismic weight (W) was based on 100% of the dead load. The response modification coefficient (R) for this type of BRB frame system was offset to 8 (ASCE 7-05, Table 12.2-1). The fundamental time periods of all frame buildings with the same total effective seismic weight were determined by the influence of the bracing system on the lateral stiffness. In this study, the fundamental (first modal) time periods (T) obtained from the preliminary 2D modal analyses were used to compute the seismic design shear forces ($V_s$) through the equivalent lateral force procedure (ASCE 7-05, Sec. 12.8.2.1). The equivalent lateral forces, which can deform a structure by conforming with the dominant mode shape, were calculated by distributing the seismic design shear force on the basis of the portion of seismic dead load at each corresponding story level (ASCE 7-05, Sec. 12.8.3). The first mode generally dominates the behavior of a frame structure when higher mode shapes contribute less than 10% of the effective seismic mass (Park *et al.*, 2011). Finally, design checks for the initial selection of the member sizes were carried out to investigate whether BRB frame models subjected to LC5 satisfied the allowable design limit in terms of inter-story drifts and P-delta effects (ASCE 7-05 Secs. 12.8.6 and 12.8.7), as obtained from elastic analyses.

### 10.6.2 Nonlinear pushover analysis results

The resulting pushover curves that were plotted to show roof story drifts versus the total base shear force are shown in Fig. 10.11. Figs. 10.11(a) and (b) show a comparison of nonlinear pushover curves for BRB frames with the same type of braced frames but different bracing systems (e.g., 6SBRB-S vs. 6SBRB-S-SMA).

Static pushover analyses were performed to clearly ascertain the important transition points in the pushover curve, such as the elastic range (or proportional limit), initial yielding, initiation of strain hardening, and ultimate strength (see Fig. 10.11(a)). For example, the results for the 6SBRB-S-SMA frame model indicate that a roof story drift of approximately 0.45% is the limit for the elastic range, that a roof story drift of about 0.53% corresponds to the yielding point, and that a roof story drift of 0.85% corresponds to the initiation of strain hardening. These static pushover curves show that the initial slope of the 6SBRB-S frame model is considerably steeper than that of the 6SBRB-S-SMA frame model, with the two models having roof yield drifts of approximately 0.32% and 0.53%, respectively. In contrast, the 6SBRB-S frame model has a smaller yield base shear strength than the 6SBRB-S-SMA frame model. It can be shown that the material behavior of the circular rods installed at the end connectors of the BRB member has a significant influence on these behavioral characteristics of braced frames. These circular rods can handle the behavior of the entire frame structure, despite the small portion that they occupy. The values of the seismic design base shear forces ($V_s$), which are shown in Fig. 10.11(a), were calculated on the basis of half the total effective seismic weight (W) of the entire building, because each frame model bears only half of the total seismic load. Note that the seismic design base shear force was the summation of the equivalent lateral forces used in the course of frame design.

The additional parameters obtained from static pushover analysis results are summarized in Table 10.5, which shows the seismic design base shear coefficients ($C_s$) defined as the base shear force divided by the effective seismic weight, the base shear forces at yielding ($V_y$), and other base shear forces when the design inter-story drift limit is reached in any story level ($V_{dl}$). Furthermore, this table presents the first modal time periods (T) for individual BRB frame models, as well as the roof story drifts corresponding to each of these base shear forces ($\Delta_y$ and $\Delta_{dl}$). The relatively long time period for the SMA braced frame models indicates that superelastic SMA bracing systems are more flexible than conventional steel bracing systems. Hence, SMA

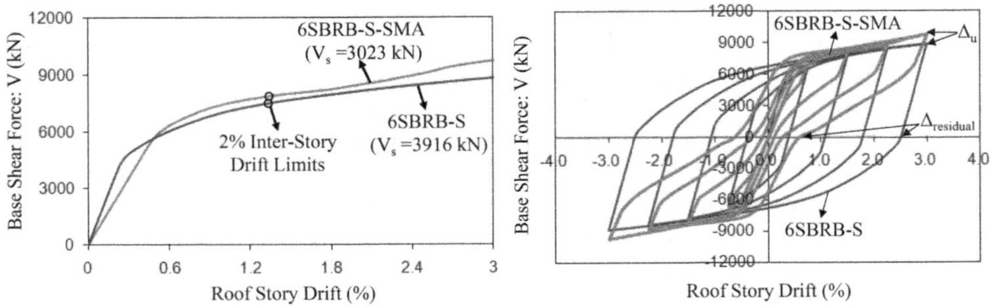


(a) Static Pushover Analysis Result (b) Cyclic Pushover Analysis Result

*Figure 10.11* Pushover test results for selected frame models.

*Table 10.5* Characteristics of 2D frame models.

| *Model ID* | *T* (sec.) | *W* (kN)[a] | $V_s$ (kN)[b] | $V_s/W$[c] | $V_y$ (kN) | $\Delta_y$ (%) | $V_y/V_s$[d] | $V_{dl}$ (kN)[e] | $\delta_{dl}$ (%) | $V_{dl}/V_s$[d] |
|---|---|---|---|---|---|---|---|---|---|---|
| 6SBRB-S | 0.82 | 26200 | 3916 | 0.149 | 4939 | 0.32 | 1.26 | 7498 | 1.33 | 1.91 |
| 6BRB-S-SMA | 1.06 | 26200 | 3023 | 0.115 | 5978 | 0.53 | 1.98 | 7879 | 1.35 | 2.61 |
| 6SBRB-V | 0.84 | 26200 | 3823 | 0.146 | 4880 | 0.32 | 1.28 | 7384 | 1.33 | 1.93 |
| 6BRB-V-SMA | 1.08 | 26200 | 2983 | 0.114 | 5893 | 0.53 | 1.98 | 7763 | 1.35 | 2.60 |
| 6SBRB-2X | 0.79 | 26200 | 4072 | 0.155 | 4912 | 0.32 | 1.21 | 7263 | 1.31 | 1.78 |
| 6BRB-2X-SMA | 1.04 | 26200 | 3120 | 0.119 | 6068 | 0.53 | 1.94 | 7687 | 1.27 | 2.46 |

[a]Effective seismic weight of the frame model equal to half of the total seismic weight of the entire building.
[b]Seismic design base shear force of the frame model equal to half of the design base force of the entire building.
[c]Seismic design base shear coefficient ($C_S = V_S/W$).
[d]Overstrength factor ($\Omega = V/V_S$).
[e]Base shear force at the design inter-story drift limit of the ASCE 7-05 (e.g. 2% for 6 story building).

braced frame models were designed using approximately 23% smaller seismic design base shear coefficients than those used for steel braced frame models. However, owing to the benefit of SMA's material behavior, BRB frame models with superelastic SMA bracing systems have relatively larger post-yield shear strength than those with steel bracing systems. The overstrength factors ($\Omega$), which were defined here as the ratio of the base shear force at each observation point to the seismic design base shear force, are also provided in this table. These factors are very useful in investigating the strength of frame structures for seismic evaluation, and they show that all BRB frame models were uniformly designed to be over twice as strong as the design strength. This indicates that the BRB frame models were overdesigned by an ultimate overstrength factor ($\Omega_o$) of 2.5. All roof story drift limits ($\Delta_{dl}$) are located on the loading path before the ultimate point (see also Fig. 10.11(a)) because the inter-story drift limit stipulated in ASCE 7-05 governed the design of these BRB frame models.

Meanwhile, the cyclic pushover analyses were performed to evaluate the strength, energy dissipation, and recentering capability according to the bracing systems used

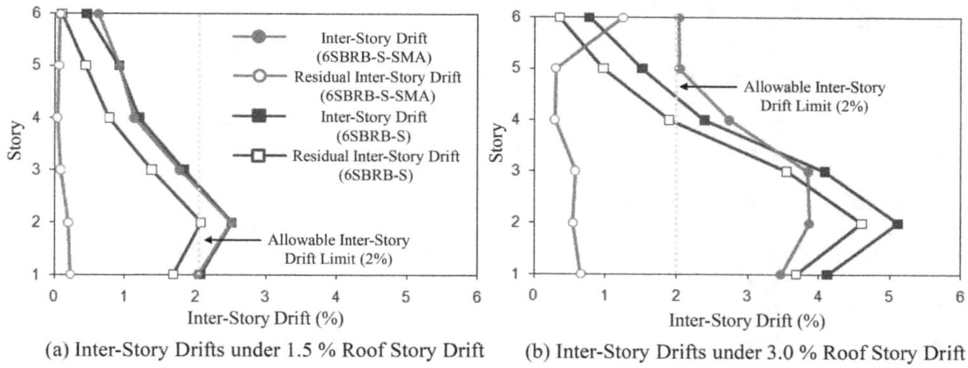


*Figure 10.12* Inter-story drifts under 1.5 (a) and 3.0% (b) roof story drifts, and residual inter-story drifts after unloading from these respective roof story drift levels.

in the BRB frame (see Fig. 10.11(b)). The same loading histories were applied to both BRB frame models being compared. For a given maximum displacement load ($\Delta_u$), the 6SBRB-S-SMA frame model benefits by being able to exploit the SMA bar connectors, which have higher ultimate strength capacities than steel bar connectors, and thus it exhibits slightly greater post-yield shear strength than the 6SBRB-S frame model. Moreover, the 6SBRB-S-SMA model shows much better recoverable roof story drifts than the 6SBRB-S frame model. As the number of cycles increased, the recoverable roof story drift, which is defined as the applied roof story drift load minus the residual roof story drift, also increased for the 6SBRB-S-SMA frame model. For the given maximum 3% roof story drift loading cycle, the recoverable roof story drift ($\Delta_{re} = \Delta_u - \Delta_{residual}$) reached as high as 2.4%, which was as much as 80% of the applied roof story drift.

The values for energy dissipation were calculated as the enclosed areas of the resulting cyclic pushover curves. The energy dissipated by the yielding of members causes the cyclic behavior curves to become plump (Hu & Leon, 2011). For the 6SBRB-S frame model, the steel end connectors help to increase the energy dissipation capacity by virtue of their plastic deformation. The occurrence of such plastic deformation, however, tends to increase the residual story drift, as is shown in the cyclic pushover curve. The recentering behavior found in the 6SBRB-S-SMA frame model reduces more or less energy dissipation, as shown by the pinching of the cyclic pushover curve during unloading, but significantly increases the degree of recoverable deformation.

### 10.6.3 Performance-based observation

Based on the results of nonlinear pushover analyses, a more comprehensive study of the behavioral characteristics of BRB frame models was conducted to assess their recentering capability in light of the availability of SMA applications. For the chevron BRB frame models (6SBRB-S and 6SBRB-S-SMA), not only inter-story drift values that are the 1.5 and 3.0% roof story drift used as the specific performance levels, but also the residual inter-story drift values obtained after unloading from these roof story drift

*Table 10.6* Summary of applied and residual inter-story drifts for all frame models.

| | | *Inter-Story Drift (%)* | | | | | | *Residual Inter-Story Drift (%)* | | | | | |
|---|---|---|---|---|---|---|---|---|---|---|---|---|---|
| | | *Story* | | | | | | *Story* | | | | | |
| *Model ID* | *V (kN)* | *1* | *2* | *3* | *4* | *5* | *6* | *1* | *2* | *3* | *4* | *5* | *6* |
| (a) Inter-Story Drifts under 1.5% Roof Story Drift and Corresponding Residual Inter-Story Drifts | | | | | | | | | | | | | |
| 6SBRB-S | 7690 | 2.07 | 2.52 | 1.84 | 1.20 | 0.92 | 0.46 | 1.69 | 2.08 | 1.38 | 0.78 | 0.44 | 0.10 |
| 6SBRB-S-SMA | 8017 | 2.04 | 2.52 | 1.77 | 1.13 | 0.92 | 0.62 | 0.23 | 0.19 | 0.09 | 0.04 | 0.06 | 0.07 |
| 6SBRB-V | 7569 | 2.06 | 2.52 | 1.83 | 1.18 | 0.92 | 0.49 | 1.69 | 2.07 | 1.36 | 0.74 | 0.43 | 0.11 |
| 6SBRB-V-SMA | 7898 | 2.04 | 2.52 | 1.76 | 1.10 | 0.92 | 0.66 | 0.27 | 0.25 | 0.12 | 0.06 | 0.08 | 0.11 |
| 6SBRB-2X | 7504 | 2.31 | 2.61 | 1.69 | 1.10 | 0.86 | 0.44 | 1.96 | 2.19 | 1.27 | 0.72 | 0.40 | 0.12 |
| 6SBRB-2X-SMA | 7921 | 2.28 | 2.62 | 1.60 | 1.02 | 0.89 | 0.59 | 0.33 | 0.26 | 0.08 | 0.06 | 0.08 | 0.09 |
| (b) Inter-Story Drifts under 3.0% Roof Story Drift and Corresponding Residual Inter-Story Drifts | | | | | | | | | | | | | |
| 6SBRB-S | 8866 | 4.12 | 5.12 | 4.08 | 2.40 | 1.53 | 0.75 | 3.68 | 4.60 | 3.57 | 1.87 | 0.98 | 0.36 |
| 6SBRB-S-SMA | 9740 | 3.47 | 3.87 | 3.85 | 2.73 | 2.03 | 2.05 | 0.65 | 0.55 | 0.57 | 0.27 | 0.31 | 1.25 |
| 6SBRB-V | 8658 | 4.09 | 5.13 | 4.07 | 2.31 | 1.55 | 0.85 | 3.65 | 4.61 | 3.53 | 1.81 | 0.95 | 0.45 |
| 6SBRB-V-SMA | 9713 | 3.40 | 4.00 | 3.97 | 2.83 | 2.00 | 1.80 | 0.65 | 0.55 | 0.57 | 0.27 | 0.31 | 1.25 |
| 6SBRB-2X | 8591 | 4.69 | 5.47 | 3.67 | 2.01 | 1.41 | 0.75 | 4.30 | 5.00 | 3.21 | 1.57 | 0.87 | 0.41 |
| 6SBRB-2X-SMA | 10017 | 3.41 | 3.99 | 3.88 | 2.76 | 2.01 | 1.95 | 0.83 | 0.81 | 0.67 | 0.29 | 0.30 | 1.18 |

levels, respectively, are presented in Fig. 10.12. The 2% allowable inter-story drift limit (ASCE, 2005; ICC, 2006) is also plotted as a dashed line in this figure.

The maximum inter-story drifts for both roof story drift levels took place in the second story. It can be shown that the plastic deformation was concentrated in the lower stories. The 6SBRB-S-SMA frame model involves smaller maximum inter-story drift than the 6SBRB-S frame model, beyond the 1.5% roof story drift level that belongs to the post-yield range of the pushover curve. This indicates that steel-braced frames are more susceptible to plastic deformation than SMA-braced frames under the strain hardening level. The findings obtained from the nonlinear pushover analyses are consistent with our investigation of the inter-story drift levels. The 6SBRB-S-SMA frame model, in particular, experienced much smaller residual inter-story drifts than the 6SBRB-S frame model. As we expected, the BRB frame models subjected to a 3.0% roof story drift underwent more plastic deformation than those subjected to 1.5% roof story drift, in that the residual inter-story drift is increased after unloading, a phenomenon that is initiated by more story drifts.

More detailed information on the inter-story drifts of all the BRB frame models that were distributed from the first to the sixth story is presented in Table 10.6. To clearly illustrate the potential self-restoring capability of individual BRB frame models on the basis of the analysis results, the recentering ratios, which are defined as the recoverable story drifts divided by the applied story drifts ($(\Delta_u - \Delta_{residual})/\Delta_u$), are additionally presented in Table 10.7. The results summarized in both tables support the findings obtained with reference to the inter-story drift data of the chevron BRB frame models (see Fig. 10.12), while also taking into account those of the other two braced-frame types (i.e., V- and 2X-BRB frame models). Most of all, these results

*Table 10.7* Summary of recentering ratios for all frame models.

| | Recentering Ratio = $100(\Delta_u - \Delta_{residual})/\Delta_u$ (%) | | | | | | |
|---|---|---|---|---|---|---|---|
| *Model ID* | *V(kN)* | *1* | *2* | *3* | *4* | *5* | *6* |
| (a) Recentering Ratios under 1.5% Roof Drift | | | | | | | |
| 6SBRB-S | 7690 | 18.45 | 17.93 | 19.94 | 22.38 | 25.55 | 28.25 |
| 6SBRB-V | 7569 | 18.04 | 18.02 | 20.21 | 22.91 | 26.20 | 29.00 |
| 6SBRB-2X | 7504 | 15.03 | 15.45 | 17.92 | 20.28 | 23.66 | 26.00 |
| **Average** | **7588** | **17.18** | **17.13** | **19.36** | **21.86** | **25.14** | **27.75** |
| 6SBRB-S-SMA | 8017 | 88.82 | 90.72 | 91.97 | 92.69 | 92.82 | 92.50 |
| 6SBRB-V-SMA | 7898 | 86.64 | 88.55 | 89.80 | 90.46 | 90.49 | 90.00 |
| 6SBRB-2X-SMA | 7921 | 85.68 | 88.08 | 89.74 | 90.32 | 90.37 | 90.00 |
| **Average** | **7945** | **87.05** | **89.12** | **90.50** | **91.16** | **91.23** | **90.83** |
| (b) Recentering Ratios under 3.0% Roof Drift | | | | | | | |
| 6SBRB-S | 8866 | 10.64 | 10.26 | 11.13 | 12.58 | 14.68 | 16.38 |
| 6SBRB-V | 8658 | 10.69 | 10.38 | 11.28 | 12.88 | 15.03 | 16.75 |
| 6SBRB-2X | 10017 | 8.36 | 8.45 | 9.70 | 11.18 | 13.34 | 14.75 |
| **Average** | **9180** | **9.90** | **9.70** | **10.70** | **12.21** | **14.35** | **15.96** |
| 6SBRB-S-SMA | 9740 | 81.17 | 83.70 | 84.18 | 85.23 | 85.22 | 80.00 |
| 6SBRB-V-SMA | 9713 | 80.79 | 83.84 | 84.42 | 85.51 | 85.43 | 80.00 |
| 6SBRB-2X-SMA | 10017 | 75.67 | 77.75 | 79.60 | 81.54 | 82.00 | 77.25 |
| **Average** | **9823** | **79.21** | **81.76** | **82.73** | **84.09** | **84.22** | **79.08** |

confirm the assertion that the inter-story drift performance of BRB frame buildings with respect to maximum and residual values mostly depends upon the establishment of SMA recentering devices as displacement control systems, rather than upon the type of braced-frame system that is used. As presented in Table 10.7, the excellent recentering ratios, whose average value exceeds 80%, are distributed over all story levels for the BRB frame models equipped with superelastic SMA bracing systems. The average recentering ratios of the BRB frame models with steel bracing systems are between only 9 and 28%. It can thus be concluded that superelastic SMA bracing systems are much more effective than conventional steel bracing systems in reducing residual deformations during unloading.

## 10.7 NONLINEAR DYNAMIC ANALYSIS RESULTS

For a comparison between the seismic responses of the SMA-braced frames and those of the steel-braced frames, nonlinear dynamic time-history analyses were performed using a total of 40 ground motions mentioned above. A comparison of the time histories for the roof displacement of both frame models, which includes 2 ground motion data that were used, is presented in Fig. 10.13. The LA17 ground motion has a relatively long duration (60 seconds) with a PGA value of 0.569 g and belongs to a seismic hazard level of 10% probability of exceedence in 50 years. Meanwhile, the LA27 ground

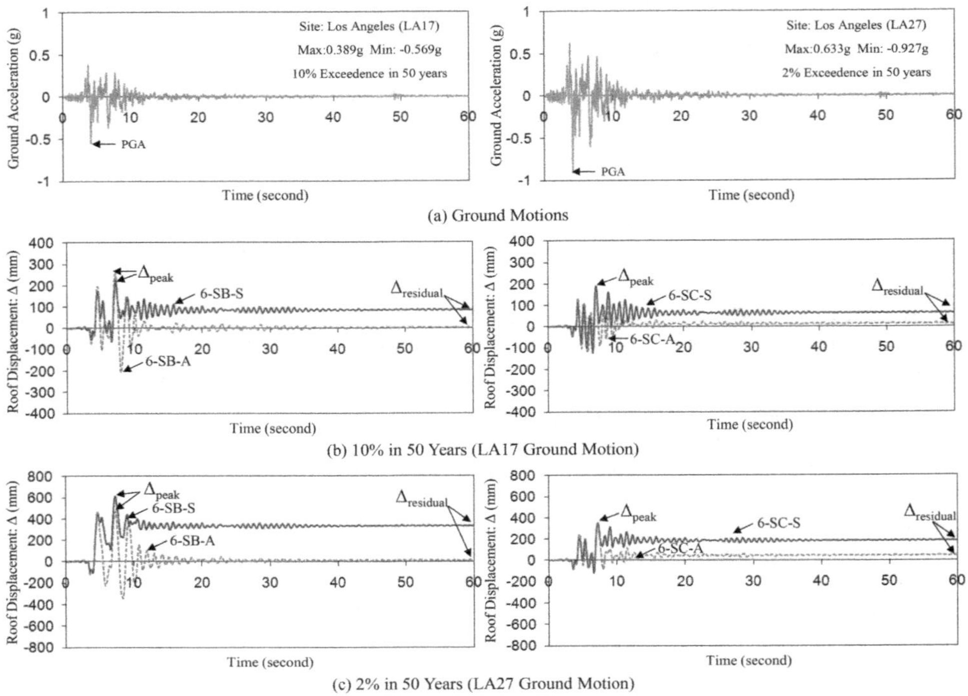


*Figure 10.13* Time vs. roof displacement histories for 6 story sample frame models under 2 scaled ground motions (LA17 and LA27).

motion has a 60-second duration with a PGA value of 0.927 g and belongs to a seismic hazard level of 2% probability of exceedence in 50 years.

For the 10%-probability-in-50-years ground motions, the BRBF models with the superelastic SMA bracing systems (e.g., 6SB-A model), whose behavior was more flexible, generally exhibited larger peak roof displacements than those with the steel bracing systems (i.e., 6SB-S model). For example, the peak roof displacements for the 6SB-A and 6SB-S frame model subjected to the LA17 ground motion were 256 mm and 225 mm, respectively, indicating the peak roof story drifts of 1.08% and 0.94%. The SMA-braced frame models subjected to the LA27 ground motion, which was considered one of the most severe seismic excitations, showed a smaller peak roof displacement than the steel-braced frame models subjected to the identical ground motion (see Fig. 10.13(c)). It can be shown that stress stiffening displayed by the superelastic SMA elements at a high large strain level enabled the SMA-braced frame models to withstand more lateral base shear force after the post-yield state had progressed quite a bit. Under the LA27 ground motion, the peak roof displacements of both BRBF models being compared (6SB-A and 6SB-S model) exceeded the 1.5% roof story drift limit (approximately 360 mm). However, the peak roof displacements for the 6SCA and 6SC-S frame models were 345 mm and 353 mm, respectively, in less than this drift limit. The occurrence time of each peak roof displacement was almost same for both frame models being compared under the same ground motion condition, and lagged behind the PGA time by about 2 to 3 seconds.

The SMA's recentering capability decreased the residual roof displacement. After the LA27 ground motion, the residual roof displacements for the 6SC-A and 6SC-S frame model were 43 mm and 185 mm, respectively. Furthermore, the residual roof displacement of the 6SB-S frame model was 330 mm, as much as 37 times larger than that of the 6SB-A frame model. Therefore, the ability of the superelastic SMA bracing systems to restore a braced frame to its original condition can be expected to considerably mitigate the permanent deformation of the entire frame building. Their excellent damping properties also help to quickly reduce the amplitude of deformations during ground motions, as well.

## 10.8 MAXIMUM INTER-STORY DRIFTS AND RESIDUAL INTER-STORY DRIFTS

According to the availability of SMA application, the recentering capability of individual frame models can be accessed through the investigation of maximum inter-story drifts and residual inter-story drifts after nonlinear dynamic time-history analyses. Average maximum inter-story drifts and residual inter-story drifts for 6 story chevron type-braced frame models under two sets of ground motions (each seismic hazard level) are presented in Fig. 10.14. The allowable inter-story drift limit (2%) stipulated in the

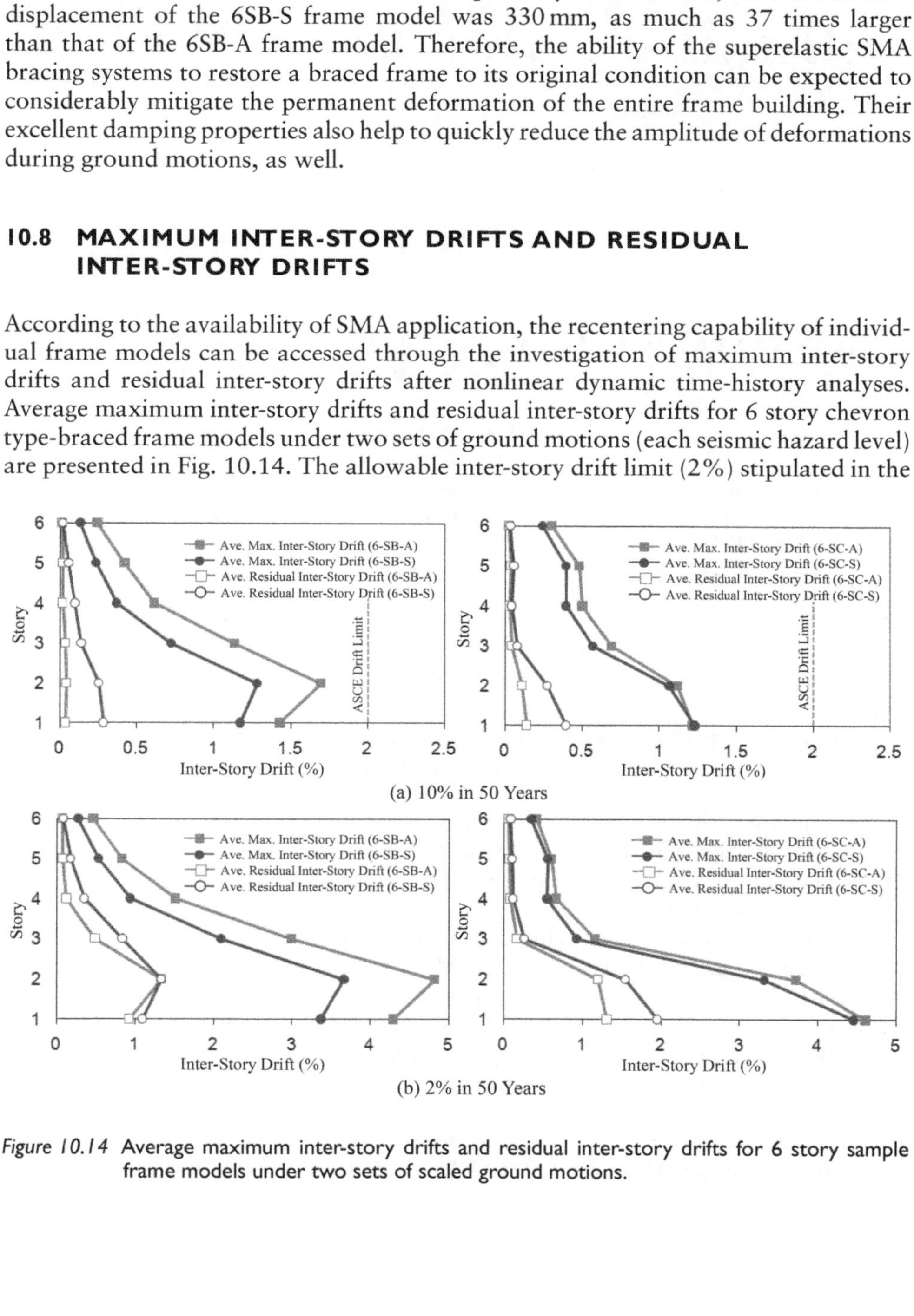


*Figure 10.14* Average maximum inter-story drifts and residual inter-story drifts for 6 story sample frame models under two sets of scaled ground motions.

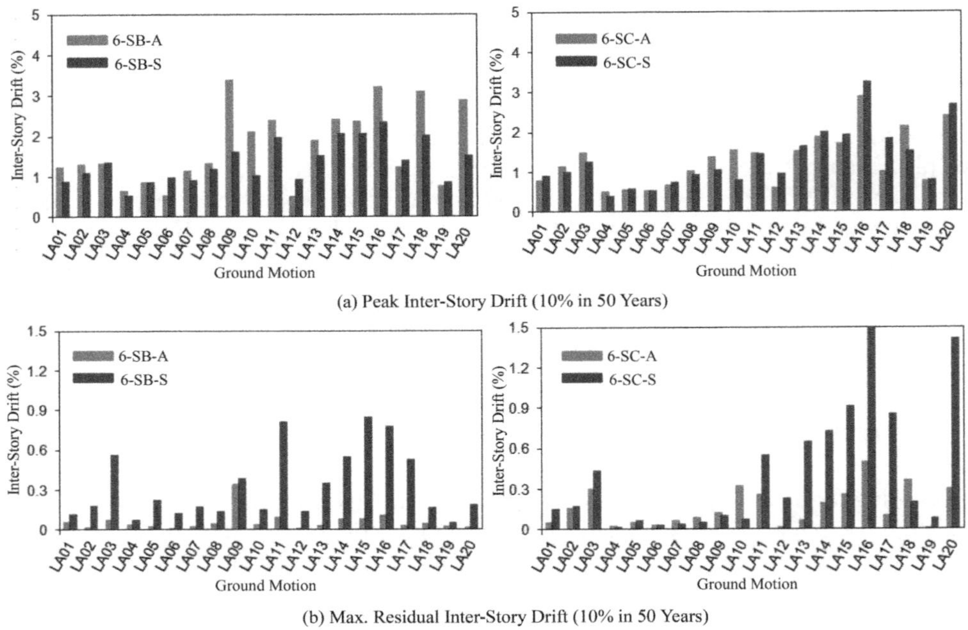


*Figure 10.15* Investigation of peak and maximum-residual inter-story drifts for 6 story sample frame models subjected to a set of design-based earthquake ground motions (DBE, 10% in 50 years).

ASCE 7-05 is also plotted as the dashed line in Fig. 10.14(a). The average maximum inter-story drifts are below this allowable limit over the whole story level. Therefore, four frame models subjected to the DLE ground motions can meet the design requirement.

The peak inter-story drifts generally occurred at first or second story levels, meaning that the plastic deformation of frame members concentrated on the lower stories during the peak motion. As we expected, the frame models subjected to 2% in 50 year ground motions suffered more damages than those subjected to 10% in 50 year ground motions in that the residual inter-story drifts were increased. It can be observed that the residual inter-story drifts initiated by more maximum inter-story drifts were also increased after seismic events.

The inter-story drifts are highly dependent on the used bracing systems. Owing to more flexible-initial behavior of the SMA materials, the frame models with the superelastic SMA bracing systems showed larger maximum inter-story drifts than those with steel bracing system throughout all story levels. However, the former model cases had smaller average residual inter-story drifts than the latter ones. The 6-SB-A model under 10% in 50 year ground motions almost recovered original conditions, especially, seeing that it had average residual inter-story drifts very close to zero.

Fig. 10.15 shows peak and maximum-residual inter-story drifts for 6 story sample frame models subjected to a set of DBE (10% in 50 years) ground motions. According

*Table 10.8* Average and standard deviation (between brackets) of maximum inter-story drifts and residual inter-story drifts for all 6 story frame models under two sets of scaled ground motions.

| | *Maximum Inter-Story Drift (%)* | | | | | | *Residual Inter-Story Drift (%)* | | | | | |
|---|---|---|---|---|---|---|---|---|---|---|---|---|
| | Story | | | | | | Story | | | | | |
| *Model ID* | *1* | 2 | 3 | 4 | 5 | 6 | *1* | 2 | 3 | 4 | 5 | 6 |
| (a) 10% in 50 Years (6F). | | | | | | | | | | | | |
| 6-SB-S | 1.17 (0.48) | **1.28** (0.49) | 0.73 (0.31) | 0.37 (0.18) | 0.24 (0.10) | 0.14 (0.05) | **0.29** (0.26) | 0.26 (0.22) | 0.14 (0.11) | 0.10 (0.11) | 0.06 (0.07) | 0.02 (0.03) |
| 6-SB-A | 1.43 (0.84) | **1.70** (0.91) | 1.14 (0.63) | 0.61 (0.23) | 0.42 (0.12) | 0.25 (0.05) | **0.04** (0.05) | 0.04 (0.07) | 0.04 (0.05) | 0.02 (0.01) | 0.02 (0.01) | 0.01 (0.01) |
| 6-VB-S | 1.17 (0.47) | **1.26** (0.51) | 0.73 (0.29) | 0.35 (0.12) | 0.25 (0.08) | 0.16 (0.05) | **0.30** (0.27) | 0.28 (0.24) | 0.14 (0.10) | 0.10 (0.10) | 0.06 (0.06) | 0.02 (0.03) |
| 6-VB-A | 1.48 (0.85) | **1.70** (0.94) | 1.06 (0.48) | 0.60 (0.20) | 0.44 (0.13) | 0.27 (0.07) | 0.03 (0.03) | **0.04** (0.04) | 0.04 (0.03) | 0.02 (0.02) | 0.02 (0.01) | 0.02 (0.01) |
| 6-XB-S | 1.39 (0.76) | **1.41** (0.73) | 0.74 (0.33) | 0.35 (0.18) | 0.25 (0.13) | 0.15 (0.08) | **0.28** (0.32) | 0.21 (0.26) | 0.10 (0.09) | 0.08 (0.10) | 0.05 (0.06) | 0.02 (0.03) |
| 6-XB-A | 1.61 (0.97) | **1.73** (0.96) | 0.97 (0.43) | 0.54 (0.18) | 0.39 (0.11) | 0.22 (0.06) | 0.03 (0.04) | **0.04** (0.06) | 0.03 (0.03) | 0.02 (0.01) | 0.02 (0.01) | 0.01 (0.01) |
| 6-SC-S | **1.23** (0.73) | 1.07 (0.51) | 0.57 (0.16) | 0.39 (0.07) | 0.40 (0.09) | 0.25 (0.06) | **0.40** (0.47) | 0.28 (0.29) | 0.08 (0.08) | 0.04 (0.05) | 0.06 (0.06) | 0.03 (0.03) |
| 6-SC-A | **1.22** (0.66) | 1.12 (0.54) | 0.69 (0.20) | 0.50 (0.10) | 0.48 (0.11) | 0.30 (0.07) | **0.14** (0.13) | 0.11 (0.12) | 0.05 (0.04) | 0.04 (0.04) | 0.04 (0.04) | 0.03 (0.03) |
| 6-VC-S | **1.17** (0.68) | 1.13 (0.58) | 0.55 (0.13) | 0.43 (0.08) | 0.42 (0.10) | 0.30 (0.07) | **0.37** (0.43) | 0.32 (0.34) | 0.07 (0.06) | 0.06 (0.05) | 0.07 (0.06) | 0.05 (0.05) |
| 6-VC-A | **1.12** (0.71) | 1.05 (0.58) | 0.66 (0.24) | 0.51 (0.16) | 0.49 (0.17) | 0.33 (0.11) | **0.12** (0.13) | 0.09 (0.12) | 0.05 (0.05) | 0.04 (0.04) | 0.04 (0.04) | 0.03 (0.03) |
| 6-XC-S | **1.16** (0.68) | 1.06 (0.50) | 0.53 (0.14) | 0.36 (0.06) | 0.37 (0.08) | 0.22 (0.05) | **0.37** (0.43) | 0.28 (0.28) | 0.08 (0.08) | 0.04 (0.04) | 0.06 (0.06) | 0.03 (0.03) |
| 6-XC-A | **1.16** (0.63) | 1.14 (0.57) | 0.66 (0.19) | 0.47 (0.10) | 0.47 (0.11) | 0.29 (0.07) | **0.14** (0.13) | 0.12 (0.13) | 0.06 (0.05) | 0.04 (0.04) | 0.04 (0.04) | 0.03 (0.03) |
| (b) 2% in 50 Years (6F). | | | | | | | | | | | | |
| 6-SB-S | 3.38 (1.73) | **3.67** (2.21) | 2.10 (1.68) | 0.95 (0.62) | 0.54 (0.28) | 0.28 (0.14) | 1.10 (1.17) | **1.34** (1.50) | 0.85 (1.08) | 0.36 (0.43) | 0.18 (0.18) | 0.09 (0.08) |
| 6-SB-A | 4.30 (1.85) | **4.82** (2.51) | 3.00 (1.50) | 1.51 (1.04) | 0.84 (0.44) | 0.47 (0.20) | 0.93 (1.40) | **1.34** (1.91) | 0.50 (0.77) | 0.13 (0.18) | 0.08 (0.08) | 0.07 (0.07) |
| 6-VB-S | 3.26 (1.73) | **3.80** (2.12) | 2.19 (1.59) | 1.00 (0.56) | 0.55 (0.27) | 0.34 (0.15) | 1.16 (1.14) | **1.42** (1.49) | 0.89 (1.07) | 0.37 (0.43) | 0.19 (0.20) | 0.11 (0.11) |
| 6-VB-A | 4.09 (2.03) | **4.68** (2.54) | 2.86 (1.53) | 1.45 (0.96) | 0.87 (0.48) | 0.53 (0.24) | 0.78 (1.37) | **1.17** (1.88) | 0.44 (0.67) | 0.16 (0.22) | 0.12 (0.12) | 0.10 (0.11) |
| 6-XB-S | 3.64 (1.97) | **3.92** (2.26) | 1.87 (1.21) | 0.80 (0.43) | 0.45 (0.22) | 0.27 (0.14) | 1.41 (1.40) | **1.63** (1.70) | 0.69 (0.86) | 0.24 (0.26) | 0.14 (0.14) | 0.10 (0.09) |
| 6-XB-A | 4.85 (2.64) | **5.44** (3.29) | 2.72 (1.27) | 1.30 (0.71) | 0.81 (0.40) | 0.49 (0.23) | 1.20 (1.72) | **1.58** (2.33) | 0.34 (0.35) | 0.11 (0.13) | 0.12 (0.12) | 0.10 (0.10) |
| 6-SC-S | **4.46** (2.93) | 3.32 (2.46) | 0.94 (0.32) | 0.56 (0.15) | 0.56 (0.16) | 0.36 (0.12) | **1.96** (1.78) | 1.55 (1.48) | 0.27 (0.29) | 0.12 (0.10) | 0.11 (0.08) | 0.09 (0.08) |
| 6-SC-A | **4.62** (3.14) | 3.73 (2.97) | 1.17 (0.75) | 0.67 (0.22) | 0.61 (0.17) | 0.42 (0.16) | **1.32** (1.85) | 1.21 (1.75) | 0.17 (0.25) | 0.08 (0.09) | 0.08 (0.08) | 0.07 (0.07) |
| 6-VC-S | **4.03** (2.61) | 3.84 (2.95) | 0.98 (0.50) | 0.56 (0.12) | 0.63 (0.19) | 0.45 (0.16) | **1.80** (1.64) | 1.80 (1.71) | 0.27 (0.28) | 0.09 (0.08) | 0.13 (0.10) | 0.11 (0.09) |
| 6-VC-A | **4.61** (3.14) | 3.80 (3.26) | 1.24 (0.95) | 0.74 (0.32) | 0.67 (0.26) | 0.49 (0.24) | **1.37** (1.94) | 1.27 (1.91) | 0.21 (0.30) | 0.12 (0.13) | 0.11 (0.12) | 0.09 (0.11) |
| 6-XC-S | **3.91** (2.93) | 3.22 (2.59) | 0.87 (0.39) | 0.50 (0.19) | 0.53 (0.20) | 0.33 (0.14) | **1.74** (1.66) | 1.50 (1.50) | 0.24 (0.25) | 0.11 (0.10) | 0.11 (0.09) | 0.09 (0.09) |
| 6-XC-A | **4.35** (2.93) | 3.97 (3.11) | 1.21 (0.80) | 0.67 (0.25) | 0.63 (0.20) | 0.43 (0.18) | **1.03** (1.59) | 1.02 (1.61) | 0.16 (0.25) | 0.08 (0.10) | 0.09 (0.10) | 0.07 (0.09) |

to each earthquake record, seismic performance and resulting damage can be investigated in this figure. The average values of these provided data are also estimated by findings to Fig. 10.14(a) (see also Table 10.8). The average peak inter-story drifts for the 6-SB-S and 6-SC-S frame models were 1.28% and 1.23%, while those for the 6-SB-A and 6-SC-A frame models were more or almost equal, 1.70% and 1.22%, respectively. The use of the superelastic SMA bracing systems to the braced frame incurred a little increase in the maximum inter-story drifts. Nevertheless, the superelastic SMA braces reduced the residual inter-story drifts by approximately 88% and 65% for the 6-SB-A and 6-SC-A frame models, respectively, as compared to the conventional steel bracing system. The average maximum-residual inter-story drifts for the 6-SB-S and 6-SC-S frame models were 0.29% and 0.40%, respectively, while SMA-braced frame models had the values of only 0.04% and 0.14% for the 6-SB-A and 6-SC-A frame models, respectively. This suggests that the ability of the superelastic SMA bracing systems to provide recentering effect leads to smaller residual inter-story drifts in a braced frame structure.

## 10.9 SEISMIC EVALUATIONS

In the following, the peak and residual responses were investigated to examine the dynamic performance of the frame models. For all frame models, the average and standard deviation values of maximum inter-story drifts and residual inter-story drifts with respect to the seismic hazard level are summarized in Tables 10.7 and 10.8. It is important to note that the frame models undergoing 2% in 50 year ground motions showed larger maximum inter-story drifts and residual inter-story drifts than those subjected to 10% in 50 year ground motions. For this reason, the scatters defined as average inter-story drifts plus one standard deviation values are expected to be larger as seismic hazard increases. A typical value of the maximum-residual inter-story drift for 6 story frame models was around 0.04% to 0.40% under the DLE seismic hazard level. This value increased to about 1.17% to 1.96% under the MCE seismic hazard level. The scatters were also larger as the building height increased (see also Fig. 10.17).

In all instances, the average residual inter-story drifts decreased with the use of the SMA bracing systems regardless of the bracing types used. Even though the SMA bracing system provided entire frame structures with recentering capability, the difference in the average residual inter-story drifts of SMA-braced frames and steel-braced frames had a tendency to decrease with the increase in seismic hazard and building height (see Tables 10.7 and 10.8). Moreover, the SMA bracing systems used in the BRBF buildings were relatively more efficient at reducing the residual inter-story drifts than those installed in the CBF buildings. Not only the buckling behavior of CB members in compression but also the onset of brace fractures in tension enabled the CBF model cases to have somewhat larger residual inter-story drifts than the BRBF model cases after analyses.

As mentioned above, the residual inter-story drifts that are greater than 0.5% represent significant damage to the frame system and consequently the building could be rendered useless by a complete loss of the structure. It is necessary to examine how many ground motion records induced the maximum-residual inter-story drift

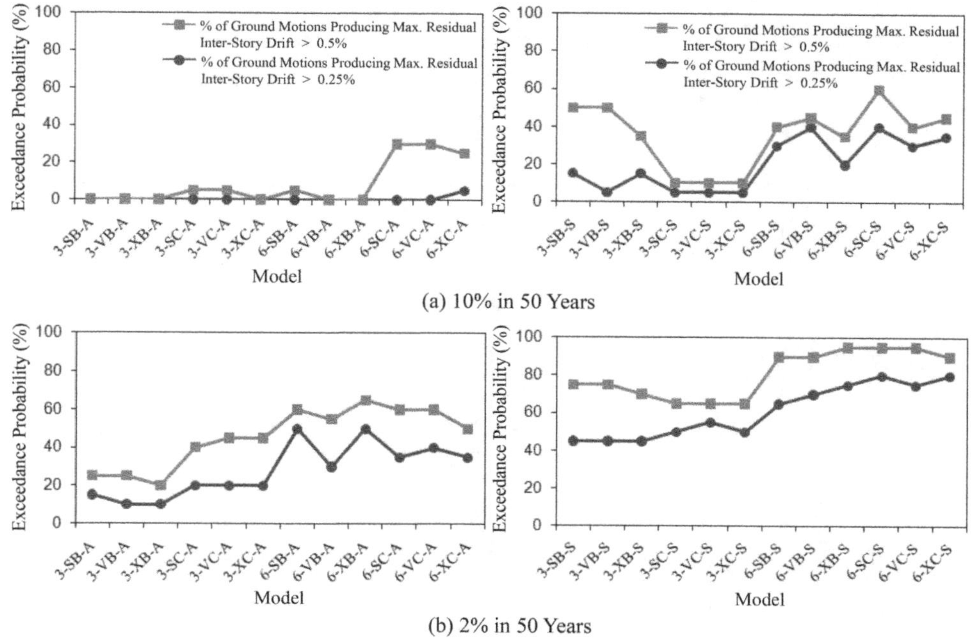


*Figure 10.16* Percentages of scaled ground motions producing a maximum-residual inter-story drift greater than 0.5% and 0.25%, respectively, for individual frame models.

exceeding the value of 0.5% on that account. For individual frame models, the percentages of scaled ground motions producing a maximum-residual inter-story drift greater than 0.5% and 0.25%, respectively, are given in Fig. 10.16. The average percentages of the ground motions that induce the maximum-residual inter-story drift exceeding the value of 0.5% are 33% and 58% for 3 and 6 story steel-braced frame models under the DLE seismic hazard level, respectively, and 70% and 93% for 3 and 6 story steel-braced frame models under the MCE seismic hazard level, respectively. On the other hand, these percentages rapidly decrease to 1.7% and 15% for 3 and 6 story SMA-braced frame models under the DLE seismic hazard level, respectively, and also decrease to 27% and 44% for 3 and 6 story SMA-braced frame models under the MCE seismic hazard level, respectively. It can be shown that the use of the SMA bracing system makes a significant contribution to a decrease in the repair cost by mitigating the permanent deformations. Meanwhile, the steel-braced frame models exhibited a maximum-residual inter-story drift that was much greater than 0.5%, which would therefore require the reconstruction of the entire building structure rather than the repair of only the damaged parts.

The maximum recoverable inter-story drifts attained by subtracting the residual inter-story drift from the peak inter-story drift are needed as well with a view to certifying that the SMA bracing system supplies excellent recentering capability to braced frames. The recentering ratios for individual frame models subjected to two sets of

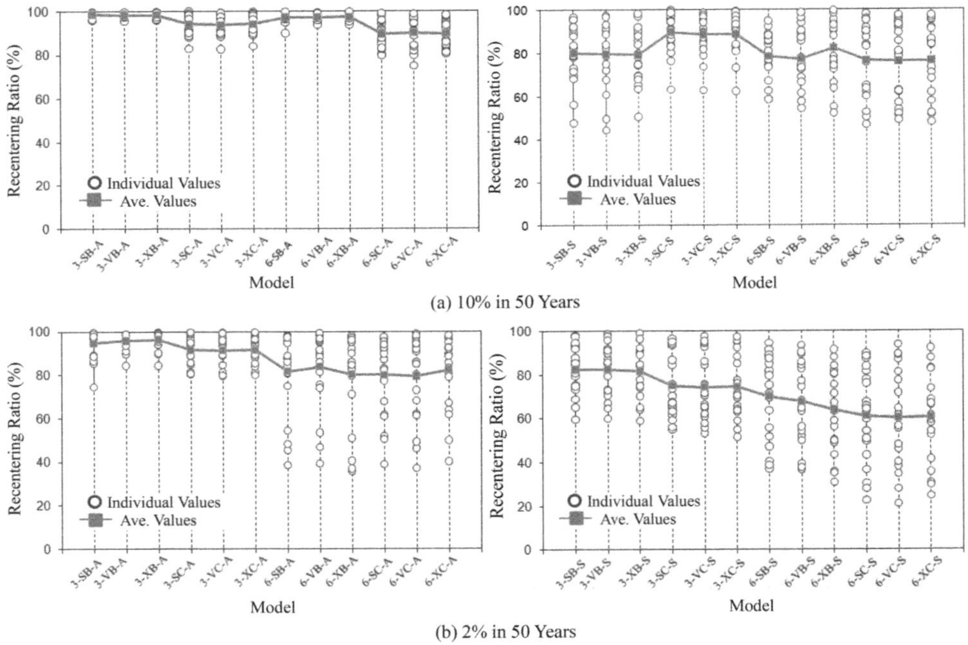


*Figure 10.17* Investigation of recentering ratios for individual frame models subjected to two sets of scaled ground motions.

ground motions are presented as circular symbols in Fig. 10.17. The corresponding average values are presented as solid lines in this figure. The recentering ratios herein were converted from the maximum recoverable inter-story drifts divided by the corresponding peak inter-story drifts. On average, the frame models with the SMA bracing systems had higher recentering ratios than those with the steel bracing systems. Above all, the 3 story BRBF models with the SMA bracing systems, when subjected to 10% in 50 year ground motions, had average recentering ratios that reached as high as 98%, which indicates that they fully reassumed their original shapes. The individual points scattered further away from average values when the residual inter-story drifts were low.

For another seismic evolution, the damage to BRB members subjected to earthquake loads should be investigated in the same manner by tracing the plastic stress hinge sequence. The plastic stress hinges shown in Figs. 10.18 and 1.19 were detected when both BRBF models being compared (6SBRB-S and 6SBRB-S-SMA) encountered the peak roof story drift during nonlinear dynamic analyses performed with long-period ground motions (LA17 and LA27). As can be seen in Figs. 10.18(b) and 10.19(b), the highest base shear demand on the BRBF model and the peak roof displacement took place at the same time. The stress contour levels were divided into 5 equal sections, based on the nominal strength of base Gr. 50 steel or the superelastic SMA materials in

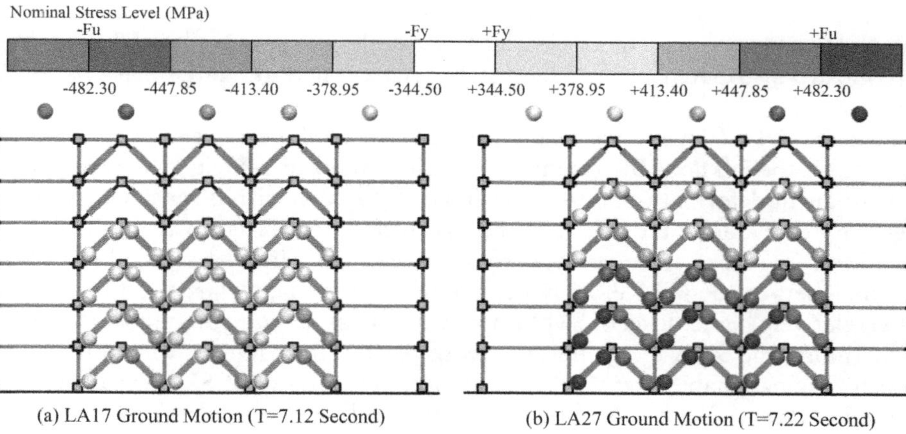


*Figure 10.18* Investigation of plastic stress hinges at peak roof drift (6SBRB-S frame model).

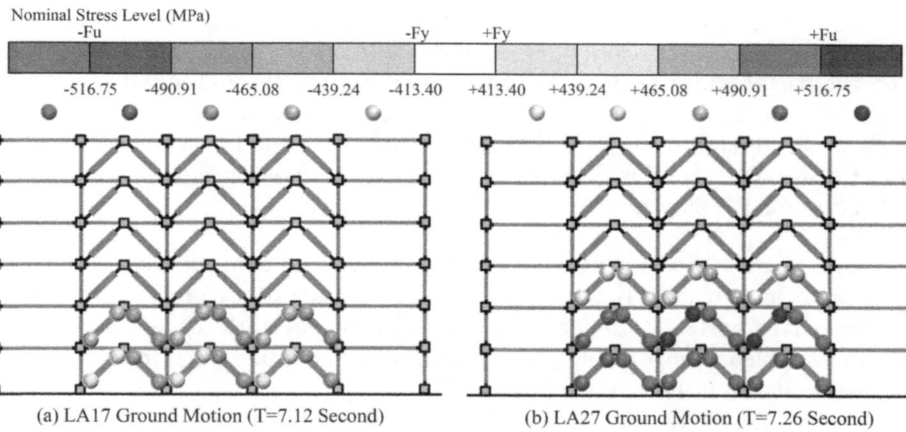


*Figure 10.19* Investigation of plastic stress hinges at peak roof drift (6SBRB-S-SMA frame model).

the BRB members. The extent of the plastic stress hinge, which is depicted as colored solid circles, was determined by examining the fiber stress measured at the integration point.

In the BRB member, the composite segments surrounded by the steel casing and mortar remained in elastic states without buckling. Instead, plastic deformations were concentrated in the extended end connectors that have relatively smaller cross-cross sections. Due to the brace configuration, two pairs of plastic stress hinges consisting of compression and tension were arranged to face each other across the center of each braced bay on both BRB members. A strong ground motion (LA27) induced more plastic stress hinges and more severe failures in the BRBF models than did a weak ground motion (LA17). Overall, the severe plastic failure hinges were mostly distributed at the lower story levels, where the large deformation was concentrated.

The 6SBRB-S frame model under the LA27 ground motion showed plastic stress hinges up to the 5th floor BRB members. The loss of brace capacity due to yielding and the permanent deformation of the steel end connectors incurred considerable residual inter-story drifts along the height of the structure. For the case of the 6SBRB-S-SMA frame model, yielding spread only up to the level of the third story. The ability of more BRB members in the 6SBRB-S-SMA frame model to remain elastic can be attributed to the relatively higher level of SMA's post-yield stress, as compared to steel materials. Furthermore, the SMA end connectors with plastic stress hinges were able to return to their original shape after a ground motion. The ability of SMA bracing systems to provide recentering can lead to smaller residual inter-story drifts in a BRB frame structure.

## 10.10 SUMMARY AND CONCLUSIONS

The seismic responses of low-to-mid rise frame buildings with steel or superelastic SMA bracing systems are mainly treated in this study. The effect of superelastic behavior on the performance of braced frames was evaluated through nonlinear dynamic analyses. On the basis of analytical results, the performance of the SMA-braced frame models was compared to that of the steel-braced frame models with respect to roof story displacements, residual inter-story drifts, and recentering ratios. The additional conclusions are as follows:

(1) Superelastic SMAs can recover substantial inelastic deformations without heat treatment only after removing stress. Their unique behavior characterized by a flag-shape hysteresis under cyclic loading provides both recentering capability and supplemental damping for the entire frame structure in case such smart materials are utilized in the braces.
(2) Meanwhile, the yielding of steel elements in the conventional steel bracing system contributes to the permanent deformations seen in the frame models. Thus, the use of the superelastic SMA materials in the bracing member resulted in considerably smaller residual inter-story drift values for two levels of seismic hazard.
(3) The pushover analyses show that the BRB frames were overdesigned to accommodate the ASCE 7-05 design drift limit. The pushover behavior of the BRB frame models, on average, was significantly improved by the use of superelastic SMA bracing systems, in terms of flexible initial slope, relatively large post-yield strength, and small permanent deformation.
(4) By virtue of the recentering system, the SMA-braced frame models proposed in this study maintained maximum inter-story drift values of less than 0.5%, which represents the upper limit at which rehabilitation is economically feasible, even after strong ground motion.

(5) Finally, our analysis results are promising for the application of the SMA braced frames, which feature moderate energy dissipation and excellent recentering capability. In the future, further work related to the experimental tests of the actual SMA bracing systems is still necessary to guarantee the analytical results with certainty.

Appendix A

# A detailed design example for CFT columns

## A.1 INTRODUCTION

The detail calculation procedures of the P-M interaction diagram for CFT beam-columns in accordance with the *AISC 2005 Specifications* are presented in this Appendix A. The basic backgrounds including the necessary notations and calculation procedures were described in Chapter 2. The calculation examples for CFT columns used in this research are summarized in the A.2 section with case-by-case procedures. The CFT column details are summarized in Table A.1. In addition, P-M interaction diagram interpolated with five points are described in A.3 section. The necessary equations to calculate these points are summarized in Table A.2 and Table A.3.

## A.2 CALCULATION EXAMPLES

### A.2.1 RCFT 16 x 16 x 500 column

Develop an axial load-moment (P-M) envelope for a HSS 16 × 16 × 500 concrete filled with $f'_c = 27.56$ MPa strength. The effective length of the member is 3812.5 mm. Assume A572 steel (Swanson 2002, Use the nominal stress value for $F_y = 378.95$ MPa and $F_u = 502.97$ MPa). The unit system is Newton and meter. Use the dimensions shown in Table A.1.

*Table A.1* Summary of the CFT columns.

| *Column ID* | *Steel section* | *d* | *T* | $F_y$ | $f'_c$ |
|---|---|---|---|---|---|
| RCFT 16 × 16 × 500 | HSS 16 × 16 × 500 | 406.4 | 12.7 | 378.95 | 27.56 |
| CCFT 18 × 500 | HSS 18 × 500 | 457.2 | 12.7 | 378.95 | 27.56 |
| RCFT 12 × 12 × 500 | HSS 12 × 12 × 500 | 304.8 | 12.7 | 378.95 | 27.56 |
| CCFT 14 × 500 | HSS 14 × 500 | 355.6 | 12.7 | 378.95 | 27.56 |
| RCFT 16 × 16 × 375 | HSS 16 × 16 × 375 | 406.4 | 9.525 | 378.95 | 27.56 |
| CCFT 18 × 375 | HSS 18 × 375 | 457.2 | 9.525 | 378.95 | 27.56 |
| RCFT 14 × 14 × 500 | HSS 14 × 14 × 500 | 355.6 | 12.7 | 378.95 | 27.56 |
| CCFT 16 × 500 | HSS 16 × 500 | 406.4 | 12.7 | 378.95 | 27.56 |

*Table A.2* Equations for the specific 5 points in the P-M interaction diagram (CCFT column).

Plastic capacities for composite field HSS about any axis

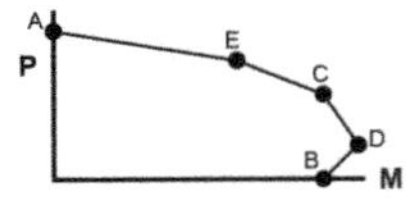


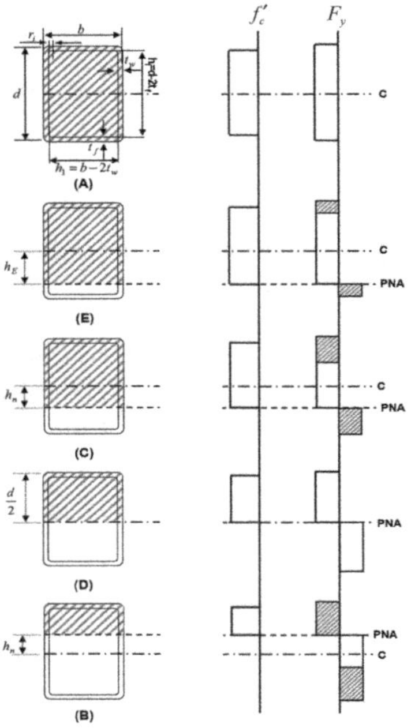


| Section | Stress distribution | Point | Defining equations |
|---|---|---|---|
| (A) | | A | $P_A = A_sF_y + 0.85f'_cA^*_c \quad M_A = 0$<br>$A_s$ = Area of steel shape $\quad A_c = h_1h_2 - 0.858r_i^2$ |
| (E) | | E | $P_E = \frac{1}{2}(0.85f'_cA_c) + 0.85f'_ch_1h_E + 4F_yt_wh_E$<br>$M_E = M_{max} - \Delta M_E \quad \Delta M_E = Z_{sE}F_y + \frac{1}{2}Z_{cE}(0.85f'_c)$<br>$Z_{sE} = bh_E^2 - Z_{cE} \quad Z_{cE} = h_1h_E^2 \quad h_E = \frac{h_n}{s} + \frac{d}{4}$ |
| (C) | | C | $P_C = 0.85f'_cA_c \quad M_c = M_B$ |
| (D) | | D | $P_D = \frac{0.85f'_cA_c}{2} \quad M_D = Z_sF_y + \frac{1}{2}Z_c(0.85f'_c)$<br>$Z_s = \frac{bd^2}{4} \quad Z_c = \frac{h_1h_2^2}{4} - 0.192r_i^3$ |
| (B) | | B | $P_B = 0 \quad M_B = M_D - Z_{sn}F_y - \frac{1}{2}Z_{cn}(0.85f'_c)$<br>$Z_{sB} = 2t_wh_n^2 \quad Z_{cB} = h_1h_n^2$<br>$h_n = \frac{0.85f'_cA_c}{2}(0.85f'_cA_ch_1 + 4t_wF_y) \le \frac{h_2}{2}$ |

**Limitation:**

1) The cross-sectional area of the steel core shall comprise at least one percent of the total composite cross section.

$$A_s = 19995\,\text{mm}^2 > (0.01)(406.4^2) = 1651.61\,\text{mm}^2\ \text{(OK)} \tag{A.1}$$

Note that $\rho = \frac{19995}{165161} = 0.121$, or 12.1% which is high ratio.

2) The slenderness of the tube wall is:

$$\left(\frac{t}{b}\right) = \frac{406.4 - 12.7}{12.7} = 31.0 \le 2.26\sqrt{\frac{E}{F_y}} = 2.26\sqrt{\frac{199810}{378.95}} = 51.90\ \text{(OK)} \tag{A.2}$$

*Table A.3* Equations for the specific 5 points in the P-M interaction diagram (CCFT column).

Plastic capacities for composite
Field round HSS bent about any axis

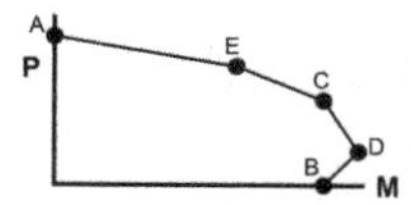


| *Section* | *Stress distribution* | *Point* | *Defining equations* |
|---|---|---|---|
| 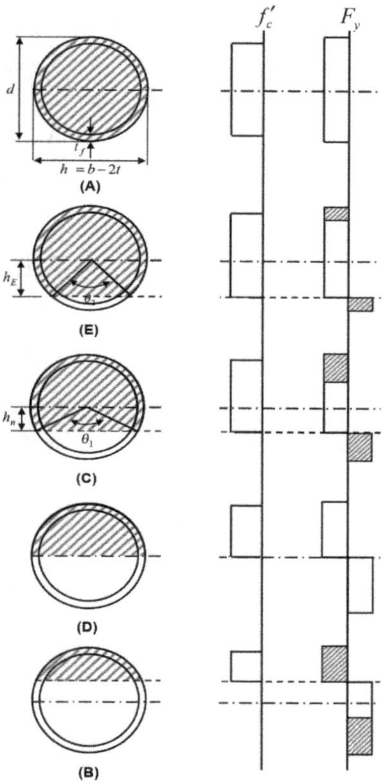 | | A | $P_A = A_sF_y + 0.85f'_cA^*_c \quad M_A = 0$<br>$A_c = \pi h^2/4 \quad A_s = \pi r_m t \quad r_m = d - t/2$<br>$P_A = A_sF_y + 0.95f'_cA_c$ (Loaded in only axial compression) |
| | | E | $P_E = (0.85f'_cA_c + F_yA_s) - \frac{1}{2}\left[F_y(d^2 - h^2) + \frac{1}{2}(0.85f'_c)h^2\right]\left[\frac{\theta_2}{2} - \sin\frac{\theta_2}{2}\right]$<br>$M_E = Z_{sE}F_y + \frac{1}{2}Z_{cE}(0.85f'_c)$<br>$h_E = \frac{h_n}{2} + \frac{d}{4} \quad Z_{sE} = \frac{d^3\sin^3\left(\frac{\theta_2}{2}\right)}{6 - Z_{cE}}$<br>$Z_{cE} = \frac{h^3\sin^3\left(\frac{\theta_2}{2}\right)}{6}$<br>$\theta_2 = \pi - 2\arcsin\left(\frac{2h_E}{h}\right)$ |
| | | C | $P_C = 0.85f'_cA_c \quad M_C = M_B$ |
| | | D | $P_D = \frac{0.85f'_cA_c}{2} \quad M_D = Z_sF_y + \frac{1}{2}Z_c(0.85f'_c)$<br>$Z_s = \frac{d^3}{6} - Z_c \quad Z_c = \frac{h^3}{6}$ |
| | | B | $P_B = 0 \quad M_B = Z_{sB}F_y + \frac{1}{2}Z_{cB}(0.85f'_c)$<br>$Z_{sB} = \frac{d^3\sin\left(\frac{\theta_1}{2}\right)}{6 - Z_B} \quad Z_{cB} = \frac{h^3\sin\left(\frac{\theta_1}{2}\right)}{6}$<br>$\theta_1 = \frac{0.026K_c - 2K_s}{0.0848K_c} + \frac{\sqrt{0.026K_c + 2K_s^2 + 0.857K_cK_s}}{0.0848K_c}$<br>$K_c = f'_ch_2 \quad K_s = F_yr_mt \quad h_n = \frac{h}{2}\sin\left(\frac{\pi - \theta}{2}\right) \le \frac{h}{2}$ |

### A.2.1.1 Point A ($M_A = 0$)

Determine the available compressive strength

$$P_o = A_sF_y + A_{sr}F_{yr} + 0.85A_cf'_c \tag{A.3}$$

$$A_c = (b - 2 \cdot t)^2 = (406.4 - 2 \cdot 12.7)^2 = 145125\,\text{mm}^2 \tag{A.4}$$

$$\begin{aligned} P_o &= A_sF_y + A_{sr}F_{yr} + 0.85A_cf'_c \\ &= 19995 \cdot 378.95 + 0 + 0.85 \cdot 145125 \cdot 27.56 \end{aligned} \tag{A.5}$$

$$P_o = 10991.5\,\text{kN} \tag{A.6}$$

$$C_3 = 0.6 + 2\left(\frac{A_s}{A_c + A_s}\right) = 0.6 + 2\left(\frac{19995}{145125 + 19995}\right) = 0.87 \tag{A.7}$$

$$I_s = \frac{d^4}{12} - \frac{(d - 2 \cdot t)^4}{12} = 516672 \cdot 10^3\ \text{mm}^4 \tag{A.8}$$

$$I_c = \frac{(d - 2 \cdot t)^4}{12} = 1755104 \cdot 10^3\ \text{mm}^4 \tag{A.9}$$

$$E_c = 57000\sqrt{f_c'} = 57000\sqrt{27560000}/1000 = 24838.450\ \text{MPa} \tag{A.10}$$

$$EI_{eff} = E_sI_s + 0.5E_sI_{sr} + C_3E_cI_c \tag{A.11}$$

$$\begin{aligned} EI_{eff} &= 199810 \cdot 517 \cdot 10^6 + 0.87 \cdot 24838.1755 \cdot 10^6 \\ &= 1.413 \cdot 10^{14}\ \text{N-mm}^2 \end{aligned} \tag{A.12}$$

$$P_e = \frac{\pi EI_{eff}}{(kL)^2} = \frac{(3.14159)^2(1.413 \cdot 10^{14})}{(1 \cdot 3810)^2} = 96088.85\ \text{kN} \tag{A.13}$$

$$\frac{P_o}{P_e} = \frac{10991.5}{96088.85} = 0.114 \leq 2.25 \quad \text{or} \quad \frac{P_e}{P_o} = \frac{96088.85}{10991.5} = 8.74 \geq 0.44 \tag{A.14}$$

∴ Use Eq. 2.1 in Chapter 2

$$P_n = P_o\left(0.658^{\left(\frac{P_O}{P_e}\right)}\right) = 10991.5(0.658^{0.114}) = 10479.75\ \text{kN} \tag{A.15}$$

$$\phi_c P_n = 0.75(10479.75) = 7858.7\ \text{kN} \tag{A.16}$$

#### A.2.1.2 Point B ($P_B = 0$)

Determine location of $h_n$

$$h_n = \frac{0.85f_c'A_c}{2(0.85f_c'h_1 + 4t_wF_y)} \leq \frac{h_2}{2}\ \text{in}^3 \tag{A.17}$$

$$\begin{aligned} h_n &= \frac{0.85 \cdot 27.56 \cdot 145125}{2(0.85 \cdot 27 \cdot (406 - 2 \cdot 12.7) + 4 \cdot 12.7 \cdot 378)} \\ &= 60\ \text{mm} \leq \frac{406}{2} = 203\ \text{mm (OK)} \end{aligned} \tag{A.18}$$

$$Z_s = 2968400\ \text{mm}^3 \tag{A.19}$$

$$Z_c = \frac{h_1h_2^2}{4} - 0.192r_i^3 = \frac{(406.4 - 2 \cdot 12.7)^3}{4} - 0 = 13837500\ \text{mm}^3 \tag{A.20}$$

$$M_D = Z_sF_y + \frac{1}{2}Z_c(0.85f_c') = 297 \cdot 10^4 \cdot 379 + \frac{1}{2} \cdot 1383 \cdot 10^4(0.85 \cdot 27)$$
$$= 1287\,\text{kN-m} \quad \text{(A.21)}$$

$$Z_{sB} = 2t_wh_n^2 = 2 \cdot 12.7 \cdot 60.198^2 = 91184\,\text{mm}^3 \quad \text{(A.22)}$$

$$Z_{cB} = h_1h_n^2 = (406.4 - 2 \cdot 12.7)(60.198)^2 = 1388588\,\text{mm}^3 \quad \text{(A.23)}$$

$$M_B = M_D - Z_{sB}F_y - \frac{1}{2}Z_{cB}(0.85f_c') \quad \text{(A.24)}$$

$$M_B = 1287 \cdot 10^6 - 91184 \cdot 378.95 - \frac{1}{2}1388588(0.85 \cdot 27.56)$$
$$= 1235.6\,\text{kN-m} \quad \text{(A.25)}$$

$$\phi_BM_B = 0.9 \cdot 1235.6 = 1112.04\,\text{kN-m} \quad \text{(A.26)}$$

#### A.2.1.3 Point C ($M_C = M_B$; $P_C = 0.85\ f_c'A_c$)

$$P_C = (0.85f_c')A_c = 145125(0.85 \cdot 27.56) = 3404\,\text{kN} \quad \text{(A.27)}$$

$$M_C = M_B = 1235.6\,\text{kN-m} \quad \text{(A.28)}$$

#### A.2.1.4 Point D

$$P_C = \frac{(0.85f_c')A_c}{2} = \frac{145125(0.85 \cdot 27.56)}{2} = 1702.125\,\text{kN} \quad \text{(A.29)}$$

$$M_D = 1287\,\text{kN-m (See computations for Point B)} \quad \text{(A.30)}$$

#### A.2.1.5 Point E

$$h_E = \frac{h_n}{2} + \frac{d}{4} = \frac{60.198}{2} + \frac{406.4}{4} = 131.826\,\text{mm} \quad \text{(A.31)}$$

$$P_E = \frac{1}{2}(0.85f_c')A_c + 0.85f_c'h_1h_E + 4F_yt_wh_E \quad \text{(A.32)}$$

$$P_E = \frac{1}{2}(0.85 \cdot 27.56) \cdot 145125 + 0.85 \cdot 27.56 \cdot 381 \cdot 131.8$$
$$+ 4 \cdot 379 \cdot 12.7 \cdot 131.8 = 5419\,\text{kN} \quad \text{(A.33)}$$

$$Z_{cE} = h_1h_E^2 = (381)(131.826)^2 = 6626420\,\text{mm}^3 \quad \text{(A.34)}$$

$$Z_{sE} = bh_E^2 - Z_{cE} = (406.4)\,(131.82)^2 - 6626420 = 441160\,\text{mm}^3 \quad \text{(A.35)}$$

*Table A.4* Calculation results for five points in P-M interaction diagram.

| Point | HSS 16 × 16 × 500 | | HSS 18 × 500 | | HSS 12 × 12 × 500 | | HSS 14 × 500 | | HSS 16 × 16 × 375 | | HSS 18 × 375 | | HSS 14 × 14 × 500 | | HSS 16 × 500 | |
|---|---|---|---|---|---|---|---|---|---|---|---|---|---|---|---|---|
| | P | M | P | M | P | M | P | M | P | M | P | M | P | M | P | M |
| (A) | 10991 | 0 | 10564 | 0 | 7458 | 0 | 7200 | 0 | 9251 | 0 | 8615 | 0 | 9167 | 0 | 8633 | |
| (B) | 0 | 1235 | 0 | 996 | 0 | 660 | 0 | 592 | 0 | 950 | 0 | 755 | 0 | 918 | 0 | 781 |
| (C) | 3404 | 1235 | 3430 | 996 | 1828 | 660 | 2006 | 592 | 3519 | 950 | 3537 | 755 | 2558 | 918 | 2674 | 781 |
| (D) | 1704 | 1286 | 1717 | 1109 | 916.7 | 681 | 1005 | 636 | 1757 | 1024 | 1766 | 888 | 1277 | 955 | 1335 | 854 |
| (E) | 5420 | 244.7 | 20363 | 517 | 3341 | 567 | 5486 | 318 | 5028 | 796 | 7071 | 376 | 4320 | 784 | 6706 | 412 |

$$\Delta M_E = Z_{sE}F_y + \frac{1}{2}Z_{cE}\left(0.85f_c'\right) = 441160 \cdot 378.95$$

$$+\frac{1}{2} \cdot 6626420 \cdot 0.85 \cdot 27.56 = 244.758\,\text{kN-m} \tag{A.36}$$

$$M_E = M_D - \Delta M_E = 1042.17\,\text{kN-m} \tag{A.37}$$

The results are summarized in Table A.4.

### A.2.2 CCFT 18 x 500 column

Develop an axial load-moment (P-M) envelope for a round HSS 18X500 concrete filled with $f_c' = 27.56$ MPa strength. The effective length of the member is 3810 mm. Assume A572 steel (Design Strength: $F_y = 378.9$ MPa and $F_u = 503$ MPa). The unit system is newton and millimeter. Use the dimensions shown in Table A.1.

**Basic Geometrical Property:**

$$d = 457.2\,\text{mm} \tag{A.38}$$

$$t = 12.7\,\text{mm} \tag{A.39}$$

$$h = d - 2t = 431.8\,\text{mm} \tag{A.40}$$

$$r_m = \frac{(d - t)}{2} = 222.25\,\text{mm} \tag{A.41}$$

$$A_s = 2\pi \cdot r_m \cdot t = 18021.3\,\text{mm}^2 \tag{A.42}$$

$$A_c = \frac{\pi h^2}{4} = 146402\,\text{mm}^2 \tag{A.43}$$

$$A_g = A_s + A_c = 164133.3\,\text{mm}^2 \tag{A.44}$$

$$E_c = 57000\sqrt{f_c'} = 57000\sqrt{27560000}/1000 = 24838.45\,\text{MPa} \tag{A.45}$$

$$I_s = \frac{\pi}{64}\left[d^4 - (d-2t)^4\right] = 438 \cdot 10^6\ \text{mm}^4 \tag{A.46}$$

$$I_c = \frac{\pi}{64}(d-2t)^4 = 1705 \cdot 10^6\ \text{mm}^4 \tag{A.47}$$

$$I_g = I_s + I_c = 2143 \cdot 10^6\ \text{mm}^4 \tag{A.48}$$

**Limits:**

3) The cross-sectional area of the steel core shall comprise at least one percent of the total composite cross section.

$$\frac{A_s}{A_s + A_c} = \frac{18021}{18021 + 146402} = 0.108 \geq 0.01\ (\text{OK}) \tag{A.49}$$

4) The slenderness of the tube wall is:

$$\begin{aligned}\left(\frac{t}{b}\right) &= \frac{457.2 - 12.7}{12.7} = 35.0 \leq 0.15\left(\frac{E_s}{F_y}\right)\\ &= 0.15\left(\frac{199810}{378.95}\right) = 79.10\ (\text{OK})\end{aligned} \tag{A.50}$$

#### *A.2.2.1 Point A ($M_A = 0$)*

Determine the available compressive strength

$$C_2 = 0.95 \tag{A.51}$$

$$C_3 = \min\left[0.9, 0.6 + 2 \cdot \left(\frac{A_s}{A_c + A_s}\right)\right] = 0.826 \tag{A.52}$$

$$P_o = A_s F_y + A_{sr} F_{yr} + C_2 A_c f'_c \tag{A.53}$$

$$EI_{eff} = E_s I_s + 0.5 E_s I_{sr} + C_3 E_c I_c \tag{A.54}$$

$$\begin{aligned}P_o &= A_s F_y + A_{sr} F_{yr} + C_2 A_c f'_c\\ &= 17731 \cdot 379 + 0 + 0.95 \cdot 146402 \cdot 27.56 = 10566.1\ \text{kN}\end{aligned} \tag{A.55}$$

$$P_o = 10566.1\ \text{kN} \tag{A.56}$$

$$EI_{eff} = E_s I_s + 0.5 E_s I_{sr} + C_3 E_c I_c \tag{A.57}$$

$$\begin{aligned}EI_{eff} &= 199810 \cdot 438 \cdot 10^6 + 0.826 \cdot 24838 \cdot 1705 \cdot 10^6\\ &= 1227 \cdot 10^8\ \text{kN-mm}^2\end{aligned} \tag{A.58}$$

$$P_e = \frac{\pi EI_{eff}}{(kL)^2} = \frac{(3.14159)^2 (1227 \cdot 10^8)}{(1 \cdot 3810)^2} = 83437.5\,\text{kN} \tag{A.59}$$

$$\frac{P_o}{P_e} = \frac{10566.1}{83437.5} = 0.127 \leq 2.25 \quad \text{or} \quad \frac{P_e}{P_o} = \frac{83437.5}{10566.1} = 7.90 \geq 0.44,$$

$$\lambda = \sqrt{\frac{P_o}{P_e}} = 0.356 \tag{A.60}$$

∴ Use Eq. 2.1 in Chapter 2

$$P_n = P_o \left(0.658^{\left(\frac{P_o}{P_e}\right)}\right) = 10566.1 \left(0.658^{0.127}\right) = 10019\,\text{kN} \tag{A.61}$$

$$\phi_c P_n = 0.75(10019) = 7514.537\,\text{kN} \tag{A.62}$$

#### A.2.2.2 *Point B ($P_B = 0$)*

From definitions of Point C in Table A.3:

$$K_c = f'_c h^2 = 27.56 \cdot 431.8^2 = 5144.2\,\text{kN} \tag{A.63}$$

$$K_s = F_y r_m t = 378 \cdot 8.75 \cdot 12.7 = 1070.78 \tag{A.64}$$

$$\theta = \frac{0.0260 \cdot K_c - 2 \cdot K_s}{0.0848 \cdot K_c} + \sqrt{\frac{(0.0260 \cdot K_c + 2 \cdot K_s)^2 + 0.857 \cdot K_c \cdot K_s}{0.0848 \cdot K_c}} \tag{A.65}$$

$$\theta = \theta_1 = 2.61\,\text{rad} = 149.62\,\text{deg} \tag{A.66}$$

$$Z_{cB} = \frac{h^3 \sin\left(\frac{\theta_1}{2}\right)}{6} = 1205 \cdot 10^4\,\text{mm}^3 \tag{A.67}$$

$$Z_{sB} = \frac{d^3 \sin\left(\frac{\theta_1}{2}\right)}{6} - Z_{cB} = 225.5 \cdot 10^4\,\text{mm}^3 \tag{A.68}$$

$$M_B = Z_{sB} F_y + \frac{1}{2} Z_{cB} \left(0.85 f'_c\right) = 996.13\,\text{kN-m} \tag{A.69}$$

$$\phi_b M_B = 896.52\,\text{kN-m} \tag{A.70}$$

#### A.2.2.3 *Point C*

From calculations for Point B (above):

$$P_C = (0.85 f'_c) A_c = 3434.2\,\text{kN} \tag{A.71}$$

$$M_C = M_B = 996.13\,\text{kN-mm} \tag{A.72}$$

#### A.2.2.4 Point D

$$P_D = \frac{0.85f'_c}{2} = 1717.12\,\text{kN} \tag{A.73}$$

$$Z_c = \frac{h^3}{6} = 1342 \cdot 10^4\ \text{mm}^3 \tag{A.74}$$

$$Z_s = \frac{d^3}{6} - Z_c = 251.2 \cdot 10^4\,\text{mm}^3 \tag{A.75}$$

$$M_D = Z_s F_y + \frac{1}{2} Z_c (0.85 f'_c) = 1109.23\,\text{kN-m} \tag{A.76}$$

#### A.2.2.5 Point E

$$h_E = \frac{h_n}{2} + \frac{d}{4} = 136.4\,\text{mm} \tag{A.77}$$

$$\theta = \theta_2 = \pi - 2\arcsin\left(\frac{2h_E}{h}\right) = 1.77\,\text{rad} = 101.41\ \text{deg} \tag{A.78}$$

$$Z_{cE} = \frac{h^3 \sin\left(\frac{\theta_1}{2}\right)}{6} = 625.9 \cdot 10^4\,\text{mm}^3 \tag{A.79}$$

$$Z_{sE} = \frac{d^3 \sin\left(\frac{\theta_1}{2}\right)}{6} - Z_{cE} = 117.1 \cdot 10^4\,\text{mm}^3 \tag{A.80}$$

$$\begin{aligned} P_E = {} & (0.85 f'_c A_c + F_y A_s) \\ & - \frac{1}{2}\left[F_y\left(d^2 - h^2\right) + \frac{1}{2}(0.85 f'_c) h^2\right]\left[\frac{\theta_2}{2} - \frac{\sin\theta_2}{2}\right] = 8024\,\text{kN} \end{aligned} \tag{A.81}$$

$$M_E = Z_{sE} F_y + \frac{1}{2} Z_{cE} (0.85 f'_c) = 517.06\,\text{kNm} \tag{A.82}$$

### A.2.3 RCFT 12 x 13 x 500 column

Develop an axial load-moment (P-M) envelope for a HSS 12 × 12 × 500 concrete filled with $f'_c = 27.56$ MPa strength. The effective length of the member is 3810 mm. Assume A572 steel (Design Strength: $F_y = 378.9$ MPa and $F_u = 503$ MPa). The unit system is newton and millimeter. Use the dimensions shown in Table A.1.

**Limitation:**

5) The cross-sectional area of the steel core shall comprise at least one percent of the total composite cross section.

$$A_s = 14835\,\mathrm{mm}^{.2} > (0.01)(304.8^2) = 928\,\mathrm{mm}^2\ (\mathrm{OK}) \tag{A.83}$$

Note that $\rho = \frac{14835}{92800} = 0.160$, or 16.0% which is high ratio.

6) The slenderness of the tube wall is:

$$\left(\frac{t}{b}\right) = \frac{304.8 - 12.7}{12.7} = 23.0 \leq 2.26\sqrt{\frac{E}{F_y}} = 2.26\sqrt{\frac{199810}{378.95}} = 51.90\ (\mathrm{OK}) \tag{A.84}$$

#### A.2.3.1 *Point A ($M_A = 0$)*

Determine the available compressive strength

$$P_o = A_sF_y + A_{sr}F_{yr} + 0.85A_cf'_c \tag{A.85}$$

$$A_c = (b - 2\cdot t)^2 = (304.8 - 2\cdot 12.7)^2 = 78045\,\mathrm{mm}^2 \tag{A.86}$$

$$\begin{aligned} P_o &= A_sF_y + A_{sr}F_{yr} + 0.85A_cf'_c = 14835\cdot 378 + 0 + 0.85\cdot 78045\cdot 27.56 \\ &= 7460\,\mathrm{kN} \end{aligned} \tag{A.87}$$

$$P_o = 7460\,\mathrm{kN} \tag{A.88}$$

$$C_3 = 0.6 + 2\left(\frac{A_s}{A_c + A_s}\right) = 0.6 + 2\left(\frac{14835}{78045 + 14835}\right) = 0.920 \tag{A.89}$$

$$I_s = \frac{d^4}{12} - \frac{(d - 2\cdot t)^4}{12} = \frac{304.8^4}{12} - \frac{(304.8 - 2\cdot 12.7)^4}{12} = 211.3\cdot 10^6\,\mathrm{mm}^4 \tag{A.90}$$

$$I_c = \frac{(d - 2\cdot t)^4}{12} = \frac{(304.8 - 2\cdot 12.7)^4}{12} = 507.5\cdot 10^6\,\mathrm{mm}^4 \tag{A.91}$$

$$E_c = 57000\sqrt{f'_c} = 57000\sqrt{27560000}/1000 = 4838.45\,\mathrm{MPa} \tag{A.92}$$

$$EI_{eff} = E_sI_s + 0.5E_sI_{sr} + C_3E_cI_c \tag{A.93}$$

$$\begin{aligned} EI_{eff} &= 199810\cdot 211.3\cdot 10^6 + 0.92\cdot 24838\cdot 507.5\cdot 10^6 \\ &= 5.39\cdot 10^8\,\mathrm{kN\text{-}mm}^2 \end{aligned} \tag{A.94}$$

$$P_e = \frac{\pi EI_{eff}}{(kL)^2} = \frac{(3.14159)^2\left(5.39\cdot 10^8\right)}{(1\cdot 3810)^2} = 36654.6\,\mathrm{kN} \tag{A.95}$$

$$\frac{P_o}{P_e} = \frac{7459.98}{36654.65} = 0.203 \leq 2.25 \quad \text{or} \quad \frac{P_e}{P_o} = \frac{36654.65}{7459.98} = 4.92 \geq 0.44 \tag{A.96}$$

Use Eq. 2.1 in Chapter 2

$$P_n = P_o\left(0.658^{\left(\frac{P_o}{P_e}\right)}\right) = 7459.98\left(0.658^{0.203}\right) = 6851.22\,\text{kN} \tag{A.97}$$

$$\phi_c P_n = 0.75\,(6851.22) = 5138.37\,\text{kN} \tag{A.98}$$

### *A.2.3.2 Point B ($P_B = 0$)*

Determine location of $h_n$

$$h_n = \frac{0.85f_c'A_c}{2(0.85f_c'h_1 + 4t_wF_y)} \leq \frac{h_2}{2} \tag{A.99}$$

$$\begin{aligned} h_n &= \frac{0.85\cdot 27.56\cdot 78045}{2(0.85\cdot 27.56\cdot(304.8-2\cdot 12.7)+4\cdot 12.7\cdot 379)} \\ &= 35.6\,\text{mm} \leq \frac{304.8}{2} = 152.4\,\text{mm} \end{aligned} \tag{A.100}$$

$$Z_s = 162.85\,\text{mm}^3 \tag{A.101}$$

$$Z_c = \frac{h_1h_2^2}{4} - 0.192r_i^3 = \frac{(406.4-2\cdot 12.7)^3}{4} - 0 = 545.7\,\text{mm}^3 \tag{A.102}$$

$$\begin{aligned} M_D &= Z_sF_y + \frac{1}{2}Z_c(0.85f_c') = 162.85\cdot 379 + \frac{1}{2}\cdot 545.7(0.85\cdot 27.56) \\ &= 681.1\,\text{kN-m} \end{aligned} \tag{A.103}$$

$$Z_{sB} = 2t_wh_n^2 = 2\cdot 12.7\cdot 35.56^2 = 32144\,\text{mm}^3 \tag{A.104}$$

$$Z_{cB} = h_1h_n^2 = (304.8-2\cdot 12.7)(35.56)^2 = 692244\,\text{mm}^3 \tag{A.105}$$

$$M_B = M_D - Z_{sB}F_y - \frac{1}{2}Z_{cB}(0.85f_c') \tag{A.106}$$

$$M_B = 681.1 - 32144\cdot 379 - \frac{1}{2}692244(0.85\cdot 27.6) = 661\,\text{kN-m} \tag{A.107}$$

$$\phi_B M_B = 0.9\cdot 661 = 594.9\,\text{kN-m} \tag{A.108}$$

### *A.2.3.3 Point C ($M_C = M_B$; $P_C = 0.85\,f_c'A_c$)*

$$P_C = (0.85f_c')A_c = 78045(0.85\cdot 27.56) = 1830.73\,\text{kN-m} \tag{A.109}$$

$$M_C = M_B = 661\,\text{kN-mm} \tag{A.110}$$

#### A.2.3.4 Point D

$$P_C = \frac{(0.85f'_c)A_c}{2} = \frac{78045\,(0.85 \cdot 27.56)}{2} = 915.37\,\text{kN} \tag{A.111}$$

$$M_D = 681.08\,\text{kN-m(See computations for Point B)} \tag{A.112}$$

#### A.2.3.5 Point E

$$h_E = \frac{h_n}{2} + \frac{d}{4} = \frac{35.56}{2} + \frac{304.8}{4} = 93.98\,\text{mm} \tag{A.113}$$

$$P_E = \frac{1}{2}(0.85f'_c)A_c + 0.85f'_c h_1 h_E + 4F_y t_w h_E \tag{A.114}$$

$$P_E = \frac{1}{2}(0.85 \cdot 4) \cdot 78045 + 0.85 \cdot 27.6 \cdot 279.4 \cdot 94 + 4 \cdot 379 \cdot 12.7 \cdot 94$$
$$= 3340.8\,\text{kN} \tag{A.115}$$

$$Z_{cE} = h_1 h_E^2 = (279.4)\,(93.4)^2 = 246.67 \cdot 10^4\,\text{mm}^3 \tag{A.116}$$

$$Z_{sE} = bh_E^2 - Z_{cE} = (279.4)(93.4)^2 - 264.67 \cdot 10^4 = 22.42 \cdot 10^4\,\text{in}^3 \tag{A.117}$$

$$\Delta M_E = Z_{sE}F_y + \frac{1}{2}Z_{cE}(0.85f'_c) = 22.4 \cdot 10^4 \cdot 379 + \frac{1}{2} \cdot 264.7 \cdot 10^4 \cdot 0.85 \cdot 27.6$$
$$= 113.83\,\text{kN-m} \tag{A.118}$$

$$M_E = M_D - \Delta M_E = 567.2\,\text{kN-m} \tag{A.119}$$

The results are summarized in Table A.4.

### A.2.4 CCFT 14 x 500 column

Develop an axial load-moment (P-M) envelope for a round HSS 14 × 500 concrete filled with $f'_c = 27.56$ MPa strength. The effective length of the member is 317.5 mm. Assume A572 steel (Design Strength: $F_y = 378.95$ MPa and $F_u = 502.97$ MPa). The unit system is newton and millimeter. Use the dimensions shown in Table A.1.

**Basic Geometrical Property:**

$$d = 355.6\,\text{mm} \tag{A.120}$$

$$t = 12.7\,\text{mm} \tag{A.121}$$

$$h = d - 2t = 330.2\,\text{mm} \tag{A.122}$$

$$r_m = \frac{(d - t)}{2} = 171.45\,\text{mm} \tag{A.123}$$

$$A_s = 2\pi \cdot r_m \cdot t = 13680.45\,\text{mm}^2 \tag{A.124}$$

$$A_c = \frac{\pi h^2}{4} = 85610.85\,\text{mm}^2 \tag{A.125}$$

$$A_g = A_s + A_c = 99291.3\,\text{mm}^2 \tag{A.126}$$

$$E_c = 57000\sqrt{f_c'} = 57000\sqrt{27560000}/1000 = 4838.45\,MPa \tag{A.127}$$

$$I_s = \frac{\pi}{64}\left[d^4 - (d-2t)^4\right] = 201.24 \cdot 10^6\,\text{mm}^4 \tag{A.128}$$

$$I_c = \frac{\pi}{64}(d-2t)^4 = 583.22\,\text{mm}^4 \tag{A.129}$$

$$I_g = I_s + I_c = 784.46 \cdot 10^6\,\text{mm}^4 \tag{A.130}$$

**Limits:**

7) The cross-sectional area of the steel core shall comprise at least one percent of the total composite cross section.

$$\frac{A_s}{A_s + A_c} = \frac{13680.45}{13680.45 + 85610.85} = 0.138 \geq 0.01\ (\text{OK}) \tag{A.131}$$

8) The slenderness of the tube wall is:

$$\begin{aligned}\left(\frac{t}{b}\right) &= \frac{355.6 - 12.7}{12.7} = 27.0 \leq 0.15\left(\frac{E_s}{F_y}\right)\\ &= 0.15\left(\frac{199810}{378.95}\right) = 79.10\ (\text{OK})\end{aligned} \tag{A.132}$$

### *A.2.4.1 Point A ($M_A = 0$)*

Determine the available compressive strength

$$C_2 = 0.95 \tag{A.133}$$

$$C_3 = \min\left[0.9, 0.6 + 2 \cdot \left(\frac{A_s}{A_c + A_s}\right)\right] = 0.876 \tag{A.134}$$

$$P_o = A_s F_y + A_{sr} F_{yr} + C_2 A_c f_c' \tag{A.135}$$

$$EI_{eff} = E_s I_s + 0.5 E_s I_{sr} + C_3 E_c I_c \tag{A.136}$$

$$\begin{aligned}P_o &= A_s F_y + A_{sr} F_{yr} + C_2 A_c f_c'\\ &= 13680 \cdot 379 + 0 + 0.95 \cdot 85611 \cdot 27.56 = 7198.32\,\text{kN}\end{aligned} \tag{A.137}$$

$$P_o = 7198.32\,\text{kN} \tag{A.138}$$

$$EI_{eff} = E_sI_s + 0.5E_sI_{sr} + C_3E_cI_c \tag{A.139}$$

$$\begin{aligned} EI_{eff} &= 199810 \cdot 201 \cdot 10^6 + 0.876 \cdot 24838 \cdot 583.2 \cdot 10^6 \\ &= 5.298 \cdot 10^{10}\,\text{kN-mm} \end{aligned} \tag{A.140}$$

$$P_e = \frac{\pi EI_{eff}}{(kL)^2} = \frac{(3.14159)^2\left(5.298 \cdot 10^{10}\right)}{(1 \cdot 3810)^2} = 36027.2\,\text{kN} \tag{A.141}$$

$$\frac{P_o}{P_e} = \frac{7198}{36027} = 0.200 \le 2.25 \quad \text{or} \quad \frac{P_e}{P_o} = \frac{36027}{7198} = 5.00 \ge 0.44,$$

$$\lambda = \sqrt{\frac{P_o}{P_e}} = 0.447 \tag{A.142}$$

∴ Use Eq. 2.1 in Chapter 2

$$P_n = P_o\left(0.658^{\left(\frac{P_o}{P_e}\right)}\right) = 7198\left(0.658^{0.2}\right) = 6619.464\,\text{kN} \tag{A.143}$$

$$\phi_cP_n = 0.75\,(6619.464) = 4964.60\,\text{kN} \tag{A.144}$$

#### *A.2.4.2 Point B ($P_B = 0$)*

From definitions of Point C in Table A.3:

$$K_c = f_c'h^2 = 27.56 \cdot 330.2^2 = 3008.2\,\text{kN} \tag{A.145}$$

$$K_s = F_yr_mt = 379 \cdot 171.45 \cdot 12.7 = 826.03\,\text{kN} \tag{A.146}$$

$$\theta = \frac{0.0260 \cdot K_c - 2 \cdot K_s}{0.0848 \cdot K_c} + \sqrt{\frac{(0.0260 \cdot K_c + 2 \cdot K_s)^2 + 0.857 \cdot K_c \cdot K_s}{0.0848 \cdot K_c}} \tag{A.147}$$

$$\theta = \theta_1 = 2.70\,\text{rad} = 154.78\,\text{deg} \tag{A.148}$$

$$Z_{cB} = \frac{h^3 \sin\left(\frac{\theta_1}{2}\right)}{6} = 558.48 \cdot 10^4\,\text{mm}^3 \tag{A.149}$$

$$Z_{sB} = \frac{d^3 \sin\left(\frac{\theta_1}{2}\right)}{6} - Z_{cB} = 139.05 \cdot 10^4\,\text{mm}^3 \tag{A.150}$$

$$M_B = Z_{sB}F_y + \frac{1}{2}Z_{cB}(0.85f_c') = 592.34\,\text{kN-m} \tag{A.151}$$

$$\phi_b M_B = 0.9 \cdot 5242.10 = 533.11 \text{ kN-m} \tag{A.152}$$

#### A.2.4.3 Point C

From calculations for Point B (above):

$$P_C = (0.85f'_c)A_c = 2008.196 \text{ kN-m} \tag{A.153}$$

$$M_C = M_B = 592.34 \text{ kN-m} \tag{A.154}$$

#### A.2.4.4 Point D

$$P_D = \frac{(0.85f'_c A_c)}{2} = 1004.1 \text{ kN} \tag{A.155}$$

$$Z_c = \frac{h^3}{6} = 600.52 \cdot 10^4 \text{ mm}^3 \tag{A.156}$$

$$Z_s = \frac{d^3}{6} - Z_c = 139.056 \cdot 10^4 \text{ mm}^3 \tag{A.157}$$

$$M_D = Z_s F_y + \frac{1}{2} Z_c (0.85f'_c) = 636.94 \text{ kN-m} \tag{A.158}$$

#### A.2.4.5 Point E

$$h_E = \frac{h_n}{2} + \frac{d}{4} = 100.584 \text{ mm} \tag{A.159}$$

$$\theta = \theta_2 = \pi - 2\arcsin\left(\frac{2h_E}{h}\right) = 1.83 \text{ rad} = 104.90 \text{ deg} \tag{A.160}$$

$$Z_{cE} = \frac{h^3 \sin^3\left(\frac{\theta_2}{2}\right)}{6} = 299.956 \cdot 10^4 \text{ mm}^3 \tag{A.161}$$

$$Z_{sE} = \frac{d^3 \sin^3\left(\frac{\theta_2}{2}\right)}{6} - Z_{cE} = 74.68 \cdot 10^4 \text{ mm}^3 \tag{A.162}$$

$$P_E = (0.85f'_c A_c + F_y A_s) - \frac{1}{2}\left[F_y(d^2 - h^2) + \frac{1}{2}(0.85f'_c)h^2\right]\left[\frac{\theta_2}{2} - \frac{\sin\theta_2}{2}\right] = 5488.9 \text{ kN} \tag{A.163}$$

$$M_E = Z_{sE} F_y + \frac{1}{2} Z_{cE}(0.85f'_c) = 318.15 \text{ kN-m} \tag{A.164}$$

The calculations for other four cases were repeated with the same procedures mentioned above, so the procedures to calculate five points for other composite column cases were omitted. Instead, all results for five points in P-M interaction diagrams are summarized in Table A.4.

Appendix B

# Design examples and failure modes

## B.1 GENERAL INTRODUCTION

Idealized strength models for each connection component provide the basic background for connection design. The strength models and design procedures were described in Chapter 3, but specific design examples for them were omitted because of the limited space. Therefore, design procedures for the smart PR-CFT connections specified in Fig. 2.26 to 2.31 are described in Section 2.5.4.

In design, brittle failure modes such as bolt fracture, weld failure, and plate fracture should be avoided in order to prevent a potential collapse of the structure. The criteria for the design strength were determined by relatively ideal failure strength models based on achieving full yielding of the beam. The available failure modes for the connection components were described in Section 2.6. The necessary design checks, including strength models and different reduction factors, are shown in Section B3.

## B.2 DESIGN EXAMPLES

Strength models and design procedures for connection components were described in Section 2.3 and Section 2.4, respectively. The detailed design procedures are described in this section.

### B.2.1 End-plate connection

The geometric details for end-plate connections were described in Figs. 2.26 and 2.27. The connection components were designed in accordance with AISC-LRFD 2001 and AISC/ANSI 358-05. The design procedures for the end-plate connection are described in this section, with the design criteria based on ideal limit states. The SI unit system (newton and millimeter) was used in this case.

#### *B.2.1.1 Check the basic information*

(1) Determine prequalified limits and geometric dimensions

Table B.1 represents a summary of the range of the geometric parameters that have satisfactorily tested. The geometric parameters are given in Fig. B.1.

(2) Determine dimensions (Refer to Table B.1 and Table B.2)

*Table B.1* Prequalification dimension limits.

| | 4E | | 4ES | | 8ES | |
|---|---|---|---|---|---|---|
| *Parameter* | *Max.* | *Min.* | *Max.* | *Min.* | *Max.* | *Min.* |
| $t_p$ | 57.15 | 12.7 | 38.1 | 12.7 | 63.5 | 19.05 |
| $b_p$ | 273.05 | 177.8 | 273.05 | 273.05 | 381 | 228.6 |
| g | 152.4 | 101.6 | 152.4 | 82.55 | 152.4 | 127 |
| $P_{r1}, P_{r2}$ | 114.3 | 38.1 | 139.7 | 44.45 | 50.8 | 44.45 |
| $P_b$ | – | – | – | – | 95.25 | 88.9 |
| d | 1397 | 635 | 609.6 | 349.25 | 914.4 | 469.9 |
| $t_{br}$ | 19.05 | 9.525 | 19.05 | 9.525 | 25.4 | 15.0622 |
| $b_{bf}$ | 234.95 | 152.4 | 228.6 | 152.4 | 311.15 | 196.85 |

Unit: millimeter.

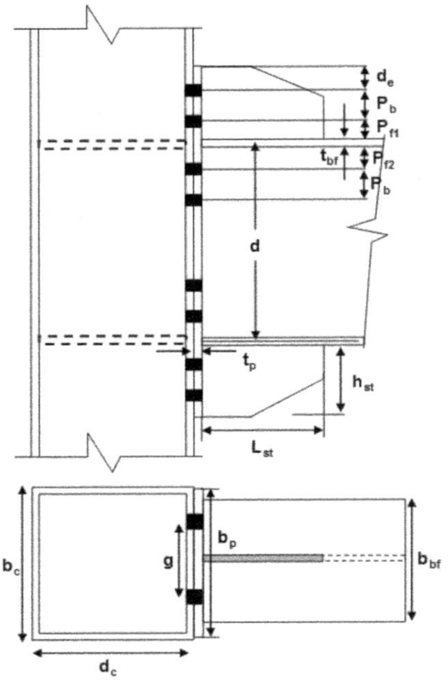


*Figure B.1* 8 bolt stiffened extended end-plate connection geometry (8ES).

Choose 8 bolt stiffened extended end-plate connection system (8ES)

| | |
|---|---|
| $t_p = 25.4$ | The thickness of the end-plate (mm) |
| $b_p = 381$ | The width of the end-plate (mm) |
| $g = 152.4$ | The horizontal distance between bars (mm) |
| $P_{f1} = P_{f2} = 44.45$ | The vertical distance between beam flange and the center of bar holes (mm) |
| $P_b = 88.8$ | The distance between the centerline of bar holes (mm) |
| $d = 622.3$ | The depth of the beam (mm) |

*Table B.2* The limit check for the geometric parameters.

| | | *Eight Bolt Stiffened (8ES)* | | |
|---|---|---|---|---|
| *Parameter* | *Max.* | *Min.* | *Design* | *Decision* |
| $t_p$ | 63.5 | 19.05 | 25.4 | OK |
| $b_p$ | 381 | 228.6 | 381 | OK |
| g | 152.4 | 127 | 152.4 | OK |
| $P_{f1}, P_{f2}$ | 50.8 | 44.45 | 44.45 | OK |
| $P_b$ | 95.25 | 88.9 | 88.9 | OK |
| d | 914.4 | 469.9 | 622.3 | OK |
| $t_{bf}$ | 25.4 | 15.0622 | 24.892 | OK |
| $b_{bf}$ | 311.15 | 196.85 | 228.6 | OK |

$t_{bf} = 24.892$ The thickness of the beam flange (mm)
$b_{bf} = 228.6$ The width of the beam flange (mm)
$t_{bw} = 14.224$ The thickness of the beam web (mm)
$d_e = 44.45$ The edge distance (mm)
$Z_x = 459.2 \cdot 10^4$ The plastic section modulus for the steel beam ($mm^3$)
Determine the position of the tension bars
$h_1 = 755.65$
$h_2 = 666.75$
$h_3 = 622.3$
$h_4 = 533.4$
where, $h_i$: the distance from the centerline of the beam compression flange to the centerline of the ith tension bar row (mm)

(3) Determine material properties
Use A572 Steel and A490M Bolt Materials (Swanson 2002)
$F_y = 378.95$ The design yield stress (MPa)
$F_{yp} = 378.95$ The design yield stress for the end plate (MPa)
$F_y = 502.97$ The design ultimate stress (MPa)

#### B.2.1.2 *Calculate the design strength*

(4) Determine the factor to consider the peak connection strength

$$C_{pr} = \frac{F_y + F_u}{2F_y} = 1.164 \leq 1.20 \text{ (OK)} \quad \text{Generally taken as } C_{pr} = 1.1 \tag{B.1}$$

(5) Determine the design strength based on the full plastic strength of the beam

$$M_{design} = C_{pr}R_yF_yZ_x = 1916.48 \text{ kN-m } (R_y = 1.1 \text{ for the material over strength factor}) \tag{B.2}$$

#### B.2.1.3 *Determine the required bar diameter*

(6) Determine the average tensile strength for bars

$\sum_{i=1}^{4} h_i = 2578.1\,mm$

$F_{nt,SMA} = 551.2\,MPa$ The nominal strength of the SMA bar

$F_{nt,Steel} = 1103.78\,MPa$ The nominal strength of the high tension steel bar

$F_{nt} = \sum_i F_{nt,i} h_i \Big/ \sum_i h_i = 798.27\,MPa$

(7) Choose the required bar diameter ($d_{b,req}$)

$$d_{b,req} = \sqrt{\frac{4M_{design}}{2\pi\pi_n F_{nt}(h_1 + h_2 + h_3 + h_4)}} = 25.019\,mm(\phi_n = 0.9 \text{ for non-ductile limit state}) \qquad (B.3)$$

Take $d_b = 25.4$ mm (OK)

#### B.2.1.4 Determine the thickness of the end-plate

(8) Calculate the yield line mechanism for the end-plate ($Y_P$ See Fig B.5)

$$s = \frac{1}{2}\sqrt{b_p g} = 120.4 \text{ mm} \qquad (B.4)$$

Use case 1 ($d_e = 1.75 \leq s = 4.74$)

$$Y_P = \frac{b_p}{2}\left[h_1\left(\frac{1}{2d_e}\right) + h2\left(\frac{1}{pf_1}\right) + h3\left(\frac{1}{pf_2}\right) + h4\left(\frac{1}{s}\right)\right] + \frac{2}{g}\left[h_1\left(d_e + \frac{p_b}{4}\right) + h_2\left(pf_1 + \frac{3p_b}{4}\right) + h_3\left(pf_2 + \frac{p_b}{4}\right) + h_4\left(s + \frac{3p_b}{4}\right) + p_b^2\right] + g \qquad (B.5)$$

$Y_P = 11756.64\,mm$

(9) Choose the required thickness of the end-plate

$$t_{p,req} = \sqrt{\frac{1.11M_{pr}}{\phi_b F_{yp} Y_p}} = 23.6728 \text{ in } (\phi_b = 1.0 \text{ for ductile limit state}) \qquad (B.6)$$

Take $t_p = 25.4\,mm$ (OK)

#### B.2.1.5 Check the shear resistance

(10) Compute the tensile axial force ($F_{fu}$)

$$F_{fu} = \frac{M_{design}}{d - t_{bf}} = 3051.45\,kN \qquad (B.7)$$

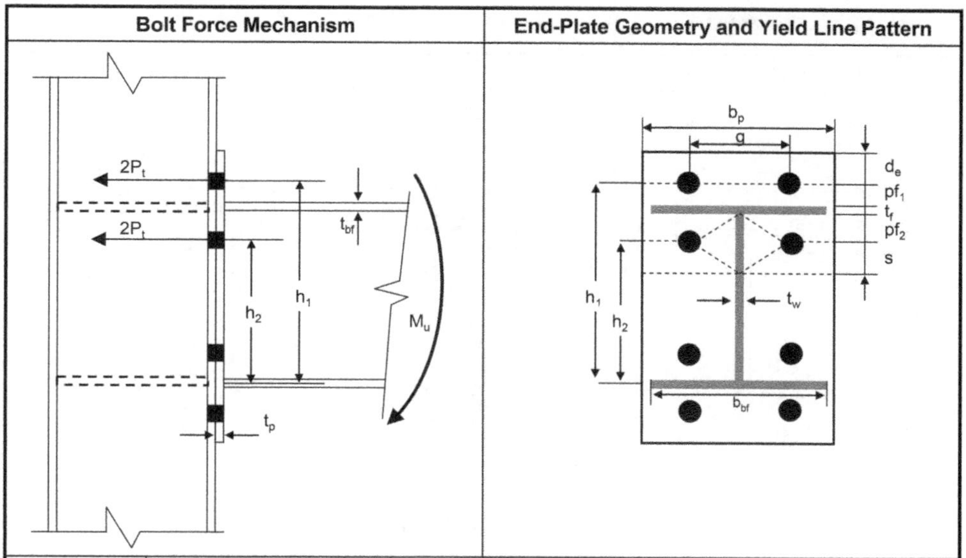


| Yield Line Equations | |
|---|---|
| | $Y_P = \frac{b_p}{2}\left[h_2\left(\frac{1}{pf_2}+\frac{1}{s}\right)+h_1\left(\frac{1}{pf_1}+\frac{1}{s}\right)-\frac{1}{2}\right]+\frac{2}{g}\left[h_2(pf_2+s)\right]$ <br> $s=\frac{1}{2}\sqrt{b_p g}$ **(If $pf_2>s$, use $pf_2=s$)** |

*Figure B.2* Geometry summary and yield line failure mechanism (4 Bolt Unstiffened, 4E).

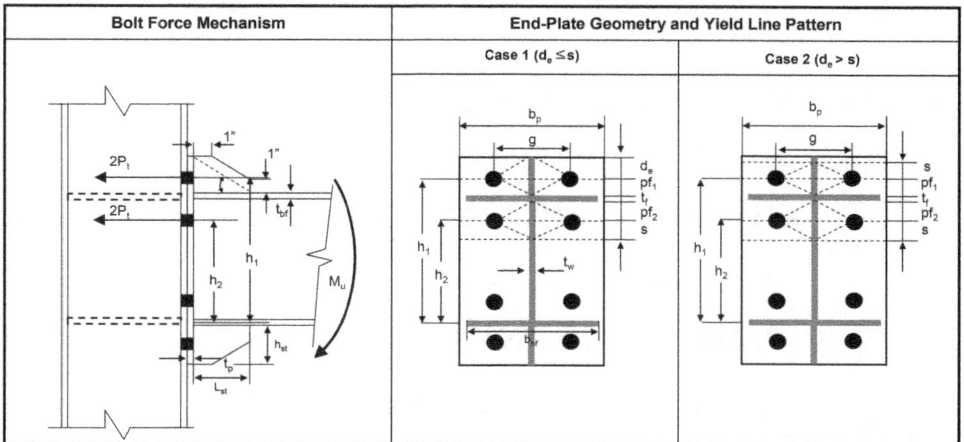


| Yield Line Equations | | |
|---|---|---|
| | Case 1 ($d_e \le s$) | $Y_P = \frac{b_p}{2}\left[h_2\left(\frac{1}{pf_2}+\frac{1}{s}\right)+h_1\left(\frac{1}{pf_1}+\frac{1}{s}\right)-\frac{1}{2}\right]+\frac{2}{g}\left[h_2(pf_2+s)+h_1(pf_1+d_e)\right]$ |
| | Case 2 ($d_e > s$) | $Y_P = \frac{b_p}{2}\left[h_2\left(\frac{1}{pf_2}+\frac{1}{s}\right)+h_1\left(\frac{1}{pf_1}+\frac{1}{s}\right)-\frac{1}{2}\right]+\frac{2}{g}\left[h_2(pf_2+s)+h_1(pf_1+s)\right]$ |
| | | $s=\frac{1}{2}\sqrt{b_p g}$ **(If $pf_2>s$, use $pf_2=s$)** |

*Figure B.3* Geometry summary and yield line failure mechanism (4 Bolt Stiffened, 4ES).

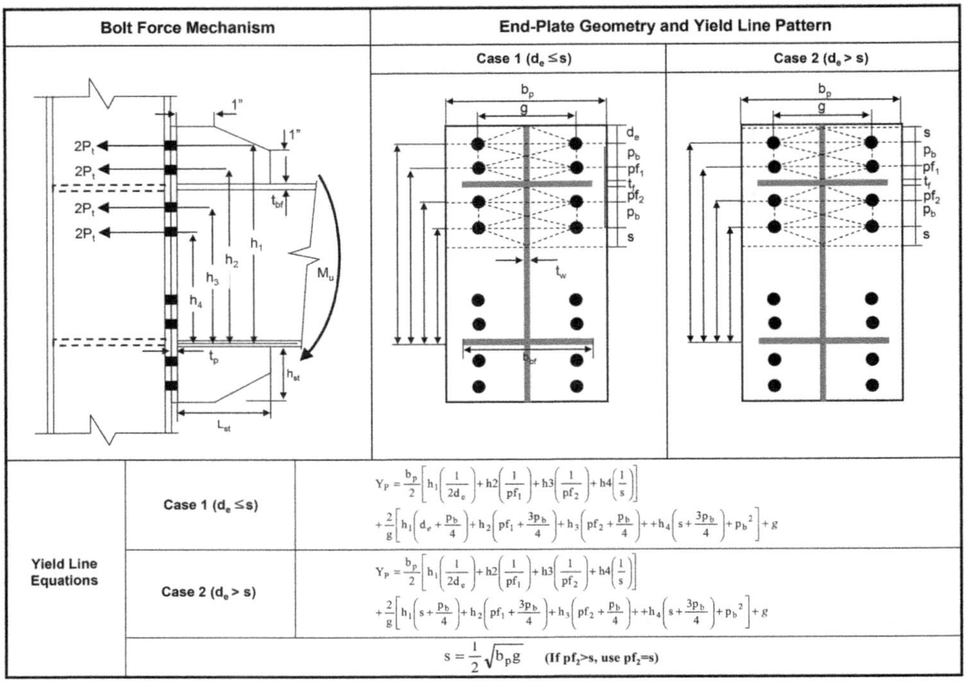


*Figure B.4* Geometry summary and yield line failure mechanism (8 Bolt Stiffened, 8ES).

(11) Check the shear resistance for the end-plate

$$\phi_n R_{n,shear} = \phi_n 0.6 F_{yp} b_p t_p = 1982.48 \text{ kN} \tag{B.8}$$

$$A_n = \left[b_p - 2\left(d_b + \frac{1}{16}\right)\right] t_p = 8223.75 \text{ mm}^2 \tag{B.9}$$

$$\phi_n R_{n,shear} = \phi_n 0.6 F_{up} A_n = 2236.6 \text{ kN} \tag{B.10}$$

$$\frac{F_{fu}}{2} = 1525.725 \leq \phi_n R_{n,shear} \text{ (OK)} \tag{B.11}$$

This connection type did not require the shear bolt system. The design check for the shear bolt is not necessary.

### B.2.1.6 *Determine the size of the end-plate stiffener*

(12) Determine the thickness of the stiffener ($t_s$)

$$t_{s,min} = t_{bw}\left(\frac{F_{yb}}{F_{ys}}\right) = 14.224 \text{ mm} \tag{B.12}$$

$$\text{Take } t_s = 14.224 \text{ in} \tag{B.13}$$

(13) Check the length of the stiffener ($L_{st}$)

$$h_{st} = \left(7 - \frac{t_{bf}}{2}\right) = 165.354 \text{ mm The height of the stiffener} \tag{B.14}$$

$$L_{st} = \frac{h_{st}}{\tan 30°} = 286.258 \text{ mm} \tag{B.15}$$

$$\frac{h_{st}}{t_s} = 20.13 \geq 0.56\sqrt{\frac{E}{F_{ys}}} = 12.86 \text{ Prevent the local buckling (OK)} \tag{B.16}$$

#### B.2.1.7 Check the rupture and bearing failure at the bars

(14) Check the tearing out failure at the end-plate

$$L_c = d_e - \left(\frac{d_b}{2} + \frac{1}{16}\right) = 30.48 \text{ mm} \tag{B.17}$$

$$R_{ni} = 1.2L_c t_p F_u = 465.03 \leq 2.4 d_b t_p F_u = 779.64 \text{ kN (OK)} \tag{B.18}$$

$$R_{no} = 1.2L_c t_p F_u = 465.03 \leq 2.4 d_b t_p F_u = 779.64 \text{ kN (OK)} \tag{B.19}$$

$$L = 4419.6 \text{ mm The length of the beam} \tag{B.20}$$

$$V_u = \frac{2M_{design}}{L} = 4709.16 \text{ kN} \tag{B.21}$$

$$V_u \leq \phi_n R_{n,bearing} = \phi_n (N_i) R_{ni} + \phi_n (N_o) R_{no} = 5611.45 \text{ kN (OK)} \tag{B.22}$$

#### B.2.1.8 Check the steel column strength

I-shape steel columns require other design strength checks such as flexural yielding of the column flange, stiffener forces, local yielding of the column web, unstiffened column web buckling strength, and unstiffened column web crippling strength. CFT column systems were used in this connection system. These columns satisfied these design strength checks and these design checks are omitted in the design procedure.

### B.2.2 T-stub connection

The geometric details for T-stub connections were described in Figs. 2.28 and 2.29. The connection components were designed in accordance with AISC-LRFD 2001 and AISC 2005 Seismic Provisions. The design procedures for the T-stub connection are described in the next section based on ideal limit states.

#### B.2.2.1 Determine the required design strength

(1) Determine the design strength ($M_{design}$) based on the full plastic strength of the steel beam in accordance with 2005 AISC Seismic Provisions.

$$R_y = 1.1 \text{ (For the types of rolled shapes and bars made by A572-Gr.50 steel)} \tag{B.23}$$

$$F_y = 378.95 \text{ MPa The design yield stress of the beam} \quad \text{(B.24)}$$

$$C_{pr} = 1.1 \text{ Factor for the design strength} \quad \text{(B.25)}$$

$$Z_x = 219.76 \cdot 10^4 \text{ mm}^3 \text{ The plastic section modulus for the steel beam} \quad \text{(B.26)}$$

$$M_{design} = C_{pr} R_y F_y Z_x = 1007.73 \text{ kNm} \quad \text{(B.27)}$$

(2) Calculate the required axial force ($P_{req}$) acting on the beam flange

$$d = 598.67 \text{ mm The depth of the beam} \quad \text{(B.28)}$$

$$P_{req} = \frac{M_{design}}{d} = 1682.1 \text{ kN} \quad \text{(B.29)}$$

#### B.2.2.2 Select the diameter of tension bars

(6) Determine the average tensile strength for bars

$$h_1 = 674.87 \text{ mm} \quad \text{(B.30)}$$

$$h_2 = 522.47 \text{ mm} \quad \text{(B.31)}$$

$$\sum_{i=1}^{2} h_i = 1197.35 \text{ mm} \quad \text{(B.32)}$$

$$F_{nt,SMA} = 356 \text{ kN} \quad \text{The nominal strength of the SMA bar} \quad \text{(B.33)}$$

$$F_{nt,Steel} = 712 \text{ kN} \quad \text{The nominal strength of the high tension steel bar} \quad \text{(B.34)}$$

$$F_{nt} = \sum_i F_{nt,i} h_i \Big/ \sum_i h_i = 671.43 \text{ MPa} \quad \text{(B.35)}$$

(7) Choose the required bar diameter ($d_{b,req}$)

$$d_{b,req} = \sqrt{\frac{4M_{design}}{4\pi\pi_n F_{nt}(h_1 + h_2)}} = 21.08 \text{ mm} \quad \text{(B.36)}$$

($\phi_n = 0.9$ for non-ductile limit state, 4 bars arrangement in each row)
For easy of the construction, the same size of the tension bars should be used.

$$\text{Take } d_b = 25.4 \text{ mm (OK)} \quad \text{(B.37)}$$

#### B.2.2.3 Layout the shear bolts and tension bars

(8) Determine the gage length and bolt spacing for the shear bolts

For the easy construction, the same grade and size of bolts should be used for the shear bolts. Ten high strength bolts (A490) were used. Use lesser maximum gage that is permitted by the beam flange. The enough edge distance more than the required distance for 1" diameter bolts was used. The details for the shear bolt arrangement are

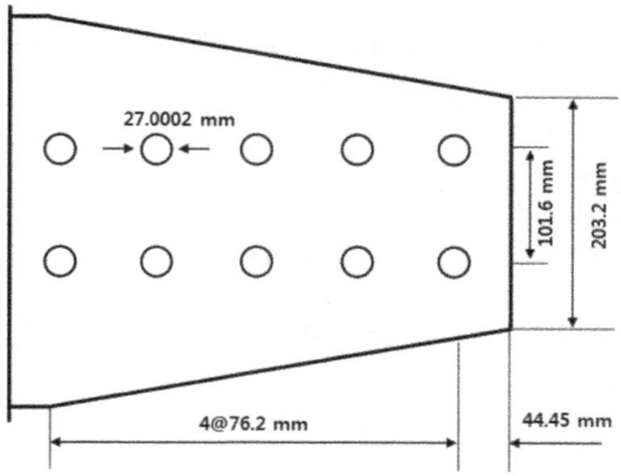


*Figure B.5* The arrangement of the shear bolts on the T-stub.

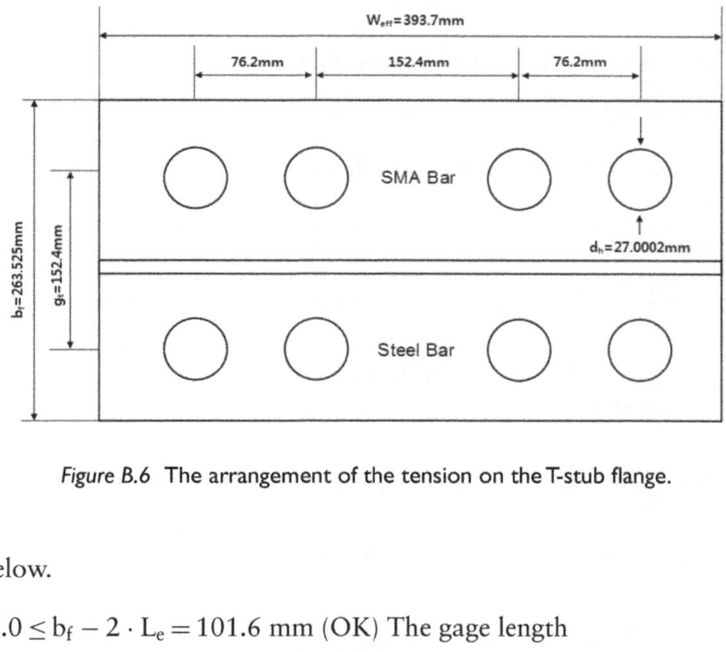


*Figure B.6* The arrangement of the tension on the T-stub flange.

given to below.

$$g_s = 4.0 \leq b_f - 2 \cdot L_e = 101.6 \text{ mm (OK) The gage length} \tag{B.38}$$

$$s = 3d_b = 76.2 \text{ mm (OK) The bolt spacing} \tag{B.39}$$

(9) Determine the arrangement for the tension bars
Use 6 inch center spacing to allow extra clearance: $W_{eff} = 393.7$ mm

### B.2.2.4 *Determine the thickness of the T-stem*

(10) Check the design strength model for the net section

$$\phi_f R_{n,net} = \phi_f F_u A_{net,stem} \text{ } (\phi_f = 0.75 \text{ for the fracture failure}) \tag{B.40}$$

(11) Determine the thickness of the T-stem based on the design strength model

$$n_{sb} = 2 \quad \text{The number of shear bolts along the T-stem width} \tag{B.41}$$

$$d_h = d_b + 1/16 = 27\ \text{mm} \tag{B.42}$$

$$t_{stem,min} = \frac{M_{design}}{\phi_f F_u \left(W_{eff} - n_{sb} d_h\right) d} = 13.1\ \text{mm}\ (F_u = F_{us} = 166\ \text{ksi, ASTM A490 Bolts}) \tag{B.43}$$

Check for $t_{stem,min}$ based on the bearing failure

$$t_{stem,min} = \frac{M_{design}}{\phi_f 2.4 m \cdot n_{sb} F_u d_b d} = 7.315\ \text{mm} \tag{B.44}$$

($m = 5\ m \cdot n_{sb}$: Total number of shear bolts)

$$t_{stem} = 14.2\ \text{mm (T-stub: Cut from W16X100, OK.)} \tag{B.45}$$

### B.2.2.5 Determine the $b_f$ and $t_f$ for the T-stub flange

(12) Determine the thickness of the T-stub flange

$$n_{tb} = 8 \quad \text{The total number of tension bars} \tag{B.46}$$

$$p = \frac{2W_{eff}}{n_{tb}} = 98.42\ \text{mm} \quad \text{The effective flange width per tension bar} \tag{B.47}$$

$$\delta = 1 - \frac{d_h}{p} = 0.726 \quad \text{The ratio of the net section area to the flange area} \tag{B.48}$$

The average bolt force should be used to estimate the bolt capacity. It causes to make design calculation easy.

$$T_{req} = \frac{P_{req}}{n_{tb}} = \text{kip} \quad \text{The required force at each tension bar} \tag{B.49}$$

Compare the required factored bar strength assuming a 40 percent prying and 8 tension bars.

$$\phi B_{req} = \frac{1.40 P_{req}}{n_{tb}} = 294.3 \leq F_{nt} = 433.6\ \text{kN (OK)} \tag{B.50}$$

$$b'_{f,req} = \left(\frac{\phi_f B_{req}}{T_{req}}\right) \cdot \left(\frac{1+\delta}{\delta}\right) \cdot 2d_b = 48.5\ \text{mm} \tag{B.51}$$

$$g_{t,req} = 2b'_{f,req} + 2t_{stem} = 125.47\ \text{mm} \tag{B.52}$$

$$\text{Take } g_t = 152.4\ \text{mm (OK)} \tag{B.53}$$

$$L_{e,min} = 31.75\ \text{mm} \quad \text{The minimum distance between the center of the bar hole and the edge} \tag{B.54}$$

$$b_{f,req} = g_t + 2L_{e,min} = 215.9 \text{ mm} \tag{B.55}$$

$$b_{f,req} = g_t + 4d_b = 254 \text{ mm} \tag{B.56}$$

$$\text{Take } b_t = 263.5 \text{ mm (OK)} \tag{B.57}$$

(13) Determine the width of the T-stub flange
Use adjusted geometric parameters ($a'$ and $b'$)

$$a' = a + \frac{d_b}{2} = 68.33 \text{ mm} \tag{B.58}$$

$$b' = b - \frac{d_b}{2} = 56.388 \text{ mm} \tag{B.59}$$

Find the range for $t_f$

$$t_{f,max} = \sqrt{\frac{4T_{req}b'}{\phi_b p F_y}} = 58.67 \text{ mm } (\phi_b = 0.75 \text{ for bolt bearing or fracture}) \tag{B.60}$$

$$t_{f,min} = \sqrt{\frac{T_{req}\left(8d_b b' - d_b\left(2d_b + b'\right)\right)}{3.75\phi_b d_b p F_y}} = 26.16 \text{ mm} \tag{B.61}$$

$$\text{Take } t_f = 25.4 \approx t_{f,min} \text{ mm (OK)} \tag{B.62}$$

#### B.2.2.6 *Check the T-stub section and failure modes*

After the T-stub section has been determined, the capacity of the T-stub section should be checked by looking at failure modes in either the flange or the stem. The failure mode checks in order to avoid brittle failure modes (net section failure at the T-stem, block shear failure, and shear bolt failure) are shown in Section B.3. The failure strength due to the brittle failures should be larger than the design strength based on the yielding of the beam.

#### B.2.2.7 *Determine the shear connection*

(14) Calculate the required shear force at the connection

$$V_u = \frac{2M_{design}}{L} = 456.13 \text{ kN} \tag{B.63}$$

$$L_{plate,max} = d - b_f = 335.3 \geq L_{plate} = 228.6 \text{ mm (OK)} \tag{B.64}$$

(15) Design the shear plate
Use four 1″ diameter A-490N bolts and two 0.56 inch thick plate would be sufficient with a capacity, $2\phi_n R = 783.2$ kN. The connection detail is given to Fig. B.8.

## B.3 FAILURE MODE CHECKS

The behavior of PR connections can be controlled by a number of different limit states including flexural yielding of the beam section, flexural yielding of the end-plates,

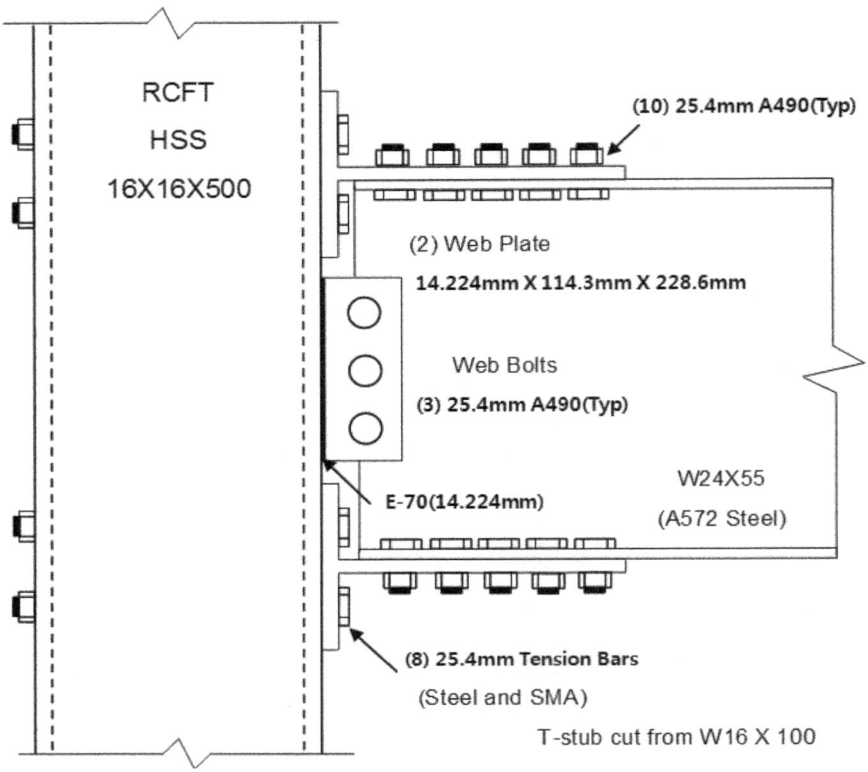


*Figure B.7* The connection details (T-stub connection).

yielding of the column panel zone, tension failure of the endplate bolts, shear failure of the end-plate bolts, or failure of various welds. The intent of the design criteria presented here is to provide sufficient strength in the components of the connections to ensure that the inelastic deformation of the connection is achieved by beam yielding.

### B.3.1 End-plate connection with RCFT columns

End-plate connection design check

Refer to AISC/ANSI 358-05 Manual

Refer to AISC-LRFD 2001
Refer to Eurocode 4 (Composite Column Design)
Refer to Steel Tip Manual (Astaneh-Asl, 1995)
Satisfy the prequalification limits: Table 6.1 in ANSI 358-05 (See Table B.1)
Use material properties for A.572 steel and A490 bolt based on the material test performed on SAC test models (Swanson, 2002).

*Table B.3* Material properties for the end-plate connection (Case 1).

| *Materials* | $F_y$ | $F_u$ |
|---|---|---|
| Beam | 378.95 MPa | 502.97 MPa |
| Column | 378.95 MPa | 502.97 MPa |
| Plate | 378.95 MPa | 502.97 MPa |
| Concrete | 22.048 MPa | 27.56 MPa |
| Steel Bar | 585.65 MPa | 1116.18 MPa |
| SMA Bar | 413.4 MPa | 551.2 MPa |

### B.3.1.1 *Given values*

Beam: W24X103
Column: HSS16X16X500
(From the standard shape in the AISC-LRFD 2001 specification)

| | |
|---|---|
| $E = 199810$ MPa | The elastic modulus for the steel material |
| $F_y = 378.95$ MPa | The yield stress for the steel material |
| $F_u = 502.97$ MPa | The ultimate stress for the steel material |
| $F_{u,steel} = 1116.18$ MPa | The ultimate strength for the steel bar |
| $F_{u,SMA} = 551.2$ MPa | The ultimate strength for the SMA bar |
| $d = 622.3$ mm | The depth of the beam |
| $t_f = 24.892$ mm | The thickness of the beam flange |
| $b_f = 228.6$ mm | The width of the beam flange |
| $t_w = 14.224$ mm | The thickness of the beam web |
| $S_{gy} = 401.8 \cdot 10^4$ mm$^3$ | The section modulus of the beam |
| $Z_{gy} = 467.4 \cdot 10^4$ mm$^3$ | The plastic section modulus of the beam |
| $h = 406.4$ mm | The height or width of the column |
| $t_c = 12.7$ mm | The thickness of the column |
| $b_p = 381$ mm | The width of the end-plate |
| $d_p = 977.9$ mm | The depth of the end-plate |
| $t_p = 25.4$ mm | The thickness of the end-plate |
| $Z_{py} = \frac{b_p d_p^2}{4} = 9115.12 \cdot 10^4$ mm$^3$ | The plastic section modulus of the end-plate |
| $d_b = 25.4$ mm | The diameter of the tension bar |
| $d_h = 26.924$ mm | The diameter of the bar hole |

Geometric parameters were illustrated in Fig. B.1.

| | |
|---|---|
| $\tau_{cu} = 44.785$ MPa | The ultimate strength of the confined concrete in panel zone (PZ) |
| $m = 2$ | The number of tension bars per row |
| $n = 8$ | The number of rows for the bar arrangement |

### B.3.1.2 *Determine the design strength*

The design strength ($M_P$) should be based on the full plastic strength of the beam.

$$M_P = Z_{gy} F_y = 1740.2 \text{ kN-m}$$

The design capacity for other component with the reduction factor ($\phi$) should be larger than the factored design strength in order to achieve the ideal failure at the connection.

An overstrength factor, taken as 1.25, was suggested in Astaneh-Asl (1995) in order to ensure that a ductile mode of behavior was reached.

### B.3.1.3 Ductile failure modes

#### B.3.1.3.1 *Slippage at the shear faying surface*

The check for the slippage strength is not available for the end-plate connections due to the lack of the shear faying surface.

#### B.3.1.3.2 *Bearing yielding around the shear bolt holes*

The check for bearing strength around the shear bolt holes is not available for the end-plate connections due to absence of the shear bolt.

#### B.3.1.3.3 *Yielding Failure of the gross section of the end-plate ($M_{P,end}$ – plate)*

$$\phi = 0.9 \quad \text{Design reduction factor for the yielding failure} \tag{B.65}$$

$$M_{P,end\text{ - }plate} = Z_{py}F_y = 34544.1 \text{ kN-m} \tag{B.66}$$

$$\phi \cdot M_{P,end\text{ - }plate} \geq 1.25 \cdot M_p \text{ (OK)} \tag{B.67}$$

Satisfy the limit to occupy the ductile failure due to yielding of the beam

### B.3.1.4 Mixed failure modes

#### B.3.1.4.1 *Local buckling at the beam flange*

$$\frac{b_f}{2 \cdot t_f} = 4.592 \tag{B.68}$$

$$\lambda_p = 0.38\sqrt{\frac{E}{F_y}} = 8.726 \text{ The compact slenderness ratio} \tag{B.69}$$

$$\frac{b_f}{2 \cdot t_f} \leq \lambda_p \text{ (OK)} \tag{B.70}$$

#### B.3.1.4.2 *Local buckling at the composite column*

$$\frac{h}{t_c} = 32 \tag{B.71}$$

$$42 \cdot \varepsilon = 42\sqrt{\frac{34.08}{F_y}} = 33.061 \text{ Local buckling check (Eurocode 4, 2004)} \tag{B.72}$$

$$\frac{h}{t_c} \leq 42 \cdot \varepsilon \text{ (OK) Compact column case} \tag{B.73}$$

*B.3.1.4.3 Shear yielding of the panel zone (PZ)*
Check shear yielding of the rectangular shaped panel zone (Wu *et al.*, 2005)

$$K_f = \frac{2Eh \cdot (t_c + t_p)^3}{(d - t_f)^2} \text{ The stiffness of the generalized column flange} \tag{B.74}$$

$$K_f = 991982.4 \text{ kN/m} \tag{B.75}$$

$$r_A = 1 - m\frac{d_h}{h - 2t_c} = 0.858 \text{ The area reduction factor due to bar} \tag{B.76}$$

$$r_C = \left(1 - 2\frac{nd_h}{d_b} + 2\frac{nd_h}{d_b r_A}\right)^{-1} = 0.897 \text{ The reduction factor} \tag{B.77}$$

$$r_w = \left(1 - \frac{nd_h}{d_b} + \frac{nd_h}{d_b r_A}\right)^{-1} = 0.946 \text{ The reduction factor for bar holes} \tag{B.78}$$

$$K_w = 2(h - 2t_c)\, t_c \frac{3E}{7} \text{ The stiffness of the column web} \tag{B.79}$$

$$K_w = 326.57 \cdot 10^5 \text{ kN/m} \tag{B.80}$$

$$A_c = (h - 2t_c)^2 = 145125 \text{ mm}^2 \text{ The area of the inside concrete} \tag{B.81}$$

Compute the yield shear strength ($V_y$) and ultimate shear strength ($V_u$) at the panel zone

$$V_y = \frac{2(h - 2t_c - md_h)\, t_c F_y}{\sqrt{3}}\left(1 + \frac{K_f}{r_w K_w}\right) + r_C \tau_{cu} A_c = 7716.3 \text{ kN} \tag{B.82}$$

$$V_u = \frac{2(h - 2t_c - md_h)\, t_c F_y}{\sqrt{3}} + r_C \tau_{cu} A_c + \frac{2h(t_c + t_p)^2 F_y}{3(d - t_f)} = 7907.65 \text{ kN} \tag{B.83}$$

$$1.25V_p = 1.25\frac{M_p}{d} = 3497.7 \text{ kN} \tag{B.84}$$

$$\phi = 0.8 \text{ The design reduction factor} \tag{B.85}$$

$$1.25V_p \le \phi V_y \text{ (OK)} \tag{B.86}$$

Satisfy the limit to occupy the ductile failure due to yielding of the beam

### B.3.1.5 Brittle (fracture) failure modes

*B.3.1.5.1 Block shear failure at the shear component*
The check for block shear failure is not applicable to the end-plate connection due to absence of a plate under direct shear force.

*B.3.1.5.2 Net section failure*
The check for the net section failure is not applicable to the end-plate connection due to absence of a plate under direct tension force.

#### *B.3.1.5.3 Fracture of the tension bars*

Apply the AISC/ANSI 358-05 specifications

$$A_b = \frac{\pi d_b^2}{4} = 506.325\ \text{mm}^2 \text{ The section area of the tension bar} \tag{B.87}$$

$$h_1 = 755.65\ \text{mm} \tag{B.88}$$

$$h_2 = 666.75\ \text{mm} \tag{B.89}$$

$$h_3 = 622.3\ \text{mm} \tag{B.90}$$

$$h_4 = 533.4\ \text{mm} \tag{B.91}$$

Here, $h_i$ ($I=1$ to 4) is the distance from the maximum bearing to each center of the bar hole (See Table B.5).

$$B_{n,steel} = F_{u,steel}A_b = 1005.7\ \text{Kn} \quad \text{The ultimate capacity for the steel tension bar} \tag{B.92}$$

$$B_{n,SMA} = F_{u,SMA}A_b = 569.6\ \text{Kn} \quad \text{The ultimate capacity for the SMA tension bar} \tag{B.93}$$

Calculate the ultimate moment capacity due to the tension bars ($M_{p,bar}$)

$$M_{p,bar} = 2B_{n,SMA}\left(h_1 + h_2\right) + 2B_{n,steel}\left(h_3 + h_4\right) = 2178.64\ \text{Kn-m} \tag{B.94}$$

$$1.25M_p \le M_{p,bar}\ (\text{OK}) \tag{B.95}$$

For the non-ductile design for the tension bars, the design factor which is 1.25 occupies the enough safety against the bolt fracture failure (Check the $d_{b,req}$, Section B2).

#### *B.3.1.5.4 Shear rupture failure at the tension bar*

$$B_s = n_b F_v A_b \text{ The capacity of the tension bar under shearing} \tag{B.96}$$

The nominal shear capacity of the SMA bar ($F_{v,SMA}$) is not provided on the specification. The check for the shear rupture failure shall be performed only at the steel tension bars with that of the steel bar ($F_{v,steel}$).

$$\phi = 0.75 \tag{B.97}$$

$$\phi F_{v,steel} = 304.5\ \text{MPa Single shear plane case (AISC 2001, 2001)} \tag{B.98}$$

$$n_b = 8 \text{ The number of steel tension bars} \tag{B.99}$$

$$\phi B_s = \phi n_b F_{v,steel} A_b = 1237.1\ \text{Kn} \tag{B.100}$$

$$L_b = 4445\ \text{mm} \quad \text{The distance from the tip of the beam to the center of the column} \tag{B.101}$$

$$\frac{1.25M_p}{L_b} = 525.1 \le \phi B_s\ (\text{OK}) \tag{B.102}$$

*B.3.1.5.5 End-plate rupture failure*

$$L_c = d_e - \frac{d_h}{2} = 30.99 \text{ mm} \tag{B.103}$$

$$1.2L_ct_pF_u = 476.15 \text{ kN} \tag{B.104}$$

$$r_i = 2.4d_ht_pF_u = 827.7 \text{ kN} \tag{B.105}$$

$$1.2L_ct_pF_u \leq 2.4d_ht_pF_u \text{ (OK)} \tag{B.106}$$

$$R_{n,tear} = 8r_i = 6626.05 \text{ kN} \tag{B.107}$$

$$\phi = 0.75 \tag{B.108}$$

$$\frac{1.25M_p}{L_b} \leq \phi R_{n,tear} \text{ (OK)} \tag{B.109}$$

*B.3.1.5.6 Fracture at the welding area*

Refer to the AISC 2001 specification (Section 16-J2)

Use submerged arc welding

$$F_{exx} = 689 \text{ MPa Electrode Strength (E100)} \tag{B.110}$$

$$\phi R_{n,w1} = 0.75F_{exx}0.6t_fb_f = 1762.2 \text{ kN} \quad \text{The welding capacity of the flange section} \tag{B.111}$$

$$\phi R_{n,w2} = 0.75F_{exx}0.6t_w\left(d - 2t_f\right) = 4423.3 \text{ kN} \quad \text{The welding capacity of the flange section} \tag{B.112}$$

$$\phi M_{p,weld} = \phi R_{n,w1}\left(d - t_f\right) + \phi R_{n,w2}\frac{d - 2t_f}{2} = 2321 \text{ kN-m} \quad \text{Moment Capacity at the welding} \tag{B.113}$$

$$1.25M_p \leq \phi M_{p,weld} \text{ (OK)} \tag{B.114}$$

Satisfy the limit to occupy the ductile failure due to yielding of the beam

## B.3.2 End-plate connection with CCFT columns

End-plate connection design check
Refer to AISC/ANSI 358-05 Manual
Refer to AISC-LRFD 2001
Refer to Eurocode 4 (Composite Column Design)
Refer to Steel Tip Manual (Astaneh-Asl, 1995)
Satisfy the prequalification limits: Table 6.1 in ANSI 358-05 (See Table B.1)
Use material properties for A.572 steel and A490 bolt based on the material test performed on SAC test models (Swanson, 2002).

### B.3.2.1 Given values

Beam: W24X103
Column: HSS18X500
(From the standard shape in the AISC-LRFD 2001 specification)

*Table B.4* Material properties for the end-plate connection (Case 2).

| Materials | $F_y$ | $F_u$ |
|---|---|---|
| Beam | 378.95 MPa | 502.97 MPa |
| Column | 378.95 MPa | 502.97 MPa |
| Plate | 378.95 MPa | 502.97 MPa |
| Concrete | 22.048 MPa | 27.56 MPa |
| Steel Bar | 585.65 MPa | 1116.18 MPa |
| SMA Bar | 413.4 MPa | 551.2 MPa |

Beam: W24X103
Column: HSS18X500
(From the standard shape in the AISC-LRFD 2001 specification)

| | |
|---|---|
| $E = 199810$ MPa | The elastic modulus for the steel material |
| $F_y = 378.95$ MPa | The yield stress for the steel material |
| $F_u = 502.97$ MPa | The ultimate stress for the steel material |
| $F_{u,steel} = 1116.18$ MPa | The ultimate strength for the steel bar |
| $F_{u,SMA} = 551.2$ MPa | The ultimate strength for the SMA bar |
| $d = 622.3$ mm | The depth of the beam |
| $t_f = 24.892$ mm | The thickness of the beam flange |
| $b_f = 228.6$ mm | The width of the beam flange |
| $t_w = 14.224$ mm | The thickness of the beam web |
| $S_{gy} = 401.8 \cdot 10^4$ mm$^3$ | The section modulus of the beam |
| $Z_{gy} = 467.4 \cdot 10^4$ mm$^3$ | The plastic section modulus of the beam |
| $h = 457.2$ mm | The height or width of the column |
| $t_c = 12.7$ mm | The thickness of the column |
| $b_p = 381$ mm | The width of the end-plate |
| $d_p = 977.9$ mm | The depth of the end-plate |
| $t_p = 25.4$ mm | The thickness of the end-plate |
| $Z_{py} = \frac{b_p d_p^2}{4} = 9115.12 \cdot 10^4$ mm$^3$ | The plastic section modulus of the end-plate |
| $d_b = 25.4$ mm | The diameter of the tension bar |
| $d_h = 26.924$ mm | The diameter of the bar hole |

Geometric parameters were illustrated in Figure B.1

| | |
|---|---|
| $\tau_{cu} = 44.785$ MPa | The ultimate strength of the confined concrete in PZ |
| $m = 2$ | The number of tension bars per row |
| $n = 8$ | The number of rows for the bar arrangement |

### B.3.2.2 *Determine the design strength*

The design strength ($M_P$) should be based on the full plastic strength of the beam. $M_P = Z_{gy} F_y = 1740.2$ kN-m

### B.3.2.3 *Ductile failure modes*

#### B.3.2.3.1 *Slippage at the shear faying surface*

The check for the slippage strength is not available for the end-plate connections due to the lack of the shear faying surface.

*B.3.2.3.2 Bearing yielding around the shear bolt holes*
The check for bearing strength around the shear bolt holes is not available for the end-plate connections due to absence of the shear bolt.

*B.3.2.3.3 Yielding failure of the gross section of the end-plate ($M_{P,end} - plate$)*

$$\phi = 0.9 \quad \text{Design reduction factor for the yielding failure} \tag{B.115}$$

$$M_{P,end\text{ - }plate} = Z_{py}F_y = 34544.1 \text{ kN-m} \tag{B.116}$$

$$\phi \cdot M_{P,end\text{ - }plate} \geq 1.25 \cdot M_p \text{ (OK)} \tag{B.117}$$

Satisfy the limit to occupy the ductile failure due to yielding of the beam

### B.3.2.4 Mixed failure modes

*B.3.2.4.1 Local buckling at the beam flange*

$$\frac{b_f}{2 \cdot t_f} = 4.592 \tag{B.118}$$

$$\lambda_p = 0.38\sqrt{\frac{E}{F_y}} = 8.726 \text{ The compact slenderness ratio} \tag{B.119}$$

$$\frac{b_f}{2 \cdot t_f} \leq \lambda_p \text{ (OK)} \tag{B.120}$$

*B.3.2.4.2 Local buckling at the composite column*

$$\frac{h}{t_c} = 36 \tag{B.121}$$

$$60 \cdot \varepsilon^2 = 60\left(\frac{34.08}{F_Y}\right) = 37.16 \text{ Local buckling check (Eurocode 4, 2004)} \tag{B.122}$$

$$\frac{h}{t_c} \leq 60 \cdot \varepsilon^2 \text{ (OK) Compact column case} \tag{B.123}$$

*B.3.2.4.3 Shear yielding of the panel zone (PZ)*
Check shear yielding of the rectangular shaped panel zone (Wu *et al.*, 2005)

$$K_f = \frac{2Eh \cdot (t_c + t_p)^3}{(d - t_f)^2} \text{ The stiffness of the generalized column flange} \tag{B.124}$$

$$K_f = 111.58 \cdot 10^4 \text{ kN/m} \tag{B.125}$$

$$r_A = 1 - m\frac{d_h}{h - 2t_c} = 0.875 \text{ The area reduction factor due to bar holes} \tag{B.126}$$

$$r_C = \left(1 - 2\frac{nd_h}{d_b} + 2\frac{nd_h}{d_b r_A}\right)^{-1} = 0.910 \text{ The reduction factor} \tag{B.127}$$

$$r_w = \left(1 - \frac{nd_h}{d_b} + \frac{nd_h}{d_b r_A}\right)^{-1} = 0.953 \text{ The reduction factor for bar holes} \quad (B.128)$$

$$K_w = 2\left(h - 2t_c\right) t_c \frac{3E}{7} \text{ The stiffness of the column web} \quad (B.129)$$

$$K_w = 370.2 \cdot 10^5 \text{ kN/mm} \quad (B.130)$$

$$A_c = \left(h - 2t_c\right)^2 = 146415 \text{ mm}^2 \text{ The area of the inside concrete} \quad (B.131)$$

Compute the yield shear strength ($V_y$) and ultimate shear strength ($V_u$) at the panel zone

$$V_y = \frac{2\left(h - 2t_c - md_h\right) t_c F_y}{\sqrt{3}} \left(1 + \frac{K_f}{r_w K_w}\right) + r_C \tau_{cu} A_c = 8317 \text{ kW} \quad (B.132)$$

$$V_u = \frac{2\left(h - 2t_c - md_h\right) t_c F_y}{\sqrt{3}} + r_C \tau_{cu} A_c + \frac{2h\left(t_c + t_p\right)^2 F_y}{3\left(d - t_f\right)} = 8357 \text{ kN} \quad (B.133)$$

$$1.25V_p = 1.25\frac{M_p}{d} = 3497.7 \text{ kN} \quad (B.134)$$

$$\phi = 0.8 \text{ The design reduction factor} \quad (B.135)$$

$$1.25V_p \le \phi V_y \text{ (OK)} \quad (B.136)$$

Satisfy the limit to occupy the ductile failure due to yielding of the beam

### B.3.2.5 Brittle (fracture) failure modes

#### *B.3.2.5.1 Block shear failure at the shear component*

The check for the block shear failure is not available for the end-plate connection due to absence of the plate under the direct shear force.

#### *B.3.2.5.2 Net section failure*

The check for the net section failure is not available for the end-plate connection due to absence of the plate under the direct shear force.

#### *B.3.2.5.3 Fracture of the tension bars*

Apply the AISC/ANSI 358-05 specifications

$$A_b = \frac{\pi d_b^2}{4} = 506.3 \text{ mm}^2 \text{ The section area of the tension bar} \quad (B.137)$$

$$h_1 = 755.65 \text{ mm} \quad (B.138)$$

$$h_2 = 666.75 \text{ mm} \quad (B.139)$$

$$h_3 = 622.3 \text{ mm} \quad (B.140)$$

$$h_4 = 533.4 \text{ mm} \quad (B.141)$$

Here, $h_i$ (i = 1 to 4) is the distance from the maximum bearing to each center of the bar hole (See Table B.5).

$$B_{n,steel} = F_{u,steel}A_b = 1183.7 \text{ kN The ultimate capacity for the steel tension bar} \tag{B.142}$$

$$B_{n,SMA} = F_{u,SMA}A_b = 569.6 \text{ kN The ultimate capacity for the SMA tension bar} \tag{B.143}$$

Calculate the ultimate moment capacity due to the tension bars ($M_{p,bar}$)

$$M_{p,bar} = 2B_{n,SMA}(h_1 + h_2) + 2B_{n,steel}(h_3 + h_4) = 20178 \text{ kN-m} \tag{B.144}$$

$$1.25M_p \leq M_{p,bar} \text{ (OK)} \tag{B.145}$$

For the non-ductile design for the tension bars, the design factor which is 1.25 occupies the enough safety against the bolt fracture failure (Check the $d_{b,req}$, Section B2).

*B.3.2.5.4 Shear Rupture Failure at the Tension Bar*

$$B_s = n_b F_v A_b \text{ The capacity of the tension bar under shearing} \tag{B.146}$$

The nominal shear capacity of the SMA bar ($F_{v,SMA}$) is not provided on the specification. The check for the shear rupture failure shall be performed only at the steel tension bars with that of the steel bar ($F_{v,steel}$).

$$\phi = 0.75 \tag{B.147}$$

$$\phi F_{v,steel} = 304.5 \text{ MPa Single shear plane case (AISC 2001, Table 7-10)} \tag{B.148}$$

$$n_b = 8 \text{ The number of steel tension bars} \tag{B.149}$$

$$\phi B_s = \phi n_b F_{v,steel} A_b = 1237.1 \text{ kN} \tag{B.150}$$

$$L_b = 4445 \text{ mm The distance from the tip of the beam to the center of the column} \tag{B.151}$$

$$\frac{1.25M_p}{L_b} = 525.1 \leq \phi B_s \text{ (OK)} \tag{B.152}$$

*B.3.2.5.5 End-plate rupture failure*

$$L_c = d_e - \frac{d_h}{2} = 30.98 \text{ mm} \tag{B.153}$$

$$1.2L_c t_p F_u = 476.15 \text{ kN} \tag{B.154}$$

$$r_i = 2.4 d_h t_p F_u = 827.7 \text{ kN} \tag{B.155}$$

$$1.2L_c t_p F_u \leq 2.4 d_h t_p F_u \text{ (OK)} \tag{B.156}$$

$$R_{n,tear} = 8r_i = 6626.05 \text{ kN} \tag{B.157}$$

$$\phi = 0.75 \tag{B.158}$$

$$\frac{1.25M_p}{L_b} \leq \phi R_{n,tear} \text{ (OK)} \tag{B.159}$$

#### *B.3.2.5.6 Fracture at the welding area*

Refer to the AISC 2001 specification (Section 16-J2)
Use submerged arc welding

$$F_{exx} = 689 \text{ MPa Electrode Strength (E100)} \tag{B.160}$$

$$\phi R_{n,w1} = 0.75F_{exx}0.6t_f b_f = 1762 \text{ kN} \quad \text{The welding capacity of the flange section} \tag{B.161}$$

$$\phi R_{n,w2} = 0.75F_{exx}0.6t_w\left(d - 2t_f\right) = 994 \text{ kN} \quad \text{The welding capacity of the flange section} \tag{B.162}$$

$$\phi M_{p,weld} = \phi R_{n,w1}\left(d - t_f\right) + \phi R_{n,w2}\frac{d - 2t_f}{2} = 2321 \text{ kN-m} \quad \text{Moment Capacity at the welding} \tag{B.163}$$

$$1.25M_p \leq \phi M_{p,weld} \text{ (OK)} \tag{B.164}$$

Satisfy the limit to occupy the ductile failure due to yielding of the beam

## B.3.3 T-stub connection with RCFT columns

T-stub connection design check
Refer to AISC-LRFD 2001
Refer to Eurocode 4 (Composite Column Design)
Refer to Steel Tip Manual (Astaneh-Asl, 1995)
Use material properties for A.572 steel and A490 bolt based on the material test performed on SAC test models (Swanson, 2001).

### *B.3.3.1 Given values*

Beam: W24X55
Column: HSS16X16X500
T-stub: Cut from W16X100
(From the standard shape in the AISC-LRFD 2001 specification)

| | |
|---|---|
| $E = 199810$ MPa | The elastic modulus for the steel material |
| $F_y = 378.95$ MPa | The yield stress for the steel material |
| $F_u = 502.97$ MPa | The ultimate stress for the steel material |
| $F_{u,steel} = 1116.18$ MPa | The ultimate strength for the steel bar |
| $F_{u,SMA} = 551.2$ MPa | The ultimate strength for the SMA bar |
| $d = 599.44$ mm | The depth of the beam |
| $t_f = 12.827$ mm | The thickness of the beam flange |
| $b_f = 178.81$ mm | The width of the beam flange |

*Table B.5* Material properties for the T-stub connection (Case 3).

| Materials | $F_y$ | $F_u$ |
|---|---|---|
| Beam | 378.95 MPa | 502.97 MPa |
| Column | 378.95 MPa | 502.97 MPa |
| T-Stub | 378.95 MPa | 502.97 MPa |
| Concrete | 22.048 MPa | 27.56 MPa |
| Steel Bar | 585.65 MPa | 1116.18 MPa |
| SMA Bar | 413.4 MPa | 551.2 MPa |
| Shear Bolt | 585.65 MPa | 1116.18 MPa |
| Web Bolts | 585.65 MPa | 1116.18 MPa |

$t_w = 10.03$ mm The thickness of the beam web
$S_{gy} = 188.6 \cdot 10^4$ mm$^3$ The section modulus of the beam
$Z_{gy} = 221.4 \cdot 10^4$ mm$^3$ The plastic section modulus of the beam
$h = 406.4$ mm The height or width of the column
$t_c = 12.7$ mm The thickness of the column
$W_{eff} = 315.5$ mm The effective width of the T-stub
$g_t = 152.4$ mm The gage length
$b_f = 263.53$ mm The height of the T-stub
$t_{tf} = 25.4$ mm The thickness of the T-stub flange
$t_{stem} = 14.22$ mm The thickness of the T-stem
$d_b = 25.4$ mm The diameter of the tension bar
$d_h = 26.92$ mm The diameter of the bar hole
$d_s = 25.4$ mm The diameter of the shear bolt
Geometric parameters were illustrated in Figure B.7
$\tau_{cu} = 44.785$ MPa The ultimate strength of the confined concrete in PZ
$m = 4$ The number of tension bars per row
$n = 4$ The number of rows for the bar arrangement
$T_m = 284.8$ The initial pretension force for the shear bolt
(AISC-LRFD 2001, Table 8.1)
$A_b = 506.32$ mm$^2$ The cross section area of shear bolt

#### B.3.3.2 *Determine the design strength*

The design strength ($M_P$) should be based on the full plastic strength of the beam.

$$M_P = Z_{gy} F_y = 839 \text{ kN-m}$$

#### B.3.3.3 *Ductile failure modes*

*Slippage at the shear faying surface*

$$u = 0.33 \quad \text{The mean slip coefficient (Class A coating)} \tag{B.165}$$

$$D_u = 1.13 \quad \text{The multiplier} \tag{B.166}$$

$$T_m = 284.8 \text{ kN} \quad \text{The specified minimum pretension (AISC Table 7.15)} \tag{B.167}$$

$$T_u = F_{u,steel}A_b = 574.05 \text{ kN} \quad \text{The required strength in tension} \tag{B.168}$$

$$N_b = 10 \quad \text{The number of shear bolts} \tag{B.169}$$

$$\phi = 1.0 \quad \text{The reduction factor for the standard hole} \tag{B.170}$$

The nominal strength for the slip resistance ($R_{n,slip}$) can be calculated.

$$R_{n,slip} = uD_uT_mN_b\left(1 - \frac{T_u}{D_uT_mN_b}\right) = 872.2 \text{ kN} \tag{B.171}$$

$$M_{p,slip} = R_{n,slip}d = 523.19 \text{ kN-m} \tag{B.172}$$

$$\phi M_{p,slip} \leq 0.8M_p = 671.22 \text{ kN-m (OK)} \tag{B.173}$$

#### *B.3.3.3.2 Bearing yielding around the shear bolt hole*

$$\phi = 0.9 \text{ Design reduction factor for the yielding failure} \tag{B.174}$$

$$R_{n,bearing} = 2.4F_yd_hN_bt_{stem} = 3497.7 \text{ kN} \tag{B.175}$$

$$M_{p,bearing} = R_{n,slip}d = 2095 \text{ kN-m} \tag{B.176}$$

$$\phi M_{p,bearing} \geq 1.25M_p = 1048.7 \text{ kN-m (OK)} \tag{B.177}$$

#### *B.3.3.3.3 Yielding failure of the gross section of the t-stem (MP,stem)*

$$\phi = 0.9 \text{ Design reduction factor for the yielding failure} \tag{B.178}$$

$$A_{stem} = W_{eff}t_{stem} = 5598 \text{ mm}^2 \text{ The gross section area of the T-stem} \tag{B.179}$$

$$R_{n,stem} = F_yA_{stem} = 2127 \text{ kN} \tag{B.180}$$

$$M_{P,stem} = R_{n,stem}d = 1273.5 \text{ kN-m} \tag{B.181}$$

$$\phi \cdot M_{P,stem} \geq 1.25 \cdot M_p \text{ (OK)} \tag{B.182}$$

Satisfy the limit to occupy the ductile failure due to yielding of the beam

### *B.3.3.4 Mixed failure modes*

#### *B.3.3.4.1 Local buckling at the beam flange*

$$\frac{b_f}{2 \cdot t_f} = 6.97 \tag{B.183}$$

$$\lambda_p = 0.38\sqrt{\frac{E}{F_y}} = 8.726 \quad \text{The compact slenderness ratio} \tag{B.184}$$

$$\frac{b_f}{2 \cdot t_f} \leq \lambda_p \quad \text{(OK)} \tag{B.185}$$

*B.3.3.4.2 Local buckling at the composite column*

$$\frac{h}{t_c} = 32 \quad \text{(B.186)}$$

$$42 \cdot \varepsilon = 42\sqrt{\frac{34.08}{F_y}} = 33.061 \quad \text{Local buckling check (Eurocode 4, 2004)} \quad \text{(B.187)}$$

$$\frac{h}{t_c} \leq 42 \cdot \varepsilon \quad \text{(OK) Compact column case} \quad \text{(B.188)}$$

*B.3.3.4.3 Shear yielding of the panel zone (PZ)*

Check shear yielding of the rectangular shaped panel zone (Wu *et al.*, 2005)

$$K_f = \frac{2Eh \cdot (t_c + t_{tf})^3}{(d - t_f)^2} \quad \text{The stiffness of the generalized column flange} \quad \text{(B.189)}$$

$$K_f = 1028.7 \cdot 10^3 \text{ kN/m} \quad \text{(B.190)}$$

$$r_A = 1 - m\frac{d_h}{h - 2t_c} = 0.717 \quad \text{The area reduction factor due to bar holes} \quad \text{(B.191)}$$

$$r_C = \left(1 - 2\frac{nd_h}{d_b} + 2\frac{nd_h}{d_b r_A}\right)^{-1} = 0.875 \quad \text{The reduction factor} \quad \text{(B.192)}$$

$$r_w = \left(1 - \frac{nd_h}{d_b} + \frac{nd_h}{d_b r_A}\right)^{-1} = 0.934 \quad \text{The reduction factor for bar holes} \quad \text{(B.193)}$$

$$K_w = 2\left(h - 2t_c\right) t_c \frac{3E}{7} \quad \text{The stiffness of the column web} \quad \text{(B.194)}$$

$$K_w = 326.6 \cdot 10^5 \text{ kN/m} \quad \text{(B.195)}$$

$$A_c = \left(h - 2t_c\right)^2 = 145125 \text{ mm}^2 \quad \text{The area of the inside concrete} \quad \text{(B.196)}$$

Compute the yield shear strength ($V_y$) and ultimate shear strength ($V_u$) at the panel zone

$$V_y = \frac{2\left(h - 2t_c - md_h\right) t_c F_y}{\sqrt{3}} \left(1 + \frac{K_f}{r_w K_w}\right) + r_C \tau_{cu} A_c = 7266.85 \text{ kN} \quad \text{(B.197)}$$

$$V_u = \frac{2\left(h - 2t_c - md_h\right) t_c F_y}{\sqrt{3}} + r_C \tau_{cu} A_c + \frac{2h\left(t_c + t_p\right)^2 F_y}{3\left(d - t_f\right)} = 7667 \text{ kN} \quad \text{(B.198)}$$

$$1.25V_p = 1.25\frac{M_p}{d} = 1748 \text{ kN} \quad \text{(B.199)}$$

$$\phi = 0.8 \quad \text{The design reduction factor} \quad \text{(B.200)}$$

$$1.25V_p \leq \phi V_y \text{ (OK)} \quad \text{(B.201)}$$

Satisfy the limit to occupy the ductile failure due to yielding of the beam

### B.3.3.5 Brittle (fracture) failure modes

#### B.3.3.5.1 Block shear failure at the shear component

Check the block shear failure at the T-stub

$$s = 101.6\ \text{mm} \quad \text{The shear bolt spacing} \tag{B.202}$$

$$A_{gt} = s \cdot t_{stem} = 1444.8\ \text{mm}^2 \quad \text{The gross area subjected to tension} \tag{B.203}$$

$$A_{nt} = \left(s - d_h\right) \cdot t_{stem} = 1061\ \text{mm}^2 \quad \text{The net section area subjected to tension} \tag{B.204}$$

$$d_e = 44.45\ \text{mm} \quad \text{The edge distance} \tag{B.205}$$

$$A_{gv} = 2(d_e + 3 \cdot 4) \cdot t_{stem} = 9933\ \text{mm}^2 \quad \text{The gross area subjected to shear} \tag{B.206}$$

$$A_{nv} = 2(d_e + 3 \cdot 4 - 4.5d_h) \cdot t_{stem} = 6482\ \text{mm}^2 \quad \text{The net section area subjected to shear} \tag{B.207}$$

Check the failure condition

$$F_u A_{nt} = 120.1 \leq 0.6 F_u A_{nv} = 1958\ \text{kN (OK)} \tag{B.208}$$

$$R_{n,block} = 0.6 F_u A_{nv} + F_y A_{gt} = 2506\ \text{kN} \tag{B.209}$$

Check the design reduction factors

$$\phi_f = 0.75 \quad \text{The design reduction factor for the fracture} \tag{B.210}$$

$$\phi_y = 0.90 \quad \text{The design reduction factor for the yielding} \tag{B.211}$$

$$\phi_f R_{n,block} = 1879\ \text{kN} \tag{B.212}$$

$$1.25\phi_y \frac{M_p}{d} = 353.95 \leq \phi_f R_{n,block}\ \text{(OK)} \tag{B.213}$$

#### B.3.3.5.2 Net section failure

$$R_{n,net} = F_u A_{stem} = 2819.5\ \text{kN} \tag{B.214}$$

$$\phi_f R_{n,net} = 2537.6 \tag{B.215}$$

$$1.25\phi_y \frac{M_p}{d} = 1575 \leq \phi_f R_{n,net}\ \text{(OK)} \tag{B.216}$$

#### B.3.3.5.3 Fracture of the tension bars

Apply the AISC/ANSI 358-05 specifications

$$A_b = \frac{\pi d_b^2}{4} = 506.3\ \text{mm}^2 \quad \text{The section area of the tension bar} \tag{B.217}$$

$$B_{n,steel} = F_{u,steel} A_b = 591.85\ \text{kN} \quad \text{The ultimate capacity for the steel tension bar} \tag{B.218}$$

$$B_{n,SMA} = F_{u,SMA}A_b = 284.8\text{ kN} \quad \text{The ultimate capacity for the SMA tension bar} \tag{B.219}$$

Calculate the ultimate moment capacity due to the tension bars ($M_{p,bar}$)

$$M_{p,bar} = 4B_{n,SMA}\left(d + \frac{g_t}{2} - \frac{t_f}{2}\right) + 4B_{n,steel}\left(d - \frac{g_t}{2} - \frac{t_f}{2}\right) = 1947\text{ kN-m} \tag{B.220}$$

$$\phi_f M_{p,bar} = 1459.9\text{ kN-m} \tag{B.221}$$

$$1.25\phi_y M_p = 943.9 \leq \phi_f M_{p,bar}\ (\text{OK}) \tag{B.222}$$

*B.3.3.5.4 Shear rupture failure at the tension bar*

$$B_s = n_b F_v A_b \quad \text{The capacity of the tension bar under shearing} \tag{B.223}$$

The nominal shear capacity of the SMA bar ($F_{v,SMA}$) is not provided on the specification. The check for the shear rupture failure shall be performed only at the steel tension bars with that of the steel bar ($F_{v,steel}$).

$$\phi = 0.75 \tag{B.224}$$

$$\phi F_{v,steel} = 304.5\text{ MPa Single shear plane case (AISC 2001, Table 7-10)} \tag{B.225}$$

$$n_b = 8 \text{ The number of steel tension bars} \tag{B.226}$$

$$\phi B_s = \phi n_b F_{v,steel} A_b = 1237\text{ kN} \tag{B.227}$$

$$L_b = 4445\text{ mm} \quad \text{The distance from the tip of the beam to the center of the column} \tag{B.228}$$

$$\frac{1.25M_p}{L_b} = 236.07 \leq \phi B_s\ (\text{OK}) \tag{B.229}$$

*B.3.3.5.5 Shear rupture failure at the shear bolts*

$$\phi = 0.75 \tag{B.230}$$

$$\phi F_v = 304.54\text{ MPa Single shear plane case (AISC 2001, Table 7-10)} \tag{B.231}$$

$$n_s = 10 \text{ The number of shear bolts} \tag{B.232}$$

$$A_s = \frac{\pi d_s^2}{4} = 506.33\text{ mm}^2 \text{ The cross section area of the shear bolts} \tag{B.233}$$

$$\phi R_{n,shear} = \phi n_s F_v A_s = 1557.5\text{ kN} \tag{B.234}$$

$$\frac{1.25\phi_y M_p}{d} = 1575.1 \approx \phi B_s\ (\text{OK}) \tag{B.235}$$

*B.3.3.5.6 T-stub rupture failure*

$$d_e = 44.45\text{ mm The edge distance (Refer to Figure B.6)} \tag{B.236}$$

$$L_c = d_e - \frac{d_h}{2} = 30.99\text{ mm} \tag{B.237}$$

$$1.2L_c t_{stem} F_u = 266.33\text{ kN} \quad (B.238)$$

$$r_i = 2.4 d_h t_{stem} F_u = 464.1\text{ kN} \quad (B.239)$$

$$1.2L_c t_p F_u \le 2.4 d_h t_p F_u \text{ (OK)} \quad (B.240)$$

$$R_{n,tear} = 10 r_i = 3712.8\text{ kN} \quad (B.241)$$

$$\phi = 0.75 \quad (B.242)$$

$$\frac{1.25M_p}{d} \le \phi R_{n,tear} \text{ (OK)} \quad (B.243)$$

*B.3.3.5.7 Shear tab failure*

Refer to AISC-LRFD 2001 (Table 10.1)
Use 4.5X9.5X0.56 double plate for shear tab.
Shear tab has enough strength to resist the applied shear force ($1.25M_p/L_b =$ 236.1 kN).
Satisfy the limit to occupy the ductile failure due to yielding of the beam

*B.3.3.5.8 Fracture at the weld area*

There are no welds used in this connection. Thus, the weld failure checks are not available for this connection system.

## B.3.4 T-stub connection with CCFT columns

T-stub connection with CCFT columns were designed with the same component as that with RCFT columns except for the panel zone and composite columns. Both cases show the same capacity against the failure modes. Thus, the procedures to estimate the identical failure strength with Case 3 are omitted in this section. Only mixed failure modes for the panel zone and composite columns will be investigated.

### B.3.4.1 Mixed failure modes

*B.3.4.1.1 Local buckling at the beam flange*

$$\frac{b_f}{2 \cdot t_f} = 6.97 \quad (B.244)$$

$$\lambda_p = 0.38\sqrt{\frac{E}{F_y}} = 8.726 \text{ The compact slenderness ratio} \quad (B.245)$$

$$\frac{b_f}{2 \cdot t_f} \le \lambda_p \text{ (OK)} \quad (B.246)$$

*B.3.4.1.2 Local buckling at the composite column*

Use HSS 18X500 size columns

$$\frac{h}{t_c} = 36 \quad (B.247)$$

$$60 \cdot \varepsilon^2 = 60\left(\frac{34.08}{F_Y}\right) = 37.16 \text{ Local buckling check (Eurocode 4, 2004)} \quad (B.248)$$

$$\frac{h}{t_c} \leq 60 \cdot \varepsilon^2 \text{ (OK) Compact column case} \quad \text{(B.249)}$$

*B.3.4.1.3 Shear yielding of the panel zone (PZ)*

Check shear yielding of the rectangular shaped panel zone (Wu *et al.*, 2005)

$$K_f = \frac{2Eh \cdot (t_c + t_p)^3}{(d - t_f)^2} \text{ The stiffness of the generalized column flange} \quad \text{(B.250)}$$

$$K_f = 1115.85 \cdot 10^3 \text{ kN/m} \quad \text{(B.251)}$$

$$r_A = 1 - m\frac{d_h}{h - 2t_c} = 0.875 \text{ The area reduction factor due to bar holes} \quad \text{(B.252)}$$

$$r_C = \left(1 - 2\frac{nd_h}{d_b} + 2\frac{nd_h}{d_b r_A}\right)^{-1} = 0.910 \text{ The reduction factor} \quad \text{(B.253)}$$

$$r_W = \left(1 - \frac{nd_h}{d_b} + \frac{nd_h}{d_b r_A}\right)^{-1} = 0.953 \text{ The reduction factor for bar holes} \quad \text{(B.254)}$$

$$K_w = 2\,(h - 2t_c)\,t_c\frac{3E}{7} \text{ The stiffness of the column web} \quad \text{(B.255)}$$

$$K_w = 370.2 \cdot 10^5 \text{ kN/m} \quad \text{(B.256)}$$

$$A_c = (h - 2t_c)^2 = 146415 \text{ mm}^2 \text{ The area of the inside concrete} \quad \text{(B.257)}$$

Compute the yield shear strength ($V_y$) and ultimate shear strength ($V_u$) at the panel zone

$$V_y = \frac{2\,(h - 2t_c - md_h)\,t_c F_y}{\sqrt{3}}\left(1 + \frac{K_f}{r_w K_w}\right) + r_C \tau_{cu} A_c = 8143.5 \text{ kN} \quad \text{(B.258)}$$

$$V_u = \frac{2\,(h - 2t_c - md_h)\,t_c F_y}{\sqrt{3}} + r_C \tau_{cu} A_c + \frac{2h\,(t_c + t_p)^2 F_y}{3\,(d - t_f)} = 8357.1 \text{ kN} \quad \text{(B.259)}$$

$$1.25V_p = 1.25\frac{M_p}{d} = 3497.7 \text{ kN} \quad \text{(B.260)}$$

$$\phi = 0.8 \text{ The design reduction factor} \quad \text{(B.261)}$$

$$1.25V_p \leq \phi V_y \text{ (OK)} \quad \text{(B.262)}$$

Satisfy the limit to occupy the ductile failure due to yielding of the beam

# Appendix C

# Detail design examples for panel zones

## C.1 INTRODUCTION

The strength models for the composite panel zone were described in Chapter 3. Based on these strength models, the calculations using the theoretical equations for the stiffness, yield shear strength, and ultimate shear strength in the panel zone are illustrated in detail in this appendix. The geometric configuration and notation are given in Fig. C.1. A rectangular concrete filled-tube panel zone was made up of steel and concrete, with both materials contributing to the stiffness and strength mechanism. The two materials can be assumed to behave independently, and strength superposition may be applied to the theoretical equations (Wu *et al.*, 2007).

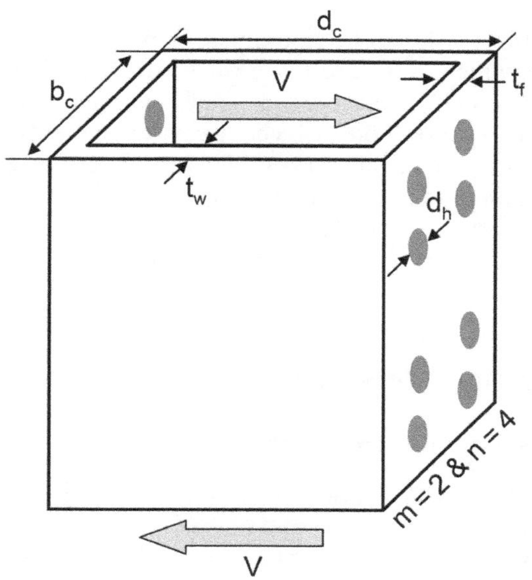


*Figure C.1* The geometric dimensions of the panel zone.

*Table C.1* Summary of the geometric dimensions for panel zone models.

| *shape ID* | *Connection type* | *Beam size* | *Column size* | $b_c = d_c$ | $t_f = t_w$ | $d_h$ |
|---|---|---|---|---|---|---|
| PZ Case1 | End-Plate | W24 × 103 | HSS16 × 16 × 500 | 406.4 mm | 12.7 mm | 27 mm |
| PZ Case2 | End-Plate | W24 × 103 | HSS18 × 500 | 457.2 mm | 12.7 mm | 27 mm |
| PZ Case3 | End-Plate | W24 × 84 | HSS16 × 16 × 500 | 406.4 mm | 12.7 mm | 27 mm |
| PZ Case4 | End-Plate | W24 × 84 | HSS18 × 500 | 406.4 mm | 12.7 mm | 27 mm |
| PZ Case5 | Clip Angle | W18 × 50 | HSS14 × 14 × 500 | 355.6 mm | 12.7 mm | 27 mm |
| PZ Case6 | Clip Angle | W18 × 50 | HSS16 × 500 | 406.4 mm | 12.7 mm | 27 mm |
| PZ Case7 | T-stub | W24 × 62 | HSS16 × 16 × 375 | 406.4 mm | 9.5 mm | 27 mm |
| PZ Case8 | T-stub | W24 × 62 | HSS18 × 375 | 457.2 mm | 9.5 mm | 27 mm |
| PZ Case9 | T-stub | W24 × 55 | HSS16 × 16 × 375 | 406.4 mm | 9.5 mm | 27 mm |
| PZ Case10 | T-stub | W24 × 55 | HSS18 × 375 | 457.2 mm | 9.5 mm | 27 mm |

Unit: kN, rad and mm.

The moment acting on the connection can be converted into equivalent axial forces at the beam flanges, which in turn result in the shear forces in the panel zone. These shear forces cause shear deformations in the panel zone webs and flexural deformation in the panel zone flanges. Bolt holes which exist in the panel zone flanges reduce the flexural strength of the panel zone flanges. The identifications and geometric dimensions for the panel zone models are summarized in Table C.1. The strength models obtained from these calculations will be assigned to the numerical joint element models in order to simulate the exact behavior of the composite panel zone.

The moment acting on the connection can be converted into equivalent axial forces at the beam flanges, which in turn result in the shear forces in the panel zone. These shear forces cause shear deformations in the panel zone webs and flexural deformation in the panel zone flanges. Bolt holes which exist in the panel zone flanges reduce the flexural strength of the panel zone flanges. The identifications and geometric dimensions for the panel zone models are summarized in Table E.1. The strength models obtained from these calculations will be assigned to the numerical joint element models in order to simulate the exact behavior of the composite panel zone.

## C.2 CALCULATION EXAMPLES

### C.2.1 End plate connection with RCFT

This panel zone model (PZ Case 1) was designed for the 6END frame model in the lower story levels. The steel beams were made up of A.572-Gr.50 steel with a 24 × 103 cross section. The RCFT columns, with a HSS16 × 16 × 500 cross section, were selected for these panel zone models. The computational procedures for the yield shear strength ($V_{y,pro}$), shear stiffness ($K_{pro}$), and ultimate strength ($V_{u,pro}$) for the panel zone model are illustrated in the step-by step process shown in the next section. The SI unit system (newton and millimeter) was used in this appendix.

### C.2.1.1 Check the basic information

(1) Determine dimensions (See Fig. C.1 and Table C.1):

| | |
|---|---|
| $b_c = 406.4$ | The width of the column (mm) |
| $d_c = 406.4$ | The depth of the column (mm) |
| $t_f = 12.7$ | The thickness of the column flange (mm) |
| $t_w = 12.7$ | The thickness of the column web (mm) |
| $d_b = 624.84$ | The depth of the H-beam (mm) |
| $t_{bf} = 24.89$ | The thickness of the H-beam flange (mm) |
| $t_{bw} = 14.22$ | The thickness of the H-beam web (mm) |
| $t_{ep} = 25.4$ | The thickness of the end-plate (mm) |
| $d_h = 27$ | The diameter of the bolt hole (mm) |
| $n = 4$ | The number of rows of bolt holes in the PZ |
| $m = 2$ | The number of bolt holes in one row |

(2) Determine material properties:

| | |
|---|---|
| $E_s = 199810$ | Elastic modulus of the steel (MPa) |
| $F_y = 378.95$ | The yield stress of the A.572-Gr. 50 steel (MPa) |
| $v_s = 0.333$ | Poisson's ratio of the steel |
| $u = 0.3$ | The friction coefficient at the interface |
| $\beta = 1.5$ | The strain hardening factor of the steel |
| $f'_c = 48.23$ | The confined compressive concrete stress including the strengthened diaphragms (MPa) |
| $E_c = 57\sqrt{1000 \cdot f'_c} = 32858$ | The elastic modulus of the concrete (MPa) |

(3) Preloading:

| | |
|---|---|
| $t = 44.5$ | The average pre-stress of each bar (kN) |
| $P = 1780$ | The axial compression acting on the CFT columns (Interior columns at the 1st story level, kN) |

### C.2.1.2 Calculation procedures

(4) Calculate the shear stiffness for two generalized column flanges ($K_f$):

$$I_{\bar{f}} = \frac{b_c \cdot (t_f + t_{ep})^3}{12} = \frac{16 \cdot (0.5 + 1)^3}{12} = 1.872 \cdot 10^6\,\text{mm}^4 \tag{C.1}$$

$$K_f = 2\frac{12 E_s I_f}{(d_b - t_{bf})^2} = 2\frac{12 \cdot 29000 \cdot 4.5}{(16 - 0.98)^2} = 24982\,\text{kN/rad} \tag{C.2}$$

(5) Calculate the shear stiffness for two column webs without the bolt hole ($K_w$):

$$G_s = \frac{E_s}{2(1 + v_s)} = \frac{29000}{2(1 + 0.333)} = 7.51 \cdot 10^4\,\text{MPa/rad} \tag{C.3}$$

$$K_w = 2(d_c - 2 \cdot t_f) t_w G_s = 2(16 - 2 \cdot 0.5) \cdot 0.5 \cdot 10878 = 7.25 \cdot 10^5\,\text{kN/rad} \tag{C.4}$$

(6) Calculate the shear stiffness of the steel column including the stiffness loss due to bar holes ($K_{wh}$):

$$K_{wh} = 2(d_c - 2 \cdot t_f - m \cdot d_h)t_w G_s$$
$$= 2(16 - 2 \cdot 0.5 - 2 \cdot 1.063) \cdot 0.5 \cdot 10878 = 6.23 \cdot 10^5 \text{ kN/rad} \quad \text{(C.5)}$$

(7) Use the superposition and calculate the shear stiffness of two column webs ($K_{w1}$):

$$K_{w1} = \left[\left(1 - \frac{nd_h}{d_b}\right) \cdot \left(\frac{1}{K_w}\right) + \left(\frac{nd_h}{d_b}\right) \cdot \left(\frac{1}{K_{wh}}\right)\right]^{-1}$$
$$= \left[\left(1 - \frac{4 \cdot 1.063}{24.6}\right) \cdot \left(\frac{1}{163166}\right) + \left(\frac{4 \cdot 1.063}{24.6}\right) \cdot \left(\frac{1}{140040}\right)\right]^{-1}$$
$$= 7.07 \cdot 10^5 \text{ kN/rad} \quad \text{(C.6)}$$

(8) Calculate the pre-stress of all tension bars (T) and friction force between end-plate and steel tube (F):

$$T = 2mnt = 2 \cdot 2 \cdot 4 \cdot 44.5 = 712 \text{ kN (Pre-stress is elastic state)} \quad \text{(C.7)}$$

$$F = 2Tu = 2 \cdot 712 \cdot 0.3 = 427.2 \text{ kN} \quad \text{(C.8)}$$

(9) Calculate the yield strength of two column webs including the loss due to the bolt holes ($V_{why}$):

$$V_{why} = \frac{2(d_c - 2t_f - m \cdot d_h)t_w F_y}{\sqrt{3}} = \frac{2(406.4 - 2 \cdot 12.7 - 2 \cdot 27) \cdot 0.5 \cdot 378.9}{\sqrt{3}}$$
$$= 1820 \text{ kN} \quad \text{(C.9)}$$

(10) Calculate the corresponding shear strain of the steel tube ($\gamma_2$):

$$\gamma_2 = \frac{V_{why}}{K_{w1}} = \frac{409}{158638} = 0.00258 \text{ rad} \quad \text{(C.10)}$$

(11) Calculate the yield strength of two column webs without bolt holes ($V_{wy}$):

$$V_{wy} = \frac{2(d_c - 2t_f)t_w F_y}{\sqrt{3}} = \frac{2(406.4 - 2 \cdot 12.7) \cdot 12.7 \cdot 378.95}{\sqrt{3}} = 3279.64 \text{ kN}$$
$$\text{(C.11)}$$

(12) Calculate the yield strength of the generalized column flanges ($V_{fy}$):

$$V_{fy} = \frac{2(t_f + t_{ep})b_c F_y}{3(d_b - t_{bf})} = \frac{2(12.7 + 25.4) \cdot 406.4 \cdot 378.95}{3(624.8 - 24.9)} = 249.2 \text{ kN} \quad \text{(C.12)}$$

(13) Calculate the axial compressive stress ($\sigma_x$) and the lateral pre-stress acted on the concrete ($\sigma_y$) in the panel zone:

$$A_c = (b_c - 2t_f)(d_b - 2t_w) = 145125\,\text{mm}^2 \tag{C.13}$$

$$A_s = b_c d_c - A_c = 19995\,\text{mm}^2 \tag{C.14}$$

$$\sigma_x = f_{cp} = \frac{-PE_c}{E_s A_s + E_c A_c} = \frac{-1780 \cdot 32858}{(199810 \cdot 19995 + 32858 \cdot 145125)} = -6.752\,\text{MPa} \tag{C.15}$$

$$\sigma_y = f_{ct} = \frac{-T}{b_c d_b} = \frac{-712}{406.4 \cdot 624.84} = -3.1\,\text{MPa} \tag{C.16}$$

(14) Calculate the ultimate shear stress of the inside confined concrete ($\tau_{cu}$):

$$f_t = \frac{7.5\sqrt{1000 f'_c}}{1000} = 4.32\,\text{MPa} \tag{C.17}$$

$$m_r = \frac{f_t}{f'_c} = 0.09 \tag{C.18}$$

$$\tau_{cu} = \frac{1}{1+m_r}\sqrt{(f'_c + \sigma_x - m_r\sigma_y)(f'_c + \sigma_y - m_r\sigma_x)} = 39.96\,\text{MPa} \tag{C.19}$$

(15) Calculate the reduction factor for the shear stiffness of the confined concrete ($r_c$):

$$r_A = 1 - \frac{md_h}{(d_c - 2t_f)} = 1 - \frac{2 \cdot 27}{(406.4 - 2 \cdot 12.7)} = 0.858 \tag{C.20}$$

$$r_c = (1 - \frac{2nd_h}{d_b} + \frac{2nd_h}{d_b r_A})^{-1} = \left(1 - \frac{2 \cdot 4 \cdot 27}{624.8} + \frac{2 \cdot 4 \cdot 27}{624.8 \cdot 0.858}\right) = 0.946 \tag{C.21}$$

(16) Calculate the ultimate shear stress of the inside confined concrete ($\overline{V}_{cu}$):

$$\overline{V}_{cu} = r_c \tau_{cu} A_c = 0.946 \cdot 39.96 \cdot 145125 = 5518\,\text{kN} \tag{C.22}$$

#### C.2.1.3 *Panel zone strength*

(17) Calculate the yield strength, shear stiffness, and the ultimate shear strength for the composite panel zone using the superposition theory

$$V_{y,pro} = (K_f + K_{w1})\gamma_2 + F + \overline{V}_{cu} = 7823.1\,\text{kN} \tag{C.23}$$

$$K_{y,pro} = K_f + K_{w1} + \frac{\left(F + \overline{V}_{cu}\right)}{\gamma_2} = 30.35 \cdot 10^5\,\text{kN/rad} \tag{C.24}$$

$$V_{y,pro} = V_{fy} + V_{wy} + F + \overline{V}_{cu} = 8303.7\,\text{kN} \tag{C.25}$$

## C.2.2 End plate connection with CCFT

This panel zone model (PZ Case 2) was designed for the 6END frame model in the lower story levels. The steel beams were made up of A.572-Gr.50 steel with a 24 × 103 cross section. The RCFT columns, with a HSS18 × 500 cross section, were selected for these panel zone models. The computational procedures for the yield shear strength (Vy,pro), shear stiffness (Kpro), and ultimate strength (Vu,pro) for the panel zone model are illustrated in the step-by step process shown in the next section. The SI unit system (newton and millimeter) was used in this appendix.

### C.2.2.1 *Check the basic information*

(1) Determine dimensions (See Fig. C.1 and Table C.1):

| | |
|---|---|
| $b_c = 457.2$ | The width of the column (mm) |
| $d_c = 457.2$ | The depth of the column (mm) |
| $t_f = 12.7$ | The thickness of the column flange (mm) |
| $t_w = 12.7$ | The thickness of the column web (mm) |
| $d_b = 624.84$ | The depth of the H-beam (mm) |
| $t_{bf} = 24.892$ | The thickness of the H-beam flange (mm) |
| $t_{bw} = 14.224$ | The thickness of the H-beam web (mm) |
| $t_{ep} = 25.4$ | The thickness of the end-plate (mm) |
| $d_h = 27$ | The diameter of the bolt hole (mm) |
| $n = 4$ | The number of rows of bolt holes in the PZ |
| $m = 2$ | The number of bolt holes in one row |

(2) Determine material properties:

| | |
|---|---|
| $E_s = 199810$ | Elastic modulus of the steel (MPa) |
| $F_y = 378.95$ | The yield stress of the A.572-Gr. 50 steel (MPa) |
| $v_s = 0.333$ | Poisson's ratio of the steel |
| $u = 0.3$ | The friction coefficient at the interface |
| $\beta = 1.5$ | The strain hardening factor of the steel |
| $f'_c = 45.474$ | The confined compressive concrete stress including the strengthened diaphragms (MPa) |
| $E_c = 57\sqrt{1000 \cdot f'_c} = 32858.41$ | The elastic modulus of the concrete (MPa) |

(3) Preloading:

| | |
|---|---|
| $t = 44.5$ | The average pre-stress of each bar (kN) |
| $P = 1780$ | The axial compression acting on the CFT columns (Interior columns at the 1st story level, kN) |

### C.2.2.2 *Calculation procedures*

(4) Calculate the shear stiffness for two generalized column flanges ($K_f$):

$$I_{\bar{f}} = \frac{b_c \cdot (t_f + t_{ep})^3}{12} = \frac{457.2 \cdot (12.7 + 25.4)^3}{12} = 2.105 \cdot 10^6 \text{ mm}^4 \quad (C.26)$$

$$K_f = 2\frac{12E_sI_f}{(d_b - t_{bf})^2} = 2\frac{12 \cdot 199810 \cdot 2.105 \cdot 10^6}{(457.2 - 24.89)^2} = 28101.75 \text{ kN/rad} \quad (C.27)$$

(5) Calculate the shear stiffness for two column webs without the bolt hole ($K_w$):

$$G_s = \frac{E_s}{2(1+v_s)} = \frac{199810}{2\,(1+0.333)} = 7.51 \cdot 10^4 \text{ MPa/rad} \tag{C.28}$$

$$K_w = 2(d_c - 2 \cdot t_f)t_w G_s = 2(457.2 - 2 \cdot 12.7) \cdot 12.7 \cdot 75.1 = 8.2325 \cdot 10^5 \text{ kN/rad} \tag{C.29}$$

(6) Calculate the shear stiffness of the steel column including the stiffness loss due to bar holes ($K_{wh}$):

$$\begin{aligned} K_{wh} &= 2(d_c - 2 \cdot t_f - m \cdot d_h)t_w G_s \\ &= 2(457.2 - 2 \cdot 12.7 - 2 \cdot 27) \cdot 12.7 \cdot 75.1 = 7.209 \cdot 10^5 \text{ kN/rad} \end{aligned} \tag{C.30}$$

(7) Use the superposition and calculate the shear stiffness of two column webs ($K_{w1}$):

$$\begin{aligned} K_{w1} &= \left[\left(1 - \frac{nd_h}{d_b}\right) \cdot \left(\frac{1}{K_w}\right) + \left(\frac{nd_h}{d_b}\right) \cdot \left(\frac{1}{K_{wh}}\right)\right]^{-1} \\ &= \left[\left(1 - \frac{4 \cdot 27}{624.84}\right) \cdot \left(\frac{1}{823250}\right) + \left(\frac{4 \cdot 27}{24.6}\right) \cdot \left(\frac{1}{720900}\right)\right]^{-1} \\ &= 8.01 \cdot 10^5 \text{ kN/rad} \end{aligned} \tag{C.31}$$

(8) Calculate the pre-stress of all tension bars (T) and friction force between end-plate and steel tube (F):

$$T = 2mnt = 2 \cdot 2 \cdot 4 \cdot 44.5 = 712 \text{ kN (Pre-stress is elastic state)} \tag{C.32}$$

$$F = 2Tu = 2 \cdot 712 \cdot 0.3 = 427.2 \text{ kN} \tag{C.33}$$

(9) Calculate the yield strength of two column webs including the loss due to the bolt holes ($V_{why}$):

$$\begin{aligned} V_{why} &= \frac{2(d_c - 2t_f - m \cdot d_h)t_w F_y}{\sqrt{3}} \\ &= \frac{2(457.2 - 2 \cdot 12.7 - 2 \cdot 27) \cdot 12.7 \cdot 378.95}{\sqrt{3}} = 2100.4 \text{ kN} \end{aligned} \tag{C.34}$$

(10) Calculate the corresponding shear strain of the steel tube ($\gamma_2$):

$$\gamma_2 = \frac{V_{why}}{K_{w1}} = \frac{2100.4}{801000} = 0.00262 \text{ rad} \tag{C.35}$$

(11) Calculate the yield strength of two column webs without bolt holes ($V_{wy}$):

$$V_{wy} = \frac{2(d_c - 2t_f)t_w F_y}{\sqrt{3}} = \frac{2(457.2 - 2 \cdot 12.7) \cdot 12.7 \cdot 378.95}{\sqrt{3}} = 2403\,\text{kN} \qquad (C.36)$$

(12) Calculate the yield strength of the generalized column flanges ($V_{fy}$):

$$V_{fy} = \frac{2(t_f + t_{ep})b_c F_y}{3(d_b - t_{bf})} = \frac{2(12.7 + 25.4) \cdot 457.2 \cdot 378.95}{3(624.8 - 24.89)} = 280.35\,\text{kN} \qquad (C.37)$$

(13) Calculate the axial compressive stress ($\sigma_x$) and the lateral pre-stress acted on the concrete ($\sigma_y$) in the panel zone:

$$A_c = (b_c - 2t_f)(d_b - 2t_w) = 186405\ \text{mm}^2 \qquad (C.38)$$

$$A_s = b_c d_c - A_c = 22575\ \text{mm}^2 \qquad (C.39)$$

$$\sigma_x = f_{cp} = \frac{-PE_c}{E_s A_s + E_c A_c} = \frac{-1780 \cdot 32858}{(199810 \cdot 22575 + 31866 \cdot 186405)} = -5.4431\ \text{MPa} \qquad (C.40)$$

$$\sigma_y = f_{ct} = \frac{-T}{b_c d_b} = \frac{-712}{457.2 \cdot 624.84} = -2.756\ \text{MPa} \qquad (C.41)$$

(14) Calculate the ultimate shear stress of the inside confined concrete ($\tau_{cu}$):

$$f_t = \frac{7.5\sqrt{1000 f_c'}}{1000} = 6.25612\,\text{MPa} \qquad (C.42)$$

$$m_r = \frac{f_t}{f_c'} = 0.092 \qquad (C.43)$$

$$\tau_{cu} = \frac{1}{1 + m_r}\sqrt{\left(f_c' + \sigma_x - m_r\sigma_y\right)\left(f_c' + \sigma_y - m_r\sigma_x\right)} = 38.1\ \text{MPa} \qquad (C.44)$$

(15) Calculate the reduction factor for the shear stiffness of the confined concrete ($r_c$):

$$r_A = 1 - \frac{md_h}{(d_c - 2t_f)} = 1 - \frac{2 \cdot 27}{(457.2 - 2 \cdot 12.7)} = 0.87 \qquad (C.45)$$

$$r_c = \left(1 - \frac{2nd_h}{d_b} + \frac{2nd_h}{d_b r_A}\right)^{-1} = \left(1 - \frac{2 \cdot 4 \cdot 27}{624.8} + \frac{2 \cdot 4 \cdot 27}{624.8 \cdot 0.87}\right) = 0.95 \qquad (C.46)$$

(16) Calculate the ultimate shear stress of the inside confined concrete ($\overline{V}_{cu}$):

$$\overline{V}_{cu} = r_c \tau_{cu} A_c = 0.95 \cdot 38.1 \cdot 186405 = 6777.35\ \text{kN} \qquad (C.47)$$

#### C.2.2.3 *Panel zone strength*

(17) Calculate the yield strength, shear stiffness, and the ultimate shear strength for the composite panel zone using the superposition theory

$$V_{y,pro} = (K_f + K_{w1})\gamma_2 + F + \overline{V}_{cu} = 9380.6\,\text{kN} \tag{C.48}$$

$$K_{y,pro} = K_f + K_{w1} + \frac{\left(F + \overline{V}_{cu}\right)}{\gamma_2} = 35.87 \cdot 10^5\ \text{kN/rad} \tag{C.49}$$

$$V_{y,pro} = V_{fy} + V_{wy} + F + \overline{V}_{cu} = 9887.9\,\text{kN} \tag{C.50}$$

### C.2.3 T-stub connection with RCFT

This panel zone model (PZ Case 9) was designed for the 6TSU frame model in the lower story levels. The steel beams were made up of A.572-Gr.50 steel with a 24 × 55 cross section. The RCFT columns, with a HSS16 × 16 × 375 cross section, were selected for these panel zone models. The computational procedures for the yield shear strength ($V_{y,pro}$), shear stiffness ($K_{pro}$), and ultimate strength ($V_{u,pro}$) for the panel zone model are illustrated in the step-by step process shown in the next section. The SI unit system (newton and millimeter) was used in this appendix.

#### C.2.3.1 *Check the basic information*

(1) Determine dimensions (See Fig. C.1 and Table C.1):

| | |
|---|---|
| $b_c = 406.4$ | The width of the column (mm) |
| $d_c = 406.4$ | The depth of the column (mm) |
| $t_f = 9.525$ | The thickness of the column flange (mm) |
| $t_w = 9.525$ | The thickness of the column web (mm) |
| $d_b = 599.44$ | The depth of the beam (mm) |
| $t_{bf} = 599.44$ | The thickness of the H-beam flange (mm) |
| $t_{bw} = 10.03$ | The thickness of the H-beam web (mm) |
| $t_{fl} = 25.4$ | The thickness of the clip angle (mm) |
| $d_h = 27$ | The diameter of the bolt hole (mm) |
| $n = 2$ | The number of rows of bolt holes in the PZ |
| $m = 4$ | The number of bolt holes in one row |

(2) Determine material properties:

| | |
|---|---|
| $E_s = 199810$ | Elastic modulus of the steel (MPa) |
| $F_y = 378.95$ | The yield stress of the A.572-Gr. 50 steel (MPa) |
| $v_s = 0.333$ | Poisson's ratio of the steel |
| $u = 0.3$ | The friction coefficient at the interface |
| $\beta = 1.5$ | The strain hardening factor of the steel |
| $f'_c = 48.23$ | The confined compressive concrete stress including the strengthened diaphragms (MPa) |
| $E_c = 57\sqrt{1000 \cdot f'_c} = 32858$ | The elastic modulus of the concrete (MPa) |

(3) Preloading:

| | |
|---|---|
| t = 44.5 | The average pre-stress of each bar (kN) |
| P = 890 | The axial compression acting on the CFT columns (Interior columns at the 4th story level, kN) |

### C.2.3.2 *Calculation procedures*

(4) Calculate the shear stiffness for two generalized column flanges ($K_f$):

$$I_{\bar{f}} = \frac{b_c \cdot (t_f + t_{fl})^3}{12} = \frac{406.4 \cdot (9.525 + 25.4)^3}{12} = 1.456 \cdot 10^6 \text{ mm}^4 \quad (C.51)$$

$$K_f = 2\frac{12E_sI_f}{(d_b - t_{bf})^2} = 2\frac{12 \cdot 199810 \cdot 1.456 \cdot 10^6}{(406.4 - 12.827)^2} = 20345.4 \text{ kN/rad} \quad (C.52)$$

(5) Calculate the shear stiffness for two column webs without the bolt hole ($K_w$):

$$G_s = \frac{E_s}{2(1 + v_s)} = \frac{199810}{2(1 + 0.333)} = 7.51 \cdot 10^4 \text{ MPa/rad} \quad (C.53)$$

$$\begin{aligned} K_w &= 2(d_c - 2 \cdot t_f)t_w G_s = 2(406 - 2 \cdot 9.525) \cdot 9.525 \cdot 75.1 \\ &= 5.607 \cdot 10^5 \text{ kN/rad} \end{aligned} \quad (C.54)$$

(6) Calculate the shear stiffness of the steel column including the stiffness loss due to bar holes ($K_{wh}$):

$$\begin{aligned} K_{wh} &= 2(d_c - 2 \cdot t_f - m \cdot d_h)t_w G_s \\ &= 2(406.4 - 2 \cdot 9.525 - 4 \cdot 27) \cdot 9.525 \cdot 75100 = 4.05 \cdot 10^5 \text{ kN/rad} \end{aligned} \quad (C.55)$$

(7) Use the superposition and calculate the shear stiffness of two column webs ($K_{w1}$):

$$\begin{aligned} K_{w1} &= \left[\left(1 - \frac{nd_h}{d_b}\right) \cdot \left(\frac{1}{K_w}\right) + \left(\frac{nd_h}{d_b}\right) \cdot \left(\frac{1}{K_{wh}}\right)\right]^{-1} \\ &= \left[\left(1 - \frac{2 \cdot 27}{599}\right) \cdot \left(\frac{1}{560700}\right) + \left(\frac{2 \cdot 27}{599.}\right) \cdot \left(\frac{1}{405000}\right)\right]^{-1} \\ &= 5.38 \cdot 10^5 \text{kN/rad} \end{aligned} \quad (C.56)$$

(8) Calculate the pre-stress of all tension bars (T) and friction force between end-plate and steel tube (F):

$$T = 2mnt = 2 \cdot 2 \cdot 4 \cdot 44.5 = 712 \text{ kN (Pre-stress is elastic state)} \quad (C.57)$$

$$F = 2Tu = 2 \cdot 712 \cdot 0.3 = 427.2 \text{ kN} \quad (C.58)$$

(9) Calculate the yield strength of two column webs including the loss due to the bolt holes ($V_{why}$):

$$V_{why} = \frac{2(d_c - 2t_f - m \cdot d_h)t_w F_y}{\sqrt{3}} = \frac{2(406.4 - 2 \cdot 9.525 - 4 \cdot 27) \cdot 9.525 \cdot 378.95}{\sqrt{3}} = 1179.25\,\text{kN} \tag{C.59}$$

(10) Calculate the corresponding shear strain of the steel tube ($\gamma_2$):

$$\gamma_2 = \frac{V_{why}}{K_{w1}} = \frac{1179.25}{538000} = 0.00218\,\text{rad} \tag{C.60}$$

(11) Calculate the yield strength of two column webs without bolt holes ($V_{wy}$):

$$V_{wy} = \frac{2(d_c - 2t_f)t_w F_y}{\sqrt{3}} = \frac{2(406 - 2 \cdot 9.525) \cdot 9.525 \cdot 378.95}{\sqrt{3}} = 1633.15\ \text{kN} \tag{C.61}$$

(12) Calculate the yield strength of the generalized column flanges ($V_{fy}$):

$$V_{fy} = \frac{2\,(t_f + t_{fl})\,b_c F_y}{3\,(d_b - t_{bf})} = \frac{2\,(9.525 + 25.4) \cdot 406.4 \cdot 378.95}{3\,(599 - 12.83)} = 213.6\ \text{kN} \tag{C.62}$$

(13) Calculate the axial compressive stress ($\sigma_x$) and the lateral pre-stress acted on the concrete ($\sigma_y$) in the panel zone:

$$A_c = (b_c - 2t_f)\ (d_b - 2t_w) = 149640\,\text{mm}^2 \tag{C.63}$$

$$A_s = b_c d_c - A_c = 15480\ \text{mm}^2 \tag{C.64}$$

$$\sigma_x = f_{cp} = \frac{-PE_c}{E_s A_s + E_c A_c} = \frac{-890 \cdot 32851}{(199810 \cdot 15480 + 32851 \cdot 149640)} = -3.658\ \text{MPa} \tag{C.65}$$

$$\sigma_y = f_{ct} = \frac{-T}{b_c d_b} = \frac{-712}{406.4 \cdot 599.4} = -3.128\ \text{MPa} \tag{C.66}$$

(14) Calculate the ultimate shear stress of the inside confined concrete ($\tau_{cu}$):

$$f_t = \frac{7.5\sqrt{1000 f'_c}}{1000} = 4.32\,\text{MPa} \tag{C.67}$$

$$m_r = \frac{f_t}{f'_c} = 0.0996 \tag{C.68}$$

$$\tau_{cu} = \frac{1}{1+m_r}\sqrt{(f_c' + \sigma_x - m_r\sigma_y)(f_c' + \sigma_y - m_r\sigma_x)} = 41.34\,\text{MPa} \tag{C.69}$$

(15) Calculate the reduction factor for the shear stiffness of the confined concrete ($r_c$):

$$r_A = 1 - \frac{md_h}{(d_c - 2t_f)} = 1 - \frac{3\cdot 27}{(406.4 - 2\cdot 9.525)} = 0.721 \tag{C.70}$$

$$r_c = \left(1 - \frac{2nd_h}{d_b} + \frac{2nd_h}{d_b r_A}\right)^{-1} = \left(1 - \frac{2\cdot 2\cdot 27}{599} + \frac{2\cdot 2\cdot 27}{599\cdot 0.721}\right) = 0.93 \tag{C.71}$$

(16) Calculate the ultimate shear stress of the inside confined concrete ($\overline{V}_{cu}$):

$$\overline{V}_{cu} = r_c\tau_{cu}A_c = 0.93\cdot 41.4\cdot 149640 = 5807.25\ \text{kN} \tag{C.72}$$

#### C.2.3.3 *Panel zone strength*

(17) Calculate the yield strength, shear stiffness, and the ultimate shear strength for the composite panel zone using the superposition theory

$$V_{y,pro} = (K_f + K_{w1})\,\gamma_2 + F + \overline{V}_{cu} = 7476\ \text{kN} \tag{C.73}$$

$$K_{y,pro} = K_f + K_{w1} + \frac{\left(F + \overline{V}_{cu}\right)}{\gamma_2} = 34.18\cdot 10^5\ \text{kN/rad} \tag{C.74}$$

$$V_{y,pro} = V_{fy} + V_{wy} + F + \overline{V}_{cu} = 8099\ \text{kN} \tag{C.75}$$

### C.2.4 T-stub connection with CCFT

This panel zone model (PZ Case 10) was designed for the 6TSU frame model in the lower story levels. The steel beams were made up of A.572-Gr.50 steel with a 24 × 55 cross section. The RCFT columns, with a HSS18 × 375 cross section, were selected for these panel zone models. The computational procedures for the yield shear strength ($V_{y,pro}$), shear stiffness ($K_{pro}$), and ultimate strength ($V_{u,pro}$) for the panel zone model are illustrated in the step-by step process shown in the next section. The SI unit system (newton and millimeter) was used in this appendix.

#### C.2.4.1 *Check the basic information*

(1) Determine dimensions (See Fig. C.1 and Table C.1):

| | |
|---|---|
| $b_c = 457.2$ | The width of the column (mm) |
| $d_c = 457.2$ | The depth of the column (mm) |
| $t_f = 9.525$ | The thickness of the column flange (mm) |
| $t_w = 9.525$ | The thickness of the column web (mm) |
| $d_b = 599.4$ | The depth of the beam (mm) |

| | |
|---|---|
| $t_{bf} = 12.83$ | The thickness of the H-beam flange (mm) |
| $t_{bw} = 10.03$ | The thickness of the H-beam web (mm) |
| $t_{fl} = 25.4$ | The thickness of the clip angle (mm) |
| $d_h = 27$ | The diameter of the bolt hole (mm) |
| $n = 2$ | The number of rows of bolt holes in the PZ |
| $m = 4$ | The number of bolt holes in one row |

(2) Determine material properties:

| | |
|---|---|
| $E_s = 199810$ | Elastic modulus of the steel (MPa) |
| $F_y = 378.95$ | The yield stress of the A.572-Gr. 50 steel (MPa) |
| $v_s = 0.333$ | Poisson's ratio of the steel |
| $u = 0.3$ | The friction coefficient at the interface |
| $\beta = 1.5$ | The strain hardening factor of the steel |
| $f'_c = 43.41$ | The confined compressive concrete stress including the strengthened diaphragms (MPa) |
| $E_c = 57\sqrt{1000 \cdot f'_c} = 31046$ | The elastic modulus of the concrete (MPa) |

(3) Preloading:

| | |
|---|---|
| $t = 44.5$ | The average pre-stress of each bar (kN) |
| $P = 890$ | The axial compression acting on the CFT columns (Interior columns at the 4th story level, kN) |

### C.2.4.2 *Calculation procedures*

(4) Calculate the shear stiffness for two generalized column flanges ($K_f$):

$$I_{\bar{f}} = \frac{b_c \cdot (t_f + t_{fl})^3}{12} = \frac{457.2 \cdot (9.525 + 25.4)^3}{12} = 1.64 \cdot 10^6 \text{ mm}^4 \tag{C.76}$$

$$K_f = 2\frac{12E_sI_f}{(d_b - t_{bf})^2} = 2\frac{12 \cdot 199810 \cdot 1.64 \cdot 10^6}{(457.2 - 12.83)^2} = 22890.8 \text{ kN/rad} \tag{C.77}$$

(5) Calculate the shear stiffness for two column webs without the bolt hole ($K_w$):

$$G_s = \frac{E_s}{2(1 + v_s)} = \frac{199810}{2(1 + 0.333)} = 7.51 \cdot 10^4 \text{ MPa/rad} \tag{C.78}$$

$$\begin{aligned} K_w &= 2(d_c - 2 \cdot t_f)t_w G_s = 2(457.2 - 2 \cdot 9.525) \cdot 9.525 \cdot 75100 \\ &= 6.319 \cdot 10^5 \text{ kN/rad} \end{aligned} \tag{C.79}$$

(6) Calculate the shear stiffness of the steel column including the stiffness loss due to bar holes ($K_{wh}$):

$$\begin{aligned} K_{wh} &= 2(d_c - 2 \cdot t_f - m \cdot d_h)t_w G_s \\ &= 2(457.2 - 2 \cdot 9.525 - 4 \cdot 27) \cdot 9.525 \cdot 75100 = 4.76 \cdot 10^5 \text{ kN/rad} \end{aligned} \tag{C.80}$$

(7) Use the superposition and calculate the shear stiffness of two column webs ($K_{w1}$):

$$K_{w1} = \left[\left(1 - \frac{nd_h}{d_b}\right)\cdot\left(\frac{1}{K_w}\right) + \left(\frac{nd_h}{d_b}\right)\cdot\left(\frac{1}{K_{wh}}\right)\right]^{-1}$$
$$= \left[\left(1 - \frac{2\cdot 27}{599.4}\right)\cdot\left(\frac{1}{631900}\right) + \left(\frac{2\cdot 27}{599.4}\right)\cdot\left(\frac{1}{476000}\right)\right]^{-1}$$
$$= 6.141\cdot 10^5\ \text{kN/rad} \tag{C.81}$$

(8) Calculate the pre-stress of all tension bars (T) and friction force between end-plate and steel tube (F):

$$T = 2mnt = 2\cdot 2\cdot 4\cdot 44.5 = 712\ \text{kN (Pre-stress is elastic state)} \tag{C.82}$$

$$F = 2Tu = 2\cdot 712\cdot 0.3 = 427.2\ \text{kN} \tag{C.83}$$

(9) Calculate the yield strength of two column webs including the loss due to the bolt holes ($V_{why}$):

$$V_{why} = \frac{2(d_c - 2t_f - m\cdot d_h)t_w F_y}{\sqrt{3}}$$
$$= \frac{2(457.2 - 2\cdot 9.525 - 4\cdot 27)\cdot 9.525\cdot 378.95}{\sqrt{3}} = 1392.85\ \text{kN} \tag{C.84}$$

(10) Calculate the corresponding shear strain of the steel tube ($\gamma_2$):

$$\gamma_2 = \frac{V_{why}}{K_{w1}} = \frac{1392.85}{614100} = 0.00226\ \text{rad} \tag{C.85}$$

(11) Calculate the yield strength of two column webs without bolt holes ($V_{wy}$):

$$V_{wy} = \frac{2(d_c - 2t_f)t_w F_y}{\sqrt{3}} = \frac{2(457.2 - 2\cdot 9.525)\cdot 9.525\cdot 378.95}{\sqrt{3}} = 2866.24\ \text{kN} \tag{C.86}$$

(12) Calculate the yield strength of the generalized column flanges ($V_{fy}$):

$$V_{fy} = \frac{2(t_f + t_{fl})b_c F_y}{3(d_b - t_{bf})} = \frac{2(9.525 + 25.4)\cdot 457.2\cdot 378.95}{3\,(599.44 - 12.83)} = 240.3\ \text{kN} \tag{C.87}$$

(13) Calculate the axial compressive stress ($\sigma_x$) and the lateral pre-stress acted on the concrete ($\sigma_y$) in the panel zone:

$$A_c = (b_c - 2t_f)\ (d_b - 2t_w) = 191565\ \text{mm}^2 \tag{C.88}$$

$$A_s = b_c d_c - A_c = 17415 \text{ mm}^2 \tag{C.89}$$

$$\sigma_x = f_{cp} = \frac{-PE_c}{E_s A_s + E_c A_c} = \frac{-890 \cdot 31046.3}{(199810 \cdot 17415 + 31046 \cdot 191565)} = -2.942 \text{ MPa} \tag{C.90}$$

$$\sigma_y = f_{ct} = \frac{-T}{b_c d_b} = \frac{-1102.4}{457.2 \cdot 599.4} = -2.77 \text{ MPa} \tag{C.91}$$

(14) Calculate the ultimate shear stress of the inside confined concrete ($\tau_{cu}$):

$$f_t = \frac{7.5\sqrt{1000 f'_c}}{1000} = 4.086 \text{ MPa} \tag{C.92}$$

$$m_r = \frac{f_t}{f'_c} = 0.0949 \tag{C.93}$$

$$\tau_{cu} = \frac{1}{1+m_r}\sqrt{(f'_c + \sigma_x - m_r\sigma_y)\,(f'_c + \sigma_y - m_r\sigma_x)} = 36.93 \text{ MPa} \tag{C.94}$$

(15) Calculate the reduction factor for the shear stiffness of the confined concrete ($r_c$):

$$r_A = 1 - \frac{md_h}{(d_c - 2t_f)} = 1 - \frac{3 \cdot 27}{(457.2 - 2 \cdot 9.525)} = 0.753 \tag{C.95}$$

$$r_c = \left(1 - \frac{2nd_h}{d_b} + \frac{2nd_h}{d_b r_A}\right)^{-1} = \left(1 - \frac{2 \cdot 2 \cdot 27}{599.4} + \frac{2 \cdot 2 \cdot 27}{599.4 \cdot 0.721}\right) = 0.944 \tag{C.96}$$

(16) Calculate the ultimate shear stress of the inside confined concrete ($\overline{V}_{cu}$):

$$\overline{V}_{cu} = r_c \tau_{cu} A_c = 0.944 \cdot 36.93 \cdot 191565 = 6701.7 \text{ kN} \tag{C.97}$$

### *C.2.4.3 Panel zone strength*

(17) Calculate the yield strength, shear stiffness, and the ultimate shear strength for the composite panel zone using the superposition theory

$$V_{y,pro} = (K_f + K_{w1})\,\gamma_2 + F + \overline{V}_{cu} = 8575.15 \text{ kN} \tag{C.98}$$

$$K_{y,pro} = K_f + K_{w1} + \frac{\left(F + \overline{V}_{cu}\right)}{\gamma_2} = 37.87 \cdot 10^5 \text{ kN/rad} \tag{C.99}$$

$$V_{y,pro} = V_{fy} + V_{wy} + F + \overline{V}_{cu} = 9211.5 \text{ kN} \tag{C.100}$$

The yield shear strength, shear stiffness, and ultimate shear strength for all panel zone models are summarized in Table C.2. These results will be used for the panel zone models in the numerical joint element models. The behavior of the panel zone in the joint element models can be simulated as shown in Fig. C.2.

*Table C.2* Theoretical results for the panel zone strength.

| *ID* | *PZ size** | $V_{ypro}$ | $K_{ypro}$ | $V_u$ | $K_t$ |
|---|---|---|---|---|---|
| PZ Case1 | 16 × 406.4 mm (12.7 mm) | $7.8 \times 10^3$ | $30.3 \times 10^5$ | $8.3 \times 10^3$ | $0.01K_{y,pro}$ |
| PZ Case2 | 18 × 457.2 mm (12.7 mm) | $9.4 \times 10^3$ | $35.8 \times 10^5$ | $9.9 \times 10^3$ | $0.01K_{y,pro}$ |
| PZ Case3 | 16 × 406.4 mm (12.7 mm) | $7.8 \times 10^3$ | $30.1 \times 10^5$ | $8.2 \times 10^3$ | $0.01K_{y,pro}$ |
| PZ Case4 | 18 × 457.2 mm (12.7 mm) | $9.5 \times 10^3$ | $35.9 \times 10^5$ | $10.0 \times 10^3$ | $0.01K_{y,pro}$ |
| PZ Case5 | 14 × 355.6 mm (12.7 mm) | $5.4 \times 10^3$ | $24.2 \times 10^5$ | $6.1 \times 10^3$ | $0.01K_{y,pro}$ |
| PZ Case6 | 16 × 406.4 mm (12.7 mm) | $7.3 \times 10^3$ | $31.1 \times 10^5$ | $8.0 \times 10^3$ | $0.01K_{y,pro}$ |
| PZ Case7 | 16 × 406.4 mm (9.5 mm) | $7.5 \times 10^3$ | $34.6 \times 10^5$ | $8.2 \times 10^3$ | $0.01K_{y,pro}$ |
| PZ Case8 | 18 × 457.2 mm (9.5 mm) | $8.6 \times 10^3$ | $38.0 \times 10^5$ | $9.2 \times 10^3$ | $0.01K_{y,pro}$ |
| PZ Case9 | 16 × 406.4 mm (9.5 mm) | $7.5 \times 10^3$ | $33.1 \times 10^5$ | $8.1 \times 10^3$ | $0.01K_{y,pro}$ |
| PZ Case10 | 18 × 457.2 mm (9.5 mm) | $8.6 \times 10^3$ | $37.8 \times 10^5$ | $9.2 \times 10^3$ | $0.01K_{y,pro}$ |

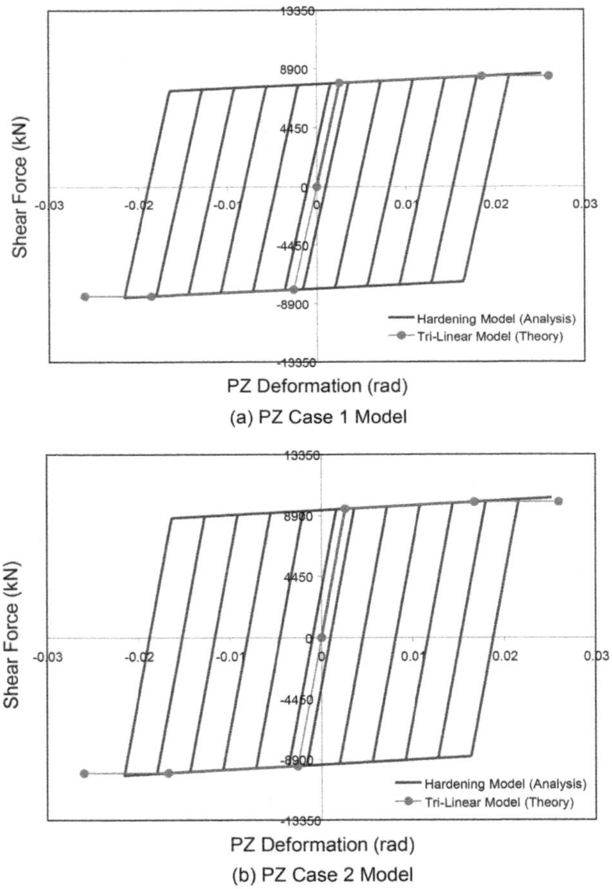


*Figure C.2* The behavioral models for composite panel zones.

Appendix D

# Earthquake ground motions

Table 8.2 provides detailed information on the records generated for Los Angeles (LA) and Seattle (SE) with probabilities of exceedence of 2% in 50 years. Ten historical ground motion pairs (a total 20 ground motions) used in this research have been derived from historical records. The detailed acceleration time history for all of the ground motions listed in the table below are shown in Figs. D.1 through D.10.

*Table D.1* Earthquake ground motions with 2% probability of exceedence in 50 years.

| *SAC Name* | *Record* | *Earthquake Magnitude* | *Distance (km)* | *Scale Factor* | *Number of Points* | *DT (sec)* | *Duration (sec)* | *PGA (g)* |
|---|---|---|---|---|---|---|---|---|
| LA21 | 1995 Kobe | 6.9 | 3.4 | 1.15 | 3000 | 0.02 | 59.98 | 1.283 |
| LA22 | 1995 Kobe | 6.9 | 3.4 | 1.15 | 3000 | 0.02 | 59.98 | 0.92069 |
| LA23 | 1989 Loma Prieta | 7 | 3.5 | 0.82 | 2500 | 0.01 | 24.99 | 0.418097 |
| LA24 | 1989 Loma Prieta | 7 | 3.5 | 0.82 | 2500 | 0.01 | 24.99 | 0.472976 |
| LA25 | 1994 Northridge | 6.7 | 7.5 | 1.29 | 2990 | 0.005 | 14.945 | 0.868544 |
| LA26 | 1994 Northridge | 6.7 | 7.5 | 1.29 | 2990 | 0.005 | 14.945 | 0.943678 |
| LA27 | 1994 Northridge | 6.7 | 6.4 | 1.61 | 3000 | 0.02 | 59.98 | 0.926758 |
| LA28 | 1994 Northridge | 6.7 | 6.4 | 1.61 | 3000 | 0.02 | 59.98 | 1.330016 |
| LA29 | 1974 Tabas | 7.4 | 1.2 | 1.08 | 2500 | 0.02 | 49.98 | 0.809218 |
| LA30 | 1974 Tabas | 7.4 | 1.2 | 1.08 | 2500 | 0.02 | 49.98 | 0.991908 |
| SE21 | 1992 Mendocino | 7.1 | 8.5 | 0.98 | 3000 | 0.02 | 59.98 | 0.7551332 |
| SE22 | 1992 Mendocino | 7.1 | 8.5 | 0.98 | 3000 | 0.02 | 59.98 | 0.4852179 |
| SE23 | 1992 Erzincan | 6.7 | 2 | 1.27 | 4156 | 0.005 | 20.775 | 0.6048157 |
| SE24 | 1992 Erzincan | 6.7 | 2 | 1.27 | 4156 | 0.005 | 20.775 | 0.5390563 |
| SE25 | 1949 Olympia | 6.5 | 56 | 4.35 | 4000 | 0.02 | 79.98 | 0.8948236 |
| SE26 | 1949 Olympia | 6.5 | 56 | 4.35 | 4000 | 0.02 | 79.98 | 0.8209028 |
| SE27 | 1965 Seattle | 7.1 | 80 | 10.04 | 4092 | 0.02 | 81.82 | 1.7549437 |
| SE28 | 1965 Seattle | 7.1 | 80 | 10.04 | 4092 | 0.02 | 81.82 | 1.3904852 |
| SE29 | 1985 Valpariso | 8 | 42 | 2.9 | 4000 | 0.025 | 99.975 | 1.6358349 |
| SE30 | 1985 Valpariso | 8 | 42 | 2.9 | 4000 | 0.025 | 99.975 | 1.5726635 |

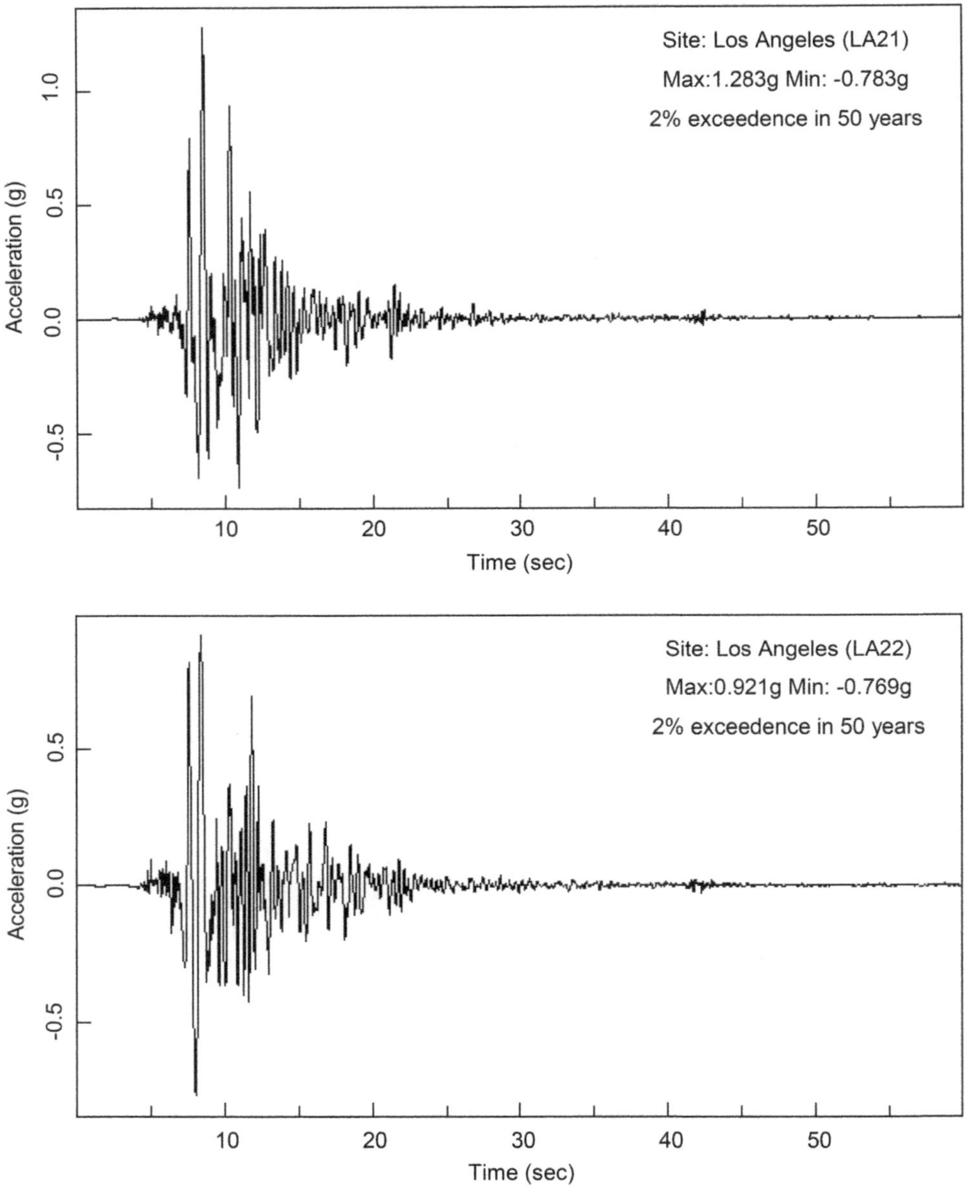


*Figure D.1* Earthquake ground motions in 1995 Kobe.

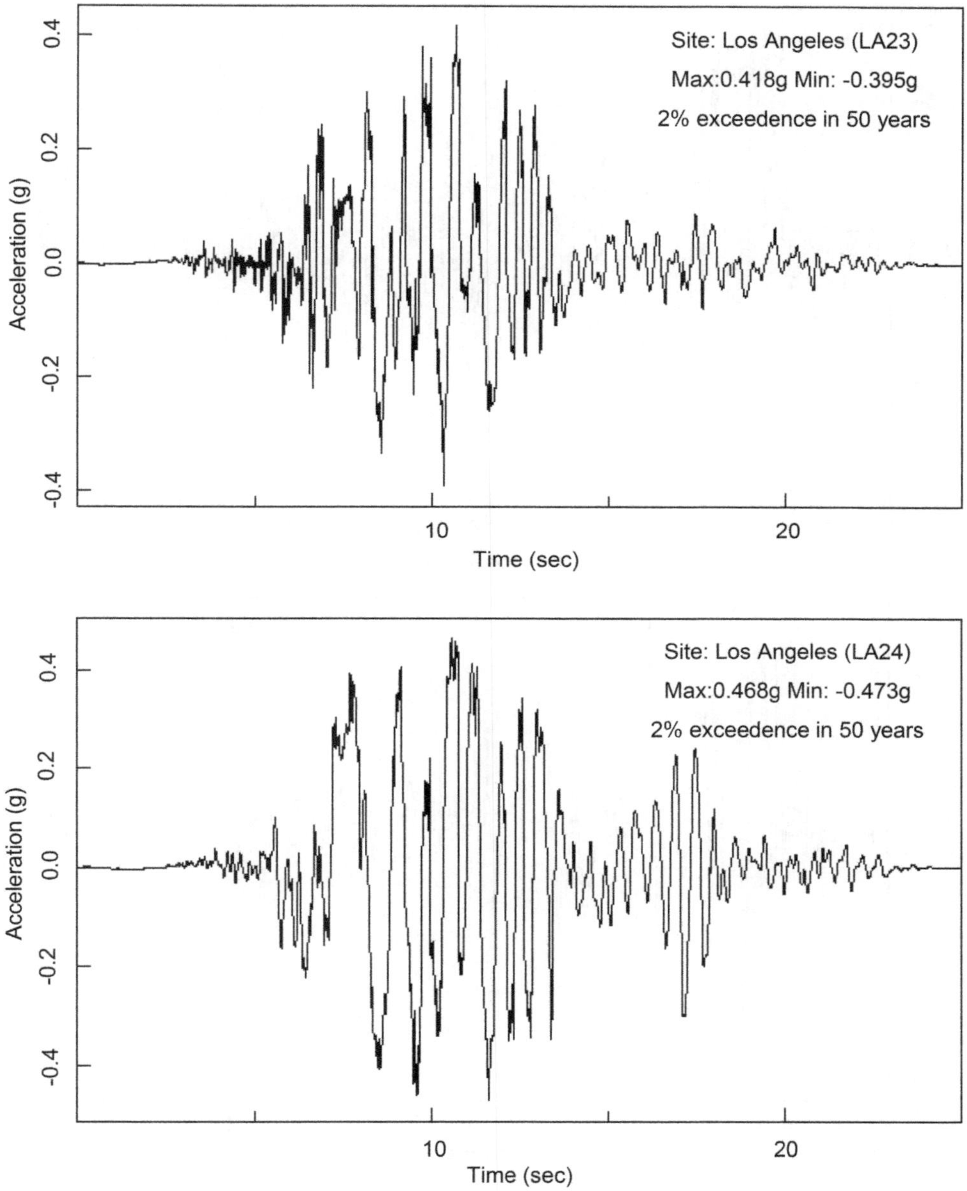


*Figure D.2* Earthquake ground motions in 1989 Loma Prieta.

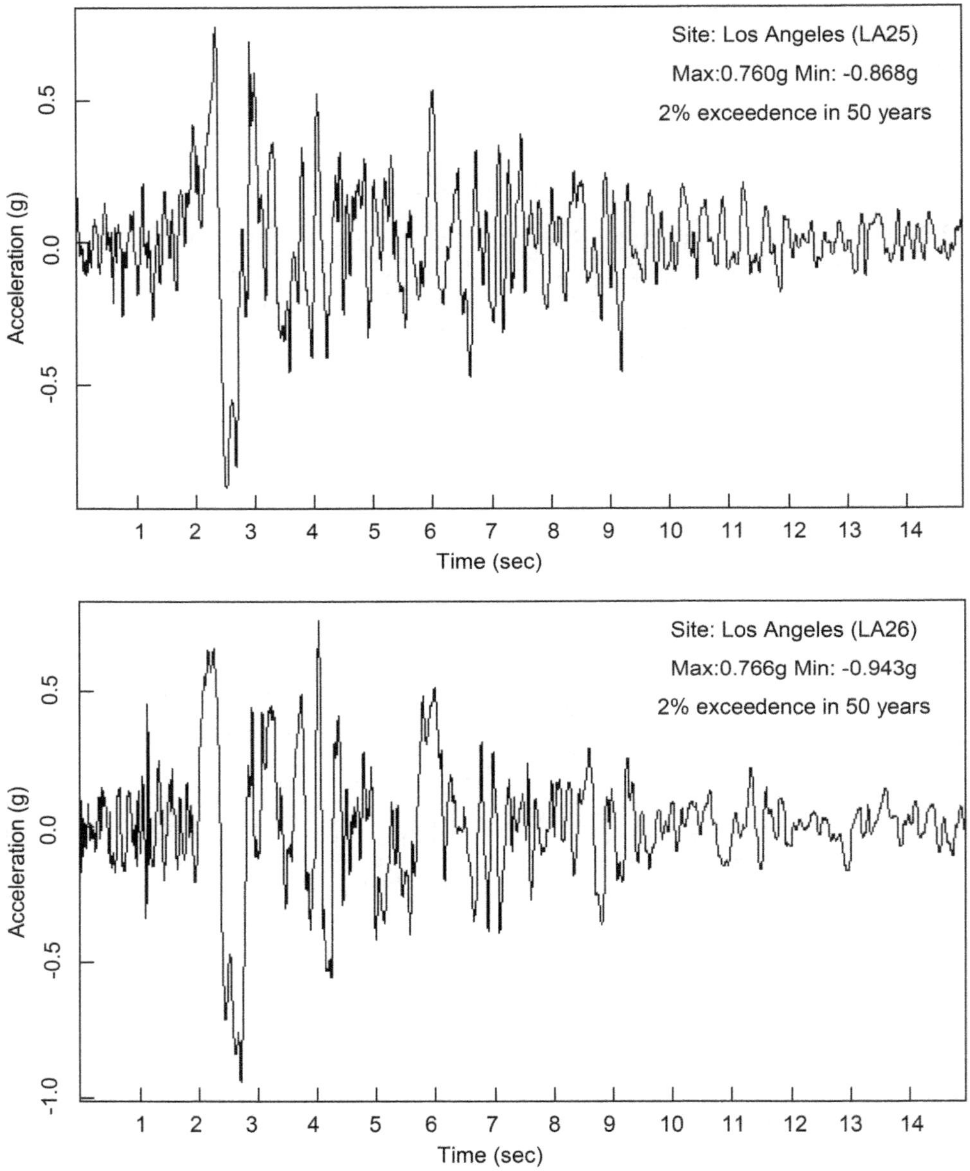


*Figure D.3* Earthquake ground motions in 1994 Northridge.

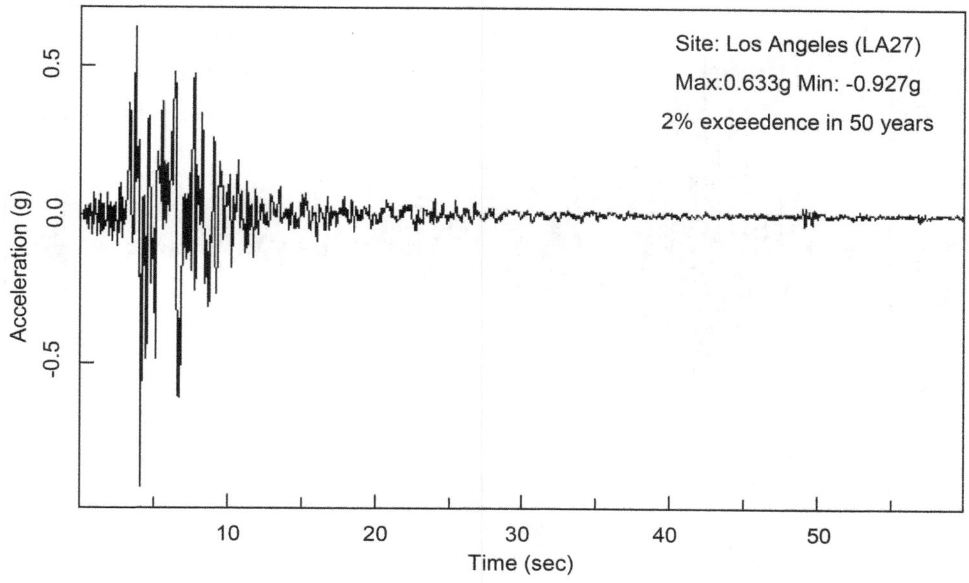


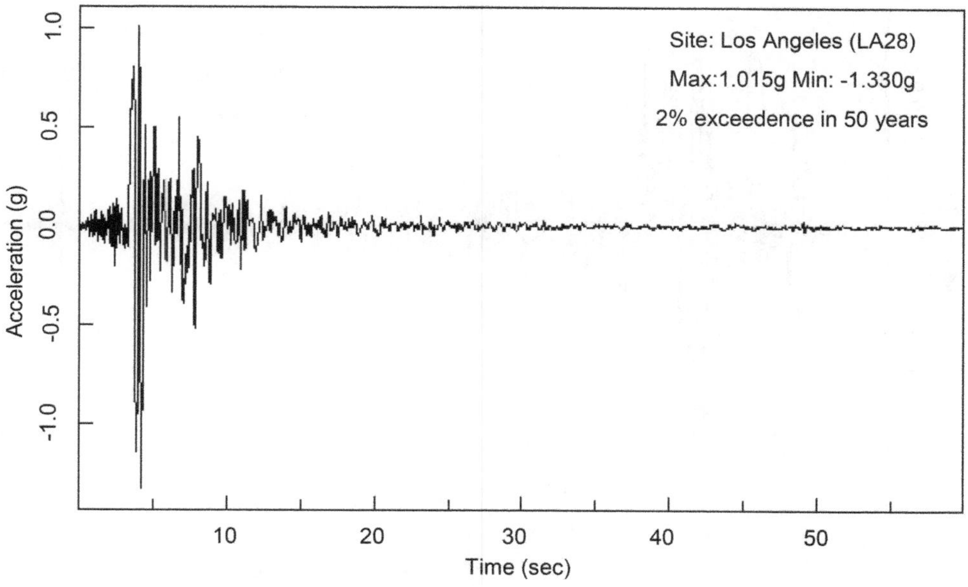


*Figure D.4* Earthquake ground motions in 1994 Northridge.

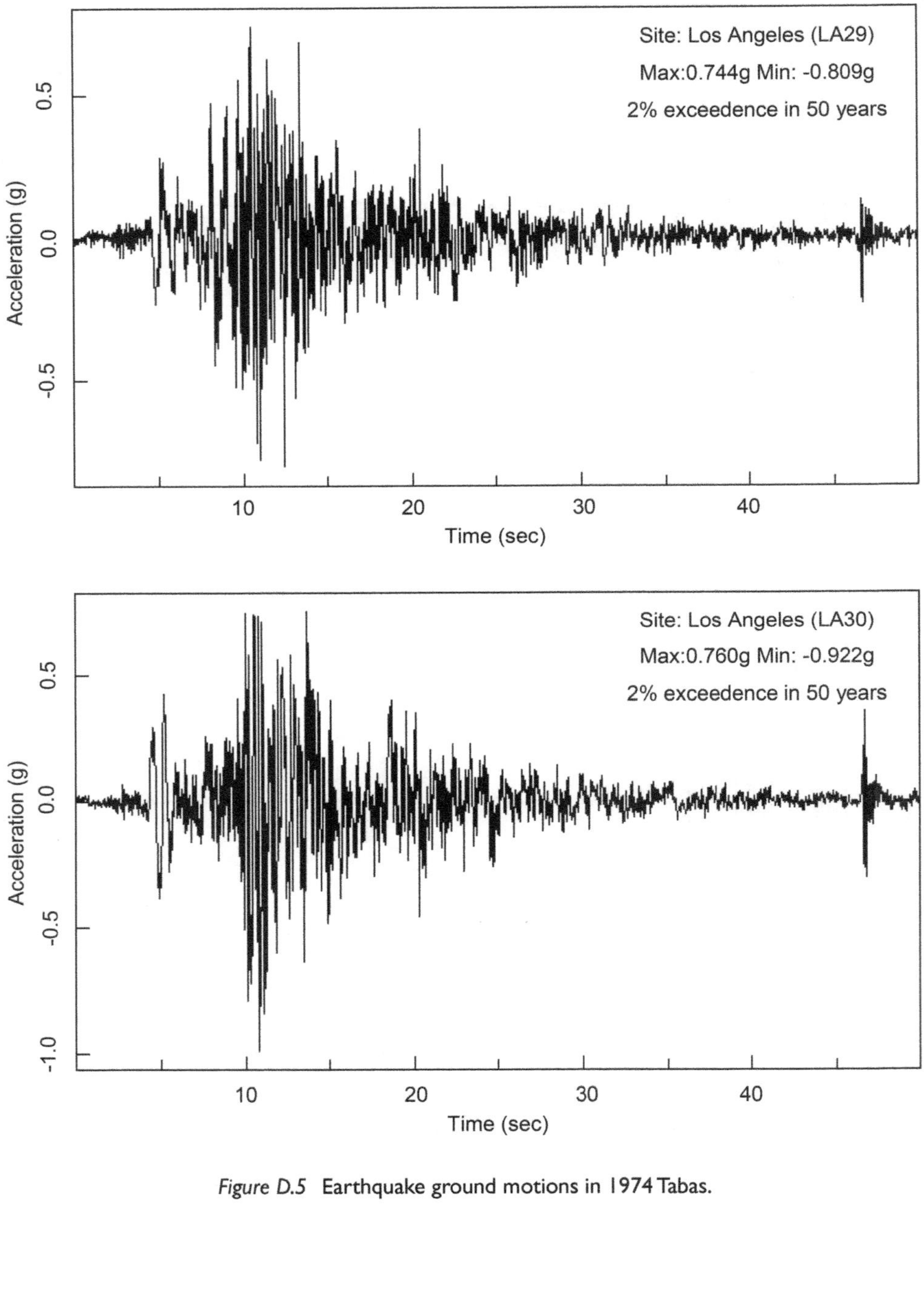


*Figure D.5* Earthquake ground motions in 1974 Tabas.

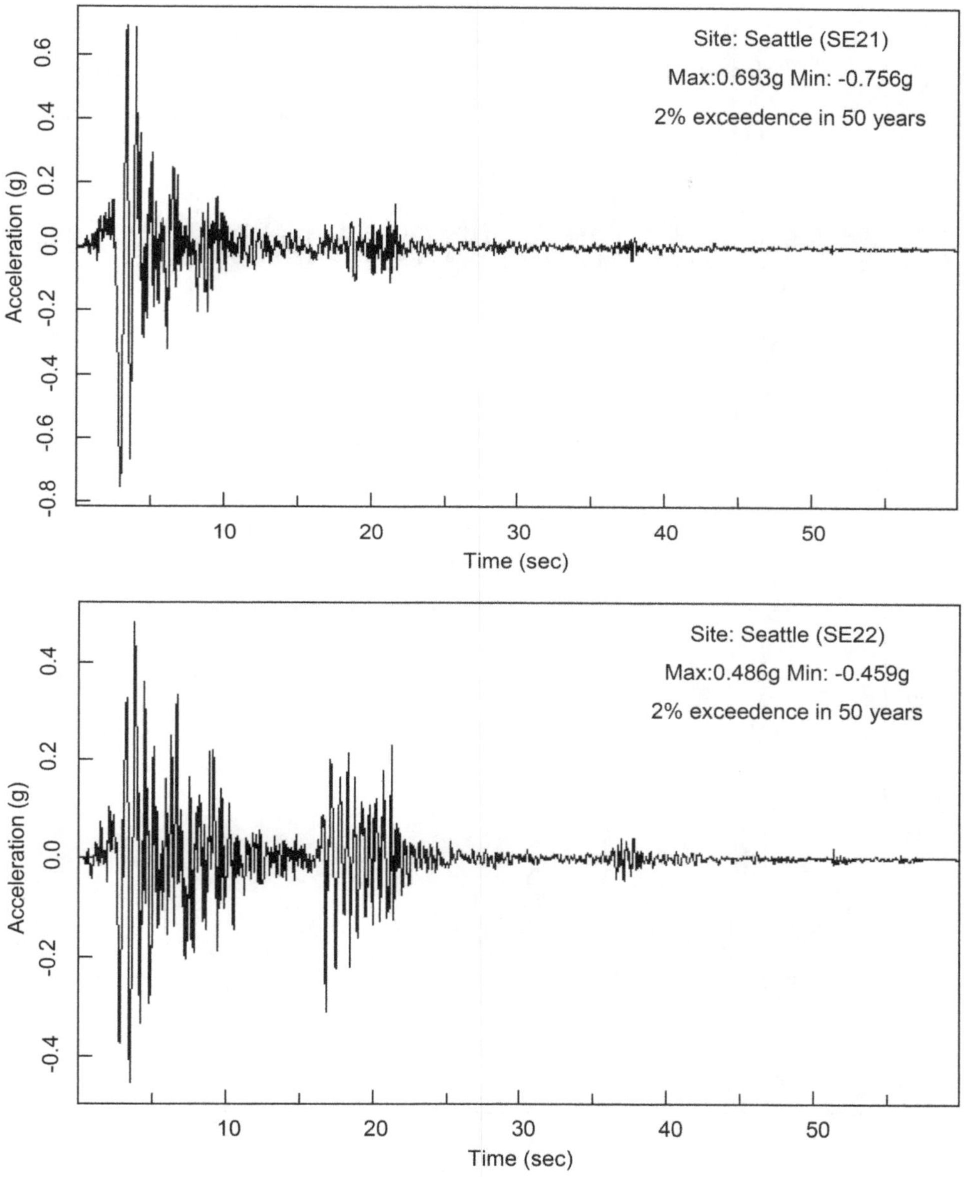


*Figure D.6* Earthquake ground motions in 1992 Mendocino.

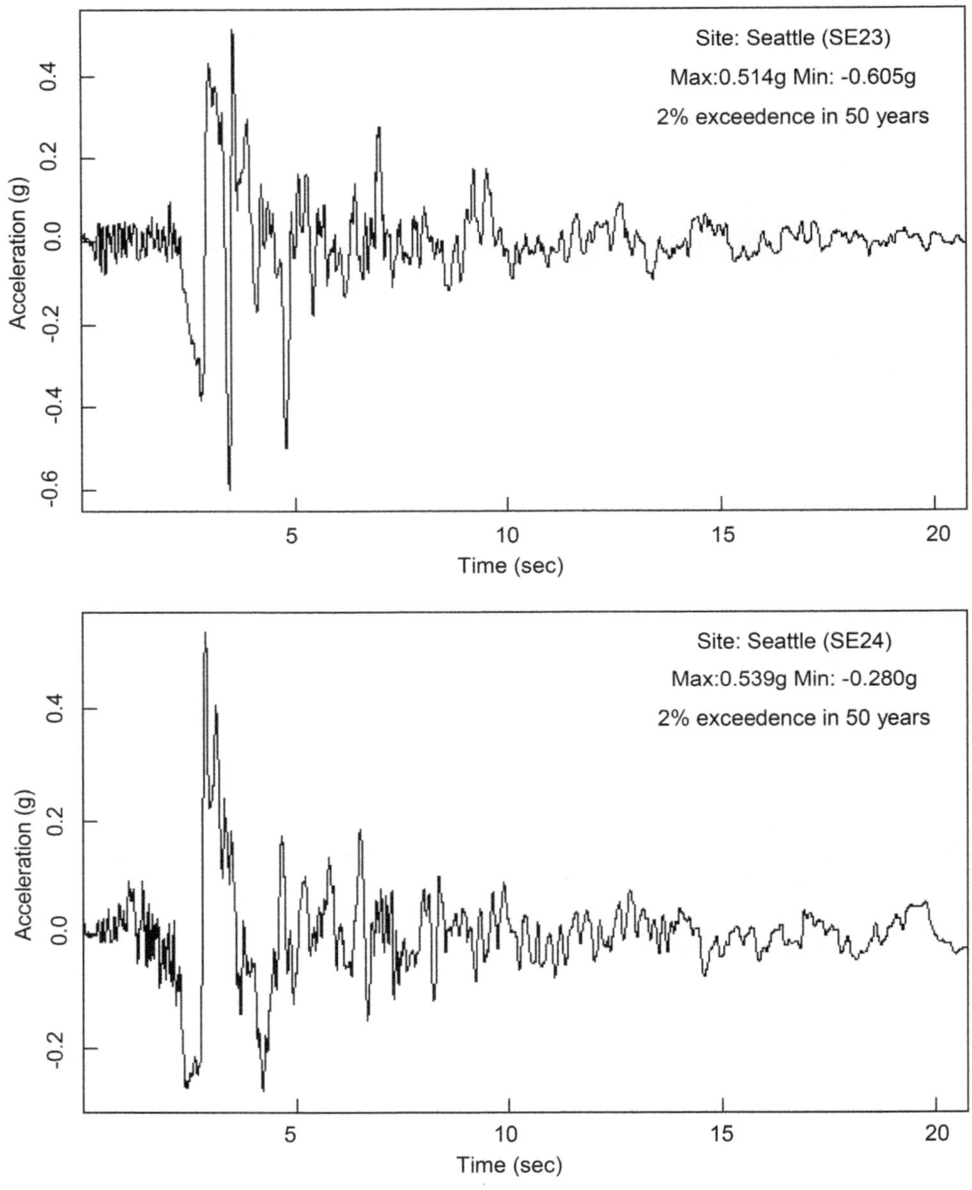


*Figure D.7* Earthquake ground motions in 1992 Erzincan.

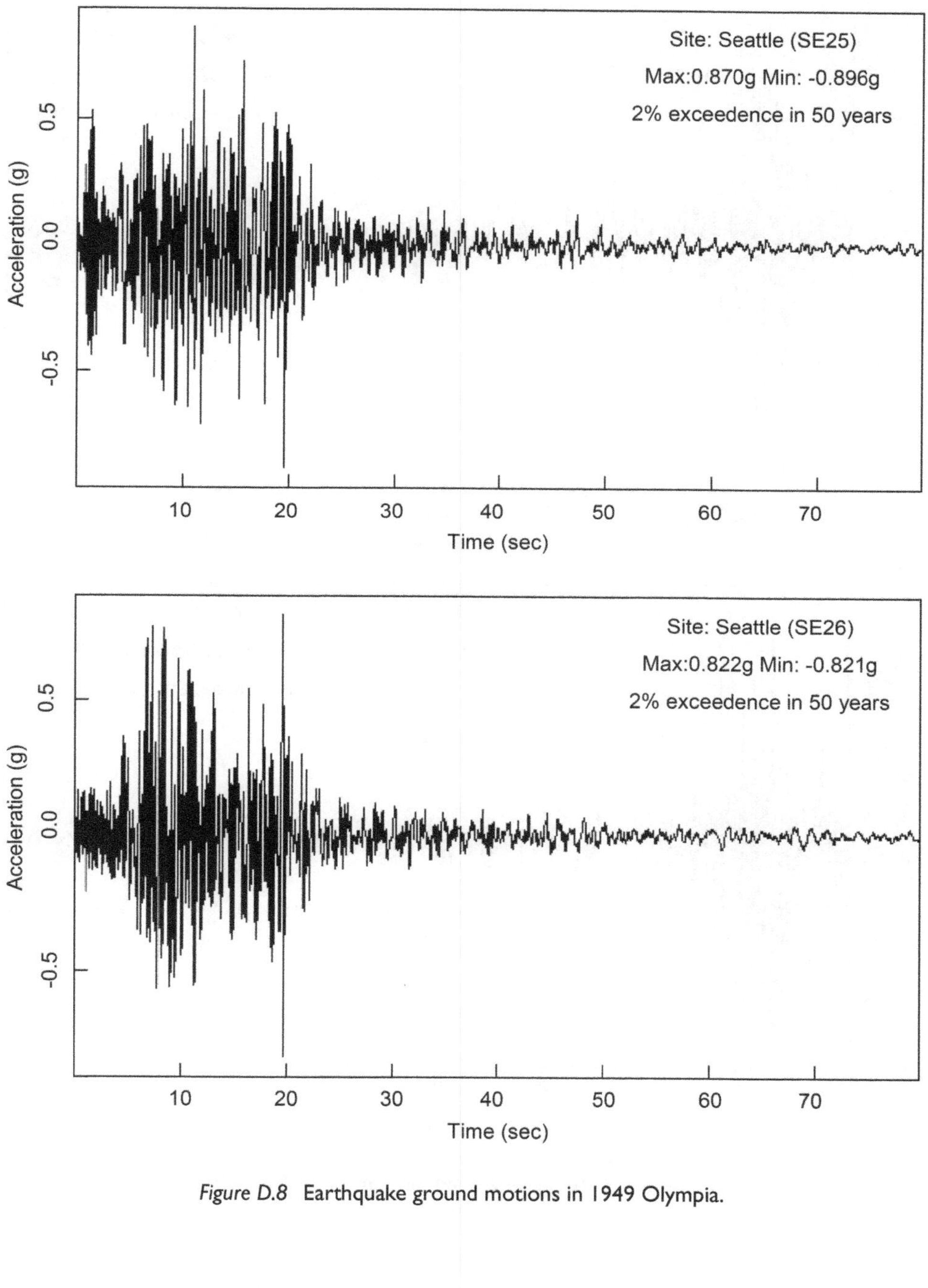


*Figure D.8* Earthquake ground motions in 1949 Olympia.

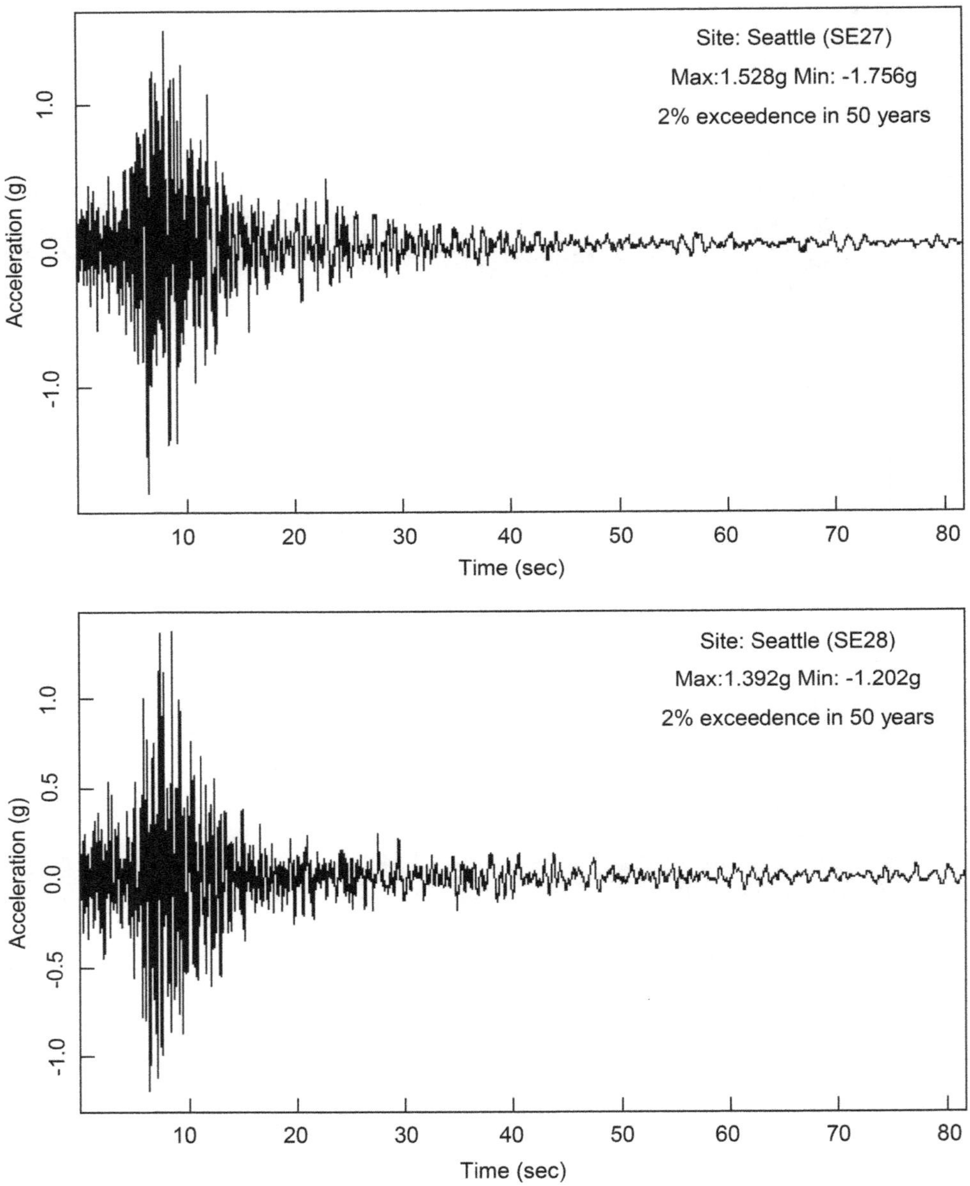


*Figure D.9* Earthquake ground motions in 1965 Seattle.

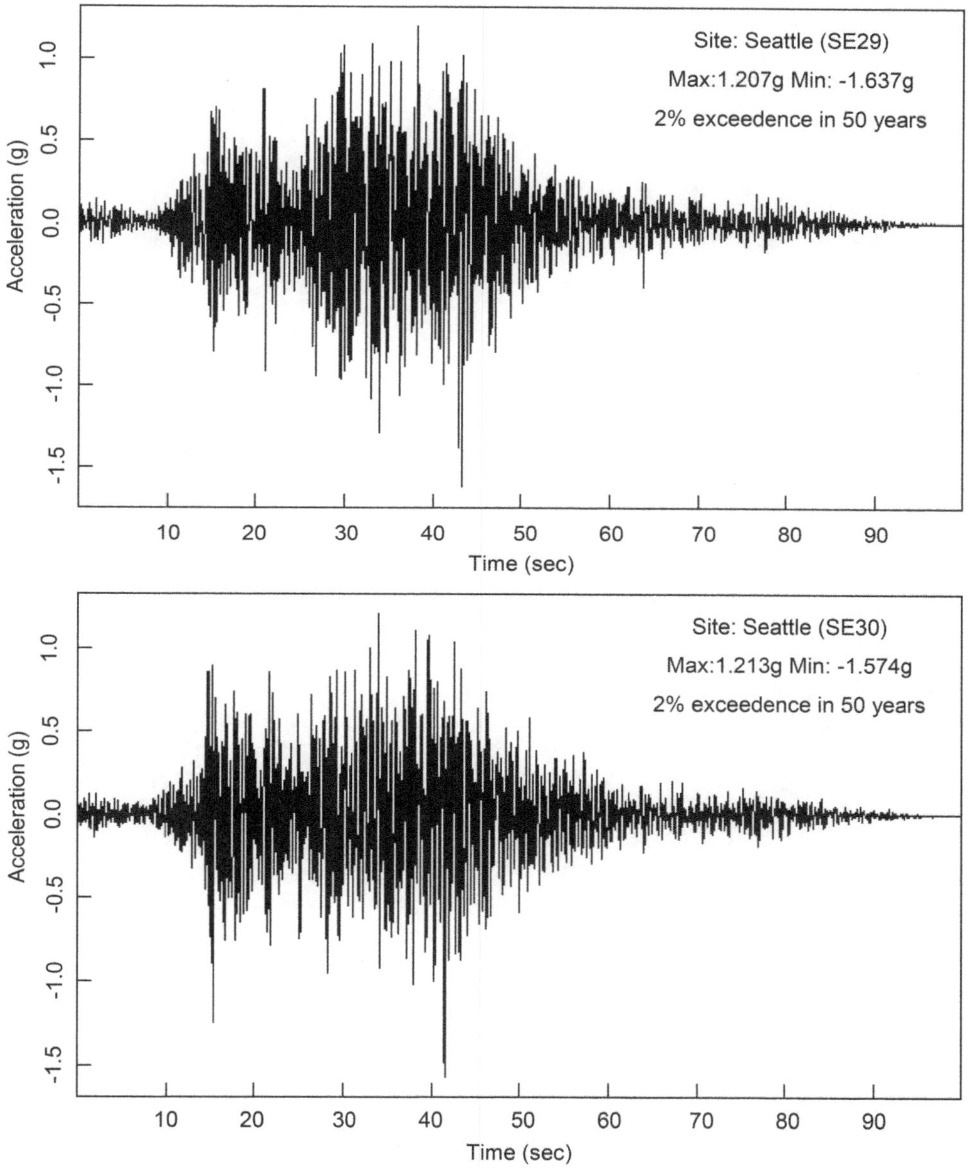


*Figure D.10* Earthquake ground motions in 1985 Valpariso.

# References


ABAQUS *ver.* 6.6-1. (2006) *Theory and User's Manual.* Pawtucket, RI, Hibbit, Karlsson & Sorensen, Inc.

Abolmaali, A., Treadway, J., Aswath, P., Lu, F.K. & McCarthy, E. (2006) Hysteresis behavior of T-stub connections with superelastic shape memory fasteners. *Journal of Constructional Steel Research*, 62 (8), 831–838.

Adey, B.T., Grondin, G.Y. & Cheng, J.J.R. (2000) Cyclic loading of end-plate moment connection. *Canadian Journal of Civil Engineering*, 27 (4), 683–701.

Aiken, I.D., Mahin, S.A. & Uriz, P. (2002) Large-scale testing of buckling-restrained braced frames. In: *Proc. Japan Passive Control Symposium*. Japan, Tokyo Institute of Technology. pp. 35–44.

Alostaz, Y.M. & Schneider, S.P. (1997) Analytical behavior of connections to concrete filled steel tubes. *Journal of Constructional Steel Research*, 40 (2), 95–127.

Altman, W.G., Azizinamini, A., Bradbrun, J.H. & Radziminski, J.B. (1982) Moment-rotation characteristics of semi-rigid steel beam-column connections. In: *Earthquake Hazard Mitigation Program, NSF, Final Report*. Columbia, SC, University of South Carolina.

American Association of State Highway and Transportation Officials (AASHTO) (2007) *AASHTO LRFD Bridge Design Specifications and 2008 Interim Revisions*. 4th edition. Washington, DC, AASHTO.

American Institute of Steel Construction (AISC) (1994) *Manual of Steel Construction, Load and Resistance Factor Design (LRFD)*. 2nd edition. Chicago, IL.

American Institute of Steel Construction (AISC) (2001) *Manual of Steel Construction, Load and Resistance Factor Design (LRFD)*. 3rd edition. Chicago, IL.

American Institute of Steel Construction (AISC) (2005a) *Prequalified Connections for Special and Intermediate Steel Moment Frames for Seismic Applications (ANSI/AISC 358-05)*. Chicago, IL.

American Institute of Steel Construction (AISC) (2005b) *Seismic Provisions for Structural Steel Buildings (ANSI/AISC 341-05)*. Chicago, IL.

American Institute of Steel Construction (AISC) (2005c). *Specification for Structural Steel Buildings (ANSI/AISC 360-05)*. Chicago, IL.

American Society of Civil Engineers (ASCE) (2002) *Minimum Design Loads for Buildings and Other Structures*. Reston, VA, ASCE/SEI No. 7-02.

American Society of Civil Engineers (ASCE) (2005) *Minimum Design Loads for Buildings and Other Structures*. Reston, VA, ASCE 7-05.

Astaneh-Asl, A. (1995) *Seismic Design of Bolted Steel Moment-Resisting Frames*. Steel Tips published by Structural Steel Educational Council, Technical Informational & Product Service, July.

Astaneh-Asl, A. (1997) *Seismic Design of Steel Column-Tree Moment-Resisting Frames*. Steel Tips published by Structural Steel Educational Council, Technical Informational & Product Service, April.

Astaneh-Asl, A. (2008) Progressive collapse of steel truss bridges the case of I 35w collaps. In: *Proc. of 7th Int. Symp. International Conference on Steel Bridges, Guimarães, Portugal.*

Auricchio, F. & Sacco, E. (1997) A one-dimensional model for superelastic shape-memory alloys with different properties between martensite and austenite. *International Journal of Non-Linear Mechanics*, 32 (6), 1101–1114.

Auricchio, F. & Taylor, R.L. (1997) Shape memory alloy: modeling and numerical simulations of the finite-strain superelastic behavior. *Computer Methods in Applied Mechanics and Engineering*, 143, 175–194.

Azizinamini, A. & Prakash, B. (1993) A tentative design guideline for a new steel beam connection detail to composite tube columns. *AISC Engineering Journal*, 3rd quarter, 103–115.

Azizinamini, A. & Schneider, S.P. (2001) Moment connections to circular concrete-filled steel tube columns. *Journal of Structural Engineering, ASCE*, 130 (2), 213–222.

Azizinamini, A., Bradbrun, J.H. & Radziminski, J.B. (1985) *Static and Cyclic Behavior of Semi-Rigid Steel Beam-Column Connections.* Technical Report. Columbia, SC, Department of Civil Engineering, University of South Carolina.

Bailey, J.R. (1970) Strength and rigidity of bolted beam-to-column connections. In: *Conference on Joints in Structures.* University of Sheffield.

Batho, C. & Rowan, H.C. (1934) Investigation on beam and stanchion connections. In: *2nd Report, Steel Structures Research Committee.* Vol. 1–2. London, Department of Scientific and Industrial Research, His Majesty's Stationary Office. pp. 61–137.

Beg, D., Zupancic, E. & Vayas, I. (2003) On the rotation capacity of moment connections. *Journal of Constructional Steel Research*, 60 (3–5), 601–620.

Bell, W.G., Chesson, E.J. & Munse, W.H. (1958) *Static Tests of Standard Riveted and Bolted Beam to Column Connections.* Urbana, IL, University of Illinois Engineering Experiment Station.

Bjorhovde, R. (1984) Effect of end-restraint on column strength-practical applications. *AISC Engineering Journal*, 20 (1), 41–45.

Bjorhovde, R. (1988) Limit states design considerations for gusset plates. *Journal of Constructional Steel Research*, 9, 61–73.

Black, C., Makris, N. & Aiken, I. (2002) *Component Testing, Stability Analysis and Characterization of Buckling-Restrained Braces.* Report No. PEER-2002/08, Pacific Earthquake Engineering Research Center. Berkeley, CA, University of California.

Black, G.R., Wenger, B.A. & Popov, E.P. (1980) *Inelastic Buckling of Steel Struts Under Cyclic Load Reversals.* UCB/EERC-80/40 Earthquake Engineering Research Center. Berkeley, CA.

Braconi, A., Salvatore, W., Tremblay, R. & Bursi, O.S. (2007) Behaviour and modelling of partial-strength beam-to-column composite joints for seismic applications. *Earthquake Engineering and Structural Dynamics*, 36, 142–161.

Buehler, W.J., Gilfrich, J.W. & Wiley, R.C. (1963) Effects of low-temperature phase changes on the mechanical properties of alloys near composition TiNi. *Journal of Applied Physics*, 34 (5), 1475–1477.

Casciati, F., Faravelli, L. & Petrini, L. (1998) Energy dissipation in shape memory alloy devices. *Computed-Aided Civil and Infrastructure Engineering*, 13, 433–442.

Chan, R.W.K. & Albermani, F. (2008) Experimental study of steel slit damper for passive energy dissipation. *Engineering Structures*, 30, 1058–1066.

Chan, S.L. & Chui, P.P.T. (2000) *Non-Linear Static and Cyclic Analysis of Steel Frame with Semi-Rigid Connections.* Elsevier Press.

Chen, W.F. & Lui, E.M. (1991) *Stability Design of Steel Frames.* Boca Raton, FL, CRC Press Inc.

Choi, E., Chung, Y.S., Choi, J.H., Kim, H.T. & Lee, H. (2010) The confining effectiveness of NiTiNb and NiTi SMA wire jackets for concrete. *Smart Materials and Structures*, 19, 1–8.

Choi, E.S., Nam, T.H. & Cho, B.S. (2005) A new concept of isolation bearings for highway steel bridges using shape memory alloys. *Canadian Journal of Civil Engineering*, 32, 957–1067.

Chopra, A.K. (1995) *Dynamics of Structures: Theory and Applications to Earthquake Engineering*. Upper Saddle River, NJ, Prentice-Hall.

Chopra, A.K. & Goel, R.K. (2002) A modal pushover analysis procedure for estimating seismic demands for buildings. *Earthquake Engineering and Structural Dynamics*, 31, 561–582.

Christopoulos, C., Filiatrault, A., Uang, C.M. & Folz, B. (2002) Posttensioned energy dissipating connections for moment-resisting steel frames. *Journal of Structural Engineering, ASCE*, 129 (9), 1111–1120.

Citipitioglu, A.M., Haj-Ali, R.M. & White, D.W. (2002) Refined 3D finite element modeling of partially-restrained connections including slip. *Journal of Constructional Steel Research*, 58, 995–1013.

Clark, P., Aiken, I., Kasai, K., Ko, E. & Kimura, I. (1999) Design procedures for buildings incorporating hysteretic damping devices. In: *Proc. 69th Annual Convention of SEAOC*. Sacramento, CA.

Clemente, I., Noe, S. & Rassati, G.A. (2004) Experimental behavior of T-stub connection components for the mechanical modeling of bare steel and composite partially-restrained beam-to-column connections. In: *Connections in Steel Structures V*. Amsterdam, June 3–4.

Coelho, A.M.G. & Silva, L.S. (2005) Ductility analysis of end plate beam-to-column joints. In: Hoffmeister, B. & Hechler, O. (eds.) *Proceeding: B. Eurosteel 2005 – 4th European Conference on Steel and Composite Structures*.

Committee European Normalization (CEN) Eurocode 3 (2003) *Design of Steel Structures – Part 1.8: Design of Joints (PREN 1993-1-8:2003)*. Stage 49 Draft, Brussels.

Computer and Structures, Inc. (CSI) (1984–2004) *SAP2000 Nonlinear Version 10.02. Structural Analysis Program*. Berkeley, CA.

Corbi, O. (2003) Shape memory alloys and their application in structural oscillations attenuation. *Simulation Modeling Practice and Theory*, 11, 387–402.

Corey, S.S. (1999) *Behavior of Full-Scale Bolted Beam-to-Column T-Stub and Clip Angle Connections Under Cyclic Loading*. Master's Thesis. Atlanta, GA, Georgia Institute of Technology.

Crisfield, M.A. (2012) *Non-Linear Finite Element Analysis of Solids and Structures*. Vol. 1. London, John Wiley and Sons Inc. pp. 152–200.

Davide, F. (2003) *Shape Alloy Devices in Earthquake Engineering: Mechanical Properties. Constitutive Modeling and Numerical Simulations*. Master's Thesis. Italy, Rose School.

Davison, J.B., Kirby, P.A. & Nethercot, D.A. (1987) Rotational stiffness characteristics of steel beam to column connections. *Journal of Constructional Steel Research*, 8, 17–54.

Deierlein, G.G. (1998) Summary of SAC case study building analysis. *Journal of Performance of Constructed Facilities, ASCE*, 12 (4), 202–212.

DesRoches, R. & Dlelamont, M. (2002) Seismic retrofit of simply supported bridges using shape memory alloys. *Engineering Structures*, 24, 325–332.

DesRoches, R., McCormick, J. & Delemont, M. (2004) Cyclic properties of superelasic shape memory alloy wires and bars. *Journal of Structural Engineering, ASCE*, 130 (1), 38–46.

Dolce, M. & Cardone, D. (2001) Mechanical behavior of shape memory alloys for seismic applications 2: Austenite NiTi wires subjected to tension. *International Journal of Mechanical Science*, 43 (11), 2657–2677.

Dolce, M., Cardone, D. & Marnetto, R. (2000) Implementation and testing of passive control devices based on shape memory alloys. *Earthquake Engineering and Structural Dynamics*, 29, 945–968.

Dolce, M., Cardone, D. & Marnetto, R. (2001) SMA re-centering devices for seismic isolation of civil structures. *Proceedings of SPIE*, 4330, 238–249.

Douty, R.T. (1964) *Strength Characteristics of High Strength Bolted Connections with Particular Application to the Plastic Design of Steel Structures*. PhD Thesis. Ithaca, NY, Cornell University.

Duerig, T., Melton, K., Stokel, D. & Wayman, C. (1990) *Engineering Aspects of Shape Memory Alloys*. London, UK, Butterworth-Heinemann.

Elremaily, A. & Azizinamini, A. (2001) Experimental behavior of steel beam to CFT column connections. *Journal of Constructional Steel Research*, 57, 1099–1119.

El-Tawil, S., Kanno, R. & Deierlein, G.G. (1996) Inelastic models for composite moment connections in RCS frames In: Buckner, C.D. & Shahrooz, B. (eds.) *Composite Construction in Steel and Concrete III*. ASCE Special Publication, SEI, American Society of Civil Engineers. pp. 197–210.

El-Tawil, S. & Deierlein, G.G. (2001a) Nonlinear analyses of mixed steel-concrete moment frames. Part I – Beam-column element formulation. *Journal of Structural Engineering, ASCE* 127 (6), 647–655.

El-Tawil, S. & Deierlein, G.G. (2001b) Nonlinear analyses of mixed steel-concrete moment frames. Part II – Implementation and verification. *Journal of Structural Engineering, ASCE* 127 (6), 656–665.

Eurocode4 (2004) *CEN: Design of Composite Steel and Concrete Structures*. Brussels, European Committee for Standardization.

Faella, C., Piluso, V. & Rizzano, G. (2000) *Structural Steel Semi-Rigid Connections – Theory, Design and Software*. Boca Raton FL, CRC Press.

Federal Emergency Management Agency (FEMA) (1995) *NEHRP Recommended Provisions for Seismic Regulations for New Buildings and Other Structures, Part 1 and 2*. 1994 edition. Report FEMA 222A and 223A. Washington, DC, Federal Emergency Management Agency.

Federal Emergency Management Agency (FEMA) (2000a) *Recommended Seismic Design Criteria for New Steel Moment-Frame Buildings*. Ref. No. FEMA-350, SAC Joint Venture, Washington, DC.

Federal Emergency Management Agency (FEMA) (2000b) *State of the Art Report on Systems Performance of Steel Moment Frames Subjected to Earthquake Ground Shaking*. Ref. No. FEMA-355C, SAC Joint Venture, Washington, DC.

Federal Highway Administration (FHWA) (2009) *Load Rating Guidance and Examples for Bolted and Riveted Gusset Plates in Truss Bridges*. FHWA-IF-09-014.

Federal Highway Administration (FHWA) (2010) *Guidelines for the Load and Resistance Factor Design and Rating of Riveted and Bolted Gusset-Plate Connections for Steel Bridges*. Project No. 12-84, Second Interim Report.

Fugazza, D. (2003) *Shape Alloy Devices in Earthquake Engineering: Mechanical Properties, Constitutive Modeling and Numerical Simulations*. Master's Thesis. Italy, Rose School.

Galambos, T.V. (1998) *Guide to Stability Design Criteria for Steel Structures, Structural Stability Research Council*. 5th edition. New York, NY, John Wiley and Sons.

Gardener, A.P. & Goldsworthy, H.M. (2005) Experimental investigation of the stiffness of critical components in a moment-resisting composite connection. *Journal of Constructional Steel Research*, 61, 709–726.

Green, T.P., Leon, R.T. & Rassati, G.A. (2004) Bidirectional tests on partially restrained, composite beam-column connections. *Journal of Structural Engineering*, 130 (2), 320–327.

Hajjar, J.F. (2002) Composite steel and concrete structural systems for seismic engineering. *Journal of Constructional Steel Research*, 58, 703–723.

Hajjar, J.F., Gourley, B.C., O'Sullivan, B.P. & Leon, R.T. (1998) Analysis of mid-rise steel frame damaged in Northridge earthquake. *Journal of Performance of Constructed Facilities, ASCE*, 12 (4), 221–231.

Hardash, S.G. & Bjorhovde, R. (1984) *Gusset Plate Design Utilizing Block-Shear Concepts.* Tucson, AZ, Department of Civil Engineering and Engineering Mechanics, The University of Arizona.

Hechtman, R.A. & Johnston, B.G. (1947) *Riveted Semi-Rigid Beam-to-Column Building Connections.* Committee of Steel Structures Research, AISC, Progress Report No. 1, November.

Historic Bridge (2013) Available from: http://www.historicbridges.org/bridges/browser/bridge browser=illinois/i94littlecalumet.

Holt, R. & Hartmann, J. (2008) *Adequacy of the U10 Gusset Plate Design for the Minnesota Bridge No. 9340 (I-35W Over the Mississippi River).* Turner-Fairbank Highway Research Center Report.

Hu, J.W. (2008) *Seismic Performance Evaluations and Analyses for Composite Moment Frames with Smart SMA PR-CFT Connections.* PhD Dissertation. Atlanta, GA, Georgia Institute of Technology.

Hu, J.W. (2013a) Design and strength evaluation of critical gusset plate in the steel bridge using new LRFD methods and advanced FE analyses. *ISIJ International*, 53 (8), 1452–1461.

Hu, J.W. (2013b) Design motivation, mechanical modeling, and nonlinear analysis for composite PR moment frames with smart SMA connection systems. *Advanced Steel Construction*, 9 (4), 334–349.

Hu, J.W. (2014a) Investigation on the cyclic response of superelastic shape memory alloy (SMA) slit damper devices simulated by quasi-static finite element (FE) analyses. *Materials*, 7 (2), 1122–1141.

Hu, J.W. (2014b) Seismic analysis and evaluation of several recentering braced frame structures. *Proceedings of the Institution of Mechanical Engineers, Part C: Journal of Mechanical Engineering Science*, 228 (5), 781–798.

Hu, J.W. & Leon, R.T. (2011) Analyses and evaluations for composite-moment frames with SMA PR-CFT connections. *Nonlinear Dynamics*, 65 (4), 732–740.

Hu, J.W. & Park, T. (2013) Continuum models for the plastic deformation of octet-truss lattice materials under multi-axial loading. *ASME Journal of Engineering Materials and Technology*, 135 (2). Available from: 10.1115/1.4023772.

Hu, J.W., Leon, R.T. & Park, T. (2009) Analytical investigation on ultimate behaviors for steel heavy clip-angle connections using FE analysis. *ISIJ International*, 50 (6), 883–892.

Hu, J.W., Kang, Y.S., Choi, D.H. & Park, T. (2010) Seismic design, performance, and behavior of composite-moment frames with steel beam-to-concrete filled tube column connections. *KSSC International Journal of Steel Structures*, 10 (2), 177–191.

Hu, J.W., Kim, D.K., Leon, R.T. & Choi, E. (2011a) Analytical studies of full-scale steel T-stub connections using delicate 3D finite element methods. *ISIJ International*, 51 (4), 619–629.

Hu, J.W., Choi, E. & Leon, R.T. (2011b) Design, analysis, and application of innovative composite PR connections between steel beams and CFT columns. *Smart Materials and Structures*, 20 (2). Available from: 10.1088/0964-1726/20/2/025019.

Hu, J.W., Leon, R.T. & Choi, E. (2011c) Investigation on the inelastic behavior of full-scale heavy clip-angle connections *International Journal of Steel Structures, KSSC* 11 (1) 1–11.

Hu, J.W., Leon, R.T. & Park, T. (2011d) Mechanical modeling of bolted T-stub connections under cyclic loads Part I: Stiffness modeling. *Journal of Constructional Steel Research*, 67, 1710–1718.

Hu, J.W., Leon, R.T. & Park, T. (2012) Mechanical models for the analysis of bolted T-stub connections under cyclic loads. *Journal of Constructional Steel Research*, 78, 45–57.

Hu, J.W., Choi, D.H. & Kim, D.K. (2013a) Inelastic behavior of smart recentering buckling-restrained braced frames with superelastic shape memory alloy bracing systems. *Proceedings of the Institution of Mechanical Engineers, Part C: Journal of Mechanical Engineering Science*, 227 (4), 806–818.

Hu, J.W., Kim, D.K. & Choi, E. (2013b) Numerical investigation on the cyclic behavior of smart recentering clip-angle connections with superelastic shape memory alloy fasteners. *Proceedings of the Institution of Mechanical Engineers, Part C: Journal of Mechanical Engineering Science*, 227 (6), 1315–1327.

Hu, T., Huang, C.S. & Chen, Z.L. (2005) Finite element analysis of CFT columns subjected to an axial compressive force and bending moment in combination. *Journal of Constructional Steel Research*, 62, 1692–1712.

Huns, B.B.S., Grondin, G.Y. & Driver, R.G. (2006) Tension and shear block failure of bolted gusset plates. *Canadian Journal of Civil Engineering* 33 (4) 395–408.

Illinois Department of Transportation IL DOT (2012) Available from: http://www.dot.state.il.us.

Indirli, M., *et al.* (2001) Demo application of shape memory alloy devices: the rehabilitation of S. Georgio church bell tower. *Proceedings of SPIE*, 4330, 262–272.

Inoue, K., Sawaisumi, S. & Higashibata, Y. (2001) Stiffening requirements for unbonded braces encased in concrete panels. *ASCE Journal of Structural Engineering*, 127 (6), 712–719.

International Code Council (ICC) (2003) *International Building Code 2003 (IBC2003)*. VA, Falls Church.

International Code Council (ICC) (2006) *International Building Code 2006 (IBC2006)*. VA, Falls Church.

Johnson, N.D. & Walpole, W.R. (1981) *Bolted End-Plate Beam-to-Column Connections Under Earthquake Type Loading*. Research Report 81-7. Christchurch, New Zealand, Department of Civil Engineering, University of Canterbury.

Kahn, M.M. & Lagoudas, D. (2002) modeling of shape memory alloy pseudoelastic spring elements using Preisach model for passive vibration isolation. *Proceedings of SPIE*, 4693, 336–347.

Kanatani, H., Tabuchi, M., Kamba, T., Hsiaolien, J. & Ishikawa, M. (1987) A study on concrete filled RHS columns to H-beam connections fabricated with HT-bolts in rigid frames. *ASCE Proceedings of the Composite Construction in Steel and Concrete*, 614–635.

Karavasilis, T.L., Krawala, S. & Hale, E. (2012) Hysteretic model for steel energy dissipation devices and evaluation of a minimal-damage seismic design approach for steel buildings. *Journal of Constructional Steel Research*, 70, 358–367.

Kennedy, D.J.L. & Hafez, M.A. (1984) A study of end-plate connections for steel beams. *Canadian Journal of Civil Engineering*, 11 (2), 139–149.

Kim, D. & Leon, R.T. (2007) Seismic performance of PR frames in mid-America earthquake region. *ASCE Journal of Structural Engineering*, 133 (12), 1808–1820.

Kim, D.H. (2003) *Seismic Performance of PR Frames in Areas of Infrequent Seismicity*. PhD Dissertation. Atlanta, GA, Georgia Institute of Technology.

Kim, D.K. (2005) *A Database for Composite Columns*. Master Thesis in School of Civil and Environmental Engineering. Atlanta, GA, Georgia Institute of Technology.

Kim, D.K., Dargush, G.F. & Hu, J.W. (2013) Cyclic damage model for E-shaped dampers in the seismic isolation system. *Journal of Mechanical Science and Technology*, 27 (8), 2275–2281.

Kim, Y.J., Shin, K.J. & Kim, W.J. (2008) Effect of stiffener details on behavior of CFT column-to-beam connection. *International Journal of Steel Structures KSSC*, 8, 119–133.

Krawinkler, H. & Popov, E.P. (1982) Seismic behavior of moment connections and joints. *ASCE Journal of the Structural Division*, 108 (2), 373–391.

Kulak, G.L. & Grondin, G.Y. (2000) *Steel Connections in the New Millennium* October 22–25 Roanoke, VA.

Kulak, G.L. & Grondin, G.Y. (2001) Block shear failure in steel members – a review of design practice. *Engineering Journal, AISC*, 38 (4), 199–203.

Kulak, G.L., Fisher, J.W. & Struik, J.H.A. (1987) *Guide to Design Criteria for Bolted and Riveted Joint*. 2nd edition. John Wiley & Sons.

Lantada, A.D., Angeles, M. & Rebollo, S. (2013) Towards low-cost effective and homogenous thermal activation of shape memory polymers. *Materials*, 6, 5447–5465.

Lee, K. & Foutch, D.A. (2002) Performance evaluation of new steel frame buildings for seismic loads. *Earthquake Engineering and Structural Dynamics*, 31, 653–670.

Leon, R.T. (1990) Semi-rigid composite connection. *Journal of Constructional Steel Research*, 15 (2), 99–120.

Leon, R.T. (1997) *Seismic Performance of Bolted and Riveted Connections, Background Reports: Metallurgy, Fracture Mechanics, Welding, Moment Connections, and Frame System Behavior*. FEMA Publication No.288. Washington, DC, Federal Emergency Management Association.

Leon, R.T. & Kim, D.H. (2004) Seismic performance of PR frames in zones in infrequency seismicity. In: *Proceeding of the 13th World Conference of Earthquake Engineering*. Paper 2696. Vancouver, BC, IAEE.

Leon, R.T. & Hu, J.W. (2008) Design of innovative SMA PR connections between steel beams and composite columns. In: *International Workshop on Connections in Steel Structures: Connections VI*.

Leon, R.T., Hajjar, J.F. & Gustafson, M.A. (1998) Seismic response of composite moment-resisting connections. I: Performance. *ASCE Journal of Structural Engineering*, 124 (8), 868–876.

Leon, R.T., DesRoches, R., Ocel, J. & Hess, G. (2001) Innovative beam column using shape memory alloys. *Proceedings of SPIE*, 4330, 227–237.

Leon, R.T., Hu, J.W. & Schrauben, C. (2005) Rotational capacity and demand in top-and-seat angle connections subjected to seismic loading. In: *Connections in Steel Structure 5*. AISC/ECCS.

Li, Y., Zhang, J., Huang, P. & Yan, B. (2008) *Proc. Int. Symp. Chinese-Croatian Joint Colloqium – Long Arch Bridges*. Brijuni Islands, China.

Liao, M., Okazaki, T., Ballarini, R., Schultz, A.E. & Galambos, T.V. (2011) Nonlinear finite-element analysis of critical gusset plates in the I-35W bridge in Minnesota. *Journal of Structural Engineering, ASCE*, 137 (1), 59–68.

Lipson, S.L. (1977) Single-angle welded-bolted beam connections. *Journal of the Structural Division, ASCE*, 103 (ST3), 559–571.

Lipson, S.L. & Antonio, M. (1980) Single-angle welded-bolted connections. *Canadian Journal of Civil Engineering*, 7, 315–324.

Liu, J. & Astaneh-Asl, A. (2000) Cyclic tests of simple connection including the effect of slabs. *Journal of Structural Engineering, ASCE*, 126 (1), 32–39.

Lowes, L.N. & Altoontash, A. (2003) Modeling reinforced-concrete beam-column joints subjected to cyclic loading. *Journal of Structural Engineering, ASCE*, 129 (12), 1686–1697.

Lubliner, J. & Auricchio, F. (1996) Generalized plasticity and shape memory alloys. *International Journal of Solids and Structures*, 33, 991–1003.

Ma, M., Wilkinson, T. & Cho, C. (2007) Feasibility study on a self-centering beam-to-column connection by using the superelastic behavior of SMAs. *Smart Materials and Structures*, 16, 1555–1563.

Mao, X.Y. & Xiao, M. (2006) Seismic behavior of confined square CFT columns. *Engineering Structures*, 28, 1378–1386.

Mazzoni, S., Mckenna, F. & Fenves, G.L. (2006) *Opensees Command Language Manual*. Berkley, Department of Civil Environmental Engineering, University of California.

McCormick, J., Barbero, L. & DesRoches, R. (2005) Effect of mechanical training on the properties of superelastic shape memory alloys for seismic applications. *Proceedings of SPIE*, 5764, 430–439.

McCormick, J., DesRoches, R., Fugazza, D. & Auricchio, F. (2007) Seismic assessment of concentrically braced steel frames with shape memory alloy braces. *Journal of Structural Engineering*, 133 (6), 862–870.

McCormick, J., Aburano, H., Ikenaga, M. & Nakashima, M.P. (2008) Permissible residual deformation levels for building structures considering both safety and human elements. In: *Proceeding of 14th World Conference Earthquake Engineering, Beijing, China: 05-06-0071.*

McCormick, J.P. (2006) *Cyclic Behavior of Shape Memory Alloys: Material Characterization and Optimization.* PhD Dissertation. Atlanta, GA, Georgia Institute of Technology.

Mehrabian, A., Ali, T. & Haldar, A. (2009) Nonlinear analysis of a steel frame. *Nonlinear Analysis*, 71, 616–623.

Miyazaki, S., Imai, T., Igo, Y. & Otsuka, K. (1986) Effect of cyclic deformation on the pseudoelasticity characteristics of Ti-Ni alloys. *Metallurgical Transactions A*, 17 (1), 115–120.

Miyazaki, S., Duerig, T., Melton, K., Stokel, D. & Wayman, C. (1990) Thermal and stress cycling effects and fatigue properties of Ni-Ti alloys. In: *Engineering Aspects of Shape Memory Alloys*. London, UK, Butterworth-Heinemann. pp. 394–413.

Morino, S., Kawanguchi, J., Yasuzaki, C. & Kanazawa, S. (1992) Behavior of concrete filled steel tubular three dimensional subassamblages. *ASCE Proceedings of the Composite Construction in Steel and Concrete*, 726–741.

National Transportation Safety Board (NTSB) (2008) *Collapse of I-35W Highway Bridge, Minneapolis, Minnesota, August 1, 2007*. Highway Accident Report NTSB/HAR-08/03, Washington, DC.

Nethercot, D.A. (1985) *Steel Beam-to-Column Connections – A Review of Test Data*. London, England, Construction Industry Research and Information Association.

Newmark, N.M. (1959) A method of computation for structural dynamics. *ASCE, Journal of the Engineering Mechanics Division*, 85 (EM3).

Ocel, J.M. (2002) *Cyclic Behavior of Steel Beam-Column Connections with Shape Memory Alloy Connection Elements*. Master's Thesis. Atlanta, GA, Georgia Institute of Technology.

Ocel, J.M., DesRoches, R., Leon, R.T., Hess, W.G., Krumme, R., Hayes, J.R. & Sweeney, S. (2004) Steel beam-column connections using shape memory alloys. *Journal of Structural Engineering, ASCE*, 130 (5), 732–740.

Ostrander, J.R. (1970) *An Experimental Investigation of End-Plate Connections*. MS Thesis. Saskatoon, SK, University of Saskatchewan.

Otani, S., Hiraishi, H. & Midorikawa, M. (2000) Development of smart systems for building structures. *Proceedings of SPIE*, 3988, 2–9.

Padgett, J.E. & DesRoches, R. (2008) Methodology for the development of analytical fragility curves for retrofitted bridges. *Earthquake Engineering and Structural Dynamics*, 37, 1157–1174.

Park, T., Hwang, W., Leon, R.T. & Hu, J.W. (2011) Damage evaluation of composite-special moment frames with concrete-filled tube columns under strong seismic loads. *KSCE Journal of Civil Engineering*, 15 (8), 1381–1394.

Penar, B.W. (2005) *Recentering Beam-Column Connections Using Shape Memory Alloys*. Master's Thesis. Atlanta, GA, Georgia Institute of Technology.

Piluso, V. & Rizzano, G. (2008) Experimental analysis and modeling of bolted T-stubs under cyclic loads. *Journal of Constructional Steel Research*, 64, 655–669.

Prion, H.G. & McLellan, A.B. (1992) Connecting steel beam to concrete-filled steel columns. In: *Proceeding of the ASCE Structures Congress on Composite Compression Members, San Antonio, TX*. pp. 918–921.

Pucinotti, R. (2001) Top-and-seat and web angle connections: prediction via mechanical model. *Journal of Constructional Steel Research*, 57 (6), 661–694.

Rassati, G.A., Leon, R.T. & Noe, S. (2004) Component modeling of partially restrained composite joints under cyclic and dynamic loading. *Journal of Structural Engineering, ASCE*, 130 (2), 343–351.

Rathburn, J.C. (1933) Elastic properties of riveted connections. *ASCE Transactions*, 101, 524–563.

Reina, P. & Normile, D. (1997) Fully braced for seismic survival. *Engineering News Record July*, 21, 34–36.

Rex, C.O. & Easterling, W.S. (1996a) *Behavior and Modeling of a Single Plate Bearing on a Single Bolt*. Report No. CE/VPI-ST 96/14. Blacksburg, VA, Virginia Polytechnic Institute and State University.

Rex, C.O. & Easterling, W.S. (1996b) *Behavior and Modeling of a Single Plate Bearing on a Single Bolt*. Report No. CE/VPI-ST 96/15. Blacksburg, VA, Virginia Polytechnic Institute and State University.

Roeder, C.W. (2000) Seismic behavior of steel braced frame connections to composite columns. In: Leon, R. & Easterling, S. (eds.) *Connections in Steel Structure 4*. Vol. 4. AISC. pp. 51–62.

Saadat, S., Noori, M., Davoodi, H., Zhou, Z., Suzuki, Y. & Masuda, A. (2001) Using NiTi SMA tendons for vibration control of coastal structures. *Smart Materials and Structures*, 10, 695–704.

Saadat, S., Salichs, J., Noori, M., Hou, Z., Davoodi, H., Bar-on, I., Suzuki, Y. & Masuda, A. (2002) An overview of vibration and seismic application of NiTi shape memory alloy. *Smart Materials and Structures*, 11, 218–229.

Sabelli, R. (2001) *Research on Improving the Design and Analysis of Earthquake-Resistant Steel-Braced Frames*. The 2000 NEHRP Professional Fellowship Report. Oakland, CA, EERI.

Sabelli, R. (2004) Recommended provisions for buckling-restrained braced frames. *AISC Engineering Journal*, 41 (4), 155–175.

Sabelli, R., Mahin, S.A. & Chang, C. (2003) Seismic demands on steel braced-frame buildings with buckling-restrained braces. *Engineering Structures*, 25 (5), 655–666.

Sabol, T.A. (2004) An assessment of seismic design practice of steel structures in the United State since the Northridge earthquake. *The Structural Design of Tall and Special Buildings*, 13 (5), 409–423.

Saiid-Saiidi, M., Sadrossadat-Zadeh, M., Ayoub, C. & Itani, A. (2007) Pilot study of behavior of concrete beams reinforced with shape memory alloys. *Journal of Materials in Civil Engineering*, 19 (6), 454–461.

Salmon, C.G. & Johnson, J.E. (1990) *Steel Structures: Design and Behavior, Emphasizing Load and Resistance Factor Design*. 3rd edition. New York, NY, Harper and Row.

SAP 2000 ver. 11 (2007) *SAP 2000 Steel Design Manual, Computer and Structures*. California, Inc. Berkeley.

Schneider, S.P. & Alostaz, Y.F. (1997) Experimental behavior of connections to concrete-filled steel tubes. *Journal of Constructional Steel Research*, 45 (3), 321–351.

Schrauben, C.S. (1999) *Behavior of Full-Scale Bolted Beam-to-Column T-Stub and Clip Angle Connections Under Cyclic Loading*. Master's Thesis. Atlanta, GA, Georgia Institute of Technology.

Sep̃ulveda, J., Boroschek, R., Herrera, R., Moroni, O. & Sarrazin, M. (2008) Steel beam-column connection using copper-based shape memory alloy dampers. *Journal of Constructional Steel Research*, 64 (4), 429–435.

Shakir-Khalil, H. (1992) Full scale test on composite connection. *ASCE Proceedings of the Composite Construction in Steel and Concrete*, 539–554.

Sherbourne, A.N. (1961) Bolted beam-to-column connections. *The Structural Engineers*, 39 (6), 203–210.

Shi, Y.J., Chan, S.L. & Wong, Y.L. (1996) Modeling for moment-rotation characteristics for end-plate connections. *Journal of Structural Engineering ASCE*, 122 (11), 1300–1306.

Smallidge, J.M. (1999) *Behavior of Bolted Beam-to-Column T-Stub Connections Under Cyclic Loading*. MS Thesis. Atlanta, GA, Georgia Institute of Technology.

Somerville, P.G., Smith, N., Punyamurthula, S. & Sun, J. (1997) *Development of Ground Motion Time Histories for Phase 2 of the FEMA/SAC Steel Project*. SAC Background Document, Report No. SAC/BD 97/04.

Sommer, W.H. (1969) *Behavior of Welded Header Plate Connections*. Master's Thesis. ON, Canada, University of Toronto.

Song, G., Ma, N. & Li, N.H. (2006) Application of shape memory alloys in civil structures. *Engineering Structures*, 28, 1266–1274.

Soong, T.T. & Spencer, B.F. (2002) Supplemental energy dissipation: state-of-the art and state-of-the practice. *Engineering Structures*, 24, 243–259.

Swanson, J.A. (1999) *Characterization of the Strength, Stiffness, and Ductility Behavior of T-Stub Connection*. PhD Dissertation. Atlanta, GA, Georgia Institute of Technology.

Swanson, J.A. (2002) Ultimate strength prying models for bolted T-stub connections. *Engineering Journal, AISC*, 39 (3), 136–147.

Swanson, J.A. & Leon, R.T. (2000) Bolted steel connections: tests on T-stub components. *Journal of Structural Engineering, ASCE*, 126 (1), 50–56.

Swanson, J.A. & Leon, R.T. (2001) Stiffness modeling of bolted T-stub connection components. *Journal of Structural Engineering, ASCE*, 127 (5), 498–505.

Swanson, J.A., Kokan, D.K. & Leon, R.T. (2002) Advanced finite element modeling of bolted T-stub connection components. *Journal of Constructional Steel Research*, 58, 1015–1031.

Tami, H. & Kitagawa, Y. (2002) Pseudoelastic behavior of shape memory alloy wires and its application to seismic resistance member for building. *Computational Material Science*, 25, 218–227.

Thermou, G.E., Elnashai, A.S., Plumier, A. & Doneaux, C. (2004) Seismic design and performance of composite frames. *Journal of Constructional Steel Research*, 60, 31–57.

Thornton, W.A. (1985) Prying action – a general treatment. *Engineering Journal, AISC*, 22 (2), 145–149.

Tobushi, H., Hayashi, S., Sugimoto, Y. & Date, K. (2009) Two-way bending properties of shape memory composite with SMA and SMP. *Materials*, 2, 1180–1192.

Tompson, L.E., Mckee, R.J. & Visintainer, D.A. (1970) *An Investigation of Rotation Characteristics of Web Shear Framed Connections using A-36 and A-41 Steels*. Rolla, MO, Department of Civil Engineers, University of Missouri-Rolla.

Torres, L.I., Lopez-Almansa, F. & Bozzo, L.M. (2004) Tension-stiffening model for cracked flexural concrete members. *Journal of Structural Engineering, ASCE*, 130 (8), 1242–1251.

Tsai, K.C. & Hsiao, P.C. (2008) Pseudo-dynamic tests of a full-scale CFT/BRB frame-Part II: Seismic performance of buckling-restrained braces and connections. *Earthquake Engineering and Structural Dynamics*, 37, 1099–1115.

Tsai, K.C., Hwang, Y.C., Weng, C.S., Shirai, T. & Nakamura, H. (2002) Experimental tests of large scale buckling restrained braces and frames. In: *Proc. Passive Control Symposium*. Japan, Tokyo Institute of Technology.

Tsai, K.C., *et al.* (2004) Pseudo dynamic tests of a full-scale CFT/BRB composite frame. In: *Proceeding of the 2004 Structure Congress*, ASCE.

Tsai, K.C., Hsiao, P.C., Wang, K.J., Weng, Y.T., Lin, M.L., Lin, K.C., Chen, C.H., Lai, J.W. & Lin, S.L. (2008) Pseudo-dynamic tests of a full-scale CFT/BRB frame-Part I: Specimen design, experiment and analysis. *Earthquake Engineering and Structural Dynamics*, 37, 1081–1098.

Uriz, P., Filippou, F.C. & Mahin, S.A. (2008) Model for cyclic inelastic buckling for steel member. *Journal of Structural Engineering, ASCE*, 134 (4), 616–628.

Viest, I.M., *et al.* (1997) *Composite Construction Design for Building*. Co-published by ASCE, McGraw-Hill, Chapter 5.3.

Wada, A., Connor, J., Kawai, H., Iwata, M. & Watanabe, A. (1992) Damage tolerant structures. In: *Proc. 5th U.S.–Japan Workshop on the Improvement of Structural Design and Construction Practices*. San Diego, CA, Applied Technology Council (ATC-15-4). pp. 27–39.

Watanabe, A., Hitomi, Y., Yaeki, E., Wada, A. & Fujimoto, M. (1988) Properties of brace encased in buckling-restraining concrete and steel tube. In: *Proc. 9th World Conference on Earthquake Engineering 5*. Tokyo-Kyoto, Japan. pp. 719–724.

Whitmore, R.E. (1952) *Experimental Investigation of Stresses in Gusset Plates*. Knoxville, University of Tennessee. Engineering Experiment Station, Bulletin, 16.

Wilde, K., Gardoni, P. & Fujino, Y. (2000) Base isolation system with shape memory alloy device for elevated highway bridges. *Engineering Structures*, 22, 222–229.

Wilson, J.C. & Wesolowsky, M.J. (2005) Shape memory alloys for seismic response modification: a state-of-the-art review. *EERI Earthquake Spectra*, 21 (2), 569–601.

Wilson, W.M. & Moore, H.F. (1917) *Tests to Determine the Rigidity of Riveted Joints of Steel Structures*. Urbana, IL, University of Illinois Engineering Experiment Station, Bulletin No. 104.

Wright, W.J. (2009) Department of Civil and Environmental Engineering, Virginia Tech.

Wu, L.Y., Chung, L.L., Tsai, S.F., Lu, C.F. & Huang, G.L. (2005) Seismic behavior of bolted beam to column connections for concrete filled steel tube. *Journal of Constructional Steel Research*, 61, 1387–1410.

Wu, L.Y., Chung, L.L., Tsai, S.F., Lu, C.F. & Huang, G.L. (2007) Seismic behavior of bidirectional bolted connections for CFT columns and H-beams. *Engineering Structures*, 29, 395–407.

Wu, Q., Yoshimura, M., Takahashi, K., Nakamura, S. & Nakamura, T. (2006) Nonlinear seismic properties of the second Saikai bridge – a concrete filled tubular (CFT) arch bridge. *Engineering Structures*, 28, 163–182.

Yamamoto, K., Akiyama, N. & Okumura, T. (1988) Buckling strengths of gusseted truss joint. *Journal of Structural Engineering, ASCE*, 114 (3), 575–590.

Yoo, J.H. Roeder, C.W. & Lehman, D.E. (2008) Analytical performance simulation of special concentrically braced frame. *Journal of Structural Engineering, ASCE*, 134 (6) 881–889

Young, C.R. & Jackson, K.B. (1934) The relative rigidity of welded and riveted connections. *Canadian Journal of Research*, 11 (1), 62–134.

Zhang, Y. & Zhu, S. (2008) Seismic response control of building structures with superelastic shape alloy wire dampers. *Journal of Engineering Mechanics*, 134 (3), 240–251.

Zhu, S. & Zhang, Y. (2008) Seismic analysis of concentrically braced frame systems with self-centering friction damping brace. *Journal of Structural Engineering*, 134 (1), 121–131.

Zoetemeijer, P. & Kolstein, M.H. (1975) *Flush End Plate Connections*. The Netherlands, Stevin Laboratory Report No. 6-75-20, Delft University of Technology.

# Index